Fundamental Biomedical Technologies

Volume 7

Series editor

Mauro Ferrari, Ph.D., Houston, TX, USA

For further volumes:
http://www.springer.com/series/7045

Aleš Prokop · Yasuhiko Iwasaki
Atsushi Harada
Editors

Intracellular Delivery II

Fundamentals and Applications

Springer

Editors
Aleš Prokop
Chemical and Biological Engineering
Vanderbilt University
Nashville, TN
USA

Atsushi Harada
Department of Applied Chemistry
Osaka Prefecture University
Osaka
Japan

Yasuhiko Iwasaki
Chemistry and Materials Engineering
Kansai University
Osaka
Japan

ISSN 1559-7083
ISBN 978-94-017-8895-3 ISBN 978-94-017-8896-0 (eBook)
DOI 10.1007/978-94-017-8896-0
Springer Dordrecht Heidelberg New York London

Library of Congress Control Number: 2014939942

Printed on acid-free paper

Springer is part of Springer Science+Business Media (www.springer.com)

Contents

Contributors

Kazunari Akiyoshi Department of Polymer Chemistry, Graduate School of Engineering, Kyoto University, Kyoto, Japan; JST-ERATO, Tokyo, Japan

Helene Andersen Nanomedicine Research Group, Faculty of Health and Medical Sciences, Department of Pharmacy, Centre for Pharmaceutical Nanotechnology and Nanotoxicology, University of Copenhagen, Copenhagen, Denmark

Eva Bártová Institute of Biophysics, Academy of Sciences of the Czech Republic, Brno, Czech Republic

Fedor Čiampor Institute of Virology, SAS, Bratislava, Slovak Republic

Pierpaolo Ceci CNR—National Research Council of Italy, Institute of Molecular Biology and Pathology, Rome, Italy

Yung Chang R&D Center for Membrane Technology, Department of Chemical Engineering, Chung Yuan Christian University, Chung-Li, Taoyuan, Taiwan

Petr Cigler Laboratory of Synthetic Nanochemistry, Institute of Organic Chemistry and Biochemistry AS CR v. v. i., Prague 6, Czech Republic

Dušan Cmarko 1st Faculty of Medicine, Institute of Cellular Biology and Pathology, Charles University in Prague, Prague, Czech Republic

Kateřina Dědková Nanotechnology Centre, VSB—Technical University of Ostrava, Ostrava, Poruba, Czech Republic

Elisabetta Falvo CNR—National Research Council of Italy, Institute of Molecular Biology and Pathology, Rome, Italy

Marcela Foglarová Contipro Biotech s.r.o, Dolní Dobrouč, Czech Republic

Eleonore Fröhlich Center for Medical Research, Medical University of Graz, Graz, Austria

Jie Han Department of Electrical and Computer Engineering, University of Alberta, Edmonton, AB, Canada

Jan Havlik Laboratory of Synthetic Nanochemistry, Institute of Organic Chemistry and Biochemistry AS CR v. v. i., Prague 6, Czech Republic

Kateřina Horská Department of Nanobiotechnology, Institute of Nanobiology and Structural Biology of GCRC, České Budějovice, Czech Republic

Jian Ji MOE Key Laboratory of Macromolecule Synthesis and Functionalization of Ministry of Education, Department of Polymer Science and Engineering, Zhejiang University, Hangzhou, China

Xiaoyan Jiang Terry Fox Laboratories, British Columbia Cancer Agency, Department of Medical Genetics, University of British Columbia, Vancouver, BC, Canada

Qiao Jin MOE Key Laboratory of Macromolecule Synthesis and Functionalization of Ministry of Education, Department of Polymer Science and Engineering, Zhejiang University, Hangzhou, China

Yoshiki Katayama Faculty of Engineering, Department of Applied Chemistry, Kyushu University, Nishi-ku, Fukuoka, Japan; Center For Future Chemistry, Kyushu University, Nishi-ku, Fukuoka, Japan

Won Jong Kim Center for Self-Assembly and Complexity, Institute for Basic Science (IBS), Department of Chemistry Polymer Research Institute, Pohang University of Science and Technology (POSTECH), Pohang, South Korea

Kateřina Knotková Contipro Biotech s.r.o, Dolní Dobrouč, Czech Republic

Tomohiro Konno Department of Bioengineering, School of Engineering, The University of Tokyo, Bunkyo-ku, Tokyo, Japan

Petr Král Department of Chemistry, University of Illinois at Chicago, Chicago, IL, USA

Gabriela Kratošová Nanotechnology Centre, VSB—Technical University of Ostrava, Ostrava, Poruba, Czech Republic

Breanne Landry Faculty of Engineering, Department of Chemical and Materials Engineering, University of Alberta, Edmonton, AB, Canada

Jinghang Liang Department of Electrical and Computer Engineering, University of Alberta, Edmonton, AB, Canada

Zdeňka Maděrová Department of Nanobiotechnology, Institute of Nanobiology and Structural Biology of GCRC, České Budějovice, Czech Republic

Deniz Meneksedag-Erol Faculty of Medicine and Dentistry and Engineering, Department of Biomedical Engineering, University of Alberta, Edmonton, AB, Canada

S. Moein Moghimi Nanomedicine Research Group, Faculty of Health and Medical Sciences, Department of Pharmacy, Centre for Pharmaceutical Nanotechnology and Nanotoxicology, University of Copenhagen, Copenhagen, Denmark

Takeshi Mori Faculty of Engineering, Department of Applied Chemistry, Kyushu University, Nishi-ku, Fukuoka, Japan; Center For Future Chemistry, Kyushu University, Nishi-ku, Fukuoka, Japan

Eva Muchova Laboratory of Synthetic Nanochemistry, Institute of Organic Chemistry and Biochemistry AS CR v. v. i., Prague 6, Czech Republic

Nobuhiro Nishiyama Chemical Resources Laboratory, Polymer Chemistry Division, Tokyo Institute of Technology, Midori-ku, Yokohama, Japan

Jindřich Novák Contipro Biotech s.r.o, Dolní Dobrouč, Czech Republic

Takashi Ohtsuki Department of Biotechnology, Okayama University, Okayama, Japan

Ladan Parhamifar Nanomedicine Research Group, Faculty of Health and Medical Sciences, Department of Pharmacy, Centre for Pharmaceutical Nanotechnology and Nanotoxicology, University of Copenhagen, Copenhagen, Denmark

Stanislav Petrík Institute for Nanomaterials, Advanced Technologies and Innovation, Technical University of Liberec, Liberec 1, Czech Republic

Marek Pokorný Contipro Biotech s.r.o, Dolní Dobrouč, Czech Republic

Kristýna Pospíšková Regional Centre of Advanced Technologies and Materials, Palacký University, Olomouc, Czech Republic

Helena Raabova Laboratory of Synthetic Nanochemistry, Institute of Organic Chemistry and Biochemistry AS CR v. v. i., Prague 6, Czech Republic

Ivan Raška 1st Faculty of Medicine, Institute of Cellular Biology and Pathology, Charles University in Prague, Prague, Czech Republic

Ivan Rehor Laboratory of Synthetic Nanochemistry, Institute of Organic Chemistry and Biochemistry AS CR v. v. i., Prague 6, Czech Republic

Eva Roblegg Department of Pharmaceutical Technology, Institute of Pharmaceutical Sciences, Karl-Franzens-University of Graz, Graz, Austria

Jana Růžičková Contipro Biotech s.r.o, Dolní Dobrouč, Czech Republic

Ivo Šafařík Department of Nanobiotechnology, Institute of Nanobiology and Structural Biology of GCRC, České Budějovice, Czech Republic; Regional Centre of Advanced Technologies and Materials, Palacký University, Olomouc, Czech Republic

Mirka Šafaříková Department of Nanobiotechnology, Institute of Nanobiology and Structural Biology of GCRC, České Budějovice, Czech Republic

Gurusamy Saravanakumar Institute for Basic Science (IBS), Center for Self-Assembly and Complexity, Department of Chemistry Polymer Research Institute, Pohang University of Science and Technology (POSTECH), Pohang, South Korea

Shin-ichi Sawada Department of Polymer Chemistry, Graduate School of Engineering, Kyoto University, Kyoto, Japan; JST-ERATO, Tokyo, Japan

Adam Schröfel 1st Faculty of Medicine, Institute of Cellular Biology and Pathology, Charles University in Prague, Prague, Czech Republic

Yu-Ju Shih Department of Chemical Engineering, R&D Center for Membrane Technology, Chung Yuan Christian University, Chung-Li, Taoyuan, Taiwan

Asako Shimoda Department of Polymer Chemistry, Graduate School of Engineering, Kyoto University, Kyoto, Japan; JST-ERATO, Tokyo, Japan

Jitka Slegerova Laboratory of Synthetic Nanochemistry, Institute of Organic Chemistry and Biochemistry AS CR v. v. i., Prague 6, Czech Republic

Chongbo Sun Faculty of Engineering, Department of Mechanical Engineering, University of Alberta, Edmonton, AB, Canada

Katarzyna Szuszkiewicz Contipro Biotech s.r.o, Dolní Dobrouč, Czech Republic

Hiroyasu Takemoto Chemical Resources Laboratory, Polymer Chemistry Division, Tokyo Institute of Technology, Midori-ku, Yokohama, Japan

Tian Tang Faculty of Engineering, Department of Mechanical Engineering, University of Alberta, Edmonton, AB, Canada

Lemmuel L. Tayo Department of Chemical Engineering, R&D Center for Membrane Technology, Chung Yuan Christian University, Chung-Li, Taoyuan, Taiwan

Ching-Wei Tsai Department of Chemical Engineering, R&D Center for Membrane Technology, Chung Yuan Christian University, Chung-Li, Taoyuan, Taiwan

Hasan Uludag Faculty of Medicine and Dentistry and Engineering, Department of Biomedical Engineering, University of Alberta, Edmonton, AB, Canada Department of Chemical and Materials Engineering, Faculty of Engineering, University of Alberta, Edmonton, AB, Canada; Faculty of Pharmacy and Pharmaceutical Sciences, University of Alberta, Edmonton, AB, Canada

Juliana Valencia-Serna Faculty of Medicine and Dentistry, Department of Biomedical Engineering, University of Alberta, Edmonton, AB, Canada

Luca Vannucci Institute of Microbiology, Academy of Sciences of the Czech Republic (ASCR), v.v.i, Prague, Czech Republic

Ivo Vávra Nanotechnology Centre, VSB—Technical University of Ostrava, Ostrava, Poruba, Czech Republic; Institute of Electrical Engineering, Slovak Academy of Sciences, Bratislava, Slovak Republic

Vladimír Velebný Contipro Biotech s.r.o, Dolní Dobrouč, Czech Republic

Lela Vuković Department of Chemistry, University of Illinois at Chicago, Chicago, IL, USA

Kazunori Watanabe Department of Biotechnology, Okayama University, Okayama, Japan

Peican Zhu Department of Electrical and Computer Engineering, University of Alberta, Edmonton, AB, Canada

Editorial and Introduction

This book features a special subsection of Nanomedicine, an application of nanotechnology to achieve breakthroughs in healthcare. The Nanomedicine exploits the improved and often novel physical, chemical, and biological properties of materials only existent at the nanometer scale. As a consequence of small scale, nanosystems in most cases are efficiently uptaken by cells and appear to act at the intracellular level. Nanotechnology has the potential to improve diagnosis, treatment and follow-up of diseases, and includes targeted drug delivery and regenerative medicine; it creates new tools and methods that impact significantly existing conservative practices. This book more specifically targets using nanotechnology in the area of drug delivery and tissue engineering, i.e., the application of various nanoparticulates based on natural or synthetic, organic or inorgarnic materials as drug carriers and tissue regenerative support, first of all to deliver substances and drugs inside cells.

During the last decade, intracellular drug delivery has become an emerging area of research in the medical and pharmaceutical field. Many therapeutic agents can be delivered to a particular compartment of a cell to achieve better activity. In Volume 1 of this series, we investigated various means of delivering cargo, via endocytosis. Various carriers have been investigated for efficient intracellular delivery, either by direct entry to cytoplasm or by escaping the endosomal compartment. These include cell-penetrating peptides, and carrier systems such as liposomes, cationic lipids and polymers, polymeric nanoparticles, etc. Various properties of these carriers, including size, surface charge, composition, and the presence of cell-specific ligands, alter their efficacy and specificity toward particular cells. Also included were various aspects of targeted intracellular delivery of therapeutics including pathways, mechanisms, and approaches.

This Volume 2, a continuation of Volume 1 (not numbered this way), is a collection of **authoritative reviews**.

The Part I of this volume deals with *Novel Nanocarrier Design and Processing*, listing some new designs and chemistry. The very first chapter deals with a survey of production methods of nanofibers, as exemplified by properietary and succesfull Nanospider^{TM} technology developed by Technical University Liberec, Czech Republic, licensed to Elmarco (Liberec, Czech Republic) (www.elmarco.com). This technology has also been licensed in several countries with applications in different fields as well as in biomedicine. It should be stressed that nanofibers are readily taken up by cells (e.g. Che et al. 2011). Other four chapters describe several different new designs for nanoparticles, with emphasis on responsiveness to different external stimuli.

Part II deals with *Nanocarrier Characterization and Function* The first chapter of this section describes, in some details, how nanoparticles (NP) enter the cells and how they are distributed within the cell interior, while the subsequent chapter describes specific problems related to delivery to mucus. Following are two chapters which cover rather physical methods of nanocarrier characterization, the rest of this section introduces novel delivery vehicles for specific sites or specific cargo.

Part III is entirely a new section; it covers *Simulation for Delivery and Function*. Future applications in nanotechnology are likely to require this level of sophisticated control in order to form precisely ordered structures, with specific chemical and physical properties. Theoretical understanding of the fundamental principles of self-assembly and the design rules for creating new self-assembling materials.

Based on a paper by Vauthier and Bouchemar (2009) two out of about ten different methods of nanoparticle production are (a) formation of polyelectrolyte complexes and (b) production of nanogels. Self-assembly processes typically, both or colloidal building blocks above combine spontaneously to form ordered structures and that without guidance or control from an outside source. Resulting from a disordered system of pre-existing components is an organized structure or pattern as a consequence of specific, local interactions among the components themselves. Self-assembly can be classified as either static or dynamic process. In *static* self-assembly, the ordered state forms as a system approaches equilibrium (thermodynamic stability). In *dynamic* self-assembly, patterns of pre-existing components organized by specific local interactions are not commonly described as "self-assembled" (characterized by the presence of long-range repulsive and short-range attractive forces), whereas they should, in fact, be denoted as "self-organized" (Wikipedia).

New computational simulation tools are required to describe the self-assembly, and to apply them to understand the structures and their thermodynamics and dynamics of both biological and synthetic self-assembling systems (Frenkel and Smit 2002). We envisage that in the future it would be possible to tailor nanoparticles to deliver cargoes at the right subcellular compartment through the use of signaling signatures and pathways. This will improve the magnitude and duration of the drug effects. It is a challenging task due to the complexity of multiple compartments such as endosomes and nuclei, which themselves are dynamic and can undergo fusion and fission and exchange their content (Csukas et al. 2011). The result is to guide further experimental efforts in determining most sensitive parameters. Moreover, there is still much room for building knowledge about the interactions of NPs with proteins and membrane structures on the cell surface. Taking advantage of computer simulations and current developments in interactomics, it would certainly be of great use to know the molecules that interact with the NPs, as well as the nature of this interaction. We emphasize this effort as the literature is relatively scarce in this direction.

The last chapter of this section seeks to emphasize an importance of theoretical background, as provided by Systems Biology, to guide the researcher in the process of discovery. That is, guide the drugs/reagents to an appropriate site. Targeting, localized and intracellular delivery present still a key challenge to effective delivery. To establish an effective fight against diseases, we have to have the ability to selectively attack specific cells, while saving the normal tissue from excessive burdens of drug toxicity. However, because many drugs are designed to simply kill specified cells, in a semi-specific fashion, the distribution of drugs in healthy organs or tissues is especially undesirable due to the potential for severe side effects. Consequently, systemic application of these drugs often causes severe side effects in other tissues (e.g., bone marrow suppression, cardiomyopathy, neurotoxicity), which greatly limits the maximal allowable dose of the drug. In addition, rapid elimination and widespread distribution into nontargeted organs and tissues requires the administration of a drug (in a suitable carrier) in large quantities, which is often not economical and sometimes complicated due to nonspecific toxicity. This vicious cycle of large doses and the concurrent toxicity is a major limitation of many current therapies. Thus, the benefit of nanocarrier design.

Part IV covers *Nanocarriers for Drug Discovery and Treatment*, listing specific applications in biology and medicine. Of a special interest should be a proprietary technology of Contipro s.r.o. (Dolni Dobrouc, Czech Republic; www.contipro.com) employing low-molecular weight hyaluronate to assemble highly biocompatible nanofibers using a technology based on needle-less electrostatic filament principle. The company's main emphasis is in wound healing and other applications.

References

Che H-L, Muthiah M, Ahn Y et al (2011) Biodegradable particulate delivery of vascular endothelial growth factor plasmid from polycaprolactone/polyethylenimine electrospun nanofibers for the treatment of myocardial infarction. J nanosci Nanotechnol 11(8):7073–7077

Csukas B, Varga M, Prokop A (2011) Simulation based analysis of nanocarrier internalization: Exciting challenges with a new computational tool, In: Prokop A (ed)., Intracellular delivery: fundamentals and applications book series: fundamental biomedical technologies Vol 5, pp 125–154

Frenkel D, Smit B (2002) Understanding molecular stimulation: From Algorithms to Applications, 2nd ed. Academic Press, New York

http://en.wikipedia.org/wiki/Self-assembly, accessed October 29, 2013

Vauthier C, Bouchemal K (2009) Methods for the Preparation and Manufacture of Polymeric Nanoparticles. *Pharm Res.* 26:1025–1058

Editorial Plan

- **Novel Nanocarrier (NC) Design and Processing**

  Silica-based
  Nanofiber design/biological
  Multifunctional NC
  Templating
  Hybrid NC
  Core-shell NC
  New Cationic NC
  Colocalization (dual-label)
  Stimuli-responsive NC
  Scale-up

- **Nanocarrier Characterization and Function**

  Physical
  Biological/toxicity
  Stem cell tracking
  Compartmental delivery/trafficking
  NC uptake
  Gene silencing
  Pharmacokinetics and compartmentalization

- **Simulation for Delivery and Function**

  Modeling of self-assembly and molecular modeling
  Payload simulation
  Simulation of release
  Simulation of ligand function/binding energy
  NC dynamics
  Regulation and simulation of NC uptake/endocytosis/exocytosis/disregulation
  NC and pharmacokinetics
  Computational understanding of nanoparticle interface and interaction

- **Nanocarriers for Drug Discovery and Treatment**

  NC imaging
  Imiging/Delivery to brain
  Spinal injury

The Editors would like to profoundly thank all contributors to this volume for their cooperation and enthusiasm, and also, for their reviewing of colleagues' chapters, which served as a basis of internal review process. Finally, we invite contributions from different researchers to this series.

In future volumes, the emphasis will be more on pharmacokinetic aspects as they control the ultimate application and utility. As pointed by Karel Petrak (personal communication on 10/28/2013), "Although I understand the importance of having 'enabling technology' available, the issue of 'promises, promises and more promises' being made about 'new delivery systems' that are never delivered only to be replaced by new promises. To me the central issue is to recognize that the systems must focus on modifying the drug's pharmacokinetics and pharmacodynamics to be optimal for the given disease target." This volume, unfortunately, does not spell out this emphasis clearly. Thus, this eminent topic is sought for future volumes.

Aleš Prokop
Yasuhiko Iwasaki
Atsushi Harada

Part I
Novel Nanocarrier Design and Processing

Proprietary Nanofiber Technologies and Scale-Up

Stanislav Petrík

Abstract An overview of scalable methods for industrial production of nanofibers is given. The theoretical principles of both nozzle- and nozzle-less electrospinning processes are discussed. Productivity limits of electrospinning and competing/complementary technologies (nano-meltblown, force-spinning, islets-in-the sea), together with their predominant potential application areas, are described. Newest developments in production methods for nanofibers are introduced, e.g. nozzle-less co-axial electrospinning and single-nanofiber preparation.

Keywords Nanofibers · Electrospinning · Co-axial · Nozzle-less · Drug delivery · Bio-medical · Force spinning · Nanofiber production

Abbreviations

Φ	Scalar velocity potential
p	Hydrostatic pressure
ρ	Liquid density
E_0	Electric field strength
γ	Surface tension
ω	Angular frequency
k	Wave number
E_c	Critical electric field intensity
λ	Spatial period ("wavelength")
a	Capillary length

S. Petrík (✉)
Institute for Nanomaterials, Advanced Technologies and Innovation, Technical University of Liberec, Studentska 1402/2 461 17 Liberec 1, Czech Republic
e-mail: stanislav.petrik@tul.cz
URL: http://cxi.tul.cz/

A. Prokop et al. (eds.), *Intracellular Delivery II*,
Fundamental Biomedical Technologies 7, DOI: 10.1007/978-94-017-8896-0_1,

1 Introduction

Nanofibers attract consistently growing attention for many applications, including bio-medical, since recent decade. Unique morphology of nanofibers, their extremely high surface area, material variability and relatively simple methods for their preparation opened huge field for both technology processes and material applications research. Number of publications related to the use of nanofibers as delivery systems exhibit probably the highest growth during last few years (Yu et al. 2009).

Electrospinning as a method for production of very fine (submicron) fibers has developed into a dominant technology of industrial production scale. Some limitations connected with the use of (often dangerous) solvents and relatively low productivity for some applications motivate developments of alternate methods which are being commercialized during recent years.

Electrospinning methods for creating nanofibers from polymer solutions have been known for decades (Kirichenko et al. 2007; Ramakrishna et al. 2005). The nozzle-less (free liquid surface) technology opened new economically viable possibilities to produce nanofiber layers in a mass industrial scale, and was developed in the past decade (Jirsak et al. 2005; Petrik and Maly 2009). Hundreds of laboratories are currently active in the research of electrospinning process, nanofiber materials, and their applications. Nanofiber nonwoven-structured layers are ideal for creating novel composite materials by combining them with usual nonwovens. The most developed application of this kind of materials is air filtration (Jaroszczyk et al. 2009). Liquid filters and separators are being developed intensively with very encouraging results. Inorganic/ceramic nanofibers attract growing interest as materials for energy generation and storage (solar and fuel cells, batteries), and catalytic materials (Kavan and Grätzel 2002; Rubacek and Duchoslav 2008; Bognitzki et al. 2001).

To fully explore the extraordinary number of application opportunities of nanofibers, the availability of reliable industrial-level production technology is essential. This chapter intends to demonstrate that some of the technologies have matured to this stage.

2 Nanofibers as Delivery Systems

Well known are several bio-medical applications utilizing nanofiber materials, often from biocompatible/degradable polymers like PLA, gelatine, collagen, chitosan. These developing applications include wound care, skin-, vessel-, bone-scaffolds, drug delivery systems and many others (Proceedings 2009).

One of the first reports about electrospinning nanofibers as delivery systems was published by Kenawy et al. (2002) Electrospun fiber mats were explored as drug delivery vehicles using tetracy-cline hydrochloride as a model drug. The mats were made either from poly (lactic acid) (PLA), poly (ethyl-ene-co-vinyl acetate)

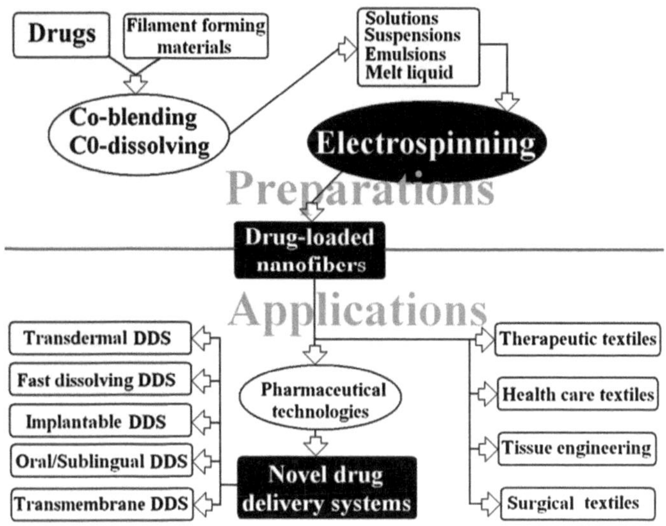

Fig. 1 Applications and preparations of electrospun drug-loaded nanofibers (Yu et al. 2009) (Courtesy of Scientific Research Publishing)

(PEVA), or from a 50:50 blend of the two from chloroform solutions. A detailed overview of delivery bio-medical applications of nanofibers was published by Yu et al. (2009). Their schematic diagram (Fig. 1) illustrates most of the opportunities the nanofiber systems offer for drug delivery, scaffold/tissue engineering, health care textiles, surgical textiles, and other systems.

The active agents (i.e. drugs) can be incorporated into nanofibers in several ways. The most common one used to be to mix functional particles into the polymer solution the nanofiber material is being prepared from. This approach often limits technological processability of the material. As many authors have proven (Buzgo et al. 2013; Mickova et al. 2012; Williams et al. 2012), co-axial (core-shell) nanofibers offer much larger potential as delivery systems, because of their capability to incorporate and protect also the agents which are not spinnable or non-dispersable in homogeneous nanofibers. Besides „trivial"technological approach based on co-axial needle electrospinning (i.e. Azarbayjani et al. (2010)), Lukas'group at the Technical University of Liberec (Vyslouzilova et al. 2010) has patented and published a nozzle-less productive electrospinning device described below.

3 Electrospinning

The electrospinning process is an interesting and well-characterized physical phenomenon and has been an attractive subject for theoretical investigations of several groups (Bognitzki et al. 2001; Doshi and Reneker 1995; Thompson et al. 2007;

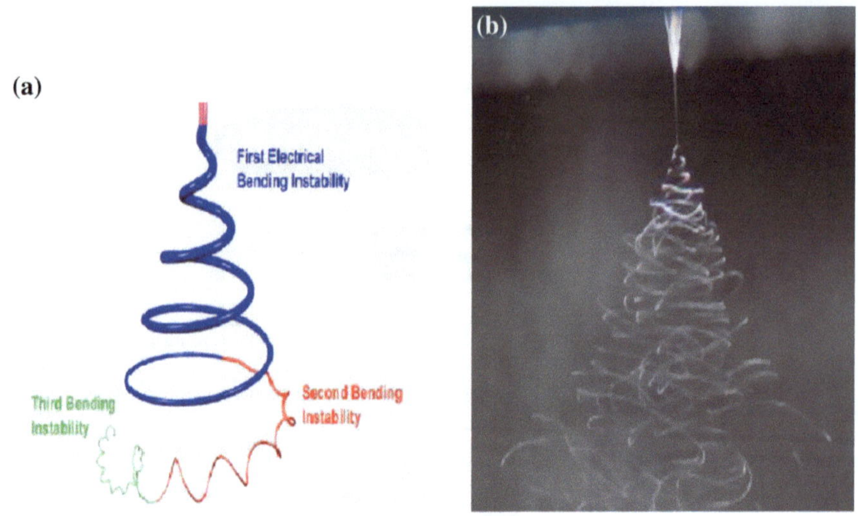

Fig. 2 The path of an electrospinning jet **a** schematic, **b** stroboscopic photograph (*Courtesy of Darrell Reneker, University of Akron*)

Shin et al. 2001; Yu et al. 2006; Hohman et al. 2001). Most work concentrates on the essentials of the process—the nanofiber formation from a liquid polymer jet in a (longitudinal) electric field. It has been theoretically described and experimentally proven that the dominant mechanism is whipping elongation occurring due to bending instability (Thompson et al. 2007; Yu et al. 2006; Hohman et al. 2001). Secondary splitting of the liquid polymer streams can occur also (Kirichenko et al. 2007), but the final thinning process is elongation.

In Fig. 2, the schematic of bending mechanism derived from physical model (a) is compared with a stroboscopic snapshot (b) (Reneker 2009).

A comprehensive analysis (electrohydrodynamic model) of the fiber formation mechanisms published by (Hohman et al. 2001) describes the regions of individual kinds of instability observed during the process. It has predicted and experimentally proven that there is a domain of the process variables where bending instability dominates, as illustrated in Fig. 3.

The efforts to scale up the electrospinning technology to an industrial production level used to be based on multiplication of the jets using multi-nozzle constructions (Kirichenko et al. 2007).

In Fig. 4, the multi-nozzle spinning head developed by NanoStatics Company is shown. The principle is based on an idea to feed multiple nozzles from a single source of the polymer solution.

Figure 5 shows the multi-nozzle spinning part of the machine being commercialized by TOPTEC Company. The device uses upwards direction of electrospinning in order to eliminate polymer droplets eventually falling from conventional down-oriented electrospinning elements.

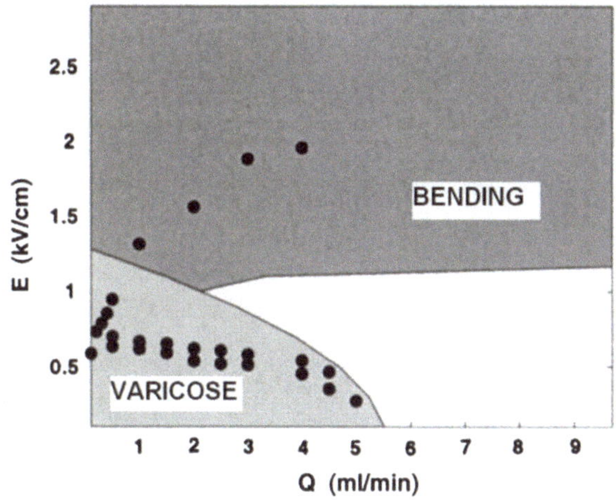

Fig. 3 Operating diagram for a PEO jet. The *upper shaded* region shows the onset of the whipping instability, the *lower* one shows the onset of the varicose instability (Hohman et al. 2001b)

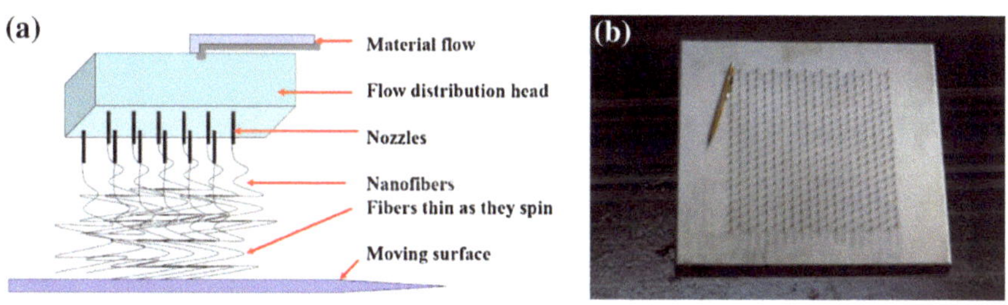

Fig. 4 Schematic (**a**), and photograph (**b**) of a multi-nozzle spinning head by NanoStatics (NanoStatics 2007)

However, the number of jets needed to reach economically acceptable productivity is very high, typically thousands. This brings into play many challenging task, generally related to reliability, quality consistency, and machine maintenance (especially cleaning). The nozzle-less electrospinning solves most of these problems due to its mechanical simplicity, however, the process itself is more complex because of its spontaneous multi-jet nature. The study by (Lukas et al. 2008) focused on the process of multi-jet generation from a free liquid surface in an electric field. They derived an expression for the critical spatial period ("wavelength")—the average distance between individual jets emerging from the liquid surface (Fig. 5). In this system, self-organization of the jets occurs, thus the number and spacing of the jets is optimal even if the technology variables (voltage, viscosity and surface tension of the solution) change. This feature leads to significant improvement of the process stability and consistent quality of the produced nanofiber layer.

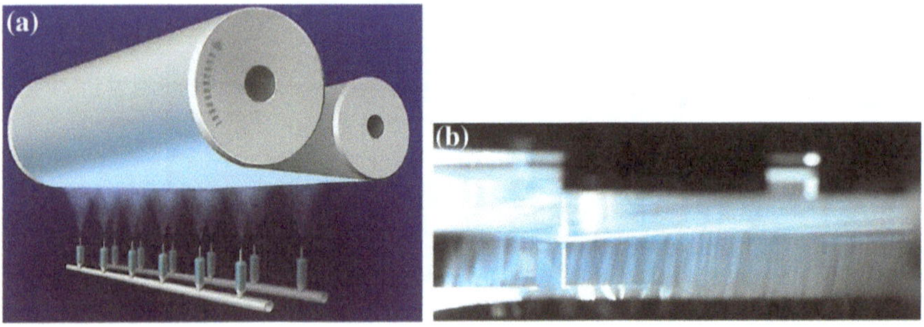

Fig. 5 Schematic (**a**), and photograph (**b**) of a multi-nozzle spinning head by TOPTEC (TOPTEC 2011)

Fig. 6 Relationship between the square of the angular frequency and the wave number for distilled water, electric field is the parameter *1* $E = E_c = 2.461\,945\,094 \times 10^6$ V/m, *2* $E = 2.4 \times 10^6$ V/m, and *3* $E = 2.5 \times 10^6$ V/m (Lukas et al. 2008) (Courtesy of D. Lukas, TU Liberec)

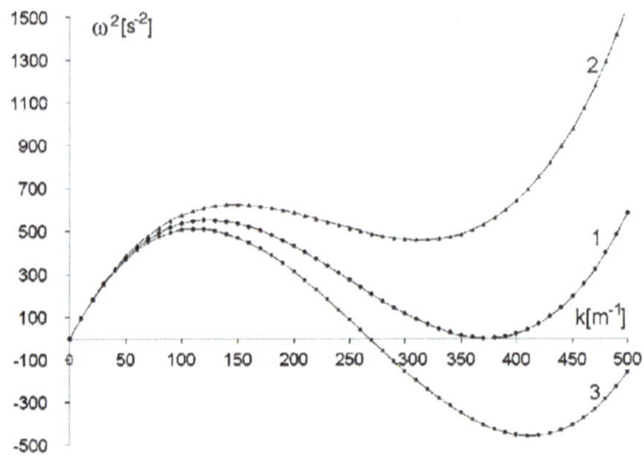

The study showed that the process can be analyzed using Euler's equations for liquid surface waves

$$\nabla\left(\rho\frac{\partial\Phi}{\partial t}+p\right)=0 \tag{1}$$

where Φ is the scalar velocity potential, p is the hydrostatic pressure, and ρ is the liquid density. They derived the dispersion law for the waves in the form

$$\omega^2=\left(\rho g+\gamma k^2-\varepsilon E_0^2 k\right)\frac{k}{\rho} \tag{2}$$

where E_0 is electric field strength, γ—surface tension.

The relationship between angular frequency ω and wave number k is in Fig. 6, electric field is the parameter. When a critical electric field intensity is reached (E_c, curve 1), ω^2 is turned to be negative, ω is then a purely imaginary value, and hence, the amplitude of the liquid surface wave

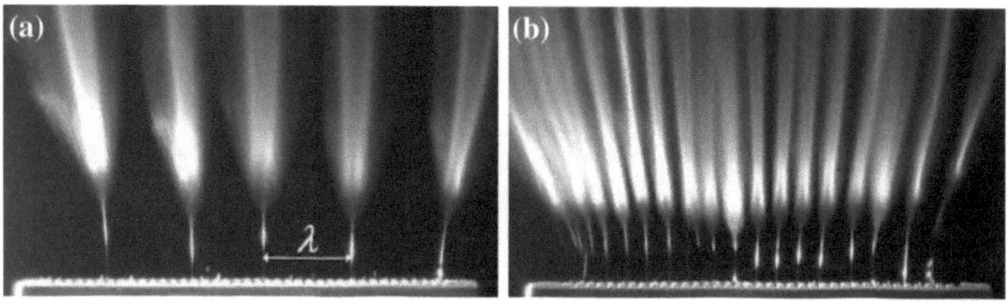

Fig. 7 **a** Free liquid surface electrospinning of Polyvinyalcohol at 32 kV, and **b** 43 kV (*Courtesy of David Lukas, Technical University of Liberec*)

$$\xi = Ae^{qt}\exp(ikx) \tag{3}$$

exponentially grows, which leads to an instability.

Critical field strength can then be expressed

$$E_c = \sqrt[4]{4\gamma\rho g/\varepsilon^2} \tag{4}$$

From this equation, they derived the expression for the critical spatial period ("wavelength")—the average distance between individual jets emerging from the liquid surface (Fig. 7).

$$\lambda_c = 2\pi/k_c = 2\pi a \tag{5}$$

and

$$\lambda = 12\pi\gamma/\left[2\varepsilon E_0^2 + \sqrt{\left(2\varepsilon E_0^2\right)^2 - 12\gamma\rho g}\right] \tag{6}$$

a is the capillary length

$$a = \sqrt{\gamma/\rho g} \tag{7}$$

The simplest realization of the nozzle-less electrospinning head is in Fig. 8a. A rotating drum is dipped into a bath of liquid polymer. The thin layer of polymer is carried on the drum surface and exposed to a high voltage electric field. If the voltage exceeds the critical value, a number of electrospinning jets are generated. One of the main advantages of nozzle-less electrospinning is that the number and location of the jets is set up naturally in their optimal positions. In the case of multi-needle spinning heads, the jet distribution is made artificially. The mismatch between "natural" jet distribution and the real mechanical structure leads to instabilities in the process, and to the production of nanofiber layers which are not homogenous.

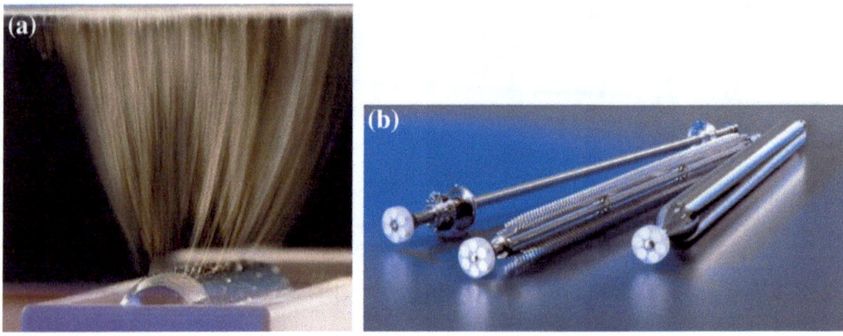

Fig. 8 **a** Free liquid surface electrospinning from a rotating electrode, and **b** various types of spinning electrodes

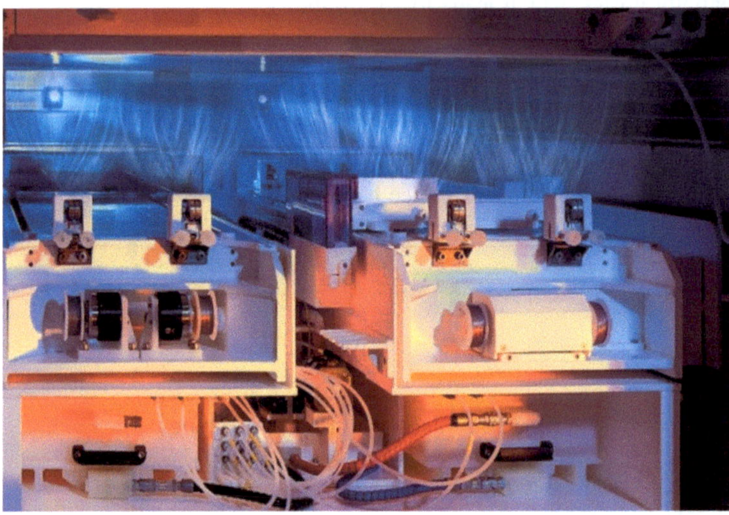

Fig. 9 Photograph of production electrospinning heads inside Nanospider^{TM} machine (2nd generation). Four strings with upwards fiber jets. (Elmarco 2013)

Several types of rotating electrodes for free liquid surface electrospinning for industrial machines have been developed (Fig. 8b). However, the drum type is still one of the most productive.

Recent significant improvement of the production nozzle-less electrospinning equipment, commercially available from Elmarco Company, is illustrated in Figs. 9 and 10. The system uses stationary string electrodes supplied with polymer solution by a proprietary moving "painting" head. This led to a dramatic decrease of solvent evaporation during the process which has to be removed from the exhaust air released from the machine. Also, the polymer solution concentration is stable, enabling to continuously run the production process for long time, typically more than 24 hours.

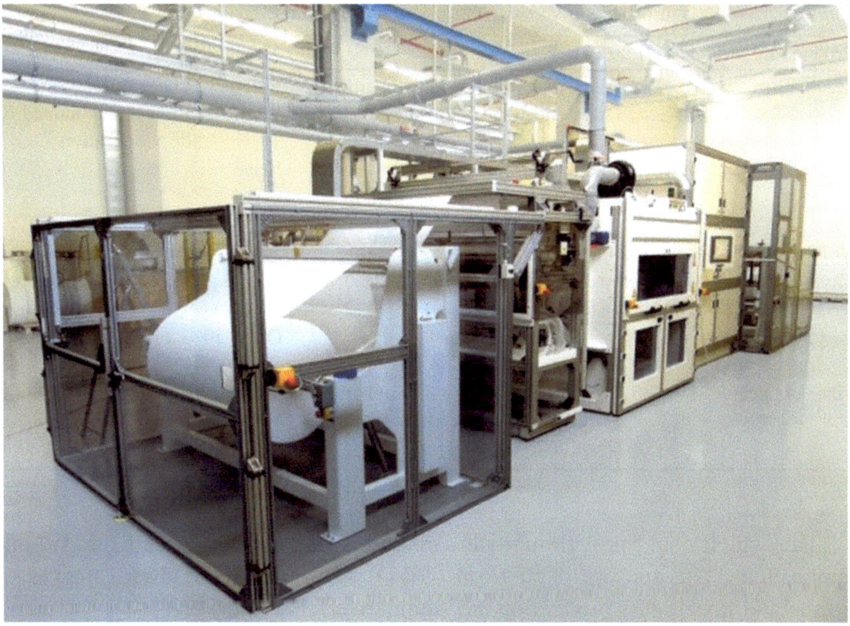

Fig. 10 Nanospider™ production line NS 8S1600U by Elmarco (2013)

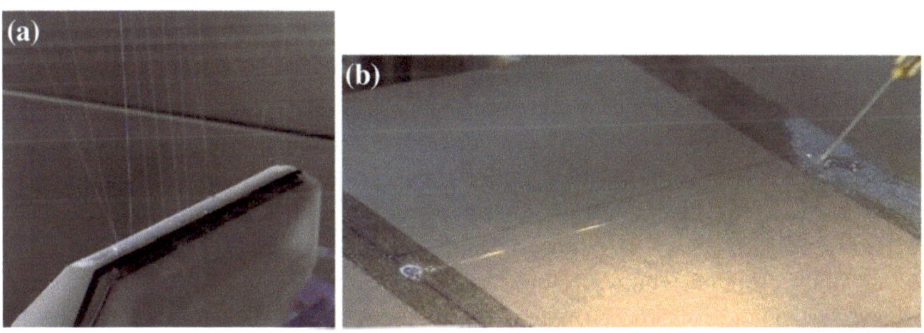

Fig. 11 Co-axial nozzle-less spinning head (**a**), single nanofiber drawing (**b**) (*Courtesy of D. Lukas, TU Liberec*)

For many applications, especially bio-medical, the co-axial (core-shell) nanofibers attract intense research interest recently. A single-needle laboratory apparatus can be relatively simple, however, a more productive equipment development is challenging. The researchers at the Technical University of Liberec (Czech Republic) successfully demonstrated the nozzle-less co-axial electrospinning head (Fig. 11) possessing the same advantages as Nanospider™ (Vyslouzilova et al. 2010). The same team works on development of a single-nanofiber production

Table 1 Comparison of nozzle versus Nozzle-less electrospinning

Production variable	Nozzle	Nozzle-less
Mechanism	Needle forces polymer downwards. Drips and issues deposited in web	Polymer is held in bath, even distribution is maintained on electrode via rotation
Hydrostatic pressure	Production variable–required to be kept level across all needles in process	None
Voltage	5–20 kV	30–120 kV
Taylor cone separation	Defined mechanically by needle distances	Nature self-optimizes distance between Taylor cones
Polymer concentration	Often 10 % of solution	Often 20 % or more of solution
Fiber diameters	80, 100, 150, 200, 250 and higher. Standard deviation likely to vary over fiber length	80, 100, 150, 200, 250 and higher. Standard deviation of ±30 %

system, which could provide unique nanofiber structures for special low-volume applications, e.g. sensors (Tsai et al. 2011). Advantages and disadvantages of nozzle and nozzle-less production are summarized in Table 1.

4 Alternate Technologies for Nanofibers

Research and development centers are very active in their efforts to further improve productivity of the manufacturing process. Novel methods for the production of sub-micron fibers are being developed. The individual methods can be considered to be complementary rather than competing. This is especially true with respect to the fiber diameter distribution and fiber layer uniformity.

Figure 12 shows the **extrusion methods** being developed by Hills Inc. (HILLS Inc 2011). Dominant technique that Hills practices to produce nanofibers is done using the Island-In-The-Sea (fibers within fibers) method. This method has the capability of making a large number of fibers within a fiber. The Hills declare they are able to spin up to 1,200 fibers within a single fiber. Using the same techniques, these filament can be produced as hollow tubes.

Centrifugal forces for elongation of liquid polymer into thin fibers are used in the approach developed by (Dauner et al. 2008), illustrated in Fig. 13. The productivity of the process they claim is high (up to 1,000 $cm^3/m\ h$), however, the fiber diameter distribution and homogeneity of the deposited nanofiber layer is not at the levels achieved by electrospinning.

Strong commercialization effort is shown recently by FibeRio Company (www.fiberio.com), developing so called "force spinning" principle into an industrial scale. The principle discovered at the University of Texas Pan American is based on high-speed rotating spinneret depositing nanofibers on the radial collector (Fig. 14)

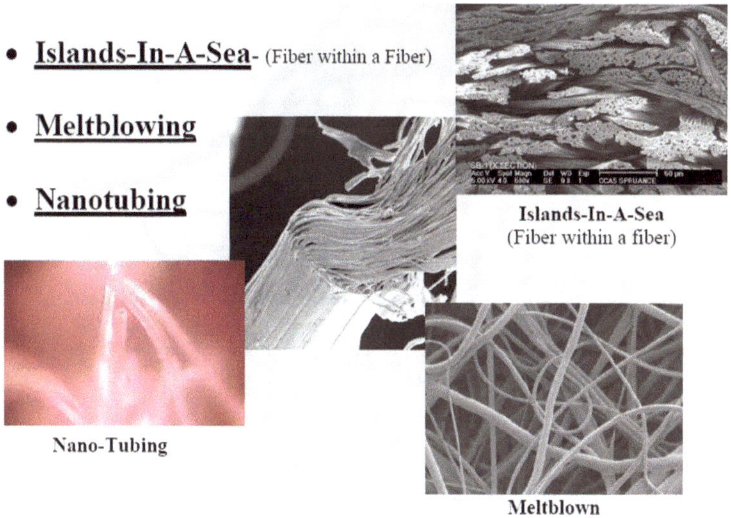

Fig. 12 Fibers made by extrusion methods (HILLS Inc 2011) (Courtesy of HILLS Inc.)

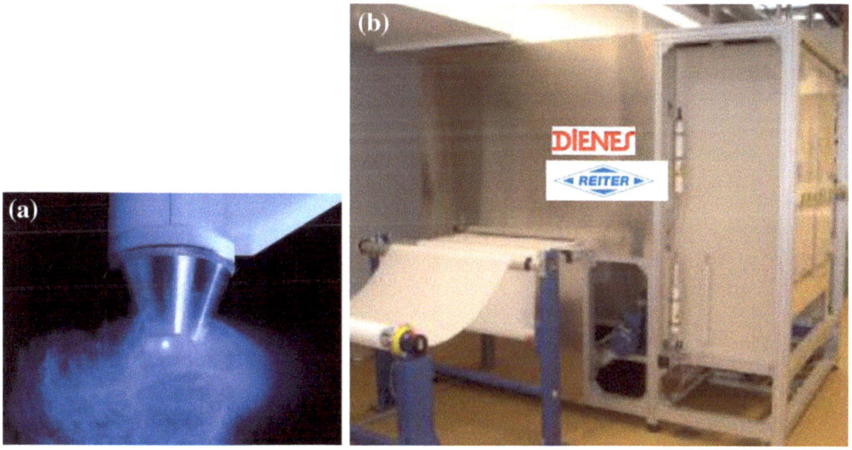

Fig. 13 Centrifugal spinning head (**a**) and a pilot production line by Dienes/Reiter (**b**) (Dauner et al. 2008)

(FibeRio 2013). The productivity of the process should be an order higher compared to electrospinning. However, mechanical design of the equipment is much more demanding (rotating parts at tens of thousands rpm, fed with liquid polymer). Also, radial deposition of the produced nanofiber materials could be challenging task for the applications, where continuous roll-to-roll deposition of thin layer to a substrate is required, e.g. at the nanofiber filtration media production.

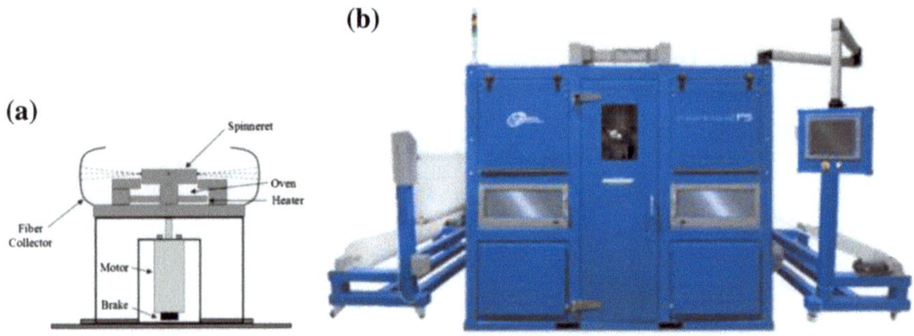

Fig. 14 Force Spinning™ technology by FibeRio: Schematic of the machine design (**a**), and photograph of the Fiber Engine® FS1000 production line (**b**) (FibeRio 2013)

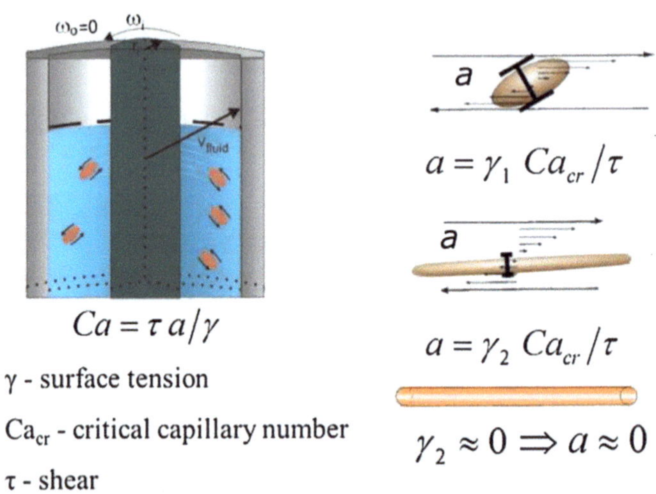

Fig. 15 Liquid shear nanofiber fabrication process (NCSU 2013)

Shear forces in a rotating liquid containing polymer droplets are the basis of XanoShear™ method developed at North Carolina State University (Alargova et al. 2004). The method is being commercialized by Xanofi Company (www.xanofi.com), offering to the market several nanofiber materials (filtration media, cell scaffolds, acoustic absorbents, etc.). Schematic of the equipment is in Fig. 15. Liquid shear nanofabrication process is a novel method to prepare nanofibers by subjecting polymer solution droplets to simultaneous shear and anti-solvent induced precipitation in viscous liquid media. The typical lab scale process involves creation of laminar shear in viscous media (glycerol + antisolvent) using a shear impeller and injection of polymer solution droplets into it. During shear process, the low interfacial tension between major component of viscous media i.e.

glycerol and polymer solution, leads to infinite stretching of polymer droplets into proto fibers and simultaneous precipitation by anti-solvent component in viscous media gives rise to solidified fibers of very thin diameter in the range 300–500 nm.

5 Conclusion

Individual production methods for nanofiber materials will likely find different areas of application. More productive extrusion technologies compromise fiber diameter and homogeneity and will likely be used in production cost sensitive applications like hygiene nonwovens, while high quality electrospinning technologies will be used in products where their high added value and need for low amounts of the material can be easily implemented (air and liquid filtration, biomedicine)..

High-quality low-cost production of nanofiber layers is essential to support the enormous amount of research results being obtained at many universities and research centers. Some of the technologies (nozzle-less electrospinning, force spinning, nano-meltblown) have matured to a level where large scale production use is common, and can be modified for practically all known polymers soluble in organic solvents and water, as well as for polymer melts. This opens commercial opportunities for hundreds of ideas developed in the academic sphere.

References

Alargova RG, Bhatt KH, Paunov VN et al (2004) Scalable synthesis of a new class of polymer Microrods by a liquid–liquid dispersion technique. Adv Mater 16:1653

Azarbayjani AF, Venugopal JR, Ramakrishna S et al (2010) Smart polymeric nanofibers for topical delivery of levothyroxine. J Pharm Pharmaceut Sci 13(3):400–410

Bognitzki M, Czado W, Frese T et al (2001) Nanostructured fibers via electrospinning. Adv Mater 13:70

Buzgo M, Jakubova R, Mickova A et al (2013) Time-regulated drug delivery system based on coaxially incorporated platelet α-granules for biomedical use. Nanomedicine 8(7):1137–1154. doi:10.2217/nnm.12.140

Dauner M, Ullrich A, Reiter F (2008) Nanofibers by centrifuge spinning to improve filter media. In: Proceedings of 10th world filtration congress (WFC10), Leipzig, Germany, 14–18 Apr 2008

Doshi J, Reneker DH (1995) Electrospinning process and applications of electrospun fibers. J Electrost 35:151

Duchoslav J, Rubacek L (2008) Electrospun TiO_2 fibers as a material for dye sensitized solar cells. In: Proceedings of nsti nanotech conference, Boston, MA 1–5 June 2008

Elmarco (2013) Pictures available at www.elmarco.com

FibeRio (2013) Pictures available at www.fiberio.com

HILLS Inc (2011) Hills Nano Technology Brochure. West Melbourne, FL

Hohman MM, Shin M, Rutledge GC et al (2001a) Electrospinning and electrically forced jets I. Stability theory. Phys Fluids 13:2201

Hohman MM, Shin M, Rutledge GC et al (2001b) Electrospinning and electrically forced jets II. Applications. Phys Fluids 13:2221

Jaroszczyk T, Petrik S, Donahue K (2009) The role of nanofiber filter media in motor vehicle air filtration. In: Proceedings of 4th biennial conference on emissions solutions in transportation, AFS, Ann Arbor, MI, 5–8 Oct 2009

Jirsak O, Sanetrnik F, Lukas D, Kotek V et al (2005). A method of nanofibers production from a polymer solution using electrostatic spinning and a device for carrying out the method. The Patent Cooperation Treaty WO 2005/024101

Kavan L, Grätzel M (2002) Facile synthesis of nanocrystalline Li4Ti5 O 12 (Spinel) exhibiting fast Li insertion. Electrochem Solid-State Lett 5:A39

Kenawy ER, Bowlin GL, Mansfield K et al (2002) Release of tetracycline hydrochloride from electrospun poly (ethylene-co-vinylacetate), poly(lactic acid), and a blend. J Control Release 81:57–64

Kirichenko V, Filatov Y, Budyka A (2007) Electrospinning of micro-and nanofibers: fundamentals in separation and filtration processes. Begell House, USA

Lukas D, Pokorny P, Sarkar A (2008) Self-organization of jets in electrospinning from free liquid surface: a generalized approach. J Appl Phys 103:084309

Mickova A, Buzgo M, Benada O et al (2012) Core/shell nanofibers with embedded liposomes as a drug delivery system. Biomacromolecules 13(4):952–962. doi:10.1021/bm2018118

NanoStatics (2007) Company brochure, available at www.nanostatics.com

NCSU (2013) Picture available at http://www.che.ncsu.edu/velevgroup/vitchuli.html

Petrik S, Maly M (2009) Production nozzle-less electrospinning nanofiber technology. Mater Res Soc Symp Proc 1240:1240-WW03-07

Proceedings of Nanofibers for the 3rd Millenium—Nano for Life Conference (2009), Prague, 11–12 Mar 2009

Ramakrishna S, Fujihara K, Teo W et al (2005) An introduction to electrospinning and nanofibers. World Scientific, Singapore

Reneker DH (2009) Personal communication

Rubacek L, Duchoslav J (2008) Electrospun nanofiber layers for applications in electrochemical devices. In: Proceedings of NSTI nanotech conference, Boston, MA, 1–5 June 2008

Shin YM, Hohman MM, Brenner MP et al (2001) Electrospinning: a whipping fluid jet generates submicron polymer fibers. Appl Phys Lett 78:1149

Thompson CJ, Chase GG, Yarin AL et al (2007) Effects of parameters on nanofiber diameter determined from electrospinning model. Polymer 48:6913

TOPTEC (2011). Company brochure, available at www.toptec.co.kr

Tsai CH, Mikes P, Andrukh T et al (2011) Nanoporous artificial proboscis for probing minute amount of liquids. Nanoscale 3:4685

Vyslouzilova L, Vodsedalkova K, Pokorny P et al (2010) Needleless co-axial electrospinning. In: Proceedings of nanofibers for the 3rd millenium—nano for life conference, Raleigh, NC, 30 Aug–1 Sept 2010

Williams GR, Chatterton NP, Nazir T, Yu D-G et al (2012) Electrospun nanofibers in drug delivery: recent developments and perspectives. Ther Deliv 3(4):515–533. doi:10.4155/tde.12.17

Yu JH, Fridrikh SV, Rutledge GC (2006) The role of elasticity in the formation of electrospun fibers. Polymer 47:4789

Yu D-G, Zhu L-M, White K et al (2009) Electrospun nanofiber-based drug delivery systems. Health 1(2):67–75. doi:10.4236/health.2009.12012

Magnetically Responsive (Nano) Biocomposites

Ivo Šafařík, Kristýna Pospíšková, Kateřina Horská, Zdeňka Maděrová and Mirka Šafaříková

Abstract Many biological materials have been successfully used in various areas of bioscience, biotechnology and environmental technology applications. Biological materials are mainly diamagnetic, which means they do not interact significantly with external magnetic field. Using various postmagnetization procedures, biological materials can be converted into magnetically responsive composite materials which can be efficiently separated using simple magnets or magnetic separation systems. The prepared magnetic biocomposites can be used for many applications, such as immobilization of target compounds, as parts of biosensors, as whole cell biocatalysts or for magnetic removal and separation of xenobiotics and biologically active compounds.

Keywords Biological materials · Postmagnetization · Magnetic composite materials · Magnetic separations

Abbreviations

DBT	Dibenzothiophene
EPEC	Enteropathogenic *Escherichia coli*
PS	Phosphatidylserine
MRI	Magnetic resonance imaging
MSCs	Mesenchymal stem cells
VTEC	Verocytotoxigenic *Escherichia coli*

I. Šafařík (✉) · K. Horská · Z. Maděrová · M. Šafaříková
Department of Nanobiotechnology, Institute of Nanobiology and Structural Biology of GCRC, Na Sádkách 7, 370 05 České Budějovice, Czech Republic
e-mail: ivosaf@yahoo.com
URL: http://www.nh.cas.cz/people/safarik/

I. Šafařík · K. Pospíšková
Regional Centre of Advanced Technologies and Materials, Palacký University, Šlechtitelů 11, 783 71 Olomouc, Czech Republic

A. Prokop et al. (eds.), *Intracellular Delivery II*, Fundamental Biomedical Technologies 7, DOI: 10.1007/978-94-017-8896-0_2,

1 Introduction

Magnetically responsive nano- and microparticles and related structures have found many important applications in various areas of biosciences, medicine, biotechnology and environmental technology (Safarik and Safarikova 2012; Safarik et al. 2012d). Different types of responses of such materials to external magnetic field enable their various applications, namely their selective separation, targeting and localization using an external magnetic field (e.g. using an appropriate magnetic separator, permanent magnet, or electromagnet), heat generation (which is caused by magnetic particles subjected to high frequency alternating magnetic field), increase of a negative T2 contrast by magnetic iron oxides nanoparticles during magnetic resonance imaging or great increase of apparent viscosity of magnetorheological fluids when subjected to a magnetic field. In addition, magnetite nanoparticles can exhibit peroxidase-like activity (Gao et al. 2007; Safarik et al. 2012d; Safarik and Safarikova 2012). Currently large amount of various magnetic nano- and micromaterials can be obtained commercially. Alternatively such materials are produced in research laboratories, using many different basic principles (Laurent et al. 2008; Safarik et al. 2011a; Li et al. 2013). Many research groups are involved in the fine tuning of procedures leading to the production of homogeneous magnetic particles with defined size, structure, composition, coating etc. (Safarik et al. 2012d).

Different types of non-magnetic particulate materials including adsorbents, catalysts, chromatography materials, carriers, microbial cells, waste biological materials etc. are available. In many cases the application potential of these materials could be improved by their modification leading to the formation of magnetically responsive materials. Such a modification can substantially simplify separation of magnetic materials from complex systems. So, in addition to the main currently used strategies focused on specific preparation of target magnetic structures, alternative strategies leading to the formation of interesting magnetically responsive materials have been developed. These procedures are based on postmagnetization of already existing diamagnetic ("non-magnetic") or paramagnetic ("weakly" magnetic) particulate materials found in nature or prepared in the laboratory or industry (Safarik et al. 2012d). The term "postmagnetization" and the whole process of magnetic modification of non-magnetic materials was invented by Mosbach (Mosbach and Andersson 1977; Griffin et al. 1981) at the end of 70s and beginning of 80s of the 20th century, when gel particles for column affinity chromatography were magnetically modified using appropriate ferrofluid (magnetic fluid); such a modification led to the preparation of magnetic derivatives with unaltered biospecificity when applied to general ligand affinity chromatography studies (Safarik et al. 2012d).

Postmagnetization usually leads to the formation of strongly magnetic composite materials, where the "original" diamagnetic or paramagnetic structure is responsible for biological, catalytic, carrier or adsorption function of the formed composite, while magnetic label (most often in the form of magnetic iron oxides nano- and microparticles, which are usually deposited on the surface or within the pores of treated materials) is responsible for the strong magnetic behavior of the formed composite materials (Safarik and Safarikova 2009; Safarik et al. 2012d). Alternatively non-magnetic materials can be modified by erbium ions which preserve their exceptionally high atomic magnetic dipole moment in various chemical structures (Zborowski et al. 1993) or by co-entrapment of nonmagnetic and magnetic particles in a gel material (Al-Dujaili et al. 1979; Safarik et al. 2012a). Postmagnetization can be applied to a broad variety of inorganic materials [e.g., clays (Bartonkova et al. 2007; Mockovciakova et al. 2010; Safarik et al. 2012b)], activated charcoal (Safarik et al. 1997, 2012a; Nakahira et al. 2007; Schwickardi et al. 2006), synthetic polymer particles (Cumbal and SenGupta 2005), biopolymer particles (Griffin et al. 1981; Mosbach and Andersson 1977; Torchilin et al. 1985), waste plant materials (Safarik et al. 2007a, 2011b, 2012c; Safarik and Safarikova 2010; Tian et al. 2011; Pospiskova and Safarik 2013), microbial cell walls (Patzak et al. 1997), whole microbial and algae cells (Safarik et al. 2007b; Pospiskova et al. 2013; Safarikova et al. 2008, 2009; Prochazkova et al. 2013) and many others.

Postmagnetization procedures can substantially increase the amount of useful magnetically responsive materials and subsequently the amount of their important applications in various areas of biosciences, biotechnology, food technology (bio)analytical chemistry, environmental chemistry and technology, removal of radionuclides, etc. (Safarik et al. 2012d).

As already mentioned, many different types of "non-magnetic" materials can be converted into the magnetic form. Table 1 shows examples of described postmagnetization procedures for the modification of inorganic materials, polymers and carbon materials. In this chapter, the attention will be mainly paid to postmagnetization of materials taken from living nature, namely plant derived materials and both prokaryotic and eukaryotic cells. Magnetic derivatives of these materials have been already successfully used e.g. as adsorbents for the removal of both organic and inorganic xenobiotics, carriers for the immobilization of biologically active compounds, whole cell biocatalysts and biosensor elements. However, the potential of magnetically responsive biocomposites is substantially greater.

2 Postmagnetization of Biomaterials

The ideal postmagnetization procedure should be cheap, easy-to-perform, scalable and tunable, leading to a stable magnetic product, both in dry state and water suspension. Preparation of magnetic derivatives of originally non-magnetic biomaterials can follow selected procedures used already for the postmagnetization of non-biological materials, as exemplified in Table 1. However, only selected procedures can be used for postmagnetization of biomaterials due to their specific characters. One of the general approaches is based on suspending the treated

Table 1 Examples of described postmagnetization procedures for the modification of inorganic materials, polymers and carbon materials

Non-magnetic material	Way of postmagnetization	Reference
Activated carbon	Impregnation of activated carbon with $Fe(NO_3)_3$ solution followed by drying at 90 °C and heating to a temperature of 700 °C under argon and benzene vapor	Schwickardi et al. (2006)
Bentonite	A suspension of bentonite in water was mixed with $FeSO_4$, then KNO_3 and KOH were added, then mixture was heated up to 90 °C	Bartonkova et al. (2007)
Bentonite	Precipitation of iron oxides from $FeSO_4$ and $FeCl_3$ by NH_4OH in the presence of bentonite in nitrogen atmosphere, followed by drying at 70 °C	Mockovciakova et al. (2010)
Carbon nanotubes	Filling of carbon nanotubes with ferrofluid followed by drying to leave deposited magnetic nanoparticles	Korneva et al. (2005)
Carbon nanotubes	Mixing of $(NH_4)_2Fe(SO_4)_2$ with hydrazine followed by the addition of carbon nanotubes, pH increase and heating	Li et al. (2010)
Charcoal	Co-entrapment of charcoal and magnetite particles in polyacrylamide gel	Al-Dujaili et al. (1979)
Charcoal	Precipitation of magnetite from $FeSO_4$ and $Fe_2(SO_4)_3$ by NaOH in the presence of charcoal, followed by aging for 24 h and heating at 473 K	Nakahira et al. (2007)
Charcoal	Precipitation of iron oxides from $FeSO_4$ and $FeCl_3$ by NaOH in the presence of charcoal, followed by drying at 100 °C for 3 h	Oliveira et al. (2002)
Charcoal	Precipitation of hydrated iron oxides from $FeSO_4$ by NaOH in the presence of charcoal, followed by heating to 100 °C for 1 h	Safarik et al. (1997)
Chromatography gels	Circulation of water based ferrofluid through a bed of chromatography gel	Mosbach and Andersson (1977), Griffin et al. (1981)
Montmorillonite	One g of powder was thoroughly mixed in a small beaker with 1 mL of water based ferrofluid stabilized with perchloric acid. This mixture was allowed to dry completely at temperatures not exceeding 50 °C	Safarik et al. (2012b)

materials in the solution of iron ions and after increase of pH and heating magnetic iron oxides particles are formed, thus modifying the treated material (Safarik et al. 1997). More gentle modification procedures are usually based on the deposition of pre-prepared magnetic iron oxides nano- or microparticles on the surface or within the pores of the treated biomaterials (Safarik et al. 2007a; Safarik and Safarikova 2014). In addition, several other alternative procedures can also be employed, such as covalent binding of magnetic particles to the treated materials (or vice versa, depending on the size), cross-linking of cells, isolated cell walls or other biomaterials with a bifunctional reagent in the presence of magnetic particles, by the biologically driven precipitation of paramagnetic compounds on the cell surface or by biospecific binding of immunomagnetic particles to the target antibody containing materials (especially cells). Also entrapment of the modified materials into synthetic polymer, biopolymer or inorganic gels containing magnetic particles can be employed, as well as labeling with erbium cation (Safarik and Safarikova 2007; Safarik et al. 2011c, 2012d).

Recently, new and efficient postmagnetization procedures have been developed. One of them employed water based magnetic fluid stabilized with perchloric acid, which was directly mixed with material to be modified and dried completely. This procedure is extremely simple and various biological, inorganic and polymer materials have been successfully transferred into their magnetic derivatives (Safarik et al. 2012b).

Another postmagnetization process is based on microwave irradiation of suspensions containing the material to be modified and iron hydroxides (formed from ferrous sulfate after increasing the pH to 10–12 by the addition of sodium hydroxide); during the microwave treatment magnetic iron oxides nano- and microparticles are formed and subsequently bound on the surface of the treated material. Using this procedure a large amount of materials has been postmagnetized (Safarik et al. 2013).

A very efficient modification of the microwave assisted postmagnetization procedure enables to modify also heat- and pH- sensitive nonmagnetic materials. At first magnetic iron oxides nano- and microparticles have been synthesized from ferrous sulfate at high pH in a microwave oven. After their washing with water the suspension of magnetic particles was mixed thoroughly with nonmagnetic material to be modified. After complete drying stable magnetically responsive materials have been formed. This procedure is extremely inexpensive, very simple, scalable and tunable. Also rather sensitive biological materials including starch grains and insoluble proteins particles have been successfully magnetized (Safarik and Safarikova 2014).

2.1 Magnetic Modification of Non-living Biomaterials

A very simple procedure leading to the formation of postmagnetized materials has been developed, using water based magnetic fluids (ferrofluids) as the modifying agent. The most frequently used magnetic fluid (stabilized with perchloric acid) can be prepared using the "classical" procedure developed by Massart (Massart 1981). This ferrofluid was usually composed of maghemite nanoparticles with the diameter bellow 15 nm (Mosiniewicz-Szablewska et al. 2007, 2010) (Fig. 1). The material to be modified was suspended in methanol and appropriate amount of ferrofluid was then added. Magnetic iron oxide nanoparticles were deposited on the surface of the treated materials after short mixing, both in the form of

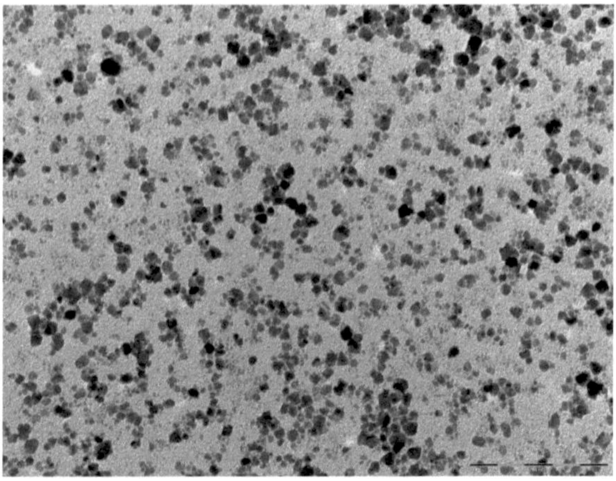

Fig. 1 TEM image of nanoparticles present in the ferrofluid stabilized with perchloric acid used for postmagnetization of biological materials (bar = 200 nm). Reproduced, with permission from (Safarik et al. 2012d)

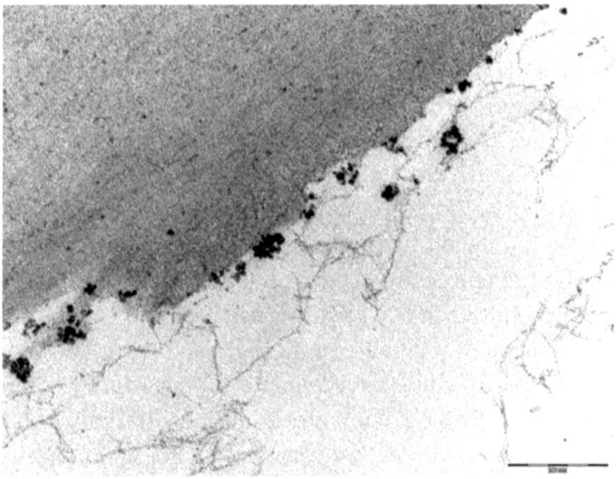

Fig. 2 Ultrathin section of magnetic sawdust particle observed in transmission electron microscopy (bar = 200 nm). Reproduced, with permission, from (Safarik et al. 2007a)

individual nanoparticles, or their aggregates (Mosiniewicz-Szablewska et al. 2007, 2010) (Fig. 2). Methanol-based procedure is especially useful for the treatment of non-living or dead biological materials, like different types of lignocellulose materials [e.g., sawdust (Safarik et al. 2007a), spent barley grains (Safarik et al. 2011b; Pospiskova and Safarik 2013), peanut husks (Safarik and Safarikova 2010) or spent coffee grounds (Safarik et al. 2012c)]. However, other materials including e.g. non-living microbial biomass or clays can be modified in this way, too. Alternatively, acetate buffer was used as a suspending medium for magnetic modification of both dead yeast cells (Safarik et al. 2007b; Safarikova et al. 2005) (Fig. 3) and algae cells (Safarikova et al. 2008). The amount of magnetic iron oxides nanoparticles deposited on the treated material surface can be influenced by the amount of ferrofluid used. Detailed magnetic characteristics of ferrofluid modified materials can be found elsewhere (Mosiniewicz-Szablewska et al. 2007, 2010). Direct magnetic modification with magnetic fluid (without the use of

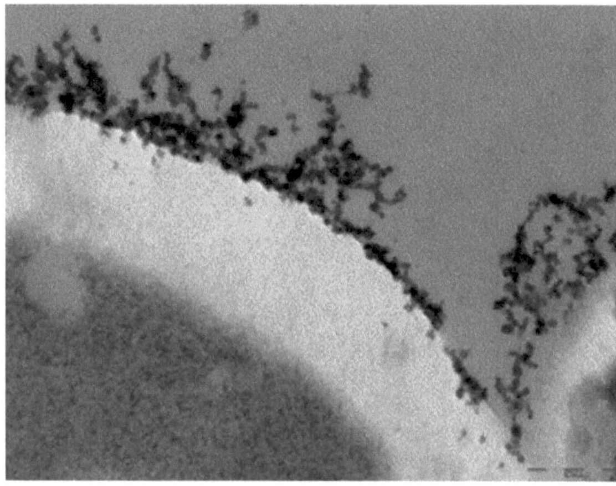

Fig. 3 TEM picture of magnetically modified dried fodder yeast (*Kluyveromyces fragilis*) cells (bar = 200 nm). Reproduced, with permission, from (Safarik et al. 2007b)

methanol or a buffer) was used for magnetization of sawdust, spent tea leaves, spent coffee grounds, spent barley grains, cellulose and starch (Safarik et al. 2012b).

As already described, microwave assisted modification was used for magnetization of large amount of biomaterials, such as microcrystalline cellulose, sawdust, spent tea leaves, spent coffee grounds, spent barley grains, powdered peanut husk or dried marine algae, in addition to other inorganic and synthetic materials (Safarik et al. 2013). The newest postmagnetization procedure employing microwave synthesized magnetic iron oxides nano- and microparticles can be used for magnetization of broad variety of materials, including sensitive biological materials (Safarik and Safarikova 2014).

2.2 Magnetic Modification of Living Biomaterials

Water-based ferrofluids have been successfully used to modify living microbial cells in order to prepare magnetically responsive whole cell biocatalysts (Safarikova et al. 2009), as shown in Fig. 4. Appropriate buffers had to be used to maintain the viability of the treated cells. Stable magnetically responsive yeast cell aggregates (Fig. 5) were formed when *S. cerevisiae* cells were mixed with magnetic iron oxides nano- and microparticles prepared by microwave assisted synthesis from ferrous sulfate (Pospiskova et al. 2013). Alternative approaches were based on the polyelectrolyte mediated deposition of magnetic iron oxides nanoparticles on the cell surface. *S. cerevisiae* yeast cells were coated with polyelectrolytes using alternating deposition of poly(allylamine hydrochloride) and poly(sodium polystyrene sulfonate) layers on the cells surface followed by deposition of non-coated magnetic nanoparticles (Fakhrullin et al. 2010a). In other experiments magnetic iron oxides nanoparticles stabilized with poly(allylamine

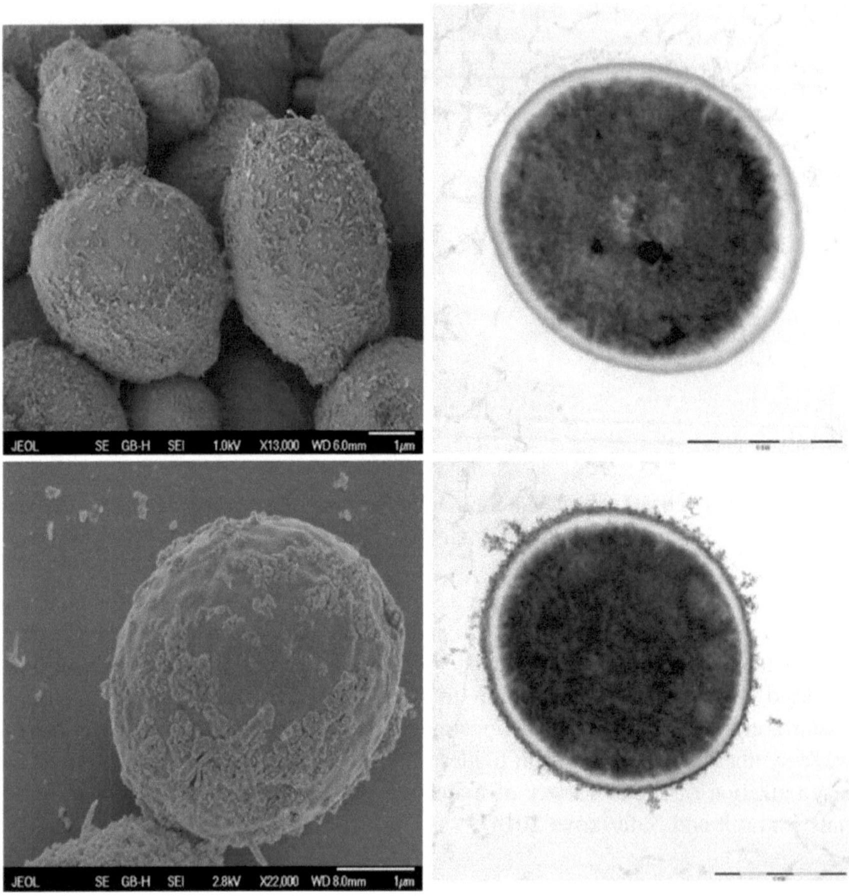

Fig. 4 *Left* SEM micrographs of ferrofluid modified *Saccharomyces cerevisiae* cells showing attached magnetic nanoparticles and their aggregates on the cell surface (bars = 1 µm). *Right* TEM micrographs of *Saccharomyces cerevisiae* cells (bars = 1 µm). *Top* Native cell. *Bottom* Ferrofluid modified cell with attached magnetic iron oxide nanoparticles on the cell wall. Reproduced, with permission, from (Safarikova et al. 2009)

hydrochloride) were used to modify *Chlorella pyrenoidosa* cells (Fakhrullin et al. 2010b) and living human cells (Dzamukova et al. 2011). The same magnetic nanoparticles were even used for magnetic modification of multicellular organisms, namely soil nematode *Caenorhabditis elegans* (Minullina et al. 2011).

Specific target cells are usually magnetically modified using immunomagnetic nano- or microparticles (Safarik and Safarikova 1999, 2012) (Fig. 6). This approach enables to detect and magnetically label cells bearing target epitopes on their surface and subsequently separate the labeled cells using flow-through or batch magnetic separation. Many immunomagnetic particles are commercially available, such as those used for the detection of microbial pathogenic bacteria

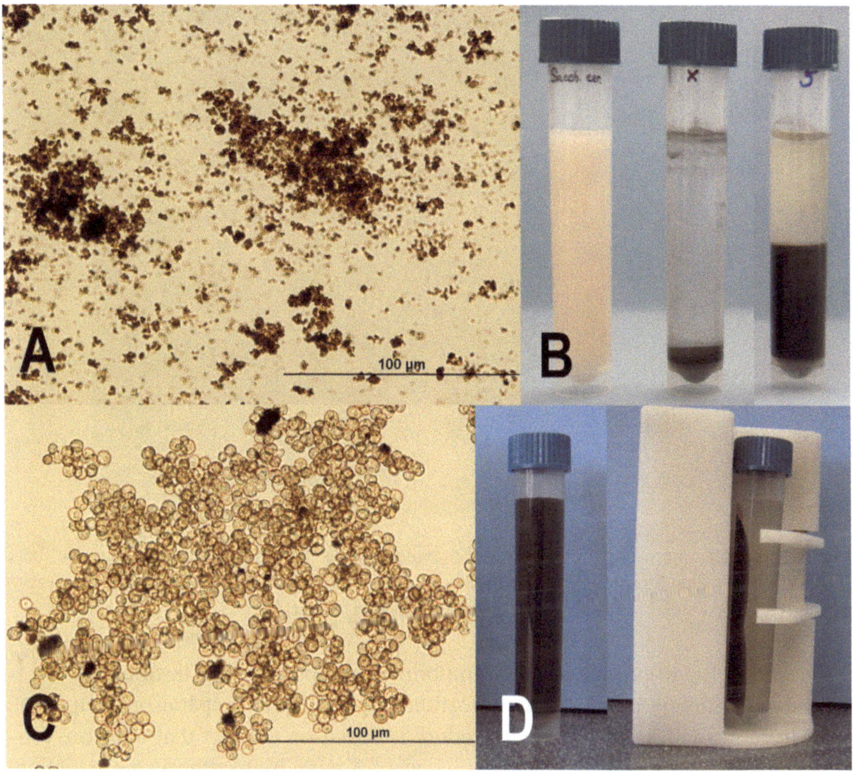

Fig. 5 Optical microscopy of magnetic iron oxides microparticles prepared by microwave assisted synthesis (**a**); process of magnetic modification of yeast cells (*left tube*—*S. cerevisiae* cells suspension; *middle tube*—sedimented iron oxides microparticles for magnetic modification; *right tube*—sedimented magnetically modified yeast cells) (**b**); optical microscopy of *S. cerevisiae* cells modified by iron oxides microparticles (**c**); magnetic separation of magnetically modified yeast cells (**d**). Reproduced, with permission, from (Pospiskova et al. 2013)

from Invitrogen, namely Dynabeads anti-*Salmonella*, Dynabeads anti *Escherichia coli* O157, Dynabeads EPEC/VTEC O26, Dynabeads EPEC/VTEC O103, Dynabeads EPEC/VTEC O111, Dynabeads EPEC/VTEC O145, Dynabeads anti-*Legionella* and Dynabeads anti-*Listeria*. The same company also produces immunomagnetic particles for the separation of parasitic protozoa, e.g. *Cryptosporidium* oocysts (Dynabeads anti-*Cryptosporidium*) and simultaneous separation of *Cryptosporidium* oocysts and *Giardia* cysts (Dynabeads GC-Combo).

Different types of lectins, such as those produced from *Triticum vulgaris* and *Agaricus bisporus*, or concanavalin A, were immobilized on magnetic microspheres and used to magnetically label specific bacterial pathogens, such as *E. coli*. Recovered cell populations were free from environmental impurities and a high percentage of the culturable cells was extracted (Payne et al. 1993; Porter and Pickup 1998; Porter et al. 1998).

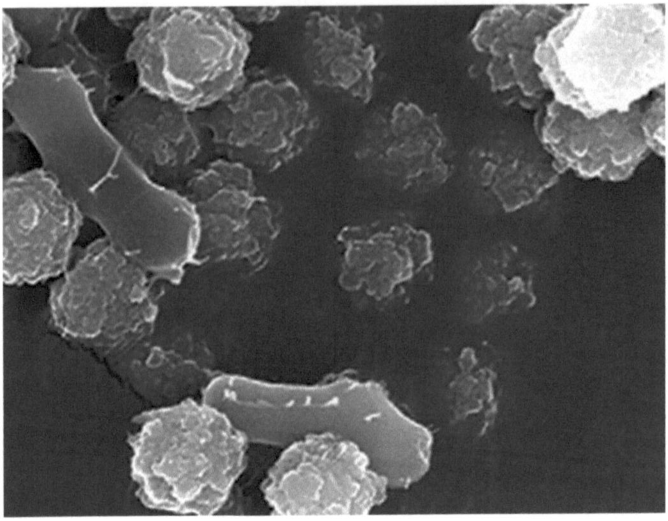

Fig. 6 Electron microscopy of *Legionella pneumophila* bound to immunomagnetic beads (Dynabeads My One Streptavidin (Invitrogen) with bound biotinylated polyclonal anti-*Legionella* antibody). Reproduced, with permission, from (Reidt et al. 2011)

Magnetic (nano)particles with immobilized annexin V have been employed for simple and efficient magnetic modification and subsequent separation of apoptotic cells from normal culture. This procedure is based on the fact that annexin V is a Ca^{2+}-dependent phospholipids-binding protein with high affinity for negatively charged phosphatidylserine (PS), which is redistributed from the inner to the outer plasma membrane leaflet in apoptotic or dead cells. Once on the cell surface, PS becomes available for binding to annexin V and any of its magnetic conjugates (Makker et al. 2008; Dirican et al. 2008; Safarik and Safarikova 2012).

Mesenchymal stem cells (MSCs), which can differentiate into multiple mesodermal tissues, were magnetically labeled using cationic magnetoliposomes (leading to the concentration of 20 pg of magnetite per cell) in order to enrich them magnetically from bone marrow. The magnetoliposomes exhibited no toxicity against MSCs in proliferation and differentiation to osteoblasts and adipocytes. During subsequent culture, substantially higher density of cells was obtained, compared to culture prepared without magnetoliposome treatment (Ito et al. 2004).

Superparamagnetic iron-oxide nanoparticles, such as MRI contrast agent Endorem or dextran-based magnetic nanoparticles MicroBeads (Miltenyi Biotec) have been used to label the stem cells. Nanoparticles can often be internalized by cells during cultivation by endocytosis (Sykova and Jendelova 2005).

The alternative magnetization procedures are based on the entrapment of nonmagnetic materials into appropriate (bio)polymer gel matrix, together with magnetic nano- or microparticles. Such a procedure can be very mild, enabling magnetic modification of living cells and subsequent employing their biological

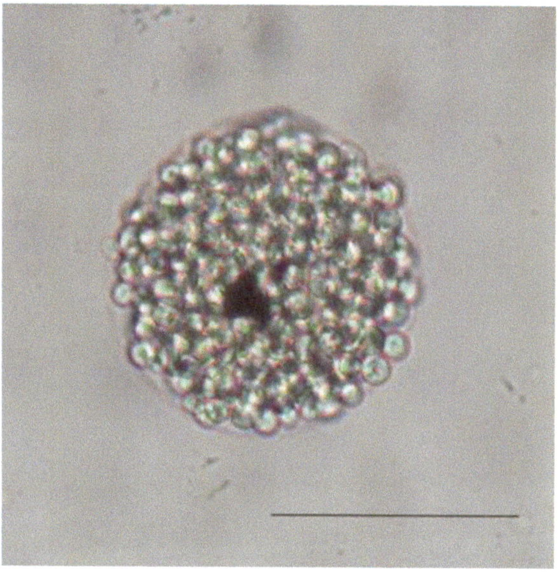

Fig. 7 Magnetically responsive alginate microbeads containing entrapped *Saccharomyces cerevisiae* cells and magnetite microparticles. The scale bar corresponds to 50 μm. Reproduced, with permission, from (Safarik et al. 2008)

activities. In a typical example magnetically responsive alginate beads containing entrapped *S. cerevisiae* cells and magnetite microparticles were prepared (Safarik et al. 2008). Larger beads (2–3 mm in diameter) were prepared by dropping the mixture into calcium chloride solution, while microbeads (the diameter of majority of particles ranged between 50 and 100 μm) were prepared using the water-in-oil emulsification process (Fig. 7).

An exceptional magnetization procedure employed erbium chloride as a magnetic label of a variety of cells. Erbium ions have a high affinity for the external cell surface and preserve their exceptionally high atomic magnetic dipole moment (9.3 Bohr magnetons) in various chemical structures. Both Gram-positive and -negative bacteria can be magnetically modified (Zborowski et al. 1993; Safarik and Safarikova 2007).

3 Application of Magnetically Modified Biological Materials

3.1 *Magnetic Plant-Derived Materials*

Plant-derived materials, such as sawdust, spent grains, spent coffee grounds, straw etc. are typical representatives of low cost (sometimes even waste) materials originating from agricultural and food industries. Plant materials have been frequently used as low-cost biosorbents for the removal of important organic and inorganic xenobiotics (Srinivasan and Viraraghavan 2010; Volesky and Holan 1995).

Fig. 8 Appearance of original lignocellulose suspension (*left*), suspension of lignocellulose after magnetic modification (*middle*) and demonstration of magnetic separation of magnetically modified lignocellulose (*right*)

Postmagnetization of such materials enables their simple magnetic manipulation (Fig. 8) and subsequently the possible development of magnetic separation-based technologies for xenobiotics removal (Safarik et al. 2011c, 2012d). Currently there are only a few examples of magnetically responsive plant-based materials applicable for xenobiotics removal. Especially, the adsorption of water soluble dyes was studied. Plant materials, modified with water-based magnetic fluid stabilized with perchloric acid or magnetized using a microwave assisted modification, was applied in laboratory adsorption experiments. In most cases analytes adsorption followed the Langmuir isotherm. This model allows calculating maximum adsorption capacities which are very important parameters for evaluation of adsorbents properties. The maximum adsorption capacities of the developed materials often exceeded values 100 mg of adsorbed dyes per one gram of magnetic biosorbent which is fully comparable with other described non-magnetic biosorbents (Srinivasan and Viraraghavan 2010; Safarik et al. 2011c, 2013).

3.2 Magnetically Modified Cells

Microbial cells, either in free or immobilized form, can be used for the preconcentration or removal of metal ions, organic and inorganic xenobiotics or biologically active compounds. Especially low cost biomaterials such as baker's yeast

5 Conclusions

Different types of biological materials, including by-products from food and agricultural industries, living and dead prokaryotic and eukaryotic cells and even multicellular organisms, can be successfully magnetized and subsequently utilized for many interesting applications, such as biosorbents for xenobiotics and biologically active compounds separation and removal, carriers for target compounds immobilization, whole cell biocatalysts and parts of sensing systems. Conversion of these originally "non-magnetic" biological materials into "smart materials" exhibiting response to external magnetic field may be one of the possible ways how to improve applicability of these materials, enabling their selective magnetic separation from difficult-to-handle systems. Of course, the same basic principles can be used for postmagnetization of many other materials, including inorganic and organic adsorbents and carriers, and other materials with interesting properties (Safarik et al. 2012d).

Despite the fact, that currently magnetically responsive biocomposites are tested mainly in laboratory experiments, there is a great potential for their large-scale applications in the near future. Postmagnetization of appropriate "non-magnetic" materials and formation of smart magnetically responsive materials will become a very useful tool for both laboratory and large scale applications. The list of available magnetic materials can increase dramatically in the near future which can lead to broader application of magnetic techniques.

Acknowledgements This research was supported by the Grant Agency of the Czech Republic (Projects No. P503/11/2263 and 13-13709S) and by the Ministry of Education of the Czech Republic (research project LD13021).

References

Al-Dujaili EAS, Forrest GC, Edwards CRW, Landon J (1979) Evaluation and application of magnetizable charcoal for separation in radioimmunoassays. Clin Chem 25(8):1402–1405

Ansari F, Grigoriev P, Libor S, Tothill LF, Ramsden JJ (2009) DBT degradation enhancement by decorating *Rhodococcus erythropolis* IGST8 with magnetic Fe_3O_4 nanoparticles. Biotechnol Bioeng 102(5):1505–1512

Bartonkova H, Mashlan M, Medrik I, Jancik D, Zboril R (2007) Magnetically modified bentonite as a possible contrast agent in MRI of gastrointestinal tract. Chem Pap 61(5):413–416

Cumbal LH, SenGupta AK (2005) Preparation and characterization of magnetically active dual-zone sorbent. Ind Eng Chem Res 44(3):600–605

Dirican EK, Ozgun OD, Akarsu S, Akin KO, Ercan O, Ugurlu M, Camsari C, Kanyilmaz O, Kaya A, Unsal A (2008) Clinical outcome of magnetic activated cell sorting of non-apoptotic spermatozoa before density gradient centrifugation for assisted reproduction. J Assist Reprod Genet 25(8):375–381

Dzamukova MR, Zamaleeva AI, Ishmuchametova DG, Osin YN, Kiyasov AP, Nurgaliev DK, Ilinskaya ON, Fakhrullin RF (2011) A direct technique for magnetic functionalization of living human cells. Langmuir 27(23):14386–14393

Fakhrullin RF, Garcia-Alonso J, Paunov VN (2010a) A direct technique for preparation of magnetically functionalised living yeast cells. Soft Matter 6(2):391–397

Fakhrullin RF, Shlykova LV, Zamaleeva AI, Nurgaliev DK, Osin YN, Garcia-Alonso J, Paunov VN (2010b) Interfacing living unicellular algae cells with biocompatible polyelectrolyte-stabilised magnetic nanoparticles. Macromol Biosci 10(10):1257–1264

Gao L, Zhuang J, Nie L, Zhang J, Zhang Y, Gu N, Wang T, Feng J, Yang D, Perrett S, Yan X (2007) Intrinsic peroxidase-like activity of ferromagnetic nanoparticles. Nat Nanotechnol 2(9):577–583

Griffin T, Mosbach K, Mosbach R (1981) Magnetic biospecific affinity adsorbents for immunoglobulin and enzyme isolation. Appl Biochem Biotechnol 6:283–292

Ino K, Ito A, Honda H (2007) Cell patterning using magnetite nanoparticles and magnetic force. Biotechnol Bioeng 97(5):1309–1317

Ito A, Hibino E, Honda H, Hata K, Kagami H, Ueda M, Kobayashi T (2004) A new methodology of mesenchymal stem cell expansion using magnetic nanoparticles. Biochem Eng J 20(2–3):119–125

Ito A, Kamihira M (2011) Tissue engineering using magnetite nanoparticles. Prog Mol Biol Transl Sci 104:355–395

Ji YQ, Hu YT, Tian Q, Shao XZ, Li JY, Safarikova M, Safarik I (2010) Biosorption of strontium ions by magnetically modified yeast cells. Sep Sci Technol 45(10):1499–1504

Korneva G, Ye HH, Gogotsi Y, Halverson D, Friedman G, Bradley JC, Kornev KG (2005) Carbon nanotubes loaded with magnetic particles. Nano Lett 5(5):879–884

Laurent S, Forge D, Port M, Roch A, Robic C, Elst LV, Muller RN (2008) Magnetic iron oxide nanoparticles: Synthesis, stabilization, vectorization, physicochemical characterizations, and biological applications. Chem Rev 108(6):2064–2110

Li JH, Hong RY, Luo GH, Zheng Y, Li HZ, Wei DG (2010) An easy approach to encapsulating Fe_3O_4 nanoparticles in multiwalled carbon nanotubes. New Carbon Mater 25(3):192–198

Li X-S, Zhu G-T, Luo Y-B, Yuan B-F, Feng Y-Q (2013) Synthesis and applications of functionalized magnetic materials in sample preparation. TrAC Trends Anal Chem 45:233–247

Makker K, Agarwal A, Sharma RK (2008) Magnetic activated cell sorting (MACS): utility in assisted reproduction. Indian J Exp Biol 46(7):491–497

Massart R (1981) Preparation of aqueous magnetic liquids in alkaline and acidic media. IEEE Trans Magn 17(2):1247–1248

Minullina RT, Osin YN, Ishmuchametova DG, Fakhrullin RF (2011) Interfacing multicellular organisms with polyelectrolyte shells and nanoparticles: A *Caenorhabtidis elegans* study. Langmuir 27(12):7708–7713

Mockovciakova A, Orolinova Z, Skvarla J (2010) Enhancement of the bentonite sorption properties. J Hazard Mater 180(1–3):274–281

Mosbach K, Andersson L (1977) Magnetic ferrofluids for preparation of magnetic polymers and their application in affinity chromatography. Nature 270(5634):259–261

Mosiniewicz-Szablewska E, Safarikova M, Safarik I (2007) Magnetic studies of ferrofluid-modified spruce sawdust. J Phys D-Appl Phys 40(21):6490–6496

Mosiniewicz-Szablewska E, Safarikova M, Safarik I (2010) Magnetic studies of ferrofluid-modified microbial cells. J Nanosci Nanotechnol 10(4):2531–2536

Nakahira A, Nagata H, Takimura M, Fukunishi K (2007) Synthesis and evaluation of magnetic active charcoals for removal of environmental endocrine disrupter and heavy metal ion. J Appl Phys 101 (9):Article No. 09J114

Oliveira LCA, Rios R, Fabris JD, Garg V, Sapag K, Lago RM (2002) Activated carbon/iron oxide magnetic composites for the adsorption of contaminants in water. Carbon 40(12):2177–2183

Olsvik O, Popovic T, Skjerve E, Cudjoe KS, Hornes E, Ugelstad J, Uhlen M (1994) Magnetic separation techniques in diagnostic microbiology. Clin Microbiol Rev 7(1):43–54

Patzak M, Dostalek P, Fogarty RV, Safarik I, Tobin JM (1997) Development of magnetic biosorbents for metal uptake. Biotechnol Tech 11(7):483–487

Payne MJ, Campbell S, Kroll RG (1993) Lectin-magnetic separation can enhance methods for the detection of *Staphylococcus aureus*, *Salmonella enteritidis* and *Listeria monocytogenes*. Food Microbiol 10(1):75–83

Porter J, Pickup RW (1998) Separation of natural populations of coliform bacteria from freshwater and sewage by magnetic-bead cell sorting. J Microbiol Methods 33(3):221–226

Porter J, Robinson J, Pickup R, Edwards C (1998) An evaluation of lectin-mediated magnetic bead cell sorting for the targeted separation of enteric bacteria. J Appl Microbiol 84(5): 722–732

Pospiskova K, Prochazkova G, Safarik I (2013) One-step magnetic modification of yeast cells by microwave-synthesized iron oxide microparticles. Lett Appl Microbiol 56(6):456–461

Pospiskova K, Safarik I (2013) Magnetically modified spent grain as a low-cost, biocompatible and smart carrier for enzyme immobilisation. J Sci Food Agric 93(7):1598–1602

Prochazkova G, Safarik I, Branyik T (2013) Harvesting microalgae with microwave synthesized magnetic microparticles. Bioresour Technol 130:472–477

Reidt U, Geisberger B, Heller C, Friedberger A (2011) Automated immunomagnetic processing and separation of *Legionella pneumophila* with manual detection by sandwich ELISA and PCR amplification of the ompS gene. J Lab Autom 16(2):157–164

Safarik I, Safarikova M (1999) Use of magnetic techniques for the isolation of cells. J Chromatogr B 722(1–2):33–53

Safarik I, Safarikova M (2007) Magnetically modified microbial cells: a new type of magnetic adsorbents. China Particuology 5(1–2):19–25

Safarik I, Safarikova M (2009) Magnetically responsive nonocomposite materials for bioapplications. Solid State Phenom 151:88–94

Safarik I, Safarikova M (2010) Magnetic fluid modified peanut husks as an adsorbent for organic dyes removal. Phys Proc 9:274–278

Safarik I, Safarikova M (2012) Magnetic nanoparticles for in vitro biological and medical applications: An overview. In: Thanh NTK (ed) Magnetic nanoparticles: from fabrication to biomedical and clinical applications. CRC Press/Taylor and Francis, Boca Raton, pp 215–242

Safarik I, Safarikova M (2014) One-step magnetic modification of non-magnetic solid materials. Int J Mater Res 105:104–107

Safarik I, Safarikova M, Forsythe SJ (1995) The application of magnetic separations in applied microbiology. J Appl Bacteriol 78(6):575–585

Safarik I, Nymburska K, Safarikova M (1997) Adsorption of water-soluble organic dyes on magnetic charcoal. J Chem Technol Biotechnol 69(1):1–4

Safarik I, Lunackova P, Mosiniewicz-Szablewska E, Weyda F, Safarikova M (2007a) Adsorption of water-soluble organic dyes on ferrofluid-modified sawdust. Holzforschung 61(3):247–253

Safarik I, Rego LFT, Borovska M, Mosiniewicz-Szablewska E, Weyda F, Safarikova M (2007b) New magnetically responsive yeast-based biosorbent for the efficient removal of water-soluble dyes. Enzyme Microb Technol 40(6):1551–1556

Safarik I, Sabatkova Z, Safarikova M (2008) Hydrogen peroxide removal with magnetically responsive *Saccharomyces cerevisiae* cells. J Agric Food Chem 56(17):7925–7928

Safarik I, Horska K, Safarikova M (2011a) Magnetic nanoparticles for biomedicine In: Prokop A (ed) Intracellular Delivery: Fundamentals and Applications. Springer, pp 363-372

Safarik I, Horska K, Safarikova M (2011b) Magnetically modified spent grain for dye removal. J Cereal Sci 53(1):78–80

Safarik I, Horska K, Safarikova M (2011c) Magnetically responsive biocomposites for inorganic and organic xenobiotics removal. In: Kotrba P, Mackova M, Macek T (eds) Microbial biosorption of metals. Springer, Berlin, pp 301–320

Safarik I, Horska K, Pospiskova K, Safarikova M (2012a) Magnetically responsive activated carbons for bio-and environmental applications. Int Rev Chem Eng 4(3):346–352

Safarik I, Horska K, Pospiskova K, Safarikova M (2012b) One-step preparation of magnetically responsive materials from non-magnetic powders. Powder Technol 229:285–289

Safarik I, Horska K, Svobodova B, Safarikova M (2012c) Magnetically modified spent coffee grounds for dyes removal. Eur Food Res Technol 234(2):345–350

Safarik I, Pospiskova K, Horska K, Safarikova M (2012d) Potential of magnetically responsive (nano)biocomposites. Soft Matter 8:5407–5413

Safarik I, Horska K, Pospiskova K, Maderova Z, Safarikova M (2013) Microwave assisted synthesis of magnetically responsive composite materials. IEEE Trans Magn 49(1):213–218

Safarikova M, Ptackova L, Kibrikova I, Safarik I (2005) Biosorption of water-soluble dyes on magnetically modified *Saccharomyces cerevisiae* subsp. *uvarum* cells. Chemosphere 59(6):831–835

Safarikova M, Pona BMR, Mosiniewicz-Szablewska E, Weyda F, Safarik I (2008) Dye adsorption on magnetically modified *Chlorella vulgaris* cells. Fresenius Environ Bull 17(4):486–492

Safarikova M, Maderova Z, Safarik I (2009) Ferrofluid modified *Saccharomyces cerevisiae* cells for biocatalysis. Food Res Int 42(4):521–524

Schwickardi M, Olejnik S, Salabas EL, Schmidt W, Schuth F (2006) Scalable synthesis of activated carbon with superparamagnetic properties. Chem Commun 38:3987–3989

Srinivasan A, Viraraghavan T (2010) Decolorization of dye wastewaters by biosorbents: a review. J Environ Manage 91(10):1915–1929

Sykova E, Jendelova P (2005) Magnetic resonance tracking of implanted adult and embryonic stem cells in injured brain and spinal cord. Ann N Y Acad Sci 1049:146–160

Tian Y, Wu M, Lin X, Huang P, Huang Y (2011) Synthesis of magnetic wheat straw for arsenic adsorption. J Hazard Mater 193:10–16

Torchilin VP, Papisov MI, Smirnov VN (1985) Magnetic Sephadex as a carrier for enzyme immobilization and drug targeting. J Biomed Mater Res 19(4):461–466

Uzun L, Saglam N, Safarikova M, Safarik I, Denizli A (2011) Copper biosorption on magnetically modified yeast cells under magnetic field. Sep Sci Technol 46(6):1045–1051

Volesky B, Holan ZR (1995) Biosorption of heavy metals. Biotechnol Prog 11(3):235–250

Yavuz H, Denizli A, Gungunes H, Safarikova M, Safarik I (2006) Biosorption of mercury on magnetically modified yeast cells. Sep Purif Technol 52(2):253–260

Zamaleeva AI, Sharipova IR, Shamagsumova RV, Ivanov AN, Evtugyn GA, Ishmuchametova DG, Fakhrullin RF (2011) A whole-cell amperometric herbicide biosensor based on magnetically functionalised microalgae and screen-printed electrodes. Anal Methods 3:509–513

Zborowski M, Tada Y, Malchesky PS, Hall GS (1993) Quantitative and qualitative analysis of bacteria in erbium(III) solution by thin-film magnetopheresis. Appl Environ Microbiol 59(4):1187–1193

Zwitterionic Nanocarriers for Gene Delivery

Yu-Ju Shih, Ching-Wei Tsai, Lemmuel L. Tayo and Yung Chang

Abstract "Gene therapy" is studied for its potential in modifying the genetic information of an organism and to treat cancerous, cardiovascular, neurological, and infectious diseases. A key limitation in the current development of human gene therapy is the lack of safe, efficient, and controllable gene delivery methods. Hundreds of clinical trials in human gene therapy have been approved worldwide since the 1980s but is met with only a small number of successes. This review discusses the hurdles and hopes of zwitterionic polybetaines in molecular cargo formulation of multi-functional gene nanocarriers.

Keywords Gene delivery · Hemocompatible gene carrier · Zwitterionic polymers · Zwitterionic nanocarriers

List of Abbreviations

ADA	Adenosine deaminase
APE	Antipolyelectrolyte effect
BCs	Block copolymers
BMA	*n*-Butyl methacrylate
CBMA	Carboxybetaine methacrylate
CST	Critical solution temperature
DMAEMA–MPC	2-(Dimethylamino)ethyl methacrylate-block-2-(methacryloyloxyethyl Phosphorylcholine)
MPC	Methacryloyloxyethyl phosphorylcholine
MPC-co-LMA2	2-Methacryloyloxyethyl phosphorylcholine-co-lauryl methacrylat
MPTMS	3-Methacryloyloxypropyl trimethoxysilane
NIPAAm	*N*-isopropylacrylamide

Y.-J. Shih · C.-W. Tsai · L. L. Tayo · Y. Chang (✉)
R&D Center for Membrane Technology and Department of Chemical Engineering, Chung Yuan Christian University, Chung-Li, Taoyuan 320, Taiwan
e-mail: ychang@cycu.edu.tw; changyung0307@gmail.com

A. Prokop et al. (eds.), *Intracellular Delivery II*, Fundamental Biomedical Technologies 7, DOI: 10.1007/978-94-017-8896-0_3,

PS	Phosphatidylserine
PC	Phosphorylcholine
PAA	Poly (acrylic acid)
PDMAEMA	Poly(2-dimethylamino)ethyl methacrylate
PBMA	Poly(butyl methacrylate)
PCBMA	Poly(carboxybetaine methacrylate)
PEG	Poly(ethylene glycol)
PLL	Poly(L-lysine)
PSD	Poly(methacryloyl sulfadimethoxine)
PMB	Poly(MPC-co-BMA)
PHEMA	Poly(*N*-(2-hydroxpropyl)methacrylamide)
PPO	Poly(propylene oxide)
PSBMA	Poly(sulfobetaine methacrylate)
PEI	Polyethyleneimine
PZ	Polyzwitterions
RES	Reticuloendothelial system
R9	Nona-arginine peptide
SAMs	Self-assembled Monolayers
SCID	Severe combined immunodeficiency
siRNA	Small interfering RNA
SBMA	Sulfobetaine methacrylate
SPR	Surface plasmon resonance

1 Introduction: Gene Therapy and Gene Delivery

Gene therapy is a bio-technique to cure a disease or improve the clinical status of the patient by the insertion, alteration, or deletion of specific genes by transferring genetic material to the specific cells or tissues. In 1990, the first case of gene therapy was conducted on a four-year old girl, Ashanti DeSilva, for treatment of severe combined immunodeficiency (SCID) with lack of adenosine deaminase (ADA) (Culver et al. 1991). Dr. William French Anderson inserted a normal copy of ADA gene into T cells collected from Ashanti's body, then the modified T cells were injected back into her body to minimize her syndrome. The repeating treatments normalized the number of T cells in her body but fail to generate new cells with functional genes. In short, the therapy did not yield a complete cure for Ashanti, but it helped mitigate the deficiency. While gene therapies were met with such successful in clinical trials, real cure via gene therapy remained elusive throughout the 1990s. Until now, trials to treat certain diseases, such as HIV infection, various cancers, Alzheimer's and Parkinson's disease, are still being continued (Ginn et al. 2013).

An important milestone in the field of gene therapy was achieved, when human genome sequence was fully identified in April 2003; however, many aspects of gene regulation are still poorly understood, such as the patient inflammation and immune responses. Another challenge faced by current gene therapy is the absence of satisfying transfer technologies. Here, we focus on the means to construct an efficient delivery system for gene transfection. On a basic level, gene delivery systems are composed of two major parts, the target gene (genetic cargo) and the delivering carrier (genetic shuttle). The target genes include plasmid DNA, oligos, and small interfering RNA (siRNA), and gene delivery carriers for gene therapy, such as transformed viral carriers, can deliver the genetic cargo into the target cells (Ushitora et al. 2010). The most commonly used DNA virus vector is the adenovirus (Vorburger and Hunt 2002), and it is the first gene therapeutic medicine (the recombined human p53 adenovirus injection) approved by China State Food and Drug Administration in 2004 (Lane et al. 2010). Although virus-based vector systems can effective deliver target genes into cells, the present viral vectors should be used with caution in human clinical trials and alternative non-viral based vectors should be developed to reduce the risks from viral based gene transfection.

Various types of non-viral based transfection technologies have been developed by scientists over time (Schenborn 2000; Guo and Huang 2012; Jafari et al. 2012; Fields et al. 2012), for instance, calcium phosphate transfection, lipid-mediated transfection (Allon et al. 2012), cell penetrating peptide-mediated transfection (Trabulo et al. 2010; Wang et al. 2011), cationic polymer transfection (Oliveira et al. 2010; Wong et al. 2011), etc. Non-viral transfection vectors are typically based on cationic lipids, peptides, or polymers which can form nano-complexes with particle size around 100 nm with the negatively charged nucleic acids (Guo and Huang 2012). The net charge of these nano-complex gene delivery systems are generally positively charged for improving the adsorption efficiency of the gene carriers toward the cell surfaces. The gene delivery systems can also be constructed through covalent conjugation or encapsulation.

In general, the delivery process has the following characteristic stages (1) circulating through the living body, (2) targeting of specific cell, (3) translocating into target cells and releasing of genetic cargo into the cell (Shoji and Nakashima 2004; Gao et al. 2007). To ensure the efficiency of the gene therapy, the gene delivery system should first be stable during the circulation stage and avoid adsorption of serum proteins, protease digestion, immune response, and renal clearance. PEGylation and zwitterionization of the cationic non-viral vectors can minimize the interaction of the genetic shuttle with blood components, reducing kidney filtration and reduce cytotoxicity (Park et al. 2006; Sinclair et al. 2013; Somasundaran et al. 2002). Secondly, the gene delivery system should be able to reduce adverse effects from targeted delivery to malignance tissue by involving the targeting domain, such as antibody, aptamer, specific cleavable peptide sequence, or environmentally sensitive moieties (Guo and Huang 2012; Wilner et al. 2012; Endsley and Ho 2012; Vessillier et al. 2012). Targeted delivery can also increase the probability and efficiency of the gene transfection towards targeted cells. Thirdly, the gene delivery system should assist with gene settles in entering cells and releasing the gene cargo after entering the cells. The initial stage of the genetic cargo delivery is dominated by two major transfection pathways. Endocytosis is one of these pathways and is usually energy-dependent or receptor-mediated. The well-known CPPs polyarginine peptide (R9), penetratin peptide, polyethyleneimine (PEI), and PEI derivatives internalize the DNA plasmid into cell via this endocytic pathway (Duchardt et al. 2007; Tian et al. 2012; Walrant et al. 2012). The transfection efficiency is, however, strongly affected by the entrapment of the "cargo" (DNA or RNA) in endosomes. Proton-sponge effect was introduced to mediate the ability of cationic polymers to escape from early endosomes. The PEI polymers can lead to an increase in proton influx followed by an enhanced accumulation of chloride ions and osmotic swelling, which is mainly attributed to its strong buffering capacity in the pH range from 5 to 7 (Nguyen and Szoka 2012). Furthermore, an extension of the proton-sponge effect involves the structure change of cationic polymers (such as polyamidoamine dendrimers) in endosome, where the gene delivery systems enters an extended structure from a collapsed state after the protonation of the amine groups with decreasing pH (Maingi et al. 2012). The electrostatic repulsion between intra-polymer chains which cause the extended polymer structure is referred to as the "umbrella effect," and structural changes leading to an increase in volume after protonation is considered optimal because of the mediation of superior transfection efficiency. The other pathway that affects the initial stages of the genetic cargo delivery is the direct penetration pathway, which is the energy-independent internalization for cargo into cell (Bode et al. 2012), and this pathway can easily avoid the problem of endosomal escapes.

The high cytotoxicity and low transfection efficiency are still major problems with non-virus based vectors today. Recently, multi-function and multi-purposed gene delivery systems were developed with high circulation stability, cell specific targeting, and facilitates endosomal escape for application in gene therapy via nano-device (nanoparticles, layer by layer nanoparticles, and liposomes) (Jabr-Milane et al. 2008; Akita et al. 2009). Such strategies can not only avoid problems of biostability, cytotoxicity, and non-specific binding, but also facilitate targeted delivery and nucleic acids release of the gene delivery systems within the cell. Therefore, the multifunctional non-viral vectors may potentially be the next generation gene delivery systems for applying gene therapy in clinical trials.

2 Zwitterionic Polybetaines: General Formulation

A fundamental understanding of nonfouling polymers' resistance towards the adsorption of plasma proteins and the adhesion of blood cells is highly desirable and critically important in the development of blood-contacting biomaterials in various applications, such as blood collection devices, antithrombogenic implants, hemodialysis membranes, drug-delivery carriers, diagnostic biosensors, and membrane bioseparation (Ratner 2000). The interactions of clotting factors, plasma proteins, and platelets with blood-contacting materials strongly affect the thrombotic reaction induced by intrinsic surface contact (Horbett 1993). To minimize the thrombogenicity of biomaterials, heparinized materials are often coated on the surfaces of blood-contacting devices or containers to prolong blood clotting time (Olsson et al. 2000; Han and Park 1995; Turk et al. 2004). Non-specific adsorption of proteins such as fibrinogen and clotting enzymes is the first interaction event to induce a full-scale platelet adhesion and activation leading to thrombosis and embolism at the blood-material interface. Hence, protein-resistant surfaces have been widely investigated in order to eliminate blood clot formation (Zhang et al. 2008a; Chang et al. 2010), and the retention of bound water molecules surrounding the functional groups of the material interfaces is now recognized to play a key role in surface resistance to nonspecific protein adsorption (Kane et al. 2003; He et al. 2008). Thus, to suppress nonspecific protein adsorption onto biomaterials, many studies have been carried out on surfaces grafted with poly(ethylene glycol) (PEG) (Chang et al. 2009a, 2011). PEGylated polymers, however, are chemically unstable in the presence of oxygen and transition-metal ions found in most biochemical solutions. Furthermore, it was demonstrated that surfaces grafted PEG brushes lose their protein repulsive properties at physiological temperatures. In this regard, it is of great advantage to have alternative nonfouling material system beyond PEG. Zwitterionic polybetaines have received growing attention for their use as blood-inert polymers because of their excellent inhibition in plasma protein adsorption, blood platelet adhesion and activation, and thrombus formation in vitro (Iwasaki and Ishihara 2005; Chen and Jiang 2008; Jiang and Cao 2010). A general characteristic of zwitterionic polymers, including polyphosphobetaine, polysulfobetaine and polycarboxylbetaine, is the cationic and anionic charged moieties on the same side chain and the overall charge neutrality (Ishihara et al. 1992; Nakabayashi and Williams 2003; Feng et al. 2006; Chang et al. 2006; Ladd et al. 2008).

In recent years, there has been a renewal of interest in the unique nonbiofouling properties of zwitterionic polymers (Georgiev et al. 2006), due to an osmotic control through the antipolyelectrolyte effect (APE). Polyzwitterions (PZ) are able to resist protein adsorption and prevent cell adhesion because of their hydrophilicity and the dipoles-dipoles moments between phospholipids and sulfobetaines side groups. Another possible reason is the higher value of the concentration ratio between "free water" and "bound water" in zwitterionic polymer-based aqueous solutions and hydrogels compared to that of other hydrophilic polymers (Georgiev et al. 2006). Whitesides et al. used Surface Plasmon Resonance spectroscopy (SPR) and Self-Assembled Monolayers (SAMs) of alkanethiols on gold to evaluate the ability to resist the nonspecific adsorption of proteins from aqueous buffer of surfaces functionalized with different combinations of charged groups (Prime and Whitesides 1993; Holmlin et al. 2001). Whether composed of zwitterionic mixed charge SAMs or zwitterionic SAMs formed by a single-component, the surface can effectively resist the protein adsorption as long as it is electrically neutral. On the other hand, all positively or all negatively charged surfaces from SAMs cannot resist proteins adsorb on the surfaces. From this result, we can anticipate an effective design of an antifouling surface from the use of zwitterionic polymers, mimicking the behavior of zwitterionic phospholipids or mixed charge SAMs composed by positive and negative moieties. In 2001, Whitesides' group came up with four basic principles, regarding the chemical structures, for the design of general antifouling formulation (1) hydrophilic, (2) hydrogen-bond acceptor, (3) no hydrogen-bond donor, and (4) electrically neutral.

In 1972, Singer and Nicolson have proposed the well-known fluid mosaic model of the cell membrane structure, enabling a satisfactory understanding of how biological systems work (Singer and Nicolson 1972). The highly polar region is in contact with the aqueous phase at the outside surface of the cell membrane while the nonpolar region is embedded in the hydrophobic domain at the inside of the cell membrane. According to phenomenon of cell membrane and natural bio-inert theory, scientists began to develop zwitterionic anti-fouling materials. Zwaal et al. reported that the inner membrane of red blood cells caused thrombogenic, while the outer layer is not. The explanation proposed was that the lipid components constituting the outside surface of a cell membrane are mainly zwitterionic phospholipids (like phosphorylcholine, PC), but the inside components are negatively charged components (like phosphatidylserine, PS). Nakabayashi then explored new monomer formulation of methacryloyloxyethyl phosphorylcholine (MPC). Over past three decades, several approaches have been used to prepare a wide variety of zwitterionic material forms, including SAMs, brushes, networks, hydrogels, polymers, micelles, nano-particels, and nano-carriers.

In 1990, Ishihara et al. improved the MPC polymerization by increasing the yield with a monomer of high purity taken as a white crystalline powder, thus enabling the development of synthetic MPC biomaterials (Olsson et al. 2000; Zhang et al. 2008b). MPC structure is composed by a methacrylate and PC head groups, as shown Fig. 1. The MPC polymerization can be well controlled in terms of chain length and molecular weight distribution by free radical polymerization. For example, the copolymer poly(MPC-co-BMA) (PMB) has been synthesized from MPC and n-butyl methacrylate (BMA) monomer (Georgiev et al. 2006). Since Nakabayashi and Ishihara et al. have improved MPC polymerization, it has been widely used in the medical field, such as biochips, contact lenses, hemocompatible devices or urinary-related medical materials.

An early study of biomimetic materials was inspired by lipid research from natural cell membranes. Chapman et al. synthesized stable phosphatidylcholine (DAPC) structure from free radical attack on a triple bond between molecules via UV or gamma ray irradiation (Lewis 2000). In 2003, Prof. Lloyd research group designed phosphorylcholine (PC)-based polymers for use in a variety of medical device applications to improve biocompatibility (Long et al. 2003). Poly(butyl methacrylate) (PBMA) and 2-methacryloyloxyethyl phosphorylcholine-co-lauryl methacrylate (MPC-co-LMA2) copolymer were spin-coated onto glass coverslips. Results showed that the adsorption of fibrinogen and albumin and the adhesion of human mononuclear cells and rabbit corneal epithelial cells decreases as the amount of MPC-co-LMA2 increased. In 2004, Armes' research group design a novel 2-(dimethylamino)ethyl methacrylate-block-2-(methacryloyloxyethyl phosphorylcholine) (DMAEMA–MPC) diblock copolymer as a new non-viral vector for gene delivery (Lam et al. 2004).

Fig. 1 The monomer structure of 2-methacryloyloxyethyl phosphorylcholine (MPC)

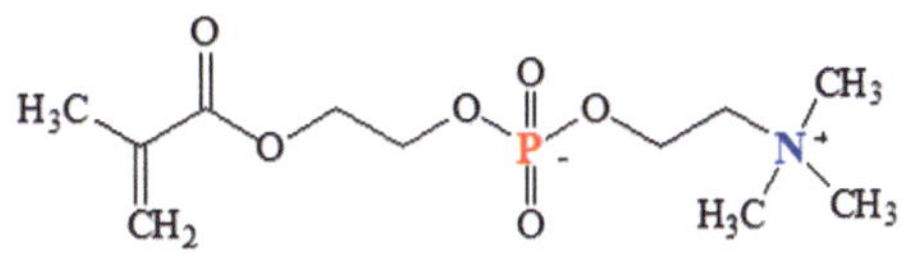

However, MPC has some disadvantages such as the rather complicated synthesis and high costs due to low yields. The other type of zwitterionic polymers which applied in antifouling field have been explored. In 2001, whiteside et al. demonstrated antifouling property of sulfobetaine-based self-assemble monolayer on the gold surface via surface plasmon resonance studies (Prime and Whitesides 1993). Sulfobetaine methacrylate (SBMA) exist a zwitterionic foumulation of electrically neutral monomer, as shown in Fig. 2. Poly(sulfobetaine methacrylate) (PSBMA), with a methacrylate main chain and an analogue of the taurine betaine as the pendant group ($CH_2CH_2N^+(CH_3)_2CH_2CH_2CH_2SO_3^-$), has become the most widely studied zwitterionic polymer owing to its ease of synthetic preparation.

In 2006, a well-defined diblock copolymer of poly(sulfobetaine methacrylate) (PSBMA) and poly(propylene oxide) (PPO) was synthesized by the sequential addition of SBMA monomer to fixed amounts of PPO using an atom transfer radical polymerization method and varying poly(SBMA) lengths (Chang et al. 2006). These copolymers were physically adsorbed onto a SPR sensor surface covered by methyl-terminated self-assembled monolayers, followed by the in situ evaluation of protein adsorption on the adsorbed copolymers. It was found that the behavior of the protein adsorption depends on the molecular weight of the copolymers. Results show that the diblock copolymers containing PSBMA can be highly protein resistant when surface SBMA densities are well controlled. In 2008, Prof. Jiang's group synthesized linear PSBMA homopolymer with an average molecular weight ranging from 20.9 to 316 kDa, via free radical polymerization at different KCl concentrations (Zhang et al. 2008b). In 2009, Chang et al. synthesized statistical copolymers of zwitterionic SBMA and nonionic *N-isopropylacrylamide* (NIPAAm) as smart biomaterial which exhibits property of double critical solution temperature (CST). In the copolymer containing 29 mol% of PSBMA had a double critical solution temperature (CST) showed extremely low protein adsorption and high anticoagulant activity in human blood and plasma. The tunable and switchable thermoresponsive phase behavior of poly(SBMA-*co*-NIPAAm), as well as its high plasma protein adsorption resistance and anticoagulant activity, suggests a potential for blood-contacting applications (Chang et al. 2009b). In 2010, Chang et al. showed the perfect hemocompatible nature of polysulfobetaines in 100 % plasma solutions (Shih and Chang 2010). It is due to zwitterionic polysulfobetaines generate a tightly bound, structured water layer around the betaine head groups via electrostatically induced hydration.

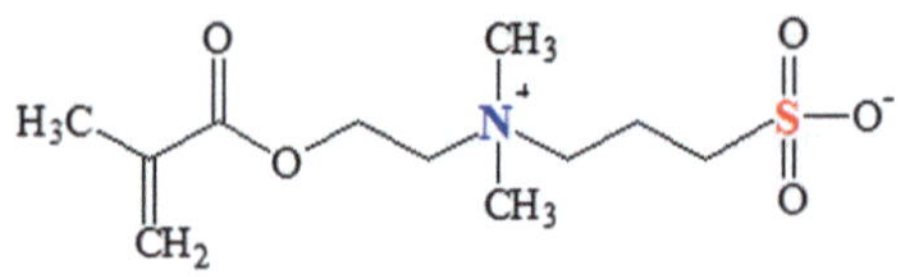

Fig. 2 Chemical structure of sulfobetaine methacrylate, SBMA

The third type of zwitterionic monomer is carboxybetaine (CBMA), as shown in Fig. 3. In recent years, Jiang's group found that zwitterionic CBMA performed a very good antifouling property comparable to SBMA. The CBMA containing carboxylic acid structure may even couple with biorecognition. The pendent group of carboxylic acid can be easily converted into other functional groups (Gao et al. 2010a). In 2010, Jiang's group prepared a biomimetic polymer with two zwitterionic poly(carboxybetaine methacrylate) (PCBMA) arms for ultra-low fouling and two adhesive catechol groups for surface anchoring. Then the carboxybetaine-based polymer was binded onto gold surfaces for the antifouling properties and evaluated using a SPR sensor. Under optimized conditions, the surface modified by PCBMA can highly resist non-specific protein adsorption. In 2012, Krishnan group prepared two different zwitterionic block copolymers (BCs) of polystyrene-*block*-poly[N-(3-dimethylamino-1-propyl) acrylamide], which function as stable, low-fouling surface modifiers in different biological environments (Wu et al. 2012). In 2012, Kitano group synthesized a random copolymer of zwitterionic monomer, CBMA and 3-methacryloyloxypropyl trimethoxysilane (MPTMS) in ethanol using 2, 2-azobisisobutyronitrile as an initiator (Suzuki et al. 2012). It was found that the copolymer-modified glass substrate became highly hydrophilic upon immersion in water, and showed resistance towards non-specific protein adsorption. The adhesion of various cells to the glass substrate was also strongly suppressed by the surface grafted PCBMA layer.

3 Multi-functional Gene Nanocarriers: Zwitterionic Cargo Design

Nowadays, several strategies are employed to address the limitations of cationic polymer vectors and to improve the gene transfection efficiencies (Haag and Kratz 2006; Wu et al. 2013; Tamura et al. 2010; Venkataraman et al. 2011; Jewell and Lynn 2008). Multifunctional nanoparticle gene carriers are now being developed to achieve excellent transfection efficiencies, and at the same time, enhance cell-specific targeting, increase cytocompatibilities, and improve other functional properties which are beneficial to the host. Aside from subtherapeutic transfection efficiencies, the issue of biocompatibility is one of the biggest challenges in the development of non-viral gene delivery systems. The current design criteria for biocompatibility in gene delivery system require both the stability of the gene

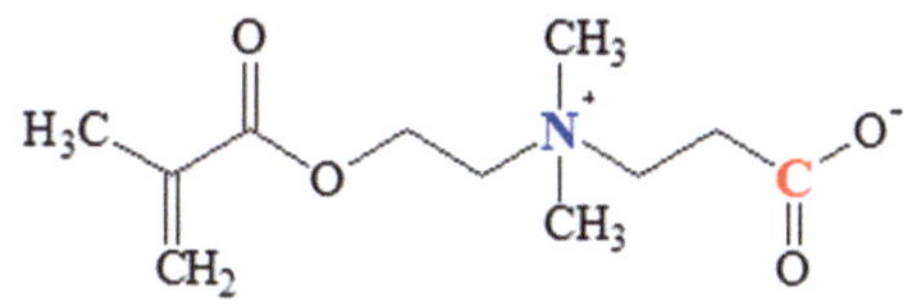

Fig. 3 Chemical structure of carboxybetaine methacrylate, CBMA

delivery particles as well as minimizing their cytotoxicity. Clinical findings have shown that cationic gene carriers can interact with blood components and other polyplexes which form embolytic entities, activate the complement system and cleared by the reticuloendothelial system. These consequences may eventually lead to premature elimination of the gene carriers and prevent the successful delivery of the therapeutic gene to its proper destination. Therefore, synthetic gene carriers should possess a multitude of functions to address the important issues in non-viral gene delivery (Venkataraman et al. 2011; Jewell and Lynn 2008; Ke and Young 2010; Sethuraman et al. 2006; Wong et al. 2007).

Multi-layering of polyplexes is one of the current approaches employed to attain multifunctional gene carriers with good transfection results, low cellular toxicity, and enhanced DNA stability. These nanoparticle gene carriers can be constructed either by layer-by-layer deposition using alternate polyanion and polycation to form tertiary and quaternary polyplexes or by encapsulation of the polycation-DNA binary polyplexes. In a recent study conducted by Ke et al., quaternary polyplexes were prepared by the sequential addition of polycations (polyethylenimine (PEI) or poly(N-(8-aminooctyl)-acrylamide) (P8Am)) for loading plasmid DNA into the core polyplexes and poly (acrylic acid) (PAA) for reversing charges to deposit additional polycation (PEI or P8Am) layer. It was found the cytotoxicity and cellular uptake of PEI core polyplexes were improved by coating a cell uptake-favorable P8Am layer. On the other hand, P8Am could not facilitate endosomal release through the proposed proton sponge effect so the PEI core was required for the P8Am-coated quaternary polyplexes to ensure efficient gene transfection results (Sethuraman et al. 2006; Wong et al. 2007).

Multi-functional properties of gene carriers can also be achieved by combining the properties of different polymers via copolymerization. Through there are several designs and arrangements available for copolymers, a compromise must be made between transfection efficiencies and cytocompatibilities. Several studies have utilized polyethylene glycol (PEG) as one of the components of block copolymers alongside with PEI and other polycations to enhance biocompatibility and increase anti-fouling effects of proteins. The incorporated PEG in gene nanocarriers can reduce the surface charge density of the polycations, hence lowering the cytotoxicity of the polymer/DNA complexes. Cationic vector-based delivery of DNA is generally complicated by interactions of the positively charged nanoparticles and serum proteins. The shielding effect and anti-fouling properties of PEG can reduce unwanted interactions, and thus, improve colloidal stability and prolong circulation time in the blood stream. More recently, zwitterionic polymers containing the pendant groups of phosphobetaine, sulfobetaine, and carboxybetaine have received a growing attention for use in the new generation of nonfouling materials because of their excellent protein resistance and have been explored for potential use as a nonfouling surface material in various biomedical applications. Both the cationic and anionic functional groups in these polymers prevent unwanted interactions with cells and other cellular components in living organisms. This general principle has become a reliable guide line in new molecular designs of nonfouling polymers with plasma protein resistance in human blood which may be of great potential in the development of biocompatible gene nanocarriers. These polymers have potential merits as components of non-viral gene carriers because of their ability to behave as biomimetic surfaces and to prevent blood clot or thrombus formation, inhibit hemolysis, and prolong circulation time of the gene carrier polyplexes in the biological system. Several studies have now utilized zwitterionic polymers such as polysulfobetaine and polycarboxybetaine in the preparation of the copolymer components of gene nanocarriers to enhance serum stability, increase gene transfection efficiency, and decrease cytotoxicity (Ke and Young 2010; Agarwal et al. 2012; Chan et al. 2007, 2008; Gaur et al. 2000; Griffiths 1991; Dai and Liu 2011; Kunath et al. 2003; Layman et al. 2009).

Another important target for multifunctional gene nanocarriers is its specific delivery of therapeutic gene cargo to cancer cells. An interesting aspect discovered in tumor physiology is that the pH of the extracellular matrix of cancer cells are relatively low compare to that of normal body pH. The acidic environment arises from the high metabolic rate of cancer cells, which leads to the production of excess lactic acid under hypoxic conditions. Consistent results over the last decades, using chemical and electrical probes, have demonstrated an acidic phenotype in the extracellular matrix of these malignant neoplasms. These results have motivated researchers to develop pH sensitive nanoparticles which would shield the polymer/DNA complex during systemic circulation and selectively deliver their therapeutic gene at tumor sites. In 2006, Sethuraman et al. (2006) has reported the use of a pH responsive sulfonamide/polyethylenimine (PEI) nanoparticles for tumor specific gene delivery. The shielding/deshielding of the nanoparticles were tested in vitro along with cell viability and transfection efficiency at normal physiological and tumor pH. The nanoparticles composed of DNA/PEI/PSD-*b*-PEG exhibited low cytotoxicity and low transfection efficiency at pH 7.4 due to shielding of PEI by PSD-*b*-PEG. At pH 6.6, the nanoparticles demonstrated high cytotoxicity and transfection, indicating PSD-*b*-PEG detachment from the nanoparticles which permits the PEI to interact with the target cells. The design of the PSD-*b*-PEG nanocarrier is able to discern the small difference in pH between normal and tumor tissues, and hence, has remarkable potential in drug targeting to tumor areas (Sethuraman et al. 2006; Agarwal et al. 2012; Griffiths 1991).

Copolymerization of zwitterionic segments with other stimuli-responsive moieties would be an important breakthrough in the development and design of multifunctional gene nanocarriers in order to achieve good transfection, increase hemocompatibility, and enhance cell-specific targeting in in vitro, in vivo, and clinical studies.

4 Zwitterionic Gene Nanocarriers in Human Whole Blood

In clinical gene therapy, ideal vectors would be administered through a noninvasive route, transducing only the desired cells within the target tissue. The basic challenge in gene therapy is the design of gene delivery nanocarriers that can control the expression of the transgene products at a therapeutic level and regulate the expression of these gene products for a defined period of time. While viral vectors show higher gene delivery efficiency than non-viral vectors, the later is more suitable in clinical trials for safety issues, such as cell toxicity and immune response. Consequently, the development of non-viral nanocarriers is one of the current trends. As we know so far, effective gene therapy vectors should have the following functions (1) carry nucleic acids into cells, (2) protect nucleic acids from decomposition by enzymes, (3) low cell toxicity, and (4) extend expression of transgenes. Most studies focused on the transfer efficiency, such as poly(L-lysine) (PLL) (Murray et al. 2001; Schwarzenberger et al. 2001), poly(ethylenimine) (PEI) (Gao et al. 2010b) and poly(2-dimethylamino)ethyl methacrylate (PDMAEMA) (Jones et al. 2004; Takeda et al. 2004) to form the polyelectrolyte complexes with DNA through in vivo charge-driven attraction. In human body, the complexes of DNA/polyelectrolyte, named polyplexes, are rapidly engulfed (phagocytosis) by cells of reticuloendothelial system (RES), because of aggregation, activation, immunisation and opsonization (Funhoff et al. 2005). Thus, it is important to prolong the circulation time as much as possible to ensure the carrier reaches the specific tissue or cell target. Previous works reported hydrophilic non-ionic polymers such as poly(ethylene glycol) (PEG) or poly(*N*-(2-hydroxpropyl)methacrylamide) (PHEMA) (Maysinger et al. 1995; Naeye et al. 2010) combined with cationic polymers to form polyplexes with DNA. With this design, it is possible to control the aggregation stage. On the other hand, some reports showed other designs of gene carriers from phosphorylcholine (PC) head groups based zwitterionic polymers. Based on the blood-inert nature of zwitterionic formulation, zwitterionic-shielded carriers might provide longer circulation time than PEGylated carriers in human body. Lam et al. prepared a new non-viral vector for gene delivery from DMAEMA-*b*-MPC diblock copolymer. The complexes with optimized copolymer shielding were found to prevent promiscuous interactions with tissues and potentially allowing for cellular specific delivery of the condensates following the attachment of a targeting ligand. This work indicated that the zwitterionic MPC-shieded carrier plays an important role in the controlled interactions with biological membranes.

In general, typical polymer-based gene carriers might trigger a series of biological reactions with human blood. The plasma protein adsorption, platelet adhesion, platelet aggregation, platelet deformation, and blood coagulation may be initiated from the carrier-blood contacting interface. In the following stages, the activated process ends up with blood clots and strongly damaged the human body. Thus, it is challenging to design a hemocompatible gene carrier. Three major concerns in carrier formulation design should be considered (1) maintain a normal function in the blood circulation without negative side effects, (2) enable specific cell/tissue/organ recognition for targeted gene delivery, (3) control and regulate transgene expression for a definite period of time.

Based on the intended medical applications, potential zwitterionic polymer such as polysulfobetaine (PSBMA) can also be present in diverse forms; they may be dissolved as unimers or micelles in aqueous medium, adsorbed or grafted onto aqueous-solid interfaces, or crosslinked in the form of physical or chemical hydrogels. While the single-protein adsorption resistance of grafted zwitterionic PSBMA brushes on aqueous-solid interfaces has been studied in great depth (Zhang et al. 2006, 2008a; Chang et al. 2010; Shih and Chang 2010), little is known about how PSBMA polymer conformations, such as zwitterionic chain lengths or associations, would influence the correlations between solution properties and blood compatibility. In 2010, Chang's group prepared a set of zwitterionic PSBMA polymers at varying molecular weights but similar molecular-weight distributions (Shih and Chang 2010). The effects of solution pH and ionic strengths on the UCST of these polymers of various molecular weights in aqueous solutions were examined in detail. The phase behavior of PSBMA in solution was also illustrated. The work also demonstrated the adsorption of plasma proteins onto the zwitterionic PSBMA suspension from human blood plasma and the anticoagulant activity of the polymers in a platelet-poor plasma solution in recalcified plasma-clotting tests. It was showed that PSBMA polymers exhibited an anticoagulant activity in 100 % human plasma and antihemolytic activity in RBC solution that depended on the molecular weights of the prepared polyzwitterions. Importantly, the PSBMA polymer with a molecular weight of about 135 kDa presented an excellent nonfouling character in human blood for plasma-protein resistance, anticoagulant activity, and antihemolytic activity, which can be attributed to the formation of a strong hydration layer due to the binding of water molecules around zwitterionic sulfobetaine groups, as shown in Fig. 4.

A pH sensitive block copolymer poly(acrylic acid)-*block*-poly(sulfobetaine methacrylate), (PAA-*b*-PSBMA) was prepared and combined with poly-(2-(Dimethylamino)ethyl methacrylate, PDMAEMA) and poly(ethylenimine) (PEI) homopolymer as the case to demonstrate a new generation of zwitterionic core-shell polyplexes. At neutral pH, the copolymer serves as a protective core by allowing the zwitterionic PSBMA moiety to provide a nonfouling surface to reduce cytotoxicity and enhance hemocompatibility of the gene carriers. PAA groups in the copolymer imparts pH sensitivity by allowing deshielding of the outer core in acidic solution. The PAA contain negative charge at pH 7.4, but became natural below its isoelectric point (pI) (pH = 5.5). To demonstrate the strategy proposed in Fig. 5, we considered a model case from tumor cells. The surrounding pH value of tumor or cancer cells is about 5.5 (Dai and Liu 2011). Thus, the PAA-*b*-PSBMA shell surrounding the polyplexe core is released from the core when close to a cancer cell through the neutralization of the PAA negative charges responsible of its adsorption. The naked polyplexe with positive charges from PDMAEMA will then be able to break the cell membrane and conduct pinocytosis. Pinocytosis ("cell-drinking", "bulk-phase pinocytosis", "non-specific, non-absorptive pinocytosis", "fluid endocytosis") is a form of endocytosis in

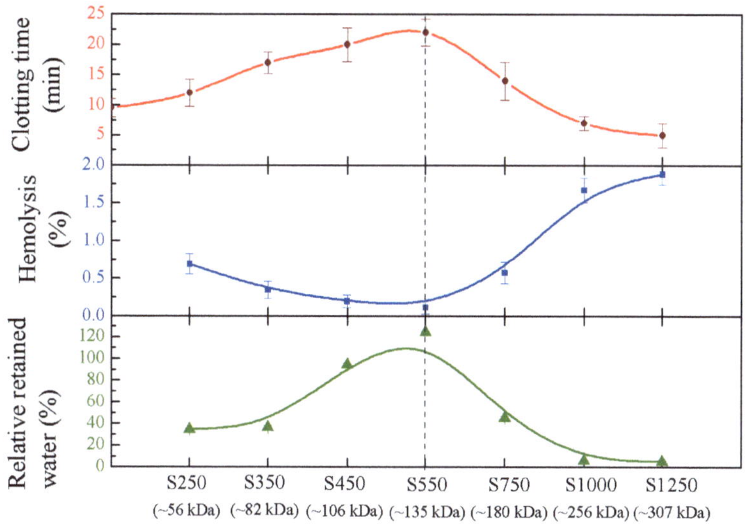

Fig. 4 Plasma-clotting time, hemolysis, and relative retained water of the different molecular weights of PSBMA polymer (Shih and Chang 2010)

which small particles are brought into the cell, forming an invagination, then suspended within small vesicles (pinocytotic vesicles) that subsequently fuse with lysosomes to hydrolyze, or to break down, the particles. Through this process, DNA can smoothly be brought to the tumor or cancer cell and achieved the treatment. From this case study, the zwitterionic core-shell formulation brings a potential molecular design in the next generation of effective nanocarriers in human whole blood with a well-controlled gene transfection.

5 Future Outlook of Zwitterionic Nano-Cargo Formulation

"Theranostics", which incorporate both therapy and diagnosis, are attracting significant attention and may revolutionize current medical treatments. Magnetic nanoparticles can work as multifunctional carriers to selectively accumulate at the target site, cure disease by certain mechanisms (either hyperthermia or drug release) and be detected using non-invasive diagnosis modality such as MRI. In this future outlook, a new nano-cargo design of multifunctional carrier is proposed to be formed from magnetic cores and surface polymeric assembling for the delivery of therapeutic nucleotides. Magnetic cores are iron oxide nanoparticles which are detectable by MRI and can be manipulated by a magnetic field, while an ideal surface polymeric assembling can carry a therapeutic nucleotides, prevent carriers cleared from the blood circulation, and provide functional groups for bio-

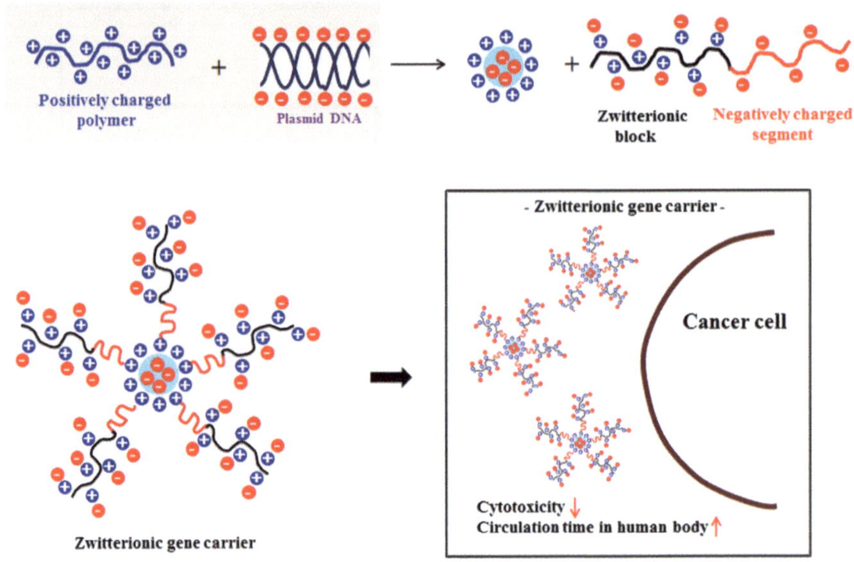

Fig. 5 Hemocompatible gene-delivery system—zwitteironic gene nanocarrier

conjugation of targeting ligands. Thus, the surface polymeric assembling plays a key role in achieving multifunctional nano-cargo carriers. Nano-cargo carriers can be designed to respond to microenvironmental differences with changes in their physiochemical properties, enabling them to perform individual delivery tasks. Cleavage of covalent bonds, disassembly of noncovalent interactions, changes of protonation, conformation, or hydrophilicity, can trigger such dynamic physicochemical adjustments. As shown in Fig. 6, zwitterionic nano-cargo formulation are ideally designed as following:

- To extend circulation in blood
  - Zwitteironization

The nano-cargo carrier has to stably bind the therapeutic nucleotides during the extracellular delivery phase and protect it against degradation in the bloodstream. Zwitteironization has been broadly explored for many liposomal and nanoparticulate carriers. In case of polymers, it can be managed through direct covalent incorporation of zwitterionic shielding into the carrier, direct zwitteironization of the nucleotides, or by zwitteironization after nano-cargo carrier formation.

- To enhance cell specificity
  - MMP cleavage
  - Targeting ligand

Nano-cargo carrier need to be shielded in the circulation and be inert against numerous possible biological interactions, but should actively interact with the target cell surface by electrostatic or ligand receptor interactions.

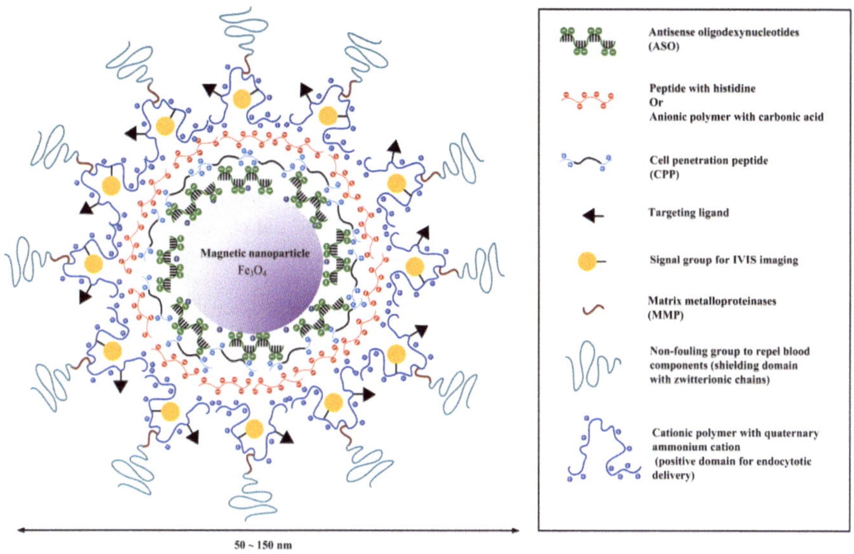

Fig. 6 Future outlook of zwitterionic nano-cargo formulation

- To improve cellular uptake
  - Controlled size distribution
  - Cationic polymers

At the intracellular site of action, the nano-cargo carrier has to disassemble to an extent that the therapeutic nucleotides is functionally accessible.

- To facilitate endosomal escape
  - Proton sponging (bioresponsive polymers)
  - Endosome membrane disruption (CPP)
- To diagnose delivery process
  - Magnetic nanoparticle for MRI
  - Signal group for IVIS (fluorescence) imaging

References

Agarwal S, Zhang Y, Maji S, Greiner A (2012) PDMAEMA based gene delivery materials. Mater Today 15(9):388–393

Akita H, Kudo A, Minoura A, Yamaguti M, Khalil IA, Moriguchi R, Masuda T, Danev R, Nagayama K, Kogure K, Harashima H (2009) Multi-layered nanoparticles for penetrating the endosome and nuclear membrane via a step-wise membrane fusion process. Biomaterials 30(15):2940–2949

Allon N, Saxena A, Chambers C, Doctor BP (2012) A new liposome-based gene delivery system targeting lung epithelial cells using endothelin antagonist. J Controlled Release 160(2):217–224

Ratner BD (2000) Blood compatibility at the close of the 20th century. Symposium proceedings, 4–6 Aug 1999, Seattle, Washington (J Biomater Sci. Polymer edition 11(11):1107–1119)

Bode SA, Thevenin M, Bechara C, Sagan S, Bregant S, Lavielle S, Chassaing G, Burlina F (2012) Self-assembling mini cell-penetrating peptides enter by both direct translocation and glycosaminoglycan-dependent endocytosis. Chem Commun 48(57):7179–7181

Chan P, Kurisawa M, Chung JE, Yang YY (2007) Synthesis and characterization of chitosan-g-poly(ethylene glycol)-folate as a non-viral carrier for tumor-targeted gene delivery. Biomaterials 28(3):540–549

Chang Y, Chen S, Zhang Z, Jiang S (2006) Highly protein-resistant coatings from well-defined diblock copolymers containing sulfobetaines. Langmuir 22(5):2222–2226

Chang Y, Liao SC, Higuchi A, Ruaan RC, Chu CW, Chen WY (2008) A highly stable nonbiofouling surface with well-packed grafted zwitterionic polysulfobetaine for plasma protein repulsion. Langmuir 24(10):5453–5458

Chang Y, Ko CY, Shih YJ, Quemener D, Deratani A, Wei TC, Wang DM, Lai JY (2009a) Surface grafting control of PEGylated poly(vinylidene fluoride) antifouling membrane via surface-initiated radical graft copolymerization. J Membr Sci 345(1–2):160–169

Chang Y, Chen WY, Yandi W, Shih YJ, Chu WL, Liu YL, Chu CW, Ruaan RC, Higuchi A (2009b) Dual-thermoresponsive phase behavior of blood compatible zwitterionic copolymers containing nonionic poly(N-isopropyl acrylamide). Biomacromolecules 10(8):2092–2100

Chang Y, Shu SH, Shih YJ, Chu CW, Ruaan RC, Chen WY (2010) Hemocompatible mixed-charge copolymer brushes of pseudozwitterionic surfaces resistant to nonspecific plasma protein fouling. Langmuir 26(5):3522–3530

Chang Y, Shih YJ, Ko CY, Jhong JF, Liu YL, Wei TC (2011) Hemocompatibility of poly(vinylidene fluoride) membrane grafted with network-like and brush-like antifouling layer controlled via plasma-induced surface PEGylation. Langmuir 27(9):5445–5455

Chen SF, Jiang SY (2008) A new avenue to nonfouling materials. Adv Mater 20(2):335

Culver KW, Anderson WF, Blaese RM (1991) Lymphocyte gene therapy. Hum Gene Ther 2(2):107–109

Dai F, Liu W (2011) Enhanced gene transfection and serum stability of polyplexes by PDMAEMA-polysulfobetaine diblock copolymers. Biomaterials 32(2):628–638

Duchardt F, Fotin-Mleczek M, Schwarz H, Fischer R, Brock R (2007) A comprehensive model for the cellular uptake of cationic cell-penetrating peptides. Traffic 8(7):848–866

Endsley AN, Ho RJ (2012) Design and characterization of novel peptide-coated lipid nanoparticles for targeting anti-HIV drug to CD4 expressing cells. AAPS J 14(2):225–235

Feng W, Brash JL, Zhu S (2006) Non-biofouling materials prepared by atom transfer radical polymerization grafting of 2-methacryloloxyethyl phosphorylcholine: separate effects of graft density and chain length on protein repulsion. Biomaterials 27(6):847–855

Fields RJ, Cheng CJ, Quijano E, Weller C, Kristofik N, Duong N, Hoimes C, Egan ME, Saltzman WM (2012) Surface modified poly(beta amino ester)-containing nanoparticles for plasmid DNA delivery. J Controlled Release 164(1):41–48

Funhoff AM, Monge S, Teeuwen R, Koning GA, Schuurmans-Nieuwenbroek NM, Crommelin DJ, Haddleton DM, Hennink WE, van Nostrum CF (2005) PEG shielded polymeric double-layered micelles for gene delivery. J Controlled Release 102(3):711–724

Gao X, Kim KS, Liu D (2007) Nonviral gene delivery: what we know and what is next. AAPS J 9(1):E92–104

Gao CL, Li GZ, Xue H, Yang W, Zhang FB, Jiang SY (2010a) Functionalizable and ultra-low fouling zwitterionic surfaces via adhesive mussel mimetic linkages. Biomaterials 31(7):1486–1492

Gao JQ, Zhao QQ, Lv TF, Shuai WP, Zhou J, Tang GP, Liang WQ, Tabata Y, Hu YL (2010b) Gene-carried chitosan-linked-PEI induced high gene transfection efficiency with low toxicity and significant tumor-suppressive activity. Int J Pharm 387(1–2):286–294

Gaur U, Sahoo SK, De TK, Ghosh PC, Maitra A, Ghosh PK (2000) Biodistribution of fluoresceinated dextran using novel nanoparticles evading reticuloendothelial system. Int J Pharm 202(1–2):1–10

Georgiev GS, Kamenska EB, Vassileva ED, Kamenova IP, Georgieva VT, Iliev SB, Ivanov IA (2006) Self-assembly, antipolyelectrolyte effect, and nonbiofouling properties of polyzwitterions. Biomacromolecules 7(4):1329–1334

Ginn SL, Alexander IE, Edelstein ML, Abedi MR, Wixon J (2013) Gene therapy clinical trials worldwide to 2012—an update. J Gene Med 15(2):65–77

Griffiths JR (1991) Are cancer cells acidic? Brit J Cancer 64(3):425–427

Guo X, Huang L (2012) Recent advances in nonviral vectors for gene delivery. Acc Chem Res 45(7):971–979

Haag R, Kratz F (2006) Polymer therapeutics: concepts and applications. Angew Chem Int Edit 45(8):1198–1215

Han G, Park B (1995) Childrens choice in conflict—application of the theory of individualism-collectivism. J Cross Cult Psychol 26(3):298–313

He Y, Chang Y, Hower JC, Zheng J, Chen S, Jiang S (2008) Origin of repulsive force and structure/dynamics of interfacial water in OEG-protein interactions: a molecular simulation study. Phys Chem Chem Phys 10(36):5539–5544

Holmlin RE, Chen XX, Chapman RG, Takayama S, Whitesides GM (2001) Zwitterionic SAMs that resist nonspecific adsorption of protein from aqueous buffer. Langmuir 17(9):2841–2850

Horbett TA (1993) Principles underlying the role of adsorbed plasma-proteins in blood interactions with foreign materials. Cardiovasc Pathol 2(3):S137–S148

Ishihara K, Oshida H, Endo Y, Ueda T, Watanabe A, Nakabayashi N (1992) Hemocompatibility of human whole blood on polymers with a phospholipid polar group and its mechanism. J Biomed Mater Res 26(12):1543–1552

Iwasaki Y, Ishihara K (2005) Phosphorylcholine-containing polymers for biomedical applications. Anal Bioanal Chem 381(3):534–546

Jabr-Milane L, van Vlerken L, Devalapally H, Shenoy D, Komareddy S, Bhavsar M, Amiji M (2008) Multi-functional nanocarriers for targeted delivery of drugs and genes. J Controlled Release 130(2):121–128

Jafari M, Soltani M, Naahidi S, Karunaratne DN, Chen P (2012) Nonviral approach for targeted nucleic acid delivery. Curr Med Chem 19(2):197–208

Jewell CM, Lynn DM (2008) Surface-mediated delivery of DNA: cationic polymers take charge. Curr Opin Colloid Interface Sci 13(6):395–402

Jiang SY, Cao ZQ (2010) Ultralow-fouling, functionalizable, and hydrolyzable zwitterionic materials and their derivatives for biological applications. Adv Mater 22(9):920–932

Jones RA, Poniris MH, Wilson MR (2004) pDMAEMA is internalised by endocytosis but does not physically disrupt endosomes. J Controlled Release 96(3):379–391

Kane RS, Deschatelets P, Whitesides GM (2003) Kosmotropes form the basis of protein-resistant surfaces. Langmuir 19(6):2388–2391

Ke JH, Young TH (2010) Multilayered polyplexes with the endosomal buffering polycation in the core and the cell uptake-favorable polycation in the outer layer for enhanced gene delivery. Biomaterials 31(35):9366–9372

Kunath K, von Harpe A, Fischer D, Peterson H, Bickel U, Voigt K, Kissel T (2003) Low-molecular-weight polyethylenimine as a non-viral vector for DNA delivery: comparison of physicochemical properties, transfection efficiency and in vivo distribution with high-molecular-weight polyethylenimine. J Controlled Release 89(1):113–125

Ladd J, Zhang Z, Chen S, Hower JC, Jiang S (2008) Zwitterionic polymers exhibiting high resistance to nonspecific protein adsorption from human serum and plasma. Biomacromolecules 9(5):1357–1361

Lam JK, Ma Y, Armes SP, Lewis AL, Baldwin T, Stolnik S (2004) Phosphorylcholine-polycation diblock copolymers as synthetic vectors for gene delivery. J Controlled Release 100(2):293–312

Lane DP, Cheok CF, Lain S (2010) p53-Based cancer therapy. Cold Spring Harb Perspect Biol 2(9):a001222

Layman JM, Ramirez SM, Green MD, Long TE (2009) Influence of polycation molecular weight on POLY(2-dimethylaminoethyl methacrylate)-mediated DNA delivery in vitro. Biomacromolecules 10(5):1244–1252

Lewis AL (2000) Phosphorylcholine-based polymers and their use in the prevention of biofouling. Colloid Surface B 18(3–4):261–275

Long SF, Clarke S, Davies MC, Lewis AL, Hanlon GW, Lloyd AW (2003) Controlled biological response on blends of a phosphorylcholine-based copolymer with poly(butyl methacrylate). Biomaterials 24(23):4115–4121

Maingi V, Kumar MV, Maiti PK (2012) PAMAM dendrimer-drug interactions: effect of pH on the binding and release pattern. J Phys Chem B 116(14):4370–4376

Maysinger D, Piccardo P, Cuello AC (1995) Microencapsulation and the grafting of genetically transformed cells as therapeutic strategies to rescue degenerating neurons of the CNS. Rev Neurosci 6(1):15–33

Murray KD, Etheridge CJ, Shah SI, Matthews DA, Russell W, Gurling HM, Miller AD (2001) Enhanced cationic liposome-mediated transfection using the DNA-binding peptide mu (mu) from the adenovirus core. Gene Ther 8(6):453–460

Naeye B, Raemdonck K, Remaut K, Sproat B, Demeester J, De Smedt SC (2010) PEGylation of biodegradable dextran nanogels for siRNA delivery. Eur J Pharm Sci 40(4):342–351

Nakabayashi N, Williams DF (2003) Preparation of non-thrombogenic materials using 2-methacryloyloxyethyl phosphorylcholine. Biomaterials 24(13):2431–2435

Nguyen J, Szoka FC (2012) Nucleic acid delivery: the missing pieces of the puzzle? Acc Chem Res 45(7):1153–1162

Oliveira AC, Neves Silva JP, Coutinho PJ, Gomes AA, Coutinho OP, Real Oliveira ME (2010) Monoolein as helper lipid for non-viral transfection in mammals. J Controlled Release 148(1):e91–e92

Olsson P, Sanchez J, Mollnes TE, Riesenfeld J (2000) On the blood compatibility of end-point immobilized heparin. J Biomater Sci Polym Ed 11(11):1261–1273

Park TG, Jeong JH, Kim SW (2006) Current status of polymeric gene delivery systems. Adv Drug Deliv Rev 58(4):467–486

Prime KL, Whitesides GM (1993) Adsorption of proteins onto surfaces containing end-attached oligo(ethylene oxide)—a model system using self-assembled monolayers. J Am Chem Soc 115(23):10714–10721

Schenborn ET (2000) Transfection technologies. Methods Mol Biol 130:91–102

Schwarzenberger P, Huang W, Oliver P, Osidipe T, Theodossiou C, Kolls JK (2001) Poly-L-lysine-based molecular conjugate vectors: a high efficiency gene transfer system for human progenitor and leukemia cells. Am J Med Sci 321(2):129–136

Sethuraman VA, Na K, Bae YH (2006) pH-responsive sulfonamide/PEI system for tumor specific gene delivery: an in vitro study. Biomacromolecules 7(1):64–70

Shih YJ, Chang Y (2010) Tunable blood compatibility of polysulfobetaine from controllable molecular-weight dependence of zwitterionic nonfouling nature in aqueous solution. Langmuir 26(22):17286–17294

Shoji Y, Nakashima H (2004) Current status of delivery systems to improve target efficacy of oligonucleotides. Curr Pharm Des 10(7):785–796

Sinclair A, Bai T, Carr LR, Ella-Menye JR, Zhang L, Jiang S (2013) Engineering buffering and hydrolytic or photolabile charge shifting in a polycarboxybetaine ester gene delivery platform. Biomacromolecules 14(5):1587–1593

Singer SJ, Nicolson GL (1972) The fluid mosaic model of the structure of cell membranes. Science 175(4023):720–731

Somasundaran T, Deo N, Somasundaran P (2002) Zwitterionic latex particles as an effective carrier for DNA. J Colloid Interface Sci 246(2):223–226

Suzuki H, Li LF, Nakaji-Hirabayashi T, Kitano H, Ohno K, Matsuoka K, Saruwatari Y (2012) Carboxymethylbetaine copolymer layer covalently fixed to a glass substrate. Colloid Surface B 94:107–113

Takeda N, Nakamura E, Yokoyama M, Okano T (2004) Temperature-responsive polymeric carriers incorporating hydrophobic monomers for effective transfection in small doses. J Controlled Release 95(2):343–355
Tamura A, Oishi M, Nagasaki Y (2010) Efficient siRNA delivery based on PEGylated and partially quaternized polyamine nanogels: enhanced gene silencing activity by the cooperative effect of tertiary and quaternary amino groups in the core. J Controlled Release 146(3):378–387
Tian H, Li F, Chen J, Huang Y, Chen X (2012) N-isopropylacrylamide-modified polyethylenimines as effective gene carriers. Macromol Biosci 12(12):1680–1688
Trabulo S, Resina S, Simoes S, Lebleu B, Pedroso de Lima MC (2010) A non-covalent strategy combining cationic lipids and CPPs to enhance the delivery of splice correcting oligonucleotides. J Controlled Release 145(2):149–158
Turk H, Haag R, Alban S (2004) Dendritic polyglycerol sulfates as new heparin analogues and potent inhibitors of the complement system. Bioconjug Chem 15(1):162–167
Ushitora M, Sakurai F, Yamaguchi T, Nakamura S, Kondoh M, Yagi K, Kawabata K, Mizuguchi H (2010) Prevention of hepatic ischemia-reperfusion injury by pre-administration of catalase-expressing adenovirus vectors. J Controlled Release 142(3):431–437
Venkataraman S, Ong WL, Ong ZY, Joachim Loo SC, Ee PL, Yang YY (2011) The role of PEG architecture and molecular weight in the gene transfection performance of PEGylated poly(dimethylaminoethyl methacrylate) based cationic polymers. Biomaterials 32(9):2369–2378
Vessillier S, Adams G, Montero-Melendez T, Jones R, Seed M, Perretti M, Chernajovsky Y (2012) Molecular engineering of short half-life small peptides (VIP, alphaMSH and gamma(3)MSH) fused to latency-associated peptide results in improved anti-inflammatory therapeutics. Ann Rheum Dis 71(1):143–149
Vorburger SA, Hunt KK (2002) Adenoviral gene therapy. Oncologist 7(1):46–59
Walrant A, Bechara C, Alves ID, Sagan S (2012) Molecular partners for interaction and cell internalization of cell-penetrating peptides: how identical are they? Nanomedicine 7(1):133–143
Wang HY, Chen JX, Sun YX, Deng JZ, Li C, Zhang XZ, Zhuo RX (2011) Construction of cell penetrating peptide vectors with N-terminal stearylated nuclear localization signal for targeted delivery of DNA into the cell nuclei. J Controlled Release 155(1):26–33
Wilner SE, Wengerter B, Maier K, de Lourdes Borba Magalhaes M, Del Amo DS, Pai S, Opazo F, Rizzoli SO, Yan A, Levy M (2012) An RNA alternative to human transferrin: a new tool for targeting human cells. Mol Ther—Nucleic Acids 1:e21
Wong SY, Pelet JM, Putnam D (2007) Polymer systems for gene delivery-past, present, and future. Prog Polym Sci 32(8–9):799–837
Wong SP, Argyros O, Howe SJ, Harbottle RP (2011) Systemic gene transfer of polyethylenimine (PEI)-plasmid DNA complexes to neonatal mice. J Controlled Release 150(3):298–306
Wu L, Jasinski J, Krishnan S (2012) Carboxybetaine, sulfobetaine, and cationic block copolymer coatings: A comparison of the surface properties and antibiofouling behavior. J Appl Polym Sci 124(3):2154–2170
Wu YZ, Ihme S, Feuring-Buske M, Kuan SL, Eisele K, Lamla M, Wang YR, Buske C, Weil T (2013) A core-shell albumin copolymer nanotransporter for high capacity loading and two-step release of doxorubicin with enhanced anti-leukemia activity. Adv Healthc Mater 2(6):884–894
Zhang Z, Chao T, Chen S, Jiang S (2006) Superlow fouling sulfobetaine and carboxybetaine polymers on glass slides. Langmuir 22(24):10072–10077
Zhang Z, Zhang M, Chen S, Horbett TA, Ratner BD, Jiang S (2008a) Blood compatibility of surfaces with superlow protein adsorption. Biomaterials 29(32):4285–4291
Zhang Z, Chao T, Jiang S (2008b) Physical, chemical, and chemical-physical double network of zwitterionic hydrogels. J Phys Chem B 112(17):5327–5332

Stimuli-Responsive Polymeric Nanocarriers as Promising Drug and Gene Delivery Systems

Gurusamy Saravanakumar and Won Jong Kim

Abstract Polymeric nanocarriers have emerged as promising drug delivery vehicles owing to their potential to selectively deliver the active agents to the disease sites through various targeting mechanisms, while minimizing the side effects. To realize more enhanced therapeutic effect, it is also highly essential to impart specific release mechanism into the nanocarriers in addition to the targeting capabilities. In this regard, stimuli-sensitive or smart polymeric nanocarriers that are responsive to inherent (e.g. pH, temperature, redox, hypoxia, enzymes, and reactive oxygen species) or external stimuli (e.g. light, ultrasound or magnetic) have received enormous attention because of their ability to control the drug release profile in a desirable fashion at the target site of action. The primary focus of this chapter is to highlight the recent advances of various stimuli-responsive nanocarriers that have been developed for a more efficient drug and gene delivery.

Keywords Stimuli-responsive nanocarriers · Drug delivery · Gene delivery · Cleavable linker · Polymeric micelles · Polymersomes · Polyplexes · Core-shell nanoparticles · Self-assembled nanoparticles · Tumor-targeting · Block copolymer

Abbreviations

AGA	Acrylglucosamine
APBA	Acrylamidephenylboronic acid
AspPBA	Aspartamidophenylboronic acid
ATRP	Atom transfer radical polymerization
BAC	*N*,*N*-Bis(acryloyl) cystamine
C18	*n*-Octadecane amine
CPT	Camptothecin

G. Saravanakumar · W. J. Kim (✉)
Center for Self-Assembly and Complexity, Institute for Basic Science (IBS), and Department of Chemistry Polymer Research Institute, Pohang University of Science and Technology (POSTECH), Pohang 790-784, South Korea
e-mail: wjkim@postech.ac.kr

A. Prokop et al. (eds.), *Intracellular Delivery II*,
Fundamental Biomedical Technologies 7, DOI: 10.1007/978-94-017-8896-0_4,

DET	*N*-(2-Aminoethyl)-2-aminoethyl group
Dex-LA	Dextran-lipoic acid
DNQ	2-Diazo-1,2-Napthoquinone
DOX	Doxorubicin
DTT	D,L-Dithiothreitol
EPR	Enhanced permeation and retention
Gal	Galactose
GI	Gastrointestinal
GSH	Glutathione
HA	Hyaluronic acid
HMAAM	Hydroxymethylacrylamide
HP	Hydrotropic polymer
Hyals	Hyaluronidases
LCST	Low critical solution temperature
MC	Merocyanine
MMPs	Matrix metalloproteinases
MTX	Methotrexate
NAS	*N*-Acryloxysuccinimide
NBC-NCA	*S*-(*O*-Nitrobenzyl)-L-cysteine) *N*-carboxyanhydride
ND	Polyester nanodendron
NIPAM	*N*-Isopropylacrylamide
NMP	Nitric oxide mediated radical polymerization
OG	Oregon green 488
P(Asp)	Poly(aspartamide)
P(HPMA-Lac_n)	Poly(2-hydroxypropyl methacrylate lactate)
P4VP	Poly(4-vinylpyridine)
PAA	Poly(acrylic acid)
PAMAM	Poly(amido amine)
PAzoMA	Polymethacrylate bearing azobenzene side groups
PBA	Phenylboronic acid
PBLG	Poly(γ-benzyl L-glutamate)
PBMA	Poly(butylmethacrylate)
PCL	Poly(caprolactone)
PDEA	Poly(diethylamino)ethyl methacrylate
PDLLA	Poly(D,L-lactide)
PDMAEMA	Poly(dimethylaminoethyl methacrylate)
PEEP	Poly(ethyl ethylene phosphate)
PEG	Poly(ethylene glycol)
PEI	Poly(ethylene imine)
PEO	Poly(ethylene oxide)
P-HA-NPs	PEGylated HA nanoparticles
PHB	Poly[(R)-3-hydroxybutyrate]
PHis	Poly(L-histidine)
PLGA	Poly(lactic-*co*-glycolic) acid

PLL	Poly(L-lysine)
PLLA	Poly(L-lactic acid)
PMAA	Poly(methacrylic acid)
PNIPAM	Poly(*N*-isopropylacrylamide)
PPS	Poly(propylene sulfide)
PS	Polystyrene
PTX	Paclitaxel
RAFT	Reversible addition-fragmentation chain transfer
ROMP	Ring opening metathesis polymerization
SP	Spiropyran
UCNPs	Upconverting nanoparticles
UCST	Upper critical solution temperature

1 Introduction

Over the past few decades, polymer-based nanocarrier systems have emerged as a versatile platform for delivery of wide range of imaging and therapeutic agents, including small-molecules drugs, proteins, and genes (Haag and Kratz 2006; Duncan 2003; Park et al. 2008). The use of polymer-based drug carriers offers several advantages over free drugs in conventional dosage forms. In particular, polymer-based carriers can enhance the solubility of poorly water-soluble drugs, improve its bioavailability, prolong circulation times, protect the drugs from harsh conditions, and release the drug in a sustained or triggered fashion at the desired site of action. Furthermore, after systemic administration, polymeric nanoparticles can passively accumulate in specific tissues such as tumors through enhanced permeation and retention (EPR) effect, attributed to leaky vasculature and lack of effective lymphatic drainage (Maeda et al. 2000). The EPR effect is also been observed in other diseases, such as chronic inflammations and infection, indicating that polymeric nanoparticles-based therapeutics may also have potential for treating these diseases as well.

More effective site-specific delivery at the cellular level can be achieved by functionalizing the surface of the polymeric nanoparticles with targeting moieties, such as antibodies, peptides or small molecular ligands like folate (Kamaly et al. 2012; Allen 2002). After accumulation at the target site, the nanoparticles functionalized with the targeting moieties can recognize, bind to, and taken up into the cells through receptor-mediated endocytosis process, thereby facilitate the release of therapeutic payloads inside the target cells. By employing a variety of natural and synthetic polymers, a number of versatile polymeric nanocarriers such as polymeric micelles, polymersomes, and polymeric nanocomplexes have been developed for effective delivery of imaging and/or therapeutic agents. For successful clinical applications, the selection of drug carrier is highly important, because it has a significant effect on the pharmacokinetics and pharmacodynamics of the transported drugs (Li and Huang 2008). One of the important prerequisites of drug carriers for use in clinical applications is that they should be biocompatible or at least they have to be degraded into small molecules with less toxicity and eliminated from the body.

In recent years, much attention has been focused on the development of stimuli-responsive or smart drug carriers that are able to release their cargo at the desired target site in a controlled or programmed fashion. Several pathological conditions are characterized by abnormal changes in the microenvironment, including change in pH, temperature, enzyme levels and oxygen concentration (Koo et al. 2011). These characteristics have been widely exploited for the development of stimuli-responsive drug carriers, which disintegrate and release their cargo in response to the local stimuli at the disease site. Alternatively, release of active agents from the drug carriers can also be manipulated by an external stimulus such as light, ultrasound or magnetic field. The developments of facile and high fidelity orthogonal transformations such as click chemistries, and the advancements in the controlled radical polymerization techniques such as atom transfer radical polymerization (ATRP), reversible addition-fragmentation chain transfer (RAFT) polymerization, nitric oxide mediated radical polymerization (NMP) and ring opening metathesis polymerization (ROMP), have enabled us to synthesis well-defined polymers to design diverse stimuli-responsive nanocarriers. In this chapter, we focus our attention on recent advances in stimuli-responsive drug delivery systems based on polymeric nanoparticles.

2 pH-Responsive Nanoparticles

The pH-responsive nanoparticles are one of the most widely investigated stimuli-responsive nanoparticles because of the changes in pH condition at the site of diseased tissue (Gao et al. 2010; Wang and Zhang 2012). The pH profile of the pathological tissues is significantly different from the normal tissues. For example, pH is lower in extracellular environment of solid tumor (6.5–7.2) and certain inflammation sites in the body compared to the healthy tissue ($\sim$7.4). The pH drop in tumor extracellular microenvironment is due to production of acidic metabolites under hypoxic conditions. This altered pH difference between the tumor and normal tissues have stimulated several investigators to develop pH-responsive nanoparticles for anticancer drug delivery. The nanoparticles employed for pH-sensitive drug targeting are generally fabricated using ionic polymers, or incorporated with acid-labile moieties in the nanoparticle structure. The nanoparticles prepared using ionic polymers undergoes solubility change (hydrophobic-hydrophilic transition) in response to pH variation and release the encapsulated drugs, while the nanoparticles with acid-labile moieties degraded in acid environment and triggered drug release.

2.1 pH-Responsive Nanoparticles Using Hydrophobic-Hydrophilic Transition

The pH-responsive nanoparticles of this category are prepared using anionic or cationic polymers (polyacids/polybases). The ionizable groups within the nanoparticles reversibly change solubility through ionization/deionization process, which enables release of drugs at the desired target site. The representative anionic polymers used for drug carriers include, poly(acrylic acid) (PAA) and poly(methyacrylate acid) (PMAA). Since PAA and PMAA become hydrophilic at normal pH (pH 7.4) and hydrophobic at low pH ($\sim$pH 1–2), they have been utilized for the development of pH-responsive block copolymer micelles for oral drug delivery. Given the broad range of pH in gastrointestinal (GI) tract, ranging from 1.0 to 2.5 in the stomach to 5.7 in the small intestine, these pH-responsive micelles could protect the drugs from harsh conditions found in stomach and enhance their absorption in the intestine (Gao et al. 2010). Further, owing to short retention time of formulations in the GI tract, it is also highly desirable to release all the drugs from the micelles in a short period after oral administration to maximize the bioavailability. To improve oral bioavailability of anticancer drug paclitaxel (PTX), Park's group prepared hydrotropic polymer (HP) micelles with AA moieties (Kim et al. 2008). In vitro release test in simulated intestinal fluid (SIF, pH = 6.5) showed a complete release of PTX within 12 h from HP micelles containing more than 20 mol % of AA. On the other hand, HP micelles without any AA moiety exhibited very slow release profile. These results indicate that HP micelles with pH-responsive AA moiety could be a promising carrier for oral delivery of PTX. Similarly, several pH-responsive PMAA-based copolymer micelles have also been investigated for small molecular drug delivery (He et al. 2010).

Many cationic polymers with ionizable amine groups, such as poly-L-lysine (PLL), poly(dimethylaminoethyl methacrylate) (PDMAEMA), poly(β-aminoester), and poly(L-histidine) (PHis) have also been used as building blocks to construct polymeric carriers, and explored for pH-sensitive drug targeting (Howard 2009; Itaka et al. 2003; Putnam et al. 2003; Green et al. 2008; De Smedt et al. 2000; Agarwal et al. 2012). The amine group of the cationic polymers undergoes protonation and deprotonation reactions during a change in pH gradients, and this events facilitate the pH-responsive nanoparticles to triggered drug release at target sites. Because of their ability to condense negatively charged nucleic acids, the cationic polymer based carriers are widely used as vectors for gene delivery. For many therapeutic agents, including nucleic acids and proteins, it is highly important to deliver them efficiently at the subcellular site of action for improved therapeutic effects. The intracellular organelles are known to maintain their own characteristic pH values, ranging from 4.5 in the lysosomes to about 8.0 in the mitochondria (Asokan and Cho 2002). This low pH and high enzymatic contents in the lysosomes may degrade biomacromolecular payload such as DNA, following their internalization by endocytosis. Therefore, to realize high transfection efficiency, it is highly

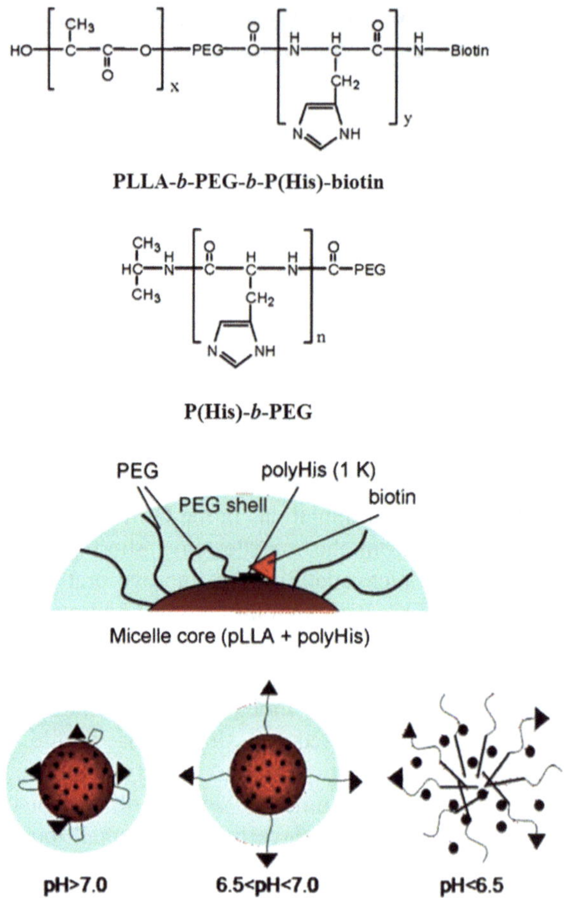

Fig. 1 Chemical structures of PLLA-*b*-PEG-*b*-P(His)-biotin and P(His)-*b*-PEG block copolymers. Schematic diagram illustrating the exposure of biotin on the surface of the pH-sensitive micelles. The biotin is anchored on the micelle core and shielded by PEG shell at pH 7.0, but exposed on the surface of the micelles for cell targeting at pH between 6.5 and 7.0, and completely disassembled at pH below 6.5. Reproduced from Lee et al. (2005, Copyright American Chemical Society) with permission

essential to escape DNA from the endosomes before degrading within lysosomes or late endosomes. The cationic polymers with high pH buffering capacity could easily facilitate endosomal escape of DNA via the proton sponge effect.

Besides gene delivery, pH-responsive cationic block copolymer micelles have also been developed for efficient tumor-targeted delivery of small molecular anticancer agents. Bae et al. developed polymeric mixed micelles that can expose biotin in response to extracellular tumor acidic pH, using the copolymers poly(L-lactic acid)-*b*-poly(ethylene glycol)-*b*-P(His)-biotin (PLLA-*b*-PEG-*b*-P(His)-biotin) and P(His)-*b*-PEG copolymer (Lee et al. 2005). The biotin is anchored on the micelle core and shielded by PEG shell at pH 7.0, but exposed on the surface of the micelles for cell targeting at pH between 6.5 and 7.0 (extracellular tumor acidic pH), and completely disassembled at pH below 6.5 (endolysosomal pH) (Fig. 1). Because of the enhanced tumor specificity and endosomal disruption characteristics, these micelles might have promising potential for tumor-targeted drug delivery. In the subsequent study, they replaced biotin with TAT (transactivator of transcription)-peptide, which has the strong capability to translocate micelles into cells (Lee et al. 2008).

2.2 pH-Responsive Nanoparticles Using Cleavable Linkers

By employing acid-labile linkers such as ester, imine, acetal, hydrazone, orthoester, β-thiopropionate, and vinylether, several pH-responsive polymeric nanocarriers have been developed for the delivery of both small molecule drugs and genes (Gao et al. 2010; Rao et al. 2011; Wei et al. 2013). In self-assembled nanostructures such as micelles or polymersomes, these cleavable linkers are generally incorporated between the hydrophilic and hydrophobic segments. The resulting self-assembled nanoparticles with acid-labile linkers induce destabilization of nanoparticles and trigger release the drugs at the target acidic extracellular tumor tissues or at the acidic intracellular compartments such as endosomes or lysosomes. To improve tumor-specific uptake and enhance intracellular delivery, a pH-responsive cleavable amphiphilic polymer with benzoic imine linker was synthesized by conjugating PEG-benzaldehyde with *n*-octadecane amine (C18) (Ding et al. 2009). Compared to the imine linker, which is prone to hydrolyzes under very weak acid conditions and unstable at physiological pH, the benzoic imine linker is stable at physiological condition (pH 7.4) but partially hydrolyzes at the tumor extracellular environment (pH $\sim$ 6.8) and completely hydrolyzes at the endosomal compartment (pH $\sim$ 5.0–6.5). Owing to the pH-dependent hydrolysis characteristic of benzoic imine, the PEG-*b*-C18 polymer formed stable micelles with neutral surface at physiological pH, but changed to positively surface charged at tumor pH and dissociated at endosomal pH (Fig. 2). This structural change suggests that these micelles could be easily taken up into the tumor cells and triggered release anticancer drugs inside the cells due to membrane disrupting capability at the endosomes. By rationally incorporating acid-labile linkers within the polymeric micelles, numerous cleavable polymeric micelles have been developed for tumor-targeted drug delivery (Tang et al. 2011; Jin et al. 2011). It is also important to note that the drug release profile of this kind of pH-responsive cleavable nanocarriers is dependent on the rate of hydrolysis of acid-labile bonds. For example, micellar aggregates and polymersomes with acid-labile β-thiopropionate linker have shown sustained release of the encapsulated molecules due to the slow acid-induced hydrolysis (Dan et al. 2010; Dan and Ghosh 2013). Thus, drug release rate from the micelles can be tuned by selecting appropriate acid-labile linkers.

In spite of promising potential of cationic polymer-nucleic acid complexes (polyplexes) for targeted gene delivery, their possible aggregation after systemic administration due to the interaction between the positively charged polyplexes and negatively charged serum protein cause a serious concern. To surmount this problem, acid-triggered dePEGylation of polyplexes strategy has been introduced. The dePEGylation not only improves the cellular uptake of the polyplexes at the acidic tumor microenvironment but also elicits endosomal destabilization and

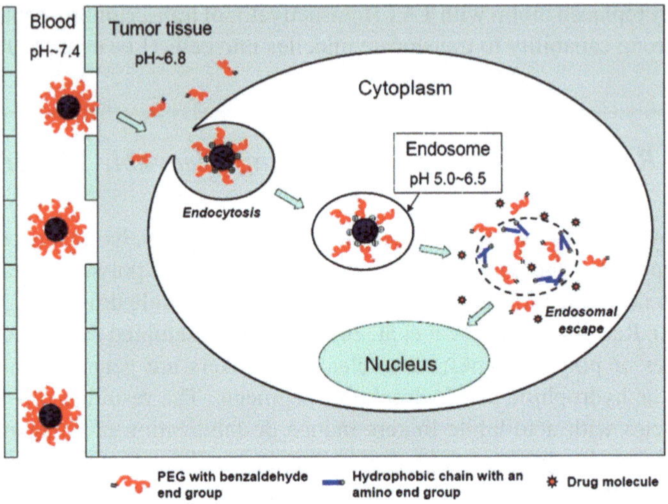

Fig. 2 Schematic illustration of cellular uptake of pH-responsive polymeric micelles with acid-labile benzoic imine. The micelles show prolonged circulation in the blood due to the PEG-shell. After accumulation at the tumor tissue via EPR effect, the surface of the micelles becomes positively charged due to the removal of PEG by partial hydrolysis of imine linker, facilitating cellular uptake through adsorptive endocytosis. Subsequently, in the more acidic endosomes, the complete hydrolysis of imine bond causes the micelles to dissociate and destabilize the endosomal membrane to release the drugs into the cytoplasm. Reproduced from Ding et al. (2009, Copyright American Chemical Society) with permission

release of genetic materials into the cytoplasm. To prepare dePEGylated polyplex micelles, PEG have been conjugated to the cationic polymers such as PLL, poly(ethylene imine) (PEI) and PDMAEMA via acid-labile linkages, hydrazone, acetal and cyclic orthoester, respectively (Walker et al. 2005; Knorr et al. 2007; Lin et al. 2007). In vitro and in vivo studies demonstrated higher gene delivery efficiency for these dePEGylated polyplex micelles than those with stable linkages.

In addition to the pH-responsive cleavable nanocarriers, many acid-triggered prodrugs have been prepared by directly conjugating small molecule anticancer drugs such as doxorubicin (DOX) or genetic drugs such as siRNA to the polymer via acid-labile linkers. Conjugation of DOX through acid-labile linkers such as hydrazone not only improves its solubility but also enhance targetability. For example, Son et al. (2003) conjugated DOX to chitosan, a natural polysaccharide, through an acid-labile *cis*-aconityl spacer. Because of the amphiphilic characteristic, the resulting DOX-chitosan conjugate under aqueous condition formed nanoparticles that enabled increase loading content of DOX through physical encapsulation. Owing to the enhanced accumulation at tumor via EPR effect and pH-responsive drug release, the DOX-loaded chitosan nanoparticles demonstrated high antitumor efficacy. Though acid-triggered prodrugs have shown improved therapeutic effects, pH-responsive nanocariers using cleavable acid-labile moieties are gaining increase attention in recent years because of easy loading and release of drugs in its native form.

3 Redox-Responsive Nanoparticles

Given the existence of a high redox potential gradient between extracellular and intracellular environments, the development of drug delivery systems based on redox-responsive polymers has garnered significant research interest. The enhanced glutathione (GSH) concentration in the cytosol and subcellular compartments ($\sim$2–10 mM) compared to the extracellular environment ($\sim$2–20 μM) has motivated researchers to incorporate reducible disulfide bonds in the polymeric nanocariers (Meng et al. 2009; Cheng et al. 2011; Son et al. 2011). The disulfide bonds present in these nanocarriers are generally preserved in the oxidizing extracellular environment during circulation, but are readily cleaved at the GSH-elevated reducing intracellular environment, triggering the cytosolic release of the drugs. Several studies have also shown that GSH is expressed in high levels in tumor tissues compared with the healthy ones (Kuppusamy et al. 2002). Thus, polymeric nanocarriers with reducible disulfide-bonds hold great potential for intracellular tumor-specific drug and gene delivery.

3.1 Redox-Responsive Block Copolymer Micelles

In recent years, tremendous effort has been directed to the development of self-assembled polymeric micelles bearing reducible disulfide bonds. Typically, disulfide bonds are incorporated at several potential locations within amphiphilic block copolymer micelles, including at the main chain of hydrophobic block, between the pendant hydrophobic segment and the main chain of the hydrophobic block, or between the hydrophilic and hydrophobic blocks. Among these three types, shell-sheddable micelles have been widely investigated as a carrier for hydrophobic anticancer drugs. Since many amphiphilic block copolymer micelles with PEG shell and biodegradable hydrophobic cores such as polyesters [such as polycaprolactone (PCL), PLLA and poly(lactic-co-glycolic) acid (PLGA)] and polypeptides (such as PBLG) have been widely established as promising drug carriers, several researchers have developed sheddable micelles by incorporating disulfide between the PEG and the hydrophobic block (Sun et al. 2009; Song et al. 2011; Saeed et al. 2011; Thambi et al. 2011). These redox-responsive micelles demonstrated triggered drug release compared to the traditional micelles without disulfide bonds, which exhibits gradual degradation kinetics and sustained release of drugs over a period of days to week through diffusion-controlled mechanism that resulted in a reduced drug efficacy. The redox-responsive shell-sheddable amphiphilic PEG-based diblock copolymers are mostly synthesized by the following two methods: polymerization of hydrophobic monomers using disulfide-containing PEG macroinitiator, or direct conjugation of functionalized PEG with hydrophobic homopolymer via disulfide functionality (Takae et al. 2008; Ryu et al. 2009). For example, a disulfide-containing diblock copolymer composed of PEG and PCL was synthesized by exchange reaction between PEG orthopyridyl disulfide and thiol end-functionalized PCL (Sun et al. 2009). The resulting PEG-SS-PCL micelles formed large aggregates rapidly in the presence of 10 mM DL-dithiothreitol (DTT) due to the shedding of PEG shell through cleavage of disulfide bonds, but the micelles were stable without the addition of DTT. The DOX-loaded in these micelles were quantitatively released within 12 h in 10 mM DTT, whereas only less than 20 % of DOX was released in the absence of DTT. In vitro cellular experiments using mouse leukemic monocyte macrophage cell line (RAW 264.7) showed faster release of DOX inside the cells for PEG-SS-PCL micelles, compared to the redox-insensitive micelles, demonstrating its potential as GSH-responsive nanocarriers for intracellular drug delivery. The drug release rate from PEG-SS-PCL micelles could be precisely controlled by fabricating micelles with a calculated amount of redox-insensitive PEG-*b*-PCL micelles (Wang et al. 2012). Interestingly, the release kinetics of another redox-responsive PEG-SS-PLLA micelles was controlled by applying high intensity focused ultrasound as an external stimulus in addition to the intracellular reducing agent GSH (Li et al. 2010b). The release mechanism involves cleavage of weak disulfide bond through solvodynamic shear produced by ultrasonic cavitation. Thus the additional stimulus could allow us to fine-tune the release kinetics of the encapsulated cargo at the desired target site in a remote and controlled way.

In addition to PEG, other hydrophilic polymers such as poly(ethyl ethylene phosphate) (PEEP) and polysaccharide such as dextran have also been investigated as shell of sheddable block copolymer micelles (Tang et al. 2009; Sun et al. 2010). Redox-responsive PEEP-SS-PCL micelles showed accelerated release of the encapsulated DOX in the presence of 10 mM GSH, mimicking the intracellular environment, than that of in the absence of GSH (Tang et al. 2009). After 2 h incubation of DOX-loaded PEEP-SS-PCL micelles in A549 cells, remarkably strong fluorescence intensity was observed in the cells pretreated with glutathione monoester (GSH-OEt, an agent used to enhance the intracellular GSH level) than the untreated one, indicating a much faster release of DOX from the micelles in a reducing environment (Fig. 3). Because of the strong affinity of PEEP shells with the cancer cells and reducing disulfide bonds, PEEP-SS-PCL micelles significantly enhanced the cytotoxicity of DOX to multidrug resistant MCF-7/ADR breast cancer cells (Wang et al. 2011).

Besides small molecular drug delivery, shell-sheddable block copolymers have also been investigated as carrier for gene delivery by replacing the hydrophobic block with cationic polymers capable of complexing with gene. An early study on these kinds of shell-sheddable polyplex micelles was carried out by Kataoka and co-workers (Takae et al. 2008). They synthesized a novel catiomer comprising disulfide bond between PEG and poly(aspartamide) (P(Asp)) with a flanking *N*-(2-aminoethyl)-2-aminoethyl (DET) group. The resulting PEG-SS-P[Asp(DET)] formed stable micelles with plasmid DNA. More importantly, PEG-SS-P[Asp(DET)] micelles demonstrated one to three orders of magnitude higher gene transfection and a more rapid gene expression, than that of micelles without disulfide bond. This high gene transfection efficiency was mainly attributed to more effective endosomal escape based on the PEG shedding in endosome. Similar kind of PEG-based

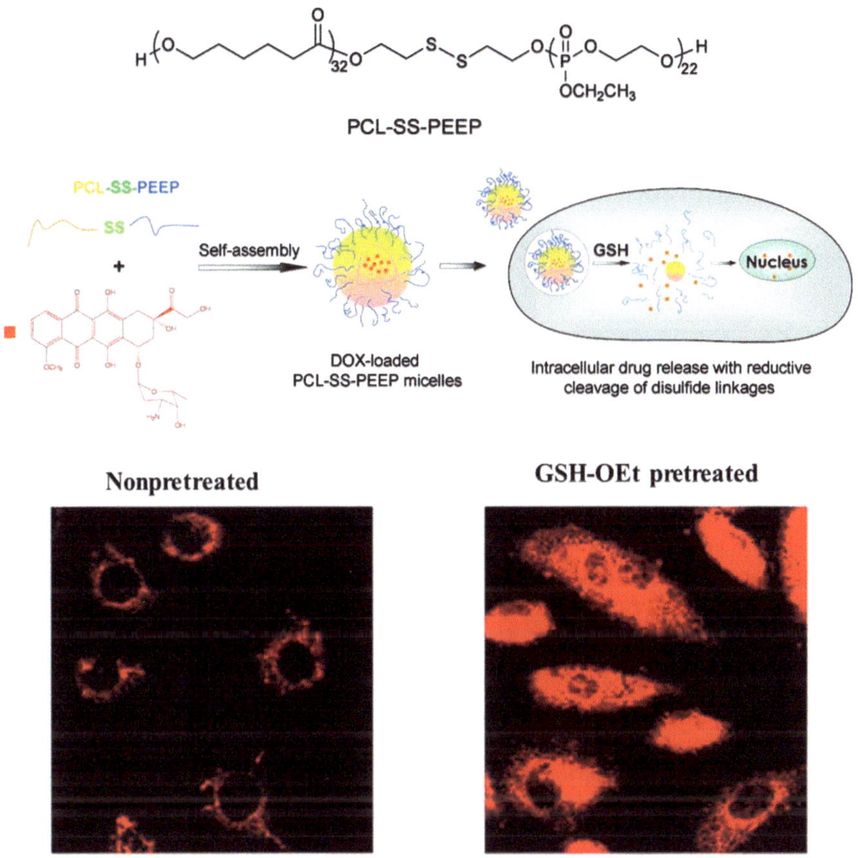

Fig. 3 Chemical structure of PEEP-SS-PCL. Schematic illustration of intracellular drug release. Confocal laser scanning microscopy (CLSM) observation of nonpretreated and GSH-OEt pretreated A549 cells after 2 h incubation of DOX-loaded PEEP-SS-PCL micelles. Reproduced from Tang et al. (2009, Copyright American Chemical Society) with permission

sheddable diblock and triblock catiomers using other cationic polymer such as poly(lysine), PDMA, PEI and chitosan oligosaccharides, have also been developed for efficient gene delivery (Cai et al. 2011; Zhu et al. 2012; Jia et al. 2013).

3.2 *Redox-Responsive Polymersomes*

Because of their ability to accommodate both hydrophilic and liphophilic active agents in the aqueous core and hydrophobic membrane respectively, polymersomes with reducible disulfide bonds could be a promising carriers for intracellular co-delivery of hydrophobic anticancer drugs and fragile hydrophilic biomolecular

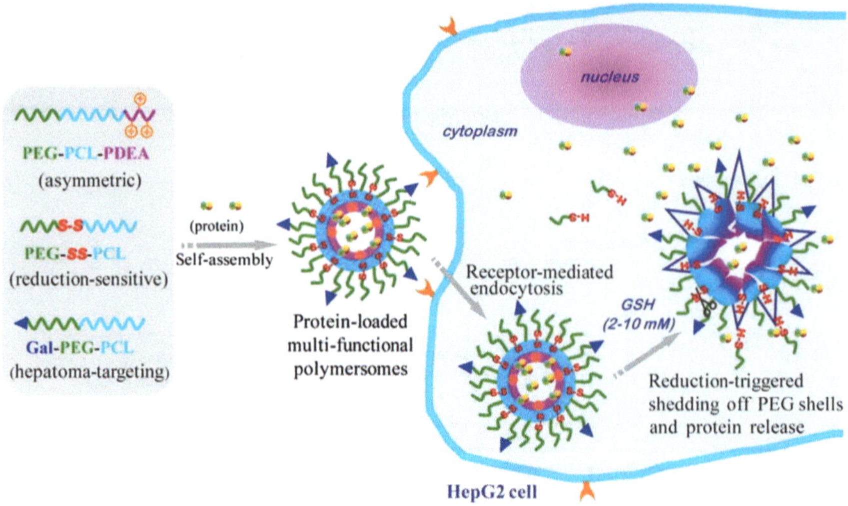

Fig. 4 Schematic illustration on hepatoma-targeting redox-responsive biodegradable chimaeric polymersomes for active loading and intracellular release of proteins. Reproduced from Wang et al. (2013, Copyright American Chemical Society) with permission

drugs. Hubbell and co-workers reported redox-responsive polymersomes based on diblock copolymers comprising disulfide bonds between hydrophilic PEG and hydrophobic poly(propylene sulfide) (PPS) (Cerritelli et al. 2007). The PEG-SS-PPS polymersomes were sensitive to the endosomal microenvironment due to the disruption of disulfide bond in the presence of intracellular concentrations of cysteine within 10 min of exposure to cells. Owing to the ability to disintegrate within the early endosomes and release their cargo before encountering harsh lysosomal environment, these polymersomes could be useful in cytoplasmic delivery of biomacromolecular drugs such as proteins, peptides and nucleic acids. Recently, novel biodegradable reduction-sensitive chimaeric polymersomes (Fig. 4) were developed based on three block copolymers, namely, PEG-SS-PCL, galactose-functionalized PEG-*b*-PCL (Gal-PEG-*b*-PCL), and asymmetric PEG-*b*-PCL-*b*-poly(diethylamino)ethyl methacrylate (PEG-*b*-PCL-*b*-PDEA) triblock copolymer, for hepatoma targeted delivery of proteins (Wang et al. 2013). The resulting polymersomes exhibited optimal loading of proteins such as bovine serum albumin, ovalbumin and cytochrome c due to the electrostatic and hydrogen bond interactions between the proteins and PDEA inner core. In particular, these Gal-decorated chimaeric polymersomes effectively delivered granzyme B, a highly potent apoptosis mediator, into HepG2 cells via receptor-mediated mechanism, indicating their potential for efficient intracellular protein delivery.

3.3 *Redox-Responsive Polymeric Nanoparticles*

Many self-assembled polymeric nanoparticles are prone to lose their structural integrity due to large dilution in the blood stream after intravenous injection. This may result in undesirable drug release before reaching the target site, causing significant toxicity to the healthy cells. Therefore, stability of the drug-loaded polymeric nanoparticles is a highly important parameter for successful in vivo applications. To improve the stability in the extracellular environment and enhance intracellular drug release, biodegradable redox-responsive nanoparticles based dextran-lipoic acid (Dex-LA) conjugate were developed (Li et al. 2009). After loading DOX, the Dex-LA nanoparticles were readily crosslinked using catalytic amount of DTT. The resulting crosslinked nanoparticles were highly robust against dilution and a high salt concentration, but released DOX under to 10 mM DTT, indicating their potential for tumor-targeted chemotherapy in vivo. Robust crosslinked redox-responsive nanogels have been prepared directly through inverse miniemulsion polymerization techniques in the presence of disulfide-functionalized dimethacrylate crosslinker (Oh et al. 2007a, b). These crosslinked nanogels demonstrated less toxicity to cells, and triggered release of encapsulated DOX and other hydrophilic drugs in the presence of GSH.

3.4 *Redox-Responsive Polyplexes*

For successful gene therapy, the nanocarriers should possess the following traits: ability to condense nucleic acids into a compact nanoparticle, protect nucleic acids from degradation by nucleases in the extracellular fluids, and capability to efficiently release nucleic acids inside the target cells. To accomplish more efficient nucleic acid delivery, several redox-responsive cationic polymers based on PEI, poly(amido amine) (PAMAM), poly(amido ethylenimines), poly(β-aminoester), and PDMAEMA have been explored (Son et al. 2010; Lin et al. 2006; Christensen et al. 2006; Zugates et al. 2006; You et al. 2007). The polyplexes prepared using these reducible cationic polymers have demonstrated enhanced DNA or siRNA transfection efficiency, and showed less cytotoxicity due to decreased charge density as a result of the intracellular reduction of disulfide bonds within the polymers. Similarly, redox-responsive complexes are also made through direct conjugation of nucleic acids to a biocompatible natural polymer such as dextran and hyaluronic acid (HA) via disulfide bond to improve the efficacy of nucleic acid delivery (Namgung and Kim 2012; Mok et al. 2007). More interestingly, cross-linking the siRNAs through cleavable disulfide bonds, followed by condensation with less cytotoxic cationic polymers, have shown improved biocompatibility and enhanced gene silencing efficiency (Lee et al. 2010; Mok et al. 2010).

4 Temperature-Responsive Nanoparticles

Several studies have shown that inflamed or neoplastic tissues could exhibit elevated temperature than the healthy tissue due to increased metabolic activity. Additionally, temperature of the body can also be conveniently tuned externally using applied signals (hyperthermia), such as magnetic field or light. Owing to these aforementioned features, increasing efforts have been focused on the development of polymeric nanocarriers that could respond to temperature (Chilkoti et al. 2002; Schmaljohann 2006). In general, the thermo-responsive polymers used for biomedical applications may display either lower critical solution temperature (LCST) or upper critical solution temperature (UCST), which are the respective critical temperature points below and above which the polymer becomes completely miscible with the water/solvent. To date, LCST-based polymers have been widely studied for drug delivery applications. Among a variety of thermo-responsive polymers, poly(*N*-isopropylacrylamide) (PNIPAM) that has a reversible sharp phase transition (LCST about 32 °C) received a great deal of interest to design and fabricate temperature-responsive carriers for drug delivery (Cammas et al. 1997; Chung et al. 1998). PNIPAM is water-soluble below their LCST, and become water-insoluble upon raising the temperature above LCST as a result of coil-to-globe transition. Furthermore, the LCST of PNIPAM can be controlled by copolymerizing NIPAM with other co-monomers. The hydrophilic co-monomers generally increase the LCST, whereas the hydrophobic co-monomers have an opposite effect. These attributes provide an excellent opportunity to fine-tune the polymers with LCST around the physiological body temperature (37 °C). Other common LCST thermo-responsive polymers are shown in Table 1. These thermo-responsive polymers have been used as building blocks to construct either core or shell of the polymeric nanoparticles. Although diverse core-shell polymeric nanoparticles have been developed using thermo-responsive polymers, we restrict our discussions in the subsequent sections mainly on the thermo-responsive block copolymer micelles.

4.1 Polymeric Micelles with Temperature-Responsive Core

Because of their change in solubility below and above LCST, PNIPAM has been utilized either as the hydrophobic segment (core) or the hydrophilic segment (shell) of the block copolymer micelles. A representative example of the former case includes PEG-*b*-PNIPAM diblock block copolymer, which is a double hydrophilic block copolymer synthesized through controlled radical polymerization such as ATRP and RAFT by using the respective PEG macroinitiator (Zhang et al. 2005; Hong et al. 2004). The PEG-*b*-PNIPAM self-assembled to form polymeric micelles with thermo-sensitive PNIPAM core and hydrophilic PEG shell above the LCST in aqueous solution. The micelle formation and the nature of micelles such as size were

Table 1 Chemical structures of representative LCST polymers. Schmaljohann (2006)

Polymer	Structure	Phase transition temperature in aqueous solution (°C)
PNIPAM	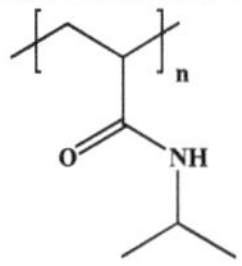	30–34
PDEAAm	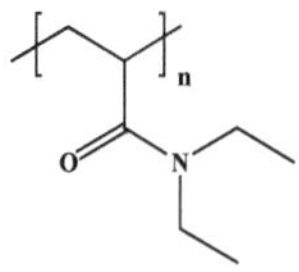	32–34
PMVE	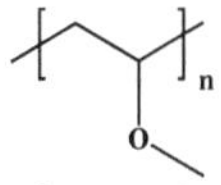	37
PVCL	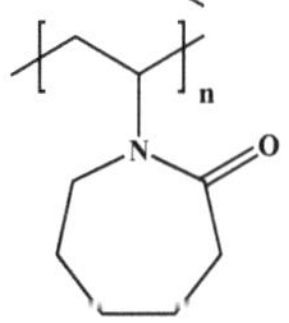	30–50[a]
P(GVGVP)	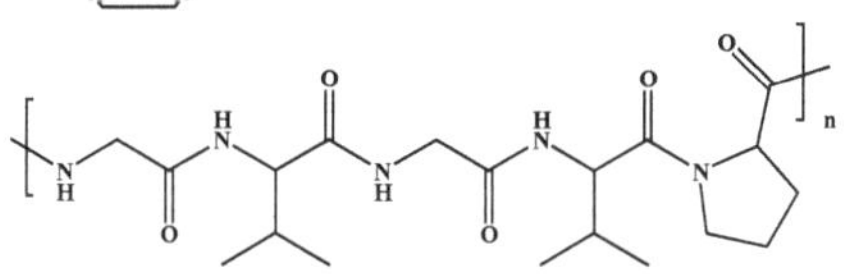	28–30

PNIPAM poly(*N*-isopropylacrylamide), *PDEAAm* poly(*N,N*-diethylacrylamide), *PMVE* poly-(methylvinylether), *PVCL* poly(*N*-vinylcaprolactam), *P(GVGVP)* elastin-like polypeptide poly(GVGVP)

[a] Dependent on the MW and concentration

depended on the concentration as well as the constituent of the block copolymer (Yan et al. 2008; Chung et al. 1998). To develop thermo-sensitive carriers for the delivery of anticancer drug methotrexate (MTX), a biotin-conjugated PEG-*b*-P(NIPAM-*co*-*N*-hydroxymethylacrylamide) (biotin-PEG-*b*-(PNIPAM-*co*-HMAAM)) copolymer was synthesized (Cheng et al. 2008). The LCST of the copolymer were adjusted by varying the molar feed ratios of NIPAM to HMAAM. The copolymer with a LCST of 41.5 °C was loaded with MTX via dialysis method. The release rate of MTX from the micelles increased markedly below LCST (37 °C) than at above LCST (43 °C) due to the temperature-induced deformation of the micellar structure. Furthermore, the biotin-conjugated micelles were efficiently taken up by the cancer cells. One of the critical problems associated with these kinds of double hydrophilic micelles with PNIPAM core are their dissociation at room temperature, making the micelles difficult in handling for practical applications. To prevent the dissociation of micelles and to increase their stability, crosslinking were performed either at the core or shell of the micelles. For example, PEG-*b*-PNIPAM micelles were crosslinked at the

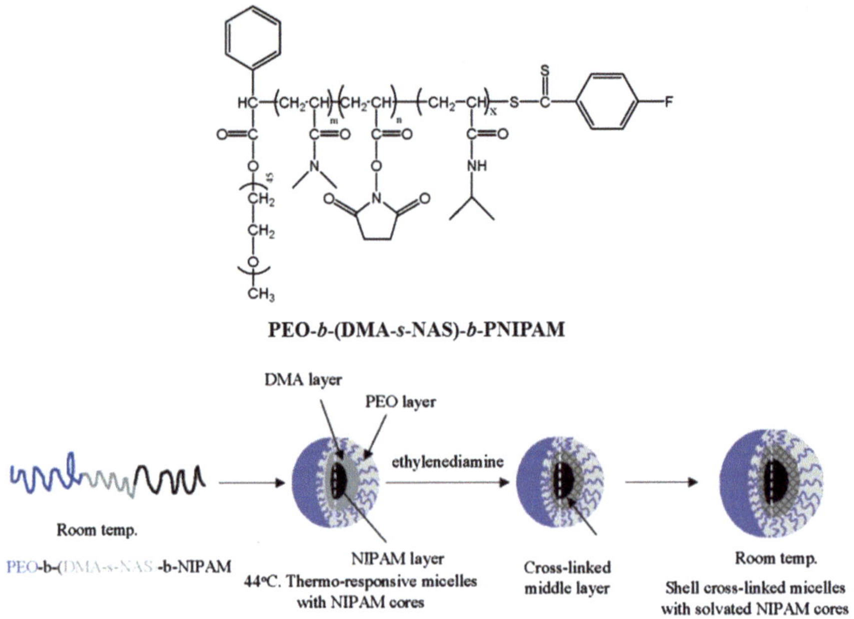

Fig. 5 Chemical structure of PEO-*b*-(DMA-*s*-NAS)-*b*-PNIPAM copolymer. Schematic illustration for the formation of crosslinked micelles. Adopted from Li et al. (2006b, Copyright American Chemical Society) with permission

PNIPAM-core using biodegradable crosslinker *N,N*-bis(acryloyl) cystamine (BAC), and evaluated as carriers for DOX (Zeng and Pitt 2005). The DOX-loaded micelles crosslinked with BAC exhibited enhanced stability at dilute concentrations than control uncrosslinked pluronic P105 micelles, which easily dissociate and released DOX at dilute concentrations. In particular, 0.5 wt% BAC crosslinked micelles were stable at room temperature for up to 2 weeks even under dilute conditions. However, the micelles were disassembled after the degradation of BAC crosslinker by the addition of reducing agent β-mercaptoethanol. These results indicate that after crosslinking thermo-responsive micelles did not disassemble at low concentrations, and can be easily handled at room temperature.

Alternatively, to crosslink the shell of the PNIPAM-core micelles, McCormick group synthesized a triblock copolymer poly(ethylene oxide)-*b*-P(*N,N*-dimethylacrylamide-*co*-*N*-acryloxysuccinimide)-*b*-PNIPAM (PEO-*b*-P(DMA-*co*-NAS)-*b*-PNIPAM comprising an activated ester block between PEO and PNIPAM blocks (Li et al. 2005). In aqueous solution at room temperature, this triblock copolymer existed as unimer but formed micelles with hydrophobic PNIPAM-core and hydrophilic PEG-outer shell and P(DMA-*co*-NAS)-inner shell above the LCST of PNIPAM (Fig. 5). The micelles were crosslinked at the activated ester inner shell using ethylenediamine as crosslinker. As expected, the crosslinked micelles were stable although the PNIPAM-core was swollen at room temperature due to solvation, whereas the uncrosslinked micelles were dissociated into unimers. Since ethylene diamine is a non-degradable crosslinker, it may acts as a permanent crosslinks and can hinders drug release. To surmount this problem, a cleavable diamine crosslinker (cystamine) was used for the preparation of shell crosslinked thermo-responsive micelles (Li et al. 2006b). This cleavable crosslinks not only facilitate the drug release, but also enables to excrete the block copolymers from the body after the micelle degradation. The shell crosslinked thermo-responsive PNIPAM-core micelles have also been prepared by reacting amine group of PLL shell of a well-defined Y-shaped miktoarm PNIPAM-*b*-P(LL)$_2$ star copolymer micelles with glutaraldehyde as a crosslinker (Li et al. 2010a). The drug release profile of these kinds of shell crosslinked micelles can be controlled by the solution temperature and the cross-linking density.

In addition to PNIPAM, other thermo-responsive polymers such as PVCL, poly(2-hydroxyethyl methacrylate lactate) and poly(2-hydroxypropyl methacrylate lactate) (p(HPMA-Lac$_n$) have also been investigated as thermo-responsive core for block copolymer micelles (Prabaharan et al. 2009; Rijcken et al. 2007a). Hennink and co-workers developed thermo-responsive PEG-*b*-p(HPMA-Lac$_n$) block copolymer micelles and investigated as a carrier of PTX, photosensitizer and superparamagnetic iron oxide nanoparticles, for chemotherapy, photodynamic therapy, and magnetic resonance imaging, respectively (Soga et al. 2005; Rijcken et al. 2007b; Talelli et al. 2009). The drug release of these thermo-responsive micelles is governed by the hydrolysis of the Lac side chains, which resulted in the destabilization of micelles.

4.2 Polymeric Micelles with Temperature-Responsive Shell

In the case of polymeric micelles with thermo-responsive shells, the hydrophobic polymers generally constitute the core-forming segments of the micelles. Therefore, unlike double hydrophilic polymeric micelles with thermo-responsive cores as discussed previously, this kind of micelles are more stable at room temperature because of the robust hydrophobic core. The hydrophilic thermo-responsive shell stabilizes the micelles and protects the loaded-drugs from enzymatic degradation during circulation, but can be switched to hydrophobic at a specific site by increasing the temperature above LCST. This increase in hydrophobicity cause shrinkage of the thermo-responsive outer shell and destabilize the micelles, thereby release the encapsulated drugs. To date, several biodegradable and non-biodegradable hydrophobic polymers such as PCL, poly(D,L-lactide) (PDLLA), poly(γ-benzyl L-glutamate) (PBLG), poly(methylmethacrylate), poly(butylmethacrylate) (PBMA), poly[(R)-3-hydroxybutyrate] (PHB) and polystyrene (PS) have been investigated as hydrophobic cores. The nature of the hydrophobic core-forming polymeric segments has a significant effect on the drug releasing characteristic of the micelles. For example, Chung et al. (2000) prepared two different block copolymer micelles, both comprising PNIPAM outer shell but different inner cores, one with PBMA and other with PS as the hydrophobic inner core. They investigated the release characteristics of the two micelles after loading DOX. The micelles having flexible PBMA inner cores with lower T_g (*ca.* 20 °C) released significant amount of DOX when heated above the LCST, while the micelles having rigid PS inner cores with high T_g (*ca.* 100 °C) did not show any enhanced DOX release. In vitro cell cytotoxic studies using bovine aorta endothelial cells demonstrated a high cytotoxicity for micelles with PBMA-cores than PS-cores. These results indicate that characteristics of hydrophobic inner cores play important roles in the physicochemical characteristics of micelles with thermo-responsive shells. It has also been reported that an increase in the length of hydrophobic biodegradable PLGA as the core-forming segment of micelles, comprising thermo-responsive P(NIAPM-*co*-*N*,*N*-dimethylacrylamide) (PNIPAM-*co*-DMAM) as the shell, could lead to a decrease in the critical association concentration and a slight increase in the DOX-loading capacity (Liu et al. 2005). The increased length of PLGA segment also caused larger polymer aggregation number per micelles, resulting in stronger PLGA-DOX interactions and thus hampering drug release. These results indicate that drug release profile from the micelles could be fine-tuned by controlling the length of the core-forming segment. Nonetheless, the DOX-loaded (PNIPAM-*co*-DMAM)-*b*-PLGA micelles demonstrated enhance drug release and high cytotoxicity against 4T1 cells at a temperature above the LCST. Unimolecular micelles with a hyperbranched polyester (Boltorn H40) as the core-forming segment and PNIPAM as the thermo-responsive shell showed a double thermal phase transition behavior, which is in contradictory to the fact that PNIPAM homopolymer exhibits a single LCST at ~32 °C in a aqueous solution (Luo et al. 2006). Temperature-dependent excimer fluorescence studies revealed that inner part of PNIPAM around the hydrophobic dendritic H40 core collapses first at lower temperature, followed by the outer part at higher temperature.

It is also important to highlight that micelles with thermo-responsive PNIPAM shells have shown enhanced intracellular uptake above their LCST. To investigate the temperature-induced intracellular uptake, Okano and co-workers prepared Oregon Green 488 (OG)-labeled thermo-responsive diblock copolymer micelles, with an LCST of 40 °C, comprising PNIPAM-*co*-DMAM as thermo-responsive segment and a biodegradable PDLLA as hydrophobic segment (Akimoto et al. 2009). The OG-labeled micelles showed a significantly higher cellular uptake above the LCST (42 °C) than below LCST (37 °C), as observed under confocal laser scanning microscopy (Fig. 6). This increase in cellular uptake above the LCST was mainly attributed to the enhanced interactions between the micelles and cell membranes mediated through the thermo-responsive phase transition of the micellar shells. Interestingly, the internalized micelles were primarily localized at the Golgi apparatus and endoplasmic reticulum compartments, bypassing lysosomes, although the exact uptake mechanism of the thermo-responsive micelles was not clearly understood (Akimoto et al. 2010). These results suggests that micelles with thermo-responsive PNIPAM-shells could be promising carriers for intracellular delivery of pH and/or enzyme-sensitive biomacromolecular drugs

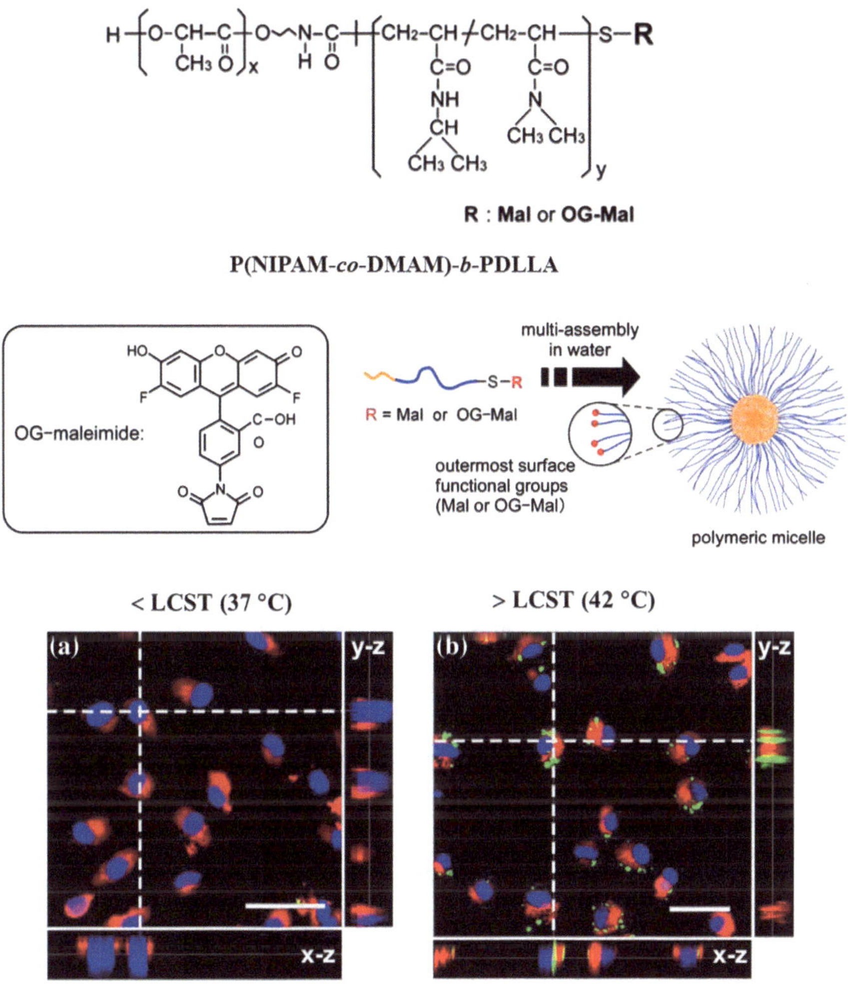

Fig. 6 Chemical structure and self-assembly of P(NIPAM-*co*-DMAM)-*b*-PDLLA micelles. Below (37 °C) and above (42 °C) LSCT. CLSM images of polymeric micelles localized within cultured cells after incubation for 9 h below the LCST (37 °C) and above the LCST (42 °C) in 10 % serum culture media. The nuclei and cytoplasm were stained with Hoechst 33258 (*blue*) and Cell Tracker Red (*red*), respectively. Green fluorescence was derived from OG-labeled micelles. Scale bars: 50 μm. Adopted from Akimoto et al. (2009, Copyright American Chemical Society) with permission

such as nucleic acids, and peptides/protein, without lysosomal degradation. Conjugation of targeting ligands such as folate at the thermo-responsive shells of the micelles has shown enhanced intracellular uptake as well as cytotoxicity above the LCST (Rezaei et al. 2012; Liu et al. 2007).

Linqi and co-workers developed thermo-responsive complex micelles with channels in the shells from the self-assembly of two diblock copolymers, namely, poly(tert-butylacrylate)-*b*-PNIPAM (P*t*BA-*b*-PNIPAM) and P*t*BA-*b*-poly(4-vinyl-pyridine) (P*t*BA-*b*-P4VP) (Li et al. 2006a). With an increase in temperature, PNIPAM collapsed and enclosed the P*t*BA core, leading to the formation of channels in the micellar shell with P4VP corona. Conversely, increase in the pH from 2.5 to 7.8 at 25 °C collapsed P4VP chains due to their deprotonation, resulting in the formation of channels on the micellar surface with PNIPAM chains stretching outside to stabilize the micelles. Interestingly, the size of the channels could be controlled by varying the environmental condition or composition of the two diblock copolymers. These complex micelles with channels could be a promising carrier for controlled release of small molecule drugs.

5 Light-Responsive Nanoparticles

The use of light as an external stimulus to trigger the release of drugs from the polymeric carriers has received much attention because it enables more efficient temporal and spatial control by tuning the light wavelength or energy (Tomatsu et al. 2011; Fomina et al. 2012). Further, a light source can be easily manipulated from outside of the body. Light-responsive polymeric nanoparticles typically consist of photochromic moieties, which upon exposure to light undergo changes through a photoreaction such as photoisomerization, photodimerization or photocleavage. These changes lead to the disruption of the nanoparticle assembly, thus, controlling the release characteristics of the cargo. The light required to induce the change of photochromic moieties strongly depends on their molecular structures. In general, polymeric nanoparticles-responsive to light in the UV, visible, and near infrared (NIR) have been widely explored for drug delivery applications. NIR light with wavelengths in the range of about 700–1,000 nm is particularly attractive for biomedical applications, because NIR has deeper penetration depth on the orders of millimeters to centimeters due to reduced absorption and scattering by biological substances and waters, and causes minimal tissue damage than UV.

Zhao et al. (2007, 2012) have developed various light-responsive block copolymer micelles. Based on the transformation brought by the photochemical events, so far, two main strategies have been utilized to design light-sensitive block copolymer micelles for drug delivery applications: (1) introducing a photochromic groups on the block copolymers that upon exposure to light disrupt the micellar structure through shifting of the hydrophilic-hydrophobic balance, and can be reassembled by exposing light of different wavelength; and (2) disassembling the micellar structure by breaking the photolabile group introduced between the hydrophobic pendant and main chain, or at the junction of hydrophobic and hydrophilic segments. In the latter case, the micellar structure cannot be restored (Fig. 7).

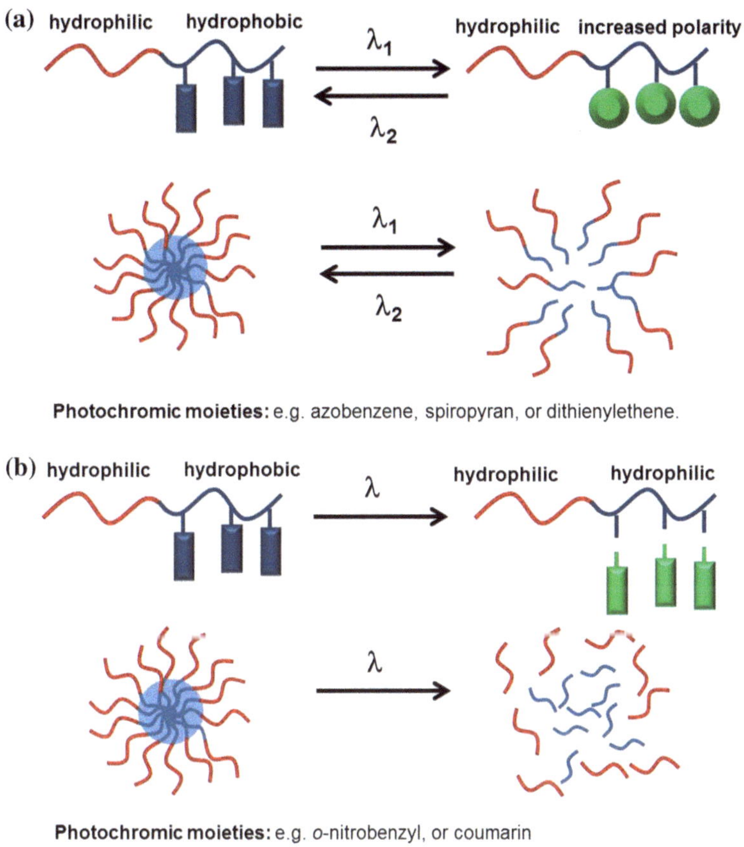

Fig. 7 Schematic illustration of two strategies used for the light-responsive block copolymer micelles. Adapted from Zhao (2012, Copyright American Chemical Society) with permission

5.1 Light-Responsive Nanoparticles Using Reversible Photoisomerization Reaction

Light-responsive polymeric nanoparticles that constitute reversible disassembly and assembly have been fabricated by employing several photochromic groups, which undergo reversible photoisomerization reaction upon exposure to light, including azobenzene, spiropyran (SP), and dithienylethene (Zhao 2012). Among these groups, azobenzene has been most widely incorporated into the block copolymer micelles. The azobenzene undergoes a reversible trans-cis photoisomerization upon UV and visible light irradiation. Zhao group has synthesized first azobenzene-containing amphiphilic diblock copolymers, which comprising of polymethacrylate bearing azobenzene side groups (PAzoMA) as the hydrophobic segment and PAA as the hydrophilic segment, through ATRP (Wang et al. 2004).

This copolymers formed aggregates, micelles or vesicles, in a selective solvent. A reversible change in the aggregate morphology was observed under alternating UV and visible light illumination as a result of trans-cis photoisomerization of azobenzene in PAzoMA block. Similarly, light-responsive vesicles were prepared using a series of amphiphilic linear-dendritic block copolymers composed of PEG segment linked to the fourth generation dendron containing azobenzene moiety and hydrophobic chains connected to the periphery of the dendron (Blasco et al. 2013). The light-induced release characteristic of the vesicles was dependent on the ratio between azobenzene moiety and hydrophobic segment. In spite of extensive investigations on the self-assembled nanostructures using azobenzene-bearing photo-responsive block copolymers, only little efforts have been made to exploit these systems for real controlled release applications. Recently, light-responsive amphiphilic polypeptide block copolymers incorporated with photochromic SP group at the hydrophobic block have been synthesized (Kotharangannagari et al. 2011). This copolymer formed flower-like micellar aggregates in the aqueous conditions. When exposed to UV light ($\lambda = 350$ nm), the aggregate disrupted and becomes water-soluble due to the isomerization of SP (hydrophobic closed form) to merocyanine (MC, hydrophilic open form). However, the MC returned to SP form under visible light irradiation ($\lambda = 590$ nm). This reversible light-induced dissolution/micellation transition indicates that it can be used as a carrier for light-regulated drug release. The size of the nanoparticles is one of the crucial factors for enhanced accumulation and deep penetration in tumor tissue, which is highly essential to improve the therapeutic effects. By exploiting the reversible photoisomerization between SP and MC, a photoswitchable nanoparticle that undergoes reversible volume changes from 150 to 40 nm upon exposure to UV light was developed to enhance tissue penetration (Tong et al. 2012). In contrast to conventional nanoparticles, where external energy sources enhance penetration by disrupting tissues, the photoswitchable nanoparticles exhibited enhanced tissue penetration and no tissue injury.

5.2 Light-Responsive Nanoparticles Using Irreversible Photo-Cleavage Reaction

For the fabrication of light-responsive micelles that undergo irreversible disassembly, photochromic molecules such as *O*-nitrobenzyl, coumarin or pyrene are incorporated in the hydrophobic pendant chain of the amphiphilic block copolymers using aryl methylester linkage (Jiang et al. 2005, 2006; Babin et al. 2009). Upon exposure of these block copolymer micelles to light, the photoreaction cleaves the photochromic molecules and converts the hydrophobic block into a hydrophilic block by forming carboxylic acid. This process generates a hydrophilic-hydrophobic imbalance and induces disassembly of the micelles, thereby triggered the encapsulated active agents. Alternatively, disassembly of the micelles has also been induced by incorporating the *O*-nitrobenzyl group at the junction of the hydrophilic and hydrophobic segments in the main backbone of the amphiphilic block copolymer (Cabane et al. 2010). Liu et al. synthesized a light-responsive polypeptide-based block copolymer having *O*-nitrobenzyl group through ring opening polymerization of S-(*O*-nitrobenzyl)-L-cysteine) *N*-carboxyanhydride (NBC-NCA) monomer using PEG as macroinitiator (Liu and Dong 2012). The potential of PNBC-*b*-PEG copolymer as a light-responsive carrier for anticancer therapy was evaluated by encapsulating DOX. The loaded nanoparticles showed light-triggered DOX-release profile under UV irradiation at 365 nm, wherein the release rate increased by increasing the irradiation time. In an attempt to improve the transfection efficiency of lipoplexes, Nagasaki et al. (2003) synthesized light-responsive cationic lipids having an *O*-nitrobenzyl moiety between its hydrophilic and hydrophobic segment. The light-responsive cationic lipid-DNA complexes showed enhanced transfection with the UV irradiation, compared to that of without UV irradiation, owing to the dissociation of the cationic lipid-DNA complex and escape of DNA from endocytic vesicles. Thayumanavan et al. synthesized light-sensitive facially amphiphilic dendrimers, which could form self-assembled aggregates in water, using a photolabile 2-nitrobenzylesters as the liphophilic units and oligoethylene glycol as the hydrophilic units. Further, they demonstrated light-induced disassembly of the dendrimers using Nile red as the liphophilic guest molecule (Yesilyurt et al. 2011). Although a number of light-responsive carrier systems using *O*-nitrobenzyl group have been developed, most of them are activated by UV light that may cause skin damage. To overcome this concern, two-photon absorption of NIR light is used to activate photoreaction of *O*-nitrobenzyl-containing nanoparticles (Jiang et al. 2006). However, the sensitivity of this process was low because of inefficient two-photon absorption. Recently, Zhao et al. addressed this issue by encapsulating upconverting nanoparticles (UCNPs) into the nitrobenzyl-based block copolymer micellar core, along with the hydrophobic payloads (Yan et al. 2011). Upon exposing these micelles to NIR light (980 nm), the UCNPs emit photons in the UV and visible regions, which are absorbed by the *O*-nitrobenzyl groups on the micelle core-forming hydrophobic block. This activated photocleavage reaction and triggered the release of hydrophobic payload (Fig. 8). Because of more efficient large-two photon absorption cross section, the micelles prepared using the coumarin chromophores have also demonstrated high sensitivity than one using *O*-nitrobenzyl (Babin et al. 2009). Nevertheless, dissociation of nanoparticles by co-loading UCNPs represents a more efficient method for potential biomedical applications.

Polymeric micelles containing light-responsive 2-diazo-1,2-napthoquinone (DNQ) group have also been synthesized and investigated their potential as drug carrier (Chen et al. 2011). DNQ is an another attractive trigger group for the design of light-responsive nanocarriers because their UV or NIR induced Wolff rearrangement can result in a drastic change in aqueous solubility through conversion of hydrophobic DNQ into hydrophilic 3-indenecarboxylic acid. Recently, light-responsive lectin-binding Janus-type dendritic PAMAM amphiphiles composed of DNQ were synthesized by connecting hydrophobic DNQ-decorated PMAMAM dendron to

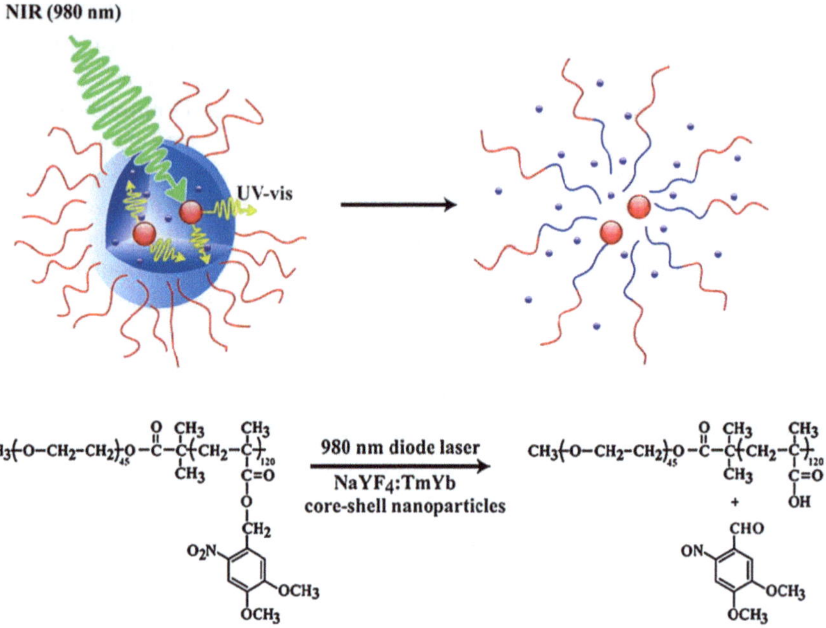

Fig. 8 Schematic illustration of NIR light-triggered dissociation of micelles using the UV light generated by the loaded upconversion nanoparticles. Reproduced from Yan et al. (2011, Copyright American Chemical Society) with permission

hydrophilic lactose-decorated PAMAM via click chemistry (Sun et al. 2012). In an aqueous condition, the dendritic amphiphiles formed self-assembled nanoparticles with DNQ-core and lactose surface. The DOX-loaded nanoparticles demonstrated NIR-triggered drug release profile, and their cytotoxicity was close to free DOX. Considering the light-triggered drug release and lectin-binding capabilities, these nanoparticles have great potential for targeted spatiotemporal drug delivery.

6 Enzyme-Responsive Nanoparticles

Enzymes play a central role in all biochemical processes that occurs in the body. Dysregulation of enzyme expression is implicated in various diseases, including cancer. Thus, enzymes are increasingly used as signals for monitoring several physiological changes, and can also be used as an effective tool for diagnose the disease early in its pathogenesis. In recent years, considerable research attention has also been focused on the design of polymeric nanoparticles that incorporate structural motifs that can be cleaved via enzymatic digestion to disassemble their structures, thereby release the drugs at the target disease site (Zelzer et al. 2013; de la Rica et al. 2012).

6.1 Enzyme-Responsive Polysaccharides-Based Nanoparticles

Nanoparticles-based on polysaccharides such as chitosan, pectin, dextran and chondroitin sulfate have been widely investigated as carriers for colon-specific drug delivery owing to the degradation of the glycosidic linkages by the colonic microflora that includes a large number of hydrolytic and reductive enzymes such as β-glycosidase, β-glucuronidase and azoreductase (Sinha and Kumria 2001; Park et al. 2010). Many malignant tumors have demonstrated elevated levels of hyaluronidases (Hyals), enzymes that are known to readily degrade polysaccharide HA into low molecular fragments HA. Since HA can bind to CD44, a receptor that is known to overexpress on various tumor cells, many researchers have investigated HA-based nanoparticles for tumor-targeted therapy and imaging (Choi et al. 2010, 2012; Saravanakumar et al. 2010). Park et al. have prepared amphiphilic HA derivative, which is capable of forming self-assembled nanoparticles, by chemically conjugating HA with 5β-cholanic acid and PEG (Choi et al. 2011). As expected, these PEGylated HA nanoparticles (P-HA-NPs) were effectively internalized into the CD44-overexpressing cancer cells (SCC7 and MDA-MB-231) via receptor mediated endocytosis, but were rarely taken by the normal fibroblasts (NIH-3T3). When incubated with Hyals, P-HA-NPs rapidly released the encapsulated anticancer drugs due to the disintegration of the nanoparticle structure by the Hyals. This characteristic of P-HA-NPs is highly advantageous for cancer therapy because of the abundant Hyals in cytosol of tumor cells, which may enable rapid release of drug from the nanoparticles after CD44-mediated cellular uptake (Fig. 9). After systemic administration in tumor-bearing mice, P-HA-NPs were effectively reached the tumor site through EPR effect and as well as specific binding of HA to CD44 on tumor cells. P-HA-NPs loaded with anticancer drug camptothecin (CPT) demonstrated high therapeutic effect in vivo, compared to the free CPT (Choi et al. 2011). Similarly, HA-cholesterol based nanoparticles encapsulated with DOX and superparmagnetic iron oxide nanoparticles also showed triggered release in the presence of Hyals (Deng et al. 2012).

6.2 Enzyme-Responsive Polypeptide-Based Nanoparticles

Besides natural polysaccharides that can be degraded by specific enzymes in particular organs or cellular environment, many synthetic polymeric nanoparticles with enzyme-degradable moieties such as peptide sequences have also been developed and investigated their potential as drug carriers. Proteases are important enzymes that play an essential role in many biological and pathological processes, and their up-regulation is associated with the diseases such as cancer, inflammation and neurodegenerative diseases. For example, matrix metalloproteinases (MMPs), a family of zinc-dependent proteases, are secreted excess in cancer cells and plays a crucial role in tumor invasion (Egeblad and Werb 2002). Owing to these features,

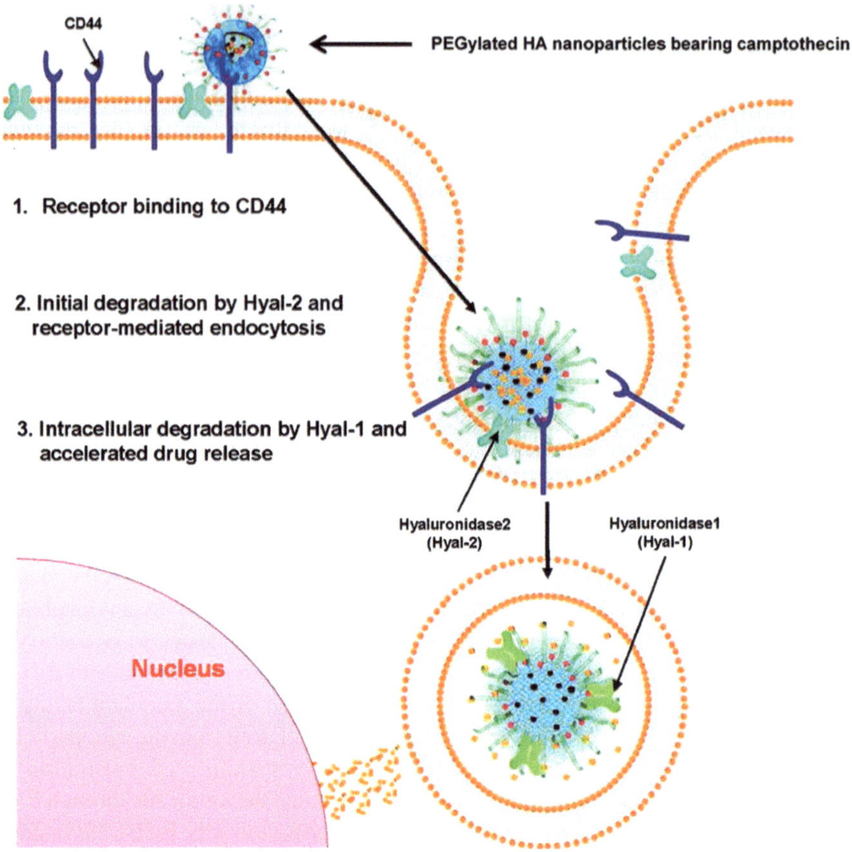

Fig. 9 Schematic illustration of cellular uptake of CPT-loaded PEGylated HA nanoparticles, and Hayls-mediated intracellular dissociation of the nanoparticles. Reproduced from Choi et al. (2011, Copyright American Chemical Society) with permission

several polymeric drug carriers with MMP-cleavable peptides have been developed to maximize the cancer therapeutics. The first class of MMP-responsive carrier systems includes simple conjugation of anticancer drugs directly to the polymers such as PEG and dextran via MMP-cleavable peptide sequences (Lee et al. 2007; Chau et al. 2004). Lee et al. conjugated DOX to PEG via MMP-specific peptide linkers GPLGV and GPLGVRG. Subsequently, they loaded free DOX into the PEG-peptide-DOX conjugates. In vivo studies demonstrated a longer half-lives and higher plasma DOX concentrations for DOX-loaded conjugated micelles compared to the bare micelles. These results indicated DOX-loaded micelles as a potent system for cancer therapy (Lee et al. 2007). In an another study, MMP-sensitive a cationic block copolymer based gene carrier was prepared by placing the GPLGVRG peptide-linker between a PEG segment and a P(Asp) segment with a flanking DET group. Because of the detachment of PEG layer at the extracellular environment in the presence of MMP-2, the resulting PEG-sheddable polyplex micelles exhibited high transfection efficiency (Li et al. 2013b). Recently, more advanced multifunctional MMP-7 triggered PEG sheddable cationic polymeric nanoparticles with a pH-dependent membrane disruptive characteristic were developed (Li et al. 2013a). In the presence of MMP-7, this smart nanoparticle effectively delivered siRNA into the cytosol, and enabled RNA interference and protein-level knockdown. In an attempt to reduce the systemic toxicities and to improve the therapeutic efficacy, anticancer drugs DOX and PTX were conjugated to a polyester nanodendron (ND) through a MMP-9 cleavable peptide linker (Samuelson et al. 2013). Although both DOX-conjugated ND (ND^{DOX}) and PTX-conjugated ND (ND^{PTX}) showed MMP-9 dependent cleavage, ND^{PTX} were highly cytotoxic upon MMP9 activation, whereas ND^{DOX} exhibited no detectable cytotoxicity.

Another PEG-sheddable block copolymer micelles were prepared using an amphiliphilic PEG-*b*-PS diblock copolymer with azobenzene linkage (Rao and Khan 2013). The block copolymer was synthesized using azobenzene linkage bearing PEG-macroinitiator via atom transfer radical polymerization. Since azo bond is susceptible to the enzyme azoreductase produced by the microbial flora present in the colon of the human intestine, these micelles were suitable for colon-specific drug delivery. Similar to the PEG-sheddable micelles, Lee et al. (2011) developed lysosomally cleavable peptide-containing polymersomes, with anti-EGFR antibody, by incorporating a peptide sequence GFLGF in between PEG and PDLLA block. This targeted cleavable polymersomes exhibited enhanced cellular uptake and relatively fast enzymatic destabilization in the presence of the lysosomal enzyme (cathepsin B). Since many tumors overexpress both cathepsin B and EGF receptor, this polymersomes might have a potential as carrier for tumor-targeted drug delivery.

7 Glucose-Responsive Nanoparticles

Diabetes mellitus is a metabolic disorder that is primarily characterized by the accumulation of glucose concentrations in the blood. Current treatment methods for diabetes are limited to insulin injection, which is painful and result in poor compliance and glucose control. For instance, while insufficient insulin injection results in hyperglycemia, an overdose of insulin injection may cause hypoglycemic coma in patients. Therefore, consistent monitoring of blood glucose, followed by appropriate insulin injections, is required to achieve normoglycemia. In this regard, significant efforts have been focused on developing a self-regulating insulin delivery system that is capable of delivering accurate levels of insulin in response to in vivo glucose level.

One of the strategies to develop such self-regulating insulin system includes fabricating glucose-responsive polymeric nanoparticles by incorporating glucose-responsive moieties such as glucose oxidase, lectin or phenylboronic acid (PBA) (Bratlie et al. 2012; Wu et al. 2011). Because of the specific and reversible interaction of PBA with diols units such as glucose, several PBA-based polymeric

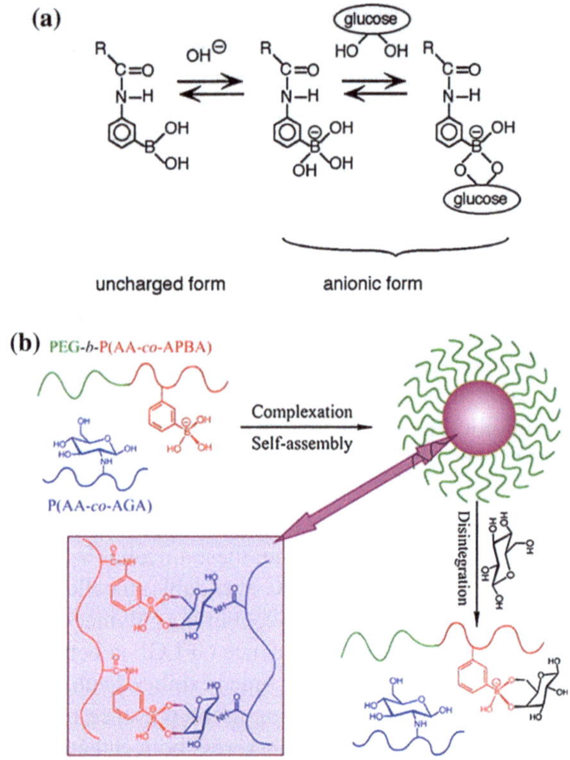

Fig. 10 **a** Equilibria of (alkylamido)phenyl boronic acid. **b** Assembly and disassembly of PBA-based glucose-responsive complex micelles. Reproduced from Kataoka et al. (1998, Copyright American Chemical Society) and Ma et al. (2012, Copyright American Chemical Society) with permission

systems have been widely developed for insulin delivery (Kataoka et al. 1998; Matsumoto et al. 2004). In an aqueous medium, PBA compounds are in equilibrium between an uncharged (hydrophobic) and charged form (hydrophilic) (Fig. 10a). The charged PBA can form a stable complex with glucose through reversible covalent bonding. Therefore, increasing the glucose concentration increases the charged forms of PBA; thereby enhance the hydrophilicity of the polymeric system with PBA moieties. Thus, the characteristic glucose-responsive equilibrium of PBA has been exploited to trigger insulin release from the polymeric system. Jin et al. (2009) synthesized glucose-responsive random amphiphilic glycopolymers that comprised of PBA and glucosamine moieties via radical polymerization. This glycopolymers could form self-assembled nanoparticles by forming crosslinks between PBA and glucosamine moieties. Insulin was loaded into the nanoparticles through nanoprecipitation method. When free glucose is added to the nanoparticles, it competes to bind with PBA and breaks the crosslink between PBA and glucosamine, thus enhancing the hydrophilicity and promoting swelling of nanoparticles, and thereby releasing its payload of insulin. The incorporation of glucosamine moieties alleviated the cytotoxicity issues associated with PBA.

To improve the glucose-sensitivity at physiological conditions, PBA-based complex micelles were prepared by self-assembly of PBA-based PEG-*b*-poly(acrylic

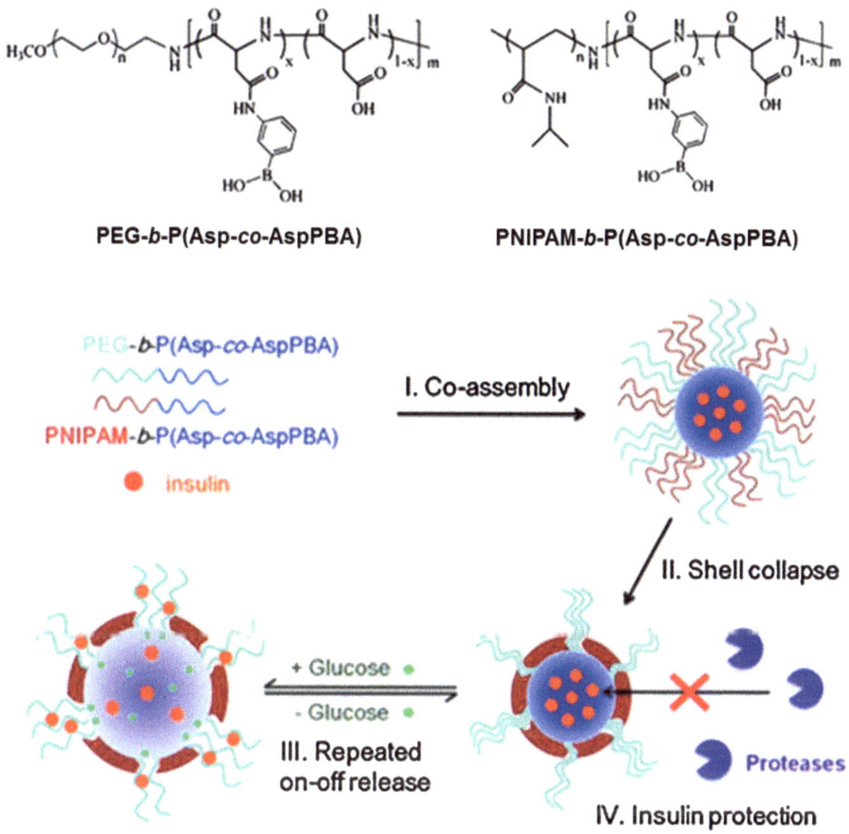

Fig. 11 Chemical structure of glucose-responsive PEG-*b*-P(Asp-*co*-AspBA) and PNIPAM-*b*-P(Asp-*co*-AspBA) block copolymers, and illustration of glucose-responsive complex micelles with on–off insulin release. Reproduced from Liu et al. (Copyright Royal Society of Chemistry) with permission

acid-*co*-acrylamidephenylboronic acid) (PEG-*b*-P(AA-*co*-APBA)) and a glycopolymer poly(acrylic acid-*co*-acrylglucosamine) (P(AA-*co*-AGA)) through covalent complexation between PBA and glycosyl (Fig. 10b) (Ma et al. 2012). The results from the study demonstrated enhanced glucose-sensitivity for complex PEG-*b*-P(AA-*co*-APBA)/P(AA-*co*-AGA) micelles (concentration of glucose 2 g/L for the polymers weight ratio 1:0.75) compared to the simple PEG-*b*-P(AA-*co*-APBA) micelles at physiological conditions. In particular, the sensitivity of complex micelles increased by decreasing the amount of PEG-*b*-P(AA-*co*-APBA) in the compositions. In another study, glucose-responsive complex micelles with on-off insulin release characteristics were prepared by self-assembly of two block copolymers, PEG-*b*-poly(aspartic acid-co-aspartamidophenylboronic acid) (PEG-*b*-P(Asp-*co*-AspPBA)) and PNIPAM-*b*-P(Asp-*co*-AspPBA) (Fig. 11) (Liu et al. 2013). By controlling the weight ratio of PEG and PNIPAM (6/4), a well-defined core-shell-corona structure PNIPAM membrane and glucose-responsive core is fabricated. The main advantages of this system include protection of insulin against protease degradation and reproducible on–off insulin release. Similarly, self-assembled polymeric nanoparticles based on poly(3-acrylamidophenylboronic acid) and poly(2-lactobionamidoethyl methacrylate) have also been developed for intracellular delivery of protein drugs (Cheng et al. 2012). Recently, Kim et al. (2012) developed monosaccharide-responsive polymersomes by employing PEG-b-poly(styreneboroxole) block copolymer synthesized via RAFT method. In particular, this polymersomes demonstrated monosaccharide-responsive disassembly under physiologically relevant pH conditions, indicating their potential as carriers for the delivery of insulin.

8 Conclusions and Outlook

Stimuli-responsive polymeric nanoparticles have emerged as a fascinating class of drug carriers that can be elegantly employed for target-specific delivery of a wide range of therapeutic agents. The stimuli-responsive drug carriers have the ability to rapidly release the drugs upon reaching the target site of action. The inherent stimuli such as pH, redox and enzymes mainly trigger drug release by altering the hydrophilic-hydrophobic balance and/or disintegration of cleavable moieties incorporated in the drug carriers. Alternatively, external stimuli such as light and ultrasound could enable to control the drug release characteristics in a remote and controlled way at the target site of action. Rationally designed carriers with multiple stimuli could be more beneficial to realize effective therapy.

Acknowledgments This work was supported by the Research Center Program of IBS (Institute for Basic Science) in Korea (CA1203-02).

References

Agarwal S, Zhang Y, Maji S, Greiner A (2012) PDMAEMA based gene delivery materials. Mater Today 15(9):388–393

Akimoto J, Nakayama M, Sakai K, Okano T (2009) Temperature-induced intracellular uptake of thermoresponsive polymeric micelles. Biomacromolecules 10(6):1331–1336

Akimoto J, Nakayama M, Sakai K, Okano T (2010) Thermally controlled intracellular uptake system of polymeric micelles possessing poly(N-isopropylacrylamide)-based outer coronas. Mol Pharm 7(4):926–935

Allen TM (2002) Ligand-targeted therapeutics in anticancer therapy. Nat Rev Cancer 2(10):750–763

Asokan A, Cho MJ (2002) Exploitation of intracellular pH gradients in the cellular delivery of macromolecules. J Pharm Sci 91(4):903–913

Babin J, Pelletier M, Lepage M, Allard J-F, Morris D, Zhao Y (2009) A new two-photon-sensitive block copolymer nanocarrier. Angew Chem Int Ed 48(18):3329–3332

Blasco E, Serrano JL, Piñol M, Oriol L (2013) Light responsive vesicles based on linear-dendritic block copolymers using azobenzene-aliphatic codendrons. Macromolecules 46(15):5951–5960

Bratlie KM, York RL, Invernale MA, Langer R, Anderson DG (2012) Materials for diabetes therapeutics. Adv Healthc Mater 1(3):267–284

Cabane E, Malinova V, Meier W (2010) Synthesis of photocleavable amphiphilic block copolymers: toward the design of photosensitive nanocarriers. Macromol Chem Phys 211(17):1847–1856

Cai X-J, Dong H-Q, Xia W-J, Wen H-Y, Li X-Q, Yu J-H, Li Y-Y, Shi D-L (2011) Glutathione-mediated shedding of PEG layers based on disulfide-linked catiomers for DNA delivery. J Mater Chem 21(38):14639–14645

Cammas S, Suzuki K, Sone C, Sakurai Y, Kataoka K, Okano T (1997) Thermo-responsive polymer nanoparticles with a core-shell micelle structure as site-specific drug carriers. J Control Release 48(2–3):157–164

Cerritelli S, Velluto D, Hubbell JA (2007) PEG-SS-PPS: reduction-sensitive disulfide block copolymer vesicles for intracellular drug delivery. Biomacromolecules 8(6):1966–1972

Chau Y, Tan FE, Langer R (2004) Synthesis and characterization of dextran-peptide-methotrexate conjugates for tumor targeting via mediation by matrix metalloproteinase II and matrix metalloproteinase IX. Bioconjug Chem 15(4):931–941

Chen C-J, Liu G-Y, Shi Y-T, Zhu C-S, Pang S-P, Liu X-S, Ji J (2011) Biocompatible micelles based on comb-like PEG derivates: formation, characterization, and photo-responsiveness. Macromol Rapid Commun 32(14):1077–1081

Cheng C, Wei H, Shi B-X, Cheng H, Li C, Gu Z-W, Cheng S-X, Zhang X-Z, Zhuo R-X (2008) Biotinylated thermoresponsive micelle self-assembled from double-hydrophilic block copolymer for drug delivery and tumor target. Biomaterials 29(4):497–505

Cheng C, Zhang X, Wang Y, Sun L, Li C (2012) Phenylboronic acid-containing block copolymers: synthesis, self-assembly, and application for intracellular delivery of proteins. New J Chem 36(6):1413–1421

Cheng R, Feng F, Meng F, Deng C, Feijen J, Zhong Z (2011) Glutathione-responsive nano-vehicles as a promising platform for targeted intracellular drug and gene delivery. J Control Release 152(1):2–12

Chilkoti A, Dreher MR, Meyer DE, Raucher D (2002) Targeted drug delivery by thermally responsive polymers. Adv Drug Deliv Rev 54(5):613–630

Choi KY, Chung H, Min KH, Yoon HY, Kim K, Park JH, Kwon IC, Jeong SY (2010) Self-assembled hyaluronic acid nanoparticles for active tumor targeting. Biomaterials 31(1):106–114

Choi KY, Saravanakumar G, Park JH, Park K (2012) Hyaluronic acid-based nanocarriers for intracellular targeting: interfacial interactions with proteins in cancer. Colloids Surf B Biointerfaces 99:82–94

Choi KY, Yoon HY, Kim J-H, Bae SM, Park R-W, Kang YM, Kim I-S, Kwon IC, Choi K, Jeong SY, Kim K, Park JH (2011) Smart nanocarrier based on PEGylated hyaluronic acid for cancer therapy. ACS Nano 5(11):8591–8599

Christensen LV, Chang C-W, Kim WJ, Kim SW, Zhong Z, Lin C, Engbersen JFJ, Feijen J (2006) Reducible poly(amido ethylenimine)s designed for triggered intracellular gene delivery. Bioconjug Chem 17(5):1233–1240

Chung JE, Yokoyama M, Aoyagi T, Sakurai Y, Okano T (1998) Effect of molecular architecture of hydrophobically modified poly(N-isopropylacrylamide) on the formation of thermoresponsive core-shell micellar drug carriers. J Control Release 53(1–3):119–130

Chung JE, Yokoyama M, Okano T (2000) Inner core segment design for drug delivery control of thermo-responsive polymeric micelles. J Control Release 65(1–2):93–103

Dan K, Ghosh S (2013) One-pot synthesis of an acid-labile amphiphilic triblock copolymer and its pH-responsive vesicular assembly. Angew Chem Int Ed 52(28):7300–7305

Dan K, Pan R, Ghosh S (2010) Aggregation and pH responsive disassembly of a new acid-labile surfactant synthesized by thiol-acrylate Michael addition reaction. Langmuir 27(2):612–617 de la Rica R, Aili D, Stevens MM (2012) Enzyme-responsive nanoparticles for drug release and diagnostics. Adv Drug Deliv Rev 64(11):967–978

De Smedt S, Demeester J, Hennink W (2000) Cationic polymer based gene delivery systems. Pharm Res 17(2):113–126

Deng L, Wang G, Ren J, Zhang B, Yan J, Li W, Khashab NM (2012) Enzymatically triggered multifunctional delivery system based on hyaluronic acid micelles. RSC Adv 2(33):12909–12914

Ding C, Gu J, Qu X, Yang Z (2009) Preparation of multifunctional drug carrier for tumor-specific uptake and enhanced intracellular delivery through the conjugation of weak acid labile linker. Bioconjug Chem 20(6):1163–1170

Duncan R (2003) The dawning era of polymer therapeutics. Nat Rev Drug Discov 2(5):347–360

Egeblad M, Werb Z (2002) New functions for the matrix metalloproteinases in cancer progression. Nat Rev Cancer 2(3):161–174

Fomina N, Sankaranarayanan J, Almutairi A (2012) Photochemical mechanisms of light-triggered release from nanocarriers. Adv Drug Deliv Rev 64(11):1005–1020

Gao W, Chan JM, Farokhzad OC (2010) pH-responsive nanoparticles for drug delivery. Mol Pharm 7(6):1913–1920

Green JJ, Langer R, Anderson DG (2008) A combinatorial polymer library approach yields insight into nonviral gene delivery. Acc Chem Res 41(6):749–759

Haag R, Kratz F (2006) Polymer therapeutics: concepts and applications. Angew Chem Int Ed 45(8):1198–1215

He E, Yue CY, Tam KC (2010) Binding and release studies of a cationic drug from a star-shaped four-arm poly(ethylene oxide)-b-poly(methacrylic acid). J Pharm Sci 99(2):782–793

Hong C-Y, You Y-Z, Pan C-Y (2004) Synthesis and characterization of well-defined diblock and triblock copolymers of poly(N-isopropylacrylamide) and poly(ethylene oxide). J Polym Sci, Part A: Polym Chem 42(19):4873–4881

Howard KA (2009) Delivery of RNA interference therapeutics using polycation-based nanoparticles. Adv Drug Deliv Rev 61(9):710–720

Itaka K, Yamauchi K, Harada A, Nakamura K, Kawaguchi H, Kataoka K (2003) Polyion complex micelles from plasmid DNA and poly(ethylene glycol)–poly(l-lysine) block copolymer as serum-tolerable polyplex system: physicochemical properties of micelles relevant to gene transfection efficiency. Biomaterials 24(24):4495–4506

Jia L, Li Z, Zhang D, Zhang Q, Shen J, Guo H, Tian X, Liu G, Zheng D, Qi L (2013) Redox-responsive catiomer based on PEG-ss-chitosan oligosaccharide-ss-polyethylenimine copolymer for effective gene delivery. Polym Chem 4(1):156–165

Jiang J, Tong X, Morris D, Zhao Y (2006) Toward photocontrolled release using light-dissociable block copolymer micelles. Macromolecules 39(13):4633–4640

Jiang J, Tong X, Zhao Y (2005) A new design for light-breakable polymer micelles. J Am Chem Soc 127(23):8290–8291

Jin X, Zhang X, Wu Z, Teng D, Zhang X, Wang Y, Wang Z, Li C (2009) Amphiphilic random glycopolymer based on phenylboronic acid: synthesis, characterization, and potential as glucose-sensitive matrix. Biomacromolecules 10(6):1337–1345

Jin Y, Song L, Su Y, Zhu L, Pang Y, Qiu F, Tong G, Yan D, Zhu B, Zhu X (2011) Oxime linkage: a robust tool for the design of pH-sensitive polymeric drug carriers. Biomacromolecules 12(10):3460–3468

Kamaly N, Xiao Z, Valencia PM, Radovic-Moreno AF, Farokhzad OC (2012) Targeted polymeric therapeutic nanoparticles: design, development and clinical translation. Chem Soc Rev 41(7):2971–3010

Kataoka K, Miyazaki H, Bunya M, Okano T, Sakurai Y (1998) Totally synthetic polymer gels responding to external glucose concentration: their preparation and application to on–off regulation of insulin release. J Am Chem Soc 120(48):12694–12695

Kim H, Kang YJ, Kang S, Kim KT (2012) Monosaccharide-responsive release of insulin from polymersomes of polyboroxole block copolymers at neutral pH. J Am Chem Soc 134(9):4030–4033

Kim S, Kim JY, Huh KM, Acharya G, Park K (2008) Hydrotropic polymer micelles containing acrylic acid moieties for oral delivery of paclitaxel. J Control Release 132(3):222–229

Knorr V, Allmendinger L, Walker GF, Paintner FF, Wagner E (2007) An acetal-based PEGylation reagent for pH-sensitive shielding of DNA polyplexes. Bioconjug Chem 18(4):1218–1225

Koo H, Huh MS, Sun I-C, Yuk SH, Choi K, Kim K, Kwon IC (2011) In vivo targeted delivery of nanoparticles for theranosis. Acc Chem Res 44(10):1018–1028

Kotharangannagari VK, Sánchez-Ferrer A, Ruokolainen J, Mezzenga R (2011) Photoresponsive reversible aggregation and dissolution of rod–coil polypeptide diblock copolymers. Macromolecules 44(12):4569–4573

Kuppusamy P, Li H, Ilangovan G, Cardounel AJ, Zweier JL, Yamada K, Krishna MC, Mitchell JB (2002) Noninvasive imaging of tumor redox status and its modification by tissue glutathione levels. Cancer Res 62(1):307–312

Lee ES, Gao Z, Kim D, Park K, Kwon IC, Bae YH (2008) Super pH-sensitive multifunctional polymeric micelle for tumor pH(e) specific TAT exposure and multidrug resistance. J Control Release 129(3):228–236

Lee ES, Na K, Bae YH (2005) Super pH-sensitive multifunctional polymeric micelle. Nano Lett 5(2):325–329

Lee GY, Park K, Kim SY, Byun Y (2007) MMPs-specific PEGylated peptide–DOX conjugate micelles that can contain free doxorubicin. Eur J Pharm Biopharm 67(3):646–654

Lee JS, Groothuis T, Cusan C, Mink D, Feijen J (2011) Lysosomally cleavable peptide-containing polymersomes modified with anti-EGFR antibody for systemic cancer chemotherapy. Biomaterials 32(34):9144–9153

Lee S-Y, Huh MS, Lee S, Lee SJ, Chung H, Park JH, Oh Y-K, Choi K, Kim K, Kwon IC (2010) Stability and cellular uptake of polymerized siRNA (poly-siRNA)/polyethylenimine (PEI) complexes for efficient gene silencing. J Control Release 141(3):339–346

Li G, Shi L, Ma R, An Y, Huang N (2006a) Formation of complex micelles with double-responsive channels from self-assembly of two diblock copolymers. Angew Chem Int Ed 45(30):4959–4962

Li H, Yu SS, Miteva M, Nelson CE, Werfel T, Giorgio TD, Duvall CL (2013a) Matrix metalloproteinase responsive, proximity-activated polymeric nanoparticles for siRNA delivery. Adv Funct Mater 23(24):3040–3052

Li J, Ge Z, Liu S (2013b) PEG-sheddable polyplex micelles as smart gene carriers based on MMP-cleavable peptide-linked block copolymers. Chem Commun 49(62):6974–6976

Li L-Y, He W-D, Li J, Zhang B-Y, Pan T-T, Sun X-L, Ding Z-L (2010a) Shell-cross-linked micelles from PNIPAM-b-(PLL)2 Y-shaped miktoarm star copolymer as drug carriers. Biomacromolecules 11(7):1882–1890

Li S-D, Huang L (2008) Pharmacokinetics and biodistribution of nanoparticles. Mol Pharm 5(4):496–504

Li Y-L, Zhu L, Liu Z, Cheng R, Meng F, Cui J-H, Ji S-J, Zhong Z (2009) Reversibly stabilized multifunctional dextran nanoparticles efficiently deliver doxorubicin into the nuclei of cancer cells. Angew Chem Int Ed 48(52):9914–9918

Li Y, Lokitz BS, Armes SP, McCormick CL (2006b) Synthesis of reversible shell cross-linked micelles for controlled release of bioactive agents. Macromolecules 39(8):2726–2728

Li Y, Lokitz BS, McCormick CL (2005) RAFT synthesis of a thermally responsive ABC triblock copolymer incorporating N-acryloxysuccinimide for facile in situ formation of shell cross-linked micelles in aqueous media. Macromolecules 39(1):81–89

Li Y, Tong R, Xia H, Zhang H, Xuan J (2010b) High intensity focused ultrasound and redox dual responsive polymer micelles. Chem Commun 46(41):7739–7741

Lin C, Zhong Z, Lok MC, Jiang X, Hennink WE, Feijen J, Engbersen JFJ (2006) Linear poly(amido amine)s with secondary and tertiary amino groups and variable amounts of disulfide linkages: Synthesis and in vitro gene transfer properties. J Control Release 116(2):130–137

Lin S, Du F, Wang Y, Ji S, Liang D, Yu L, Li Z (2007) An acid-labile block copolymer of PDMAEMA and PEG as potential carrier for intelligent gene delivery systems. Biomacromolecules 9(1):109–115

Liu G, Dong C-M (2012) Photoresponsive poly(S-(o-nitrobenzyl)-l-cysteine)-b-PEO from a l-Cysteine N-Carboxyanhydride monomer: synthesis, self-assembly, and phototriggered drug release. Biomacromolecules 13(5):1573–1583

Liu G, Ma R, Ren J, Li Z, Zhang H, Zhang Z, An Y, Shi L (2013) A glucose-responsive complex polymeric micelle enabling repeated on–off release and insulin protection. Soft Matter 9(5):1636–1644

Liu S-Q, Wiradharma N, Gao S-J, Tong YW, Yang Y-Y (2007) Bio-functional micelles self-assembled from a folate-conjugated block copolymer for targeted intracellular delivery of anticancer drugs. Biomaterials 28(7):1423–1433

Liu SQ, Tong YW, Yang Y-Y (2005) Incorporation and in vitro release of doxorubicin in thermally sensitive micelles made from poly(N-isopropylacrylamide-co-N, N-dimethylacrylamide)-b-poly(d, l-lactide-co-glycolide) with varying compositions. Biomaterials 26(24): 5064–5074

Luo S, Xu J, Zhu Z, Wu C, Liu S (2006) Phase transition behavior of unimolecular micelles with thermoresponsive poly(N-isopropylacrylamide) coronas. J Phys Chem B 110(18):9132–9139

Ma R, Yang H, Li Z, Liu G, Sun X, Liu X, An Y, Shi L (2012) Phenylboronic acid-based complex micelles with enhanced glucose-responsiveness at physiological pH by complexation with glycopolymer. Biomacromolecules 13(10):3409–3417

Maeda H, Wu J, Sawa T, Matsumura Y, Hori K (2000) Tumor vascular permeability and the EPR effect in macromolecular therapeutics: a review. J Control Release 65(1–2):271–284

Matsumoto A, Yoshida R, Kataoka K (2004) Glucose-responsive polymer gel bearing phenylborate derivative as a glucose-sensing moiety operating at the physiological pH. Biomacromolecules 5(3):1038–1045

Meng F, Hennink WE, Zhong Z (2009) Reduction-sensitive polymers and bioconjugates for biomedical applications. Biomaterials 30(12):2180–2198

Mok H, Lee SH, Park JW, Park TG (2010) Multimeric small interfering ribonucleic acid for highly efficient sequence-specific gene silencing. Nat Mater 9(3):272–278

Mok H, Park JW, Park TG (2007) Antisense oligodeoxynucleotide-conjugated hyaluronic acid/ protamine nanocomplexes for intracellular gene inhibition. Bioconjug Chem 18(5): 1483–1489

Nagasaki T, Taniguchi A, Tamagaki S (2003) Photoenhancement of transfection efficiency using novel cationic lipids having a photocleavable spacer. Bioconjug Chem 14(3):513–516

Namgung R, Kim WJ (2012) A highly entangled polymeric nanoconstruct assembled by siRNA and its reduction-triggered siRNA release for gene silencing. Small 8(20):3209–3219

Oh JK, Siegwart DJ, H-i Lee, Sherwood G, Peteanu L, Hollinger JO, Kataoka K, Matyjaszewski K (2007a) Biodegradable nanogels prepared by atom transfer radical polymerization as potential drug delivery carriers: synthesis, biodegradation, in vitro release, and bioconjugation. J Am Chem Soc 129(18):5939–5945

Oh JK, Siegwart DJ, Matyjaszewski K (2007b) Synthesis and biodegradation of nanogels as delivery carriers for carbohydrate drugs. Biomacromolecules 8(11):3326–3331

Park JH, Lee S, Kim J-H, Park K, Kim K, Kwon IC (2008) Polymeric nanomedicine for cancer therapy. Prog Polym Sci 33(1):113–137

Park JH, Saravanakumar G, Kim K, Kwon IC (2010) Targeted delivery of low molecular drugs using chitosan and its derivatives. Adv Drug Deliv Rev 62(1):28–41

Prabaharan M, Grailer JJ, Steeber DA, Gong S (2009) Thermosensitive micelles based on folate-conjugated poly(N-vinylcaprolactam)-block-poly(ethylene glycol) for tumor-targeted drug delivery. Macromol Biosci 9(8):744–753

Putnam D, Zelikin AN, Izumrudov VA, Langer R (2003) Polyhistidine–PEG: DNA nanocomposites for gene delivery. Biomaterials 24(24):4425–4433

Rao J, Khan A (2013) Enzyme sensitive synthetic polymer micelles based on the azobenzene motif. J Am Chem Soc 135(38):14056–14059

Rao NV, Mane S, Kishore A, Das Sarma J, Shunmugam R (2011) Norbornene derived doxorubicin copolymers as drug carriers with pH responsive hydrazone linker. Biomacromolecules 13(1):221–230

Rezaei SJT, Nabid MR, Niknejad H, Entezami AA (2012) Folate-decorated thermoresponsive micelles based on star-shaped amphiphilic block copolymers for efficient intracellular release of anticancer drugs. Int J Pharm 437(1–2):70–79

Rijcken CJ, Snel CJ, Schiffelers RM, van Nostrum CF, Hennink WE (2007a) Hydrolysable core-crosslinked thermosensitive polymeric micelles: synthesis, characterisation and in vivo studies. Biomaterials 28(36):5581–5593

Rijcken CJF, Hofman J-W, van Zeeland F, Hennink WE, van Nostrum CF (2007b) Photosensitiser-loaded biodegradable polymeric micelles: preparation, characterisation and in vitro PDT efficacy. J Control Release 124(3):144–153

Ryu J-H, Park S, Kim B, Klaikherd A, Russell TP, Thayumanavan S (2009) Highly ordered gold nanotubes using thiols at a cleavable block copolymer interface. J Am Chem Soc 131(29):9870–9871

Saeed AO, Magnusson JP, Moradi E, Soliman M, Wang W, Stolnik S, Thurecht KJ, Howdle SM, Alexander C (2011) Modular construction of multifunctional bioresponsive cell-targeted nanoparticles for gene delivery. Bioconjug Chem 22(2):156–168

Samuelson LE, Scherer RL, Matrisian LM, McIntyre JO, Bornhop DJ (2013) Synthesis and in vitro efficacy of MMP9-activated nanodendrons. Mol Pharm 10(8):3164–3174

Saravanakumar G, Choi KY, Yoon HY, Kim K, Park JH, Kwon IC, Park K (2010) Hydrotropic hyaluronic acid conjugates: synthesis, characterization, and implications as a carrier of paclitaxel. Int J Pharm 394(1–2):154–161

Schmaljohann D (2006) Thermo- and pH-responsive polymers in drug delivery. Adv Drug Deliv Rev 58(15):1655–1670

Sinha VR, Kumria R (2001) Polysaccharides in colon-specific drug delivery. Int J Pharm 224(1–2):19–38

Soga O, van Nostrum CF, Fens M, Rijcken CJF, Schiffelers RM, Storm G, Hennink WE (2005) Thermosensitive and biodegradable polymeric micelles for paclitaxel delivery. J Control Release 103(2):341–353

Son S, Namgung R, Kim J, Singha K, Kim WJ (2011) Bioreducible polymers for gene silencing and delivery. Acc Chem Res 45(7):1100–1112

Son S, Singha K, Kim WJ (2010) Bioreducible BPEI-SS-PEG-cNGR polymer as a tumor targeted nonviral gene carrier. Biomaterials 31(24):6344–6354

Son YJ, Jang JS, Cho YW, Chung H, Park RW, Kwon IC, Kim IS, Park JY, Seo SB, Park CR, Jeong SY (2003) Biodistribution and anti-tumor efficacy of doxorubicin loaded glycol-chitosan nanoaggregates by EPR effect. J Control Release 91(1–2):135–145

Song N, Liu W, Tu Q, Liu R, Zhang Y, Wang J (2011) Preparation and in vitro properties of redox-responsive polymeric nanoparticles for paclitaxel delivery. Colloids Surf B Biointerfaces 87(2):454–463

Sun H, Guo B, Cheng R, Meng F, Liu H, Zhong Z (2009) Biodegradable micelles with sheddable poly(ethylene glycol) shells for triggered intracellular release of doxorubicin. Biomaterials 30(31):6358–6366

Sun H, Guo B, Li X, Cheng R, Meng F, Liu H, Zhong Z (2010) Shell-sheddable micelles based on dextran-SS-poly(ε-caprolactone) diblock copolymer for efficient intracellular release of doxorubicin. Biomacromolecules 11(4):848–854

Sun L, Ma X, Dong C-M, Zhu B, Zhu X (2012) NIR-responsive and lectin-binding doxorubicin-loaded nanomedicine from janus-type dendritic PAMAM amphiphiles. Biomacromolecules 13(11):3581–3591

Takae S, Miyata K, Oba M, Ishii T, Nishiyama N, Itaka K, Yamasaki Y, Koyama H, Kataoka K (2008) PEG-detachable polyplex micelles based on disulfide-linked block catiomers as bioresponsive nonviral gene vectors. J Am Chem Soc 130(18):6001–6009

Talelli M, Rijcken CJF, Lammers T, Seevinck PR, Storm G, van Nostrum CF, Hennink WE (2009) Superparamagnetic iron oxide nanoparticles encapsulated in biodegradable thermosensitive polymeric micelles: toward a targeted nanomedicine suitable for image-guided drug delivery. Langmuir 25(4):2060–2067

Tang L-Y, Wang Y-C, Li Y, Du J-Z, Wang J (2009) Shell-detachable micelles based on disulfide-linked block copolymer as potential carrier for intracellular drug delivery. Bioconjug Chem 20(6):1095–1099

Tang R, Ji W, Panus D, Palumbo RN, Wang C (2011) Block copolymer micelles with acid-labile ortho ester side-chains: synthesis, characterization, and enhanced drug delivery to human glioma cells. J Control Release 151(1):18–27

Thambi T, Yoon HY, Kim K, Kwon IC, Yoo CK, Park JH (2011) Bioreducible block copolymers based on poly(Ethylene Glycol) and poly(γ-Benzyl l-Glutamate) for intracellular delivery of camptothecin. Bioconjug Chem 22(10):1924–1931

Tomatsu I, Peng K, Kros A (2011) Photoresponsive hydrogels for biomedical applications. Adv Drug Deliv Rev 63(14–15):1257–1266

Tong R, Hemmati HD, Langer R, Kohane DS (2012) Photoswitchable nanoparticles for triggered tissue penetration and drug delivery. J Am Chem Soc 134(21):8848–8855

Walker GF, Fella C, Pelisek J, Fahrmeir J, Boeckle S, Ogris M, Wagner E (2005) Toward synthetic viruses: endosomal pH-triggered deshielding of targeted polyplexes greatly enhances gene transfer in vitro and in vivo. Mol Ther 11(3):418–425

Wang G, Tong X, Zhao Y (2004) Preparation of azobenzene-containing amphiphilic diblock copolymers for light-responsive micellar aggregates. Macromolecules 37(24):8911–8917

Wang W, Sun H, Meng F, Ma S, Liu H, Zhong Z (2012) Precise control of intracellular drug release and anti-tumor activity of biodegradable micellar drugs via reduction-sensitive shell-shedding. Soft Matter 8(14):3949–3956

Wang X-Q, Zhang Q (2012) pH-sensitive polymeric nanoparticles to improve oral bioavailability of peptide/protein drugs and poorly water-soluble drugs. Eur J Pharm Biopharm 82(2): 219–229

Wang X, Sun H, Meng F, Cheng R, Deng C, Zhong Z (2013) Galactose-decorated reduction-sensitive degradable chimaeric polymersomes as a multifunctional nanocarrier to efficiently chaperone apoptotic proteins into hepatoma cells. Biomacromolecules 14(8):2873–2882

Wang Y-C, Wang F, Sun T-M, Wang J (2011) Redox-responsive nanoparticles from the single disulfide bond-bridged block copolymer as drug carriers for overcoming multidrug resistance in cancer cells. Bioconjug Chem 22(10):1939–1945

Wei H, Zhuo R-X, Zhang X-Z (2013) Design and development of polymeric micelles with cleavable links for intracellular drug delivery. Prog Polym Sci 38(3–4):503–535

Wu Q, Wang L, Yu H, Wang J, Chen Z (2011) Organization of glucose-responsive systems and their properties. Chem Rev 111(12):7855–7875

Yan B, Boyer J-C, Branda NR, Zhao Y (2011) Near-infrared light-triggered dissociation of block copolymer micelles using upconverting nanoparticles. J Am Chem Soc 133(49):19714–19717

Yan J, Ji W, Chen E, Li Z, Liang D (2008) Association and aggregation behavior of poly(ethylene oxide)-b-poly (N-isopropylacrylamide) in aqueous solution. Macromolecules 41(13):4908–4913

Yesilyurt V, Ramireddy R, Thayumanavan S (2011) Photoregulated release of noncovalent guests from dendritic amphiphilic nanocontainers. Angew Chem 123(13):3094–3098

You Y-Z, Manickam DS, Zhou Q-H, Oupický D (2007) Reducible poly(2-dimethylaminoethyl methacrylate): synthesis, cytotoxicity, and gene delivery activity. J Control Release 122(3): 217–225

Zelzer M, Todd SJ, Hirst AR, McDonald TO, Ulijn RV (2013) Enzyme responsive materials: design strategies and future developments. Biomaterials Science 1(1):11–39

Zeng YI, Pitt WG (2005) Poly(ethylene oxide)-b-poly(N-isopropylacrylamide) nanoparticles with cross-linked cores as drug carriers. J Biomater Sci Polym Ed 16(3):371–380

Zhang W, Shi L, Wu K, An Y (2005) Thermoresponsive micellization of poly(ethylene glycol)-b-poly(N-isopropylacrylamide) in water. Macromolecules 38(13):5743–5747

Zhao Y (2007) Rational design of light-controllable polymer micelles. Chem Rec 7(5):286–294

Zhao Y (2012) Light-responsive block copolymer micelles. Macromolecules 45(9):3647–3657

Zhu C, Zheng M, Meng F, Mickler FM, Ruthardt N, Zhu X, Zhong Z (2012) Reversibly shielded DNA polyplexes based on bioreducible PDMAEMA-SS-PEG-SS-PDMAEMA triblock copolymers mediate markedly enhanced nonviral gene transfection. Biomacromolecules 13(3):769–778

Zugates GT, Anderson DG, Little SR, Lawhorn IEB, Langer R (2006) Synthesis of poly(β-amino ester)s with thiol-reactive side chains for DNA delivery. J Am Chem Soc 128(39):12726–12734

Photo-Responsive Polymeric Nanocarriers for On-Demand Drug Delivery

Jian Ji and Qiao Jin

Abstract Polymeric assemblies self-assembled from amphiphilic copolymers are the topic of intense research and have emerged as versatile drug nanocarriers in the past few decades. To enhance the bioavailability of drugs at the target disease site, the use of stimuli-responsive nanocarriers with triggered release properties is highly desired. Among all the available chemical and physical stimuli, photo has attracted much attention since it can be localized in time and space, and it can also be triggered from outside of the system. In this chapter, we highlight the recent progress of photo-responsive assemblies as drug nanocarriers. Different types of photo-responsive polymers were classified. A wide variety of photo-responsive moieties, including UV and NIR responsive polymers, as well as synthetic routes were introduced to the drug nanocarriers. Finally, we suggest possible future developments of photo-responsive polymeric nanocarriers for biomedical applications, especially for drug delivery.

Keywords Photo-responsive · Drug delivery · NIR · UV · Micelle · Nanocarrier

Abbreviations

UV	Ultra violet
NIR	Near-infrared
NIPAAm	N-isopropylacrylamide
OEGMA	Oligo(ethylene glycol) monomethyl ether methacrylate
DOX	Doxorubicin
NBA	O-nitrobenzyl acrylate

J. Ji (✉) · Q. Jin
MOE Key Laboratory of Macromolecule Synthesis and Functionalization of Ministry of Education, Department of Polymer Science and Engineering, Zhejiang University, Hangzhou 310027, China
e-mail: jijian@zju.edu.cn

Q. Jin
e-mail: jinqiao@zju.edu.cn

A. Prokop et al. (eds.), *Intracellular Delivery II*,
Fundamental Biomedical Technologies 7, DOI: 10.1007/978-94-017-8896-0_5,

DENBMA	5-(2′-(dimethylamino)ethoxy)-2-nitrobenzyl methacrylate
RAFT	Reversible addition–fragmentation chain transfer polymerization
BSA	Bovine serum albumin
DNQ	2-Diazo-1,2-naphthoquinone

1 Introduction

There are a lot of anticancer drugs that have been developed for the treatment of cancers, but their clinical outcomes are always disappointing due to severe side effects. The lack of tumor selectivity and recurrence of cancers with intrinsic or acquired drug resistance will also decrease the therapeutic efficacy of the drugs. Various nanocarriers, including micelles, vesicles, nanoparticles, nanogels, and liposomes have been developed to address these problems (Wei et al. 2013; Sailor and Park 2012; Li et al. 2013). Generally, the drugs can either be physically encapsulated into the hydrophobic inner cores of nanocarriers or chemically conjugated to the polymer chains to form polymer-drug conjugates. Compared to conventional formulation technology, the polymeric nanocarriers have many advantages, including: (1) accumulation passively in solid tumors by enhanced permeation and retention (EPR) effects; (2) avoiding clearance from the reticuloendothelial system; (3) reducing the multiple drug resistance; (4) improving drug pharmacokinetics and biodistribution in vivo (Elsabahy and Wooley 2012; Cabral et al. 2011).

As soon as the drug nanocarriers arrive at the target disease site, a triggered release of encapsulated drugs will be highly desired. As a result, stimuli-responsive nanocarriers, which can show sharp and eventually reversible responses to various environmental changes have found ever-increasing opportunities. Various chemical and physical stimuli, such as pH, redox, temperature, enzyme, have been used to construct stimuli-responsive nanocarriers (Gil and Hudson 2004; Srinivas et al. 2008; Rapoport 2007). This will increase the feasibility of obtaining local high-dose therapy in cancerous tissues and at intracellular compartments. Especially, taking advantage of internal stimuli, such as low pH values of endosome and lysosome (5.0–5.5), lower extracellular pH (6.5) and glutathione (GSH) concentration gradients in tumor tissues, various stimuli-responsive nanocarriers have been deigned that can release the endogenous triggered on-demand drug release (Du et al. 2011; Christie et al. 2011). For the above-mentioned endogenous triggered drug release (pH, redox potential, enzyme), it is still a considerable challenge to realize accurately controlled release due to the complex and still not fully understood physiological environment in cells, tissue and body. Thus, it would be favorable to develop a new kind of externally triggered release of drugs in vivo and/or in vitro. Photo is quite advantageous since it can be localized in time and space, and it can also be triggered from outside of the system (Gohy and Zhao 2013; Liu et al. 2013; Habault et al. 2013; Pasparakis et al. 2012; Jin et al. 2010).

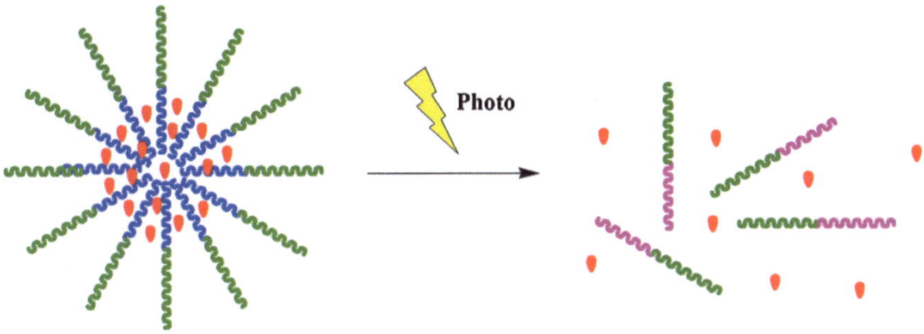

Fig. 1 Photo-responsive drug release from polymeric micelles

Furthermore, the wavelength and energy of photo can be facile adjusted. The basic design of photo-responsive drug delivery system is shown in Fig. 1. Hence, the advantages of photo-responsive polymeric micelles may include: (1) realizing spatiotemporal drug release with a high local concentration in the diseased sites, and reduce the overall injected dose and systemic toxicity; (2) realizing pulsed drug release; (3) controlling the targeting properties of nanocarriers.

2 Classify of Photo-Responsive Polymers

With the development and applications of photo-responsive materials in diverse fields including catalysis, sensors, templates, etc., various photo-responsive polymers have been designed. Depending on different premises, photo-responsive polymers can be classified in different ways. Based on the reversibility, photo-responsive polymers can be classified as reversible and irreversible photo-responsive polymers. Meanwhile, they can also be classified into UV, visible light and NIR responsive polymers depending on the light source. Additionally, based on the reaction mechanisms, photo-responsive polymers can be classified as photo-cleavable polymers and photo-isomerized polymers. In the presence of photo irradiation, photo-cleavable polymers will be degraded or the side group of the polymers will be cleaved (Fig. 2). On the contrary, photo-isomerized polymer will be isomerized to another polymer under photo irradiation (Fig. 3). Table 1 summarizes different types of photo-responsive groups, which might be helpful for the rational design of photo-responsive polymers in biomedical applications. The self-assembly and intracellular on-demand drug delivery of photo-responsive polymers will be discussed in detail in this chapter.

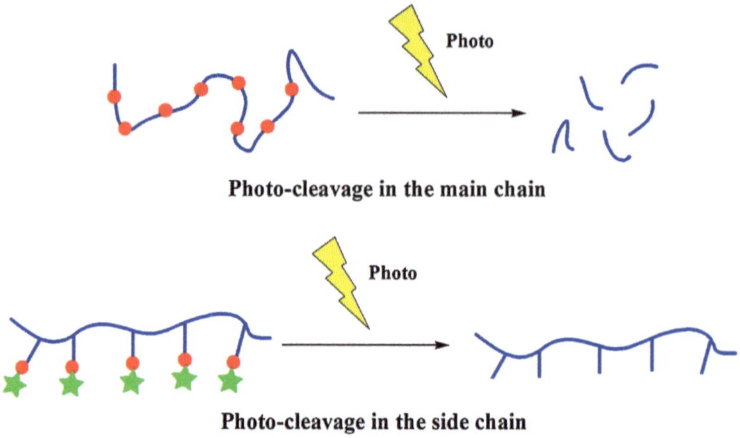

Fig. 2 Schematic illustration of photo-cleavage reaction

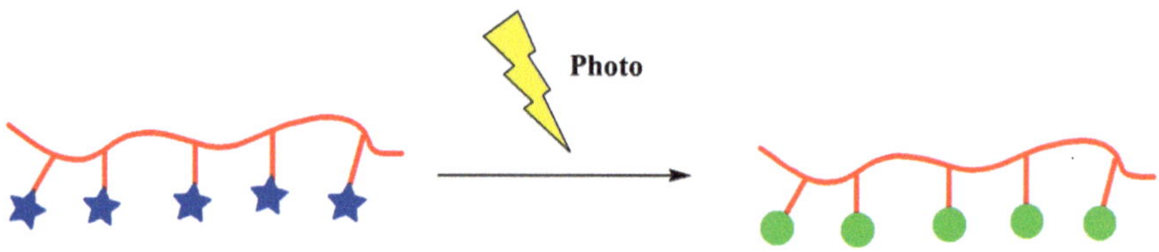

Fig. 3 Schematic illustration of photo-isomerization reaction

3 UV Responsive Polymeric Nanocarriers for Drug Delivery

UV light is widely used for the construction of photo-responsive materials since UV light apparatus with various wavelengths and intensity is cheap and easily set up in laboratory. What's more, most of the photo-responsive groups are UV light sensitive. In general, low-power photo irradiation of long wavelength UV light (365 nm) within 30 min is safe to body and imposes little toxicity to cells. However, it is enough for most of the photo reactions and photo-responsive drug release can be achieved, which opens a door for the design of UV responsive nanocarriers. Since photo-responsive pyrene-containing amphiphilic diblock copolymers were first introduced as drug delivery vehicles in 2005 by Yue Zhao (Jiang et al. 2005), there has been considerable interest in photo-responsive polymeric micelles for on-demand drug-delivery.

Table 1 Characteristic features of photo-responsive groups

Photo-responsive group	Light source	Reversibility	Effect of photo	Polymer design
O-nitrobenzyl ester	UV, NIR	No	Photocleavage	Side-chain (Jiang et al. 2006), main-chain (Han et al. 2011)
Azobenzene	UV	Yes	Photoisomerization	Side-chain (Jin et al. 2010)
Spiropyran	UV	Yes	Photoisomerization	Side-chain (Jin et al. 2010)
Coumarinyl ester	UV, NIR	No	Photoleavage	Side-chain (Babin et al. 2009)
2-diazo-1,2-naphthoquinone	UV, NIR	No	Photocleavage	Side-chain (Chen et al. 2011)
Pyrenylmethyl ester	UV	No	Photocleavage	Side-chain (Jiang et al. 2005)

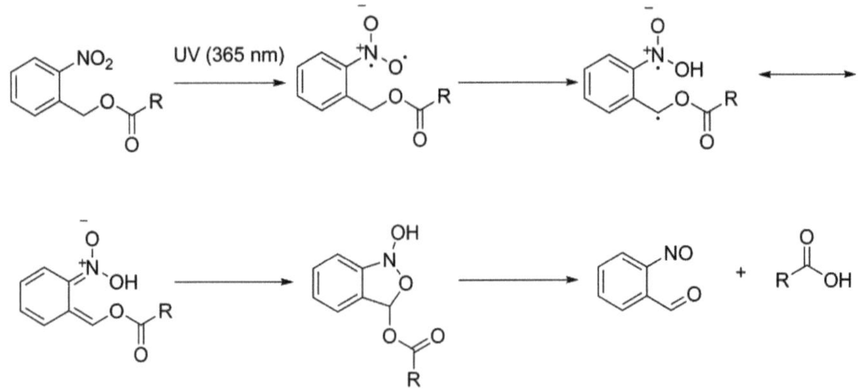

Fig. 4 The mechanism of photocleavage of o-nitrobenzyl ester

3.1 O-Nitrobenzyl Containing Drug Nanocarriers

As a most widely studied photo-labile group, the o-nitrobenzyl group is utilized frequently in polymer and materials science. o-Nitrobenzyl ester has become one of the most popular photo-labile protecting groups since first described by Schofield and co-workers (Barltrop et al. 1966). The photocleavage of o-nitrobenzyl ester is an intramolecular rearrangement process and the mechanism of the photocleavage of o-nitrobenzyl ester to corresponding o-nitrosobenzaldehyde upon UV irradiation is shown in Fig. 4. It is very important that this kind of photoreaction doesn't need protonic solvent and can carry out both in solution and solid state. Moreover, the photocleavage of o-nitrobenzyl can occur via one-photon UV (365 nm) or two-photon (700 nm) irradiation. The photo-cleavable o-nitrobenzyl group can be conjugated either in the side chain or in the main chain. If it is in the side chain, the hydrophobic o-nitrobenzyl containing block will transform to hydrophilic carboxylic acid containing block under UV irradiation, which will result in the hydrophobicity-to-hydrophilicity transition. If o-nitrobenzyl group is in the main chain, the polymer backbone will be degraded into small molecules under UV irradiation.

The first example of photo-responsive o-nitrobenzyl containing micelles was reported by Zhao and co-workers (Jiang et al. 2006). They synthesized amphiphilic block copolymer poly(ethylene oxide)-b-poly(2-nitrobenzyl methacrylate) (PEO-b-PNBMA) which can self-assemble into micelles. The photolysis of 2-nitrobenzyl groups resulted in the cleavage of 2-nitrosobenzaldehyde from the polymer, which transformed the hydrophobic PNBMA into the hydrophilic PMAA and triggered the micellar disruption. Using Nile red as a hydrophobic model compound, they investigated the photo-controlled release of Nile red from PEO-b-PNBMA micelles. The release of Nile red was monitored by fluorescence spectra as the fluorescence intensity decreased and the maximum emission wavelength red shifted upon UV irradiation.

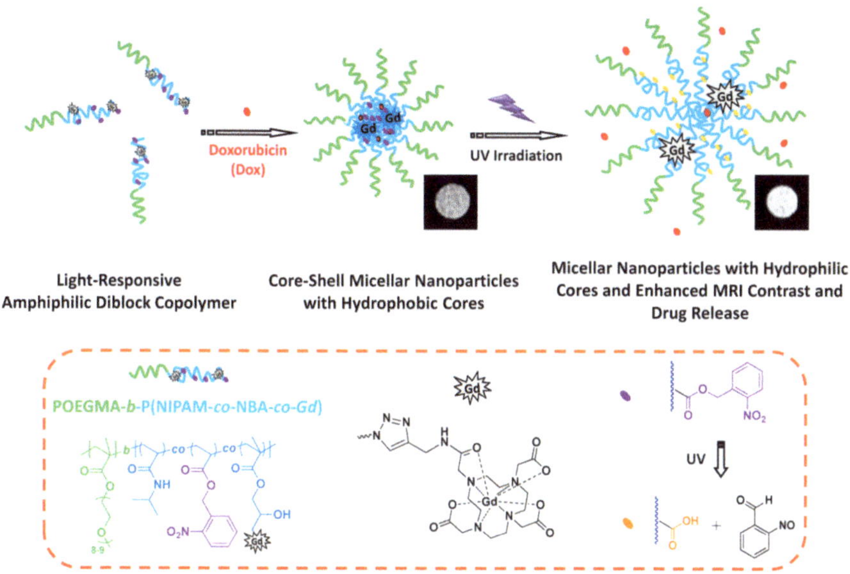

Fig. 5 Schematic illustration for the fabrication of photo-responsive polymeric micelles of POEGMA-b-P(NIPAAm-co-NBA-co-Gd) amphiphilic diblock o-nitrobenzyl containing copolymer (Li et al. 2012). Reproduced by permission of the American Chemical Society

Similarly, Liu and co-workers fabricated amphiphilic o-nitrobenzyl containing diblock copolymer micelles exhibiting photo-triggered hydrophobic-hydrophilic transition within micellar cores and the concomitant enhancement of magnetic resonance (MR) imaging contrast performance and release rate of physically encapsulated hydrophobic drugs (Li et al. 2012). POEGMA-b-P(NIPAAm-co-NBA-co-Gd) diblock copolymers covalently labeled with Gd3+ complex were synthesized via the combination of consecutive RAFT polymerizations and "click" postfunctionalization. Self-assembled micelles were further used to encapsulate anticancer drug DOX. In the presence of UV irradiation, hydrophobic NBA moieties transformed into hydrophilic carboxyl derivatives, resulting in the changes of micellar microstructural changes and the core swelling. As a result, the microenvironment surrounding Gd3+ complexes was subjected to a transition from being hydrophobic to hydrophilic, which led to the enhancement of MR imaging contrast performance. At the same time, the release rate of encapsulated Dox was also enhanced (Fig. 5). In another research, the same group fabricated photo-degradable, protein-polyelectrolyte complex-coated, mesoporous silica nanoparticles (MSNs) and realized controlled co-release of protein and model drugs (Wan et al. 2013). Random copolymers composed of OEGMA and a photocleavable o-nitrobenzyl-containing monomer, DENBMA, were first conjugated to MSNs by RAFT and then quaternary aminated to obtain positively charged P(OEGMA-co-TENBMA) which exhibits photo-induced charge conversion characteristics. Rhodamine B (RhB) was encapsulated into the nanopores of

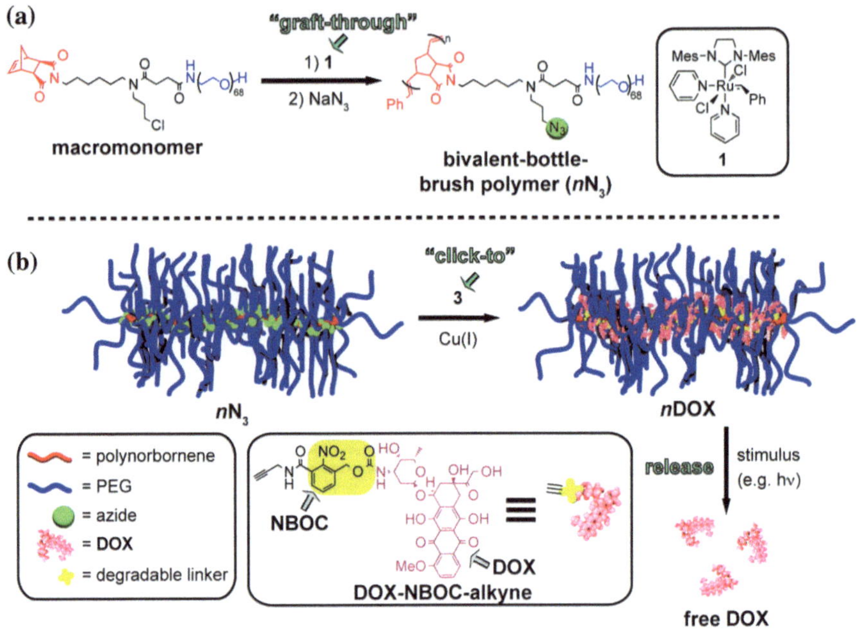

Fig. 6 **a** Synthesis of the polymer-drug conjugates; **b** Photo-triggered release of DPOX from the polymer-drug conjugates (Johnson et al. 2011). Reproduced by permission of the American Chemical Society

the MSNs. Meanwhile, Negatively charged BSA was loaded on the surface of the hybrid MSNs by electrostatic interactions. Upon UV irradiation, positively charged P(OEGMA-co-TENBMA) were transformed to negatively charged, which led to the disruption of protein-polyelectrolyte complex on MSNs and co-release of BSA and RhB by electrostatic repulsion.

In another interesting research, Grubbs and co-workers synthesized photo-responsive o-nitrobenzyl containing polymer-drug conjugates (Johnson et al. 2011). A class of DOX conjugated bottle-brush copolymer nanoconjugates were prepared by ring-opening metathesis polymerization and click chemistry. DOX was conjugated to the backbone by photocleavable o-nitrobenzyl groups (Fig. 6). After UV irradiation, DOX was cleaved from the backbone and the free DOX could kill the cancer cells effectively.

Besides photo-responsive liner polymers, dendritic amphiphilic o-nitrobenzyl containing Nanocontainers were also reported by Yesilyurt et al. (2011). They designed and synthesized photo-responsive amphiphilic dendrimers that can form dendritic micelles in water. Hydrophobic guest molecules were loaded into the dendritic micelles. Upon UV irradiation, Hydrophobic o-nitrobenzyl ester groups were converted to hydrophilic carboxylic acid segments, which destroyed the hydrophilic-hydrophobic balance and resulted in the supramolecular disassembly of the dendritic micelles (Fig. 7). 88 % of the encapsulated guest molecules were

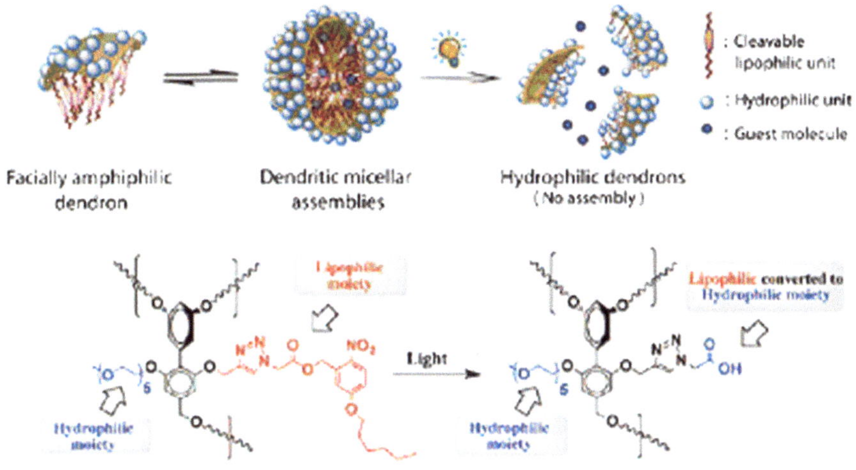

Fig. 7 Photo-induced disassembly of dendritic micellar assemblies after photo irradiation (Yesilyurt et al. 2011). Reproduced by permission of John Wiley and Sons

released in only 200s upon UV irradiation, compared to the 5 % guest release upon irradiation of a control dendrimer, which lacked the photolabile o-nitrobenzyl groups.

Combining photo with other stimuli can significantly broaden the scope of applications of such drug nanocarriers. Very recently, Zhu and co-workers reported chitosan-based nanocarriers with pH and photo dual response for anticancer drug delivery (Meng et al. 2013). The pH and photo-responsive cross-linked polymeric micelles were prepared by the self-assembly of amphiphilic glycol chitosan-o-nitrobenzyl succinate conjugates and then cross-linking with glutaraldehyde, which was synthesized by grafting photo-responsive o-nitrobenzyl group onto hydrophilic glycol chitosan. Anticancer drug camptothecin (CPT) can be loaded into the micelles. The in vitro drug release experiments showed the encapsulated CPT could be quickly released at low pH with UV irradiation (Fig. 8). At the same time, the CPT-loaded micelles exhibited better cytotoxicity against cancer cells under UV irradiation.

The o-nitrobenzyl groups mentioned above are all in the side chains. Zhao and co-workers also conjugated o-nitrobenzyl groups in the main chains. In one their research (Han et al. 2011), they synthesized an amphiphilic ABA triblock copolymer PEO-b-PUNB-b-PEO of which the middle block PUNB is a hydrophobic polyurethane containing o-nitrobenzyl groups (Fig. 9). The resulting micelles possess a photodegradable core and can be disrupted very fast upon UV irradiation. As a result, more than 70 % of encapsulated guest molecules can be released after only 10 s of UV irradiation (365 nm). What's more, they prepared block copolymer micelles with a dual-stimuli degradable core for fast or slow drug release (Han et al. 2012). ABA triblock copolymer PEO-b-poly(disulfide-alt-nitrobenzene)-b-PEO was synthesized by two-step click chemistry of which the

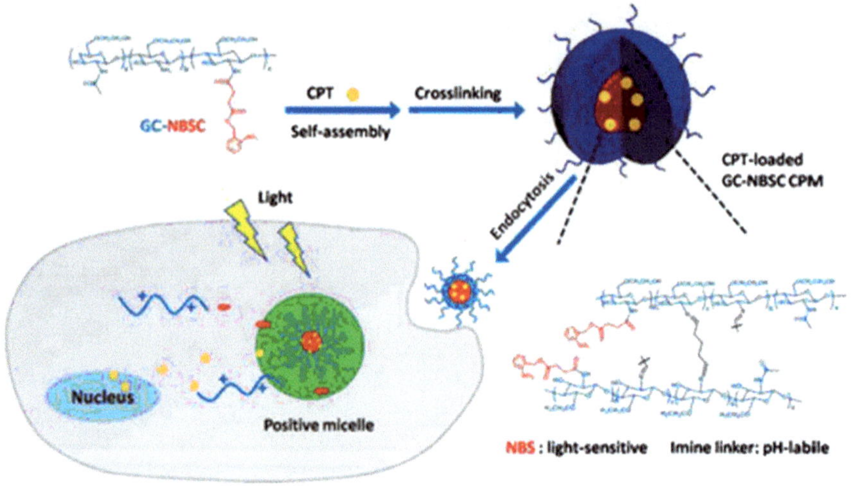

Fig. 8 Schematic illustration for the preparation of CPT-loaded micelles and intracellular drug release triggered by pH and UV Light (Meng et al. 2013). Reproduced by permission of the American Chemical Society

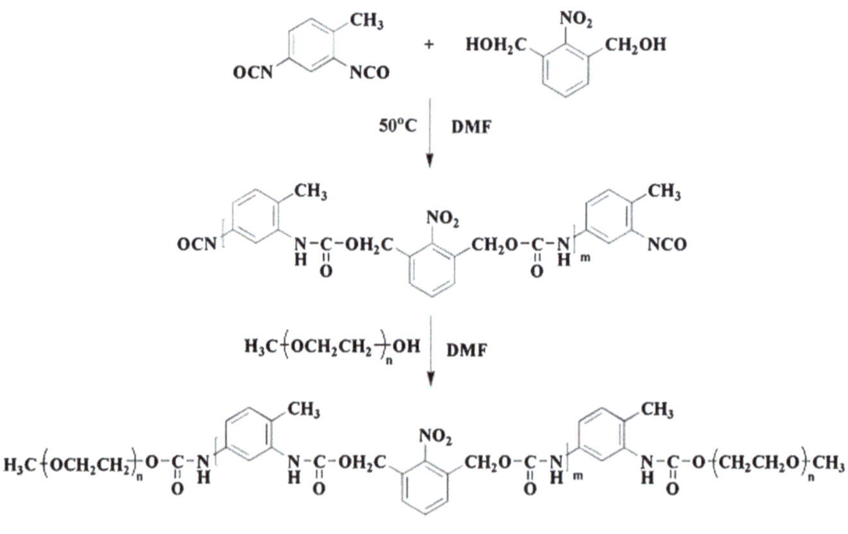

Fig. 9 Synthesis of photo-degradable triblock copolymer of PEO-b-PUNB-b-PEO (Han et al. 2011). Reproduced by permission of the American Chemical Society

middle block is composed of redox-responsive disulfide and photo-responsive o-nitrobenzyl groups. With this design, the micelles be disrupted under the effect of either a reducing agent DTT that breaks the disulfide bonds or UV light that cuts the o-nitrobenzyl ester groups. This feature makes it possible to have either burst

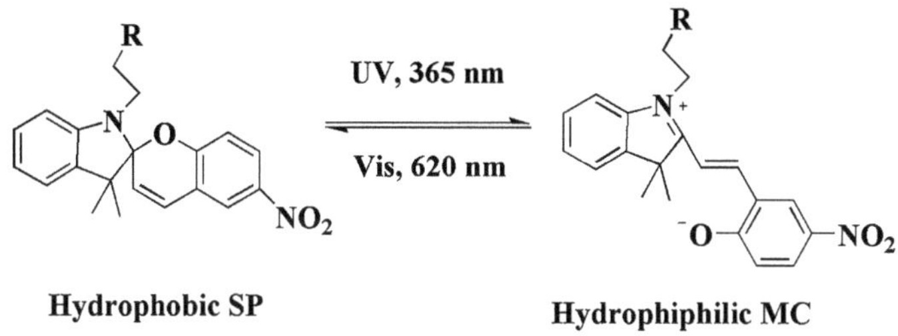

Fig. 10 Photo-responsive reversible spiropyran-merocyanine (SP-MC) transition

release of encapsulated guest molecules by UV irradiation, or slow release by the action of a reducing agent DTT in the micellar solution. The two stimuli could also be utilized in combination to generate on-demand release rate profiles.

3.2 Spiropyran-Containing Drug Nanocarriers

Photochromic spiropyran molecules are well known for the reversible spiropyran-merocyanine (SP-MC) transition (Goldburt et al. 1984). The SP molecule is colorless, nonpolar, hydrophobic, and in a "closed" form under visible light irradiation (620 nm). Exposure of the SP molecule to UV (365 nm) irradiation induces ring-opening isomerization, giving the colored, polar, hydrophilic, and zwitterionic MC form (Fig. 10). Recently, the spiropyran-merocyanine (SP-MC) chemistry has been introduced to construct photo-responsive nanocarriers. Lee and co-workers synthesized a diblock copolymer PEO-b-PSPMA, where SPMA was spiropyran-containing methacrylate monomer (Lee et al. 2007). The SP units respond to photo and undergo a reversible isomerization between hydrophobic SP and hydrophilic MC, which results in the assembly and disassembly of the micelles. This micelle system was successfully applied to the efficient encapsulation, release, and partial re-encapsulation of hydrophobic guest molecules. Similarly, Ji and co-workers synthesized amphiphilic spiropyran-containing hyperbranched polyphosphate HPHEEP-SP which can self-assemble to biocompatible micelles (Chen et al. 2012). Model drug coumarin 102 was then encapsulated into the micelles successfully. Photo-controlled release and re-encapsulation was also realized.

3.3 Azobenzene-Containing Drug Nanocarriers

Azobenzene can undergo trans-cis photoisomerization in response to UV and visible light which results in large structural change as reflected in the spatial requirement and absorption spectra (Tamai and Miyasaka 2000). At the same time, trans-azobenzene is an excellent guest for inclusion complexation with α- and

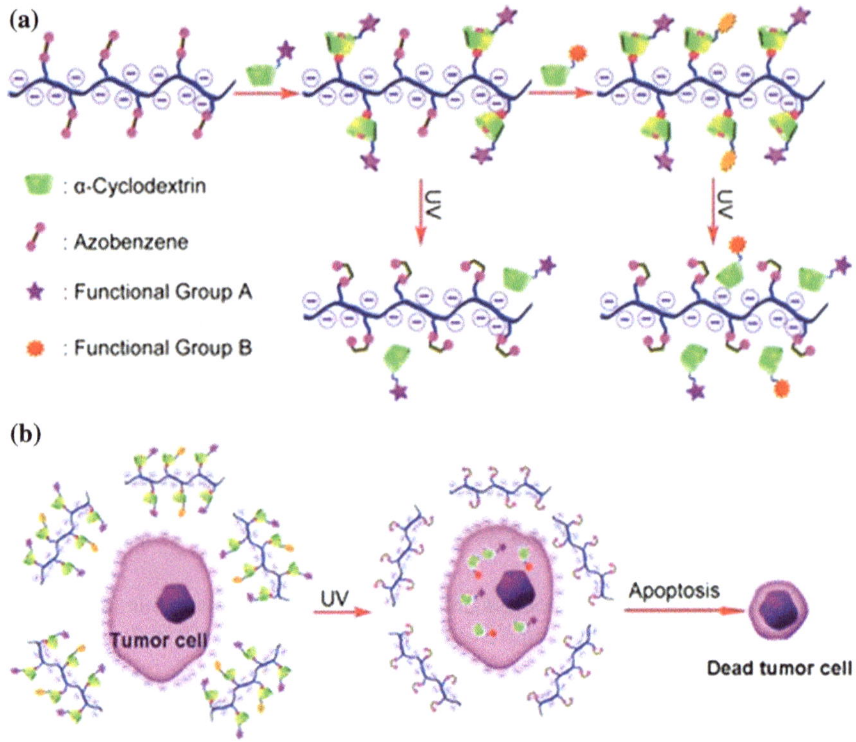

Fig. 11 Schematic illustration of the photo responsive polyanionic template loading (**a**) and release (**b**) of α-CD modified drug and functional groups (Xiao et al. 2011). Reproduced by permission of John Wiley and Sons

β-cyclodextrin (CD). In contrast, cis-azobenzene cannot form inclusion complexation with CD because of the mismatch between the host and the guest. This photo-switchable host-guest interaction might be a nice inspiration for the design of smart drug nanocarriers (Wang et al. 2007). Zhang and co-workers fabricated cellular-uptake shielding "Plug and Play" template for photo triggered drug release (Xiao et al. 2011). The polyanionic template consisted of polyacrylic acid (PAA) and azobenzene. The α-CD modified drug or functional groups can be loaded onto the template via host-guest interaction. Upon UV irradiation, the α-CD modified drug can be released (Fig. 11). Since the surface of cells is negatively charged, the negatively charged drug carriers will be repelled by cells and cannot be uptaken by normal cells. In the presence of UV irradiation, the loaded drug can be released from the drug carrier quickly and thereafter endocytosed by target cells to achieve the desirable cure effect. Azobenzene-containing photo-responsive amphiphilic linear-dendritic block copolymers were reported by Oriol and co-workers recently (Blasco et al. 2013). The release of encapsulated guest molecules can be triggered by low intensity UV illumination, which was related to the trans-cis photoisomerization of azobenzene.

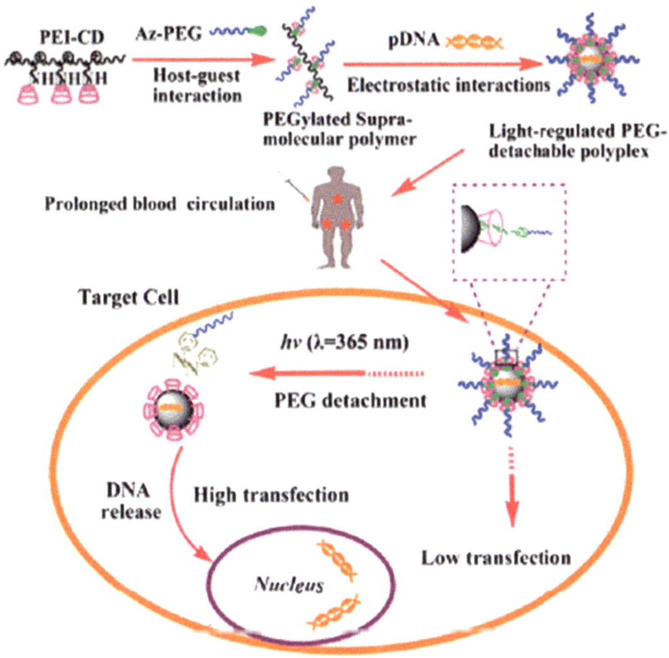

Fig. 12 Formation of PEGylated polyplexes via host–guest interactions and the detachment of PEG upon UV irradiation (Li et al. 2012). Reproduced by permission of the Royal Society of Chemistry

Photo-responsive azobenzene-containing polymers can also be used in gene delivery. Wang and co-workers constructed intracellular PEG-detachable polyplexes to facilitate nuclear entry by photo-responsive host–guest interactions (Li et al. 2012). They synthesized β-CD-modified branched polyethylenimine (PEI-CD) and azobenzene modified PEG (AZ-PEG) was grafted to PEI-CD via host–guest interactions. By this way, PEGylated supramolecular PEI (Az-PEG/PEI-CD) was prepared. It was used to associate with DNA and Az-PEG/PEI-CD/DNA polyplexes were obtained. PEGylated polyplexes provided excellent biocompatibility during circulation. After they were internalized by cells, photo-regulated dePEGylation facilitated DNA release and nuclear entry, thus resulting in efficient transfection (Fig. 12).

4 NIR Responsive Polymeric Nanocarriers for Drug Delivery

As is known, UV light can be easily absorbed by water and biomolecules. As a result, UV light can only penetrate shallow tissue, which greatly limits the applications of UV light in vivo. What's more, UV light is somewhat harmful

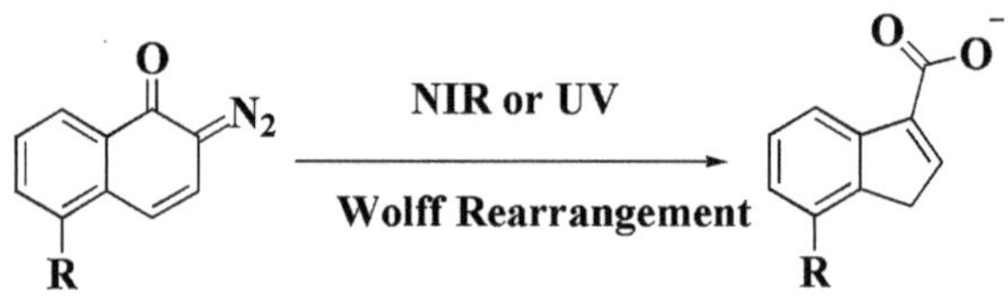

Fig. 13 Wolff rearrangement of DNQ

to the tissues and cells. To overcome this drawback, an increasing number of drug-delivery materials sensitive to near infrared (NIR) light have been reported in recent years. Compared to UV light, NIR between 700 and 1,000 nm can penetrate several millimeters up to centimeter depth of tissues and shows fewer risks of damage to the irradiated area, enabling the NIR-sensitive nanomedicines great potential as a noninvasive clinical therapy (Fomina et al. 2012; Weissleder 2001).

4.1 DNQ-Containing Drug Nanocarriers

2-Diazo-1,2-naphthoquinone (DNQ) is one of the most widely used NIR responsive groups in scientific research. Under UV (365 nm) or NIR (800 nm) irradiation, the hydrophobic DNQ undergoes a Wolff rearrangement to afford hydrophilic 3-indenecarboxylic acid with a pKa of 4.5 (Fig. 13) (Almstead et al. 1994). Therefore, the DNQ-containing polymers provide a versatile platform for the design of NIR-responsive drug nanocarriers. Fréchet and co-workers first synthesized amphiphilic copolymers of DNQ-terminated PEO (PEO-DNQ) (Goodwin et al. 2005). PEO-DNQ can self-assemble into micelles for the encapsulation of hydrophobic Nile Red. The photo-responsive release of Nile Red was monitored by fluorescence spectra under NIR irradiation (795 nm). The disadvantage of this system was that PEO-DNQ had a relatively high critical micelle concentration (CMC) of 0.15 mg/mL and exhibited a rather high toxicity towards cells. In order to overcome this disadvantage, they synthesized DNQ-containing linear-dendritic copolymers, which showed much lower CMC (12 μg/mL) and was less toxic than PEO-DNQ. Meanwhile, the NIR-responsive release of guest molecules was also realized (Mynar et al. 2007). Notably, Ji and co-workers designed a series of DNQ modified polymers, including hyperbranched polyphosphate, comb-like PEG, and dextran to explore the NIR-responsive drug release (Chen et al. 2011a, b; Liu et al. 2012). Especially, they constructed DNQ grafted dextran (Dex-DNQ), which was used for the NIR-responsive intracellular drug delivery (Liu et al. 2012). Anticancer drug DOX can be easily encapsulated into Dex-DNQ micelles withloading content of 24 %. In vitro cell viability studies, the micelles exhibited higher intracellular DOX release under NIR irradiation at 808 nm, which resulted in significant growth inhibition of HepG2 cancer cells (Fig. 14).

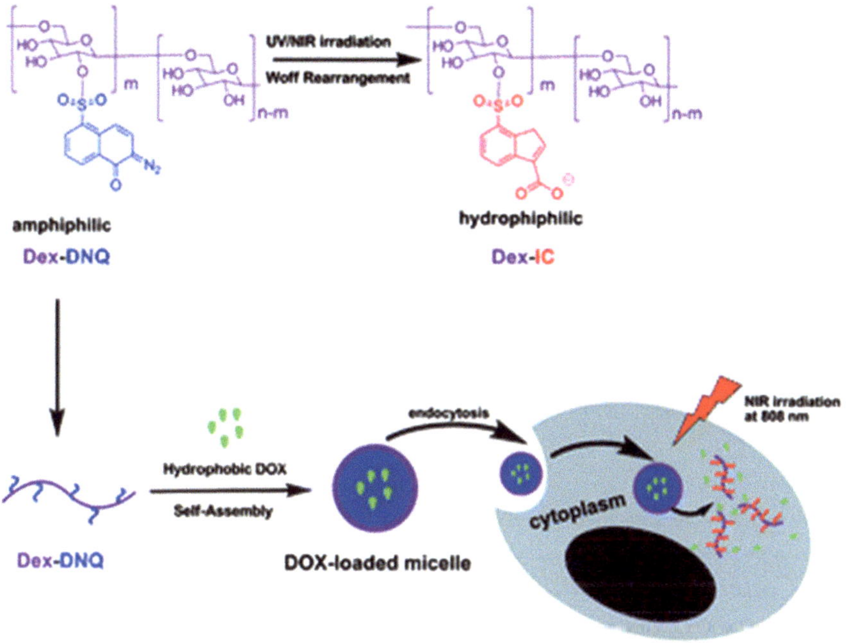

Fig. 14 Schematic illustration of the self-assembly and NIR-responsive intracellular release of DOX (Liu et al. 2012). Reproduced by permission of the Royal Society of Chemistry

Importantly, DNQ was introduced into multifunctional drug nanocarriers. Dong and co-workes prepared NIR-responsive sugar-targeted nanocarriers from degradable and dendritic amphiphiles (Sun et al. 2011). They synthesized DNQ grafted dendronic poly(amido amine)-b-poly(ε-caprolactone) (D3-PCL-DNQ), D3-PCL-DNQ micelles can be used as nanocarriers for the loading of DOX. Upon NIR irradiation, DOX can be released in a controlled manner by changing thc photo irradiation time, which was induced by the gradual disruption of micelles in aqueous solution. Furthermore, the sugar-coated micelles demonstrated specific binding with the lectins Concanavalin A and Ricinus communis agglutinin, respectively, which madc thcm uscful as targctcd drug-delivery vesicles. In another research, the same group fabricated NIR-responsvie and lectin-binding nanocarriers from Janus-type dendritic PAMAM amphiphiles (Sun et al. 2012). Janus-type dendritic poly(amido amine) (PAMAM) amphiphiles Dm-Lac-D3DNQ were synthesized by connecting hydrophobic DNQ-decorated PAMAM dendron D3 and hydrophilic lactose (Lac)-decorated PAMAM dendrons Dm via click chemistry. Amphiphilic Dm-Lac-D3DNQ can self-assemble into the DNQ-cored micelles dangled by densely free Lac groups (Fig. 15). After 30 min of NIR irradiation (808 nm), most of the micelles were disassembled. The DOX-loaded nanomedicine exhibited a NIR-triggered drug release profile and a related cytotoxicity that was close to free DOX. Meanwhile, the micelles showed binding with RCA_{120} lectin, which was related to the generation of Lac-decorated dendrons.

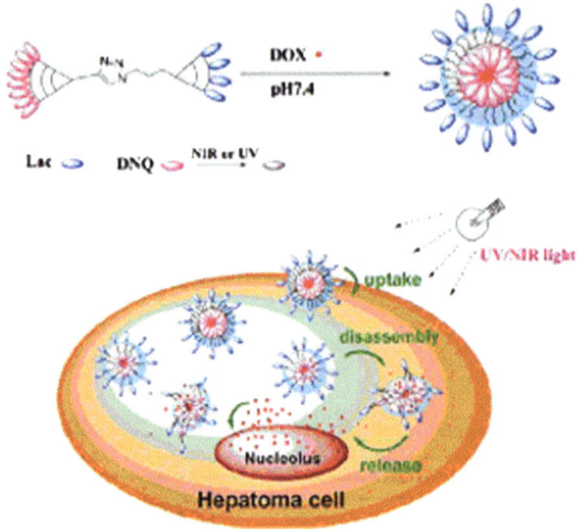

Fig. 15 Self-assembly of DOX-loaded nanocarriers of Dm-Lac-D3DNQ and NIR-responsive intracellular DOX release (Sun et al. 2012). Reproduced by permission of the American Chemical Society

4.2 Coumarin-Containing Drug Nanocarriers

The coumarin family has thousands of derivatives and has been widely used in various areas. The coumarin units are well known for the reversible dimerization and de-dimerization under UV irradiation with different wavelength (365 and 254 nm) (Trenor et al. 2004). Thus, coumarin units are widely used in photo cross-linking. On the other hand, some special coumarin derivatives can also be photo-cleaved unde one-photon UV or two-photon NIR irradiation if the coumarin derivatives are linked along the side chain of polymer by ester units (Fig. 16). Zhao and co-workers first synthesized coumarin-containing NIR-responsive nanocarriers. The amphiphilic NIR-responsive block copolymer is composed of hydrophilic PEO block and hydrophobic poly-([7-(diethylamino)coumarin-4-yl] methyl methacrylate) (PDEACMM) block (Babin et al. 2009). Upon UV or NIR irradiation, the DEACMM groups were cleaved from the backbone, which converted the ester groups to carboxylic acid groups and the hydrophobic PDEACMM to hydrophilic poly(methacrylic acid) (PMA). After 285 min of NIR exposure, micelles appeared to be highly degraded, accompanying with the release of guest molecules. In order to improve the biocompatibility and biodegradability of the micelles, Zhao and co-workers a new kind of polypeptide copolymer (Kumar et al. 2012). The NIR responsive copolymer was composed of PEO and poly(L-glutamic acid) bearing a number of 6-bromo-7-hydroxycoumarin-4-ylmethyl groups (PEO-b-P(LGA-co-COU)). The coumarin compound, 6-bromo-7-hydroxycoumarin-4-ylmethyl, was selected as the NIR-responsive group because this chromophore is known to have a large two-photon absorption cross-section for NIR light. The PEO-b-P(LGA-co-COU) micelles could be disrupted by 794 nm NIR excitation via two-photon absorption. The NIR-responsive release of an antibacterial drug

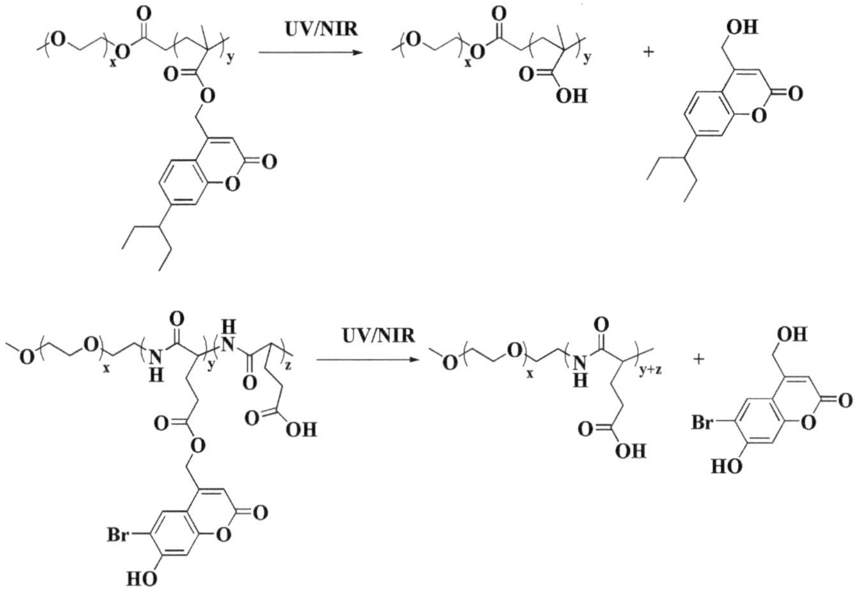

Fig. 16 NIR responsive coumarin derivatives PEO-b-PDEACMM and PEO-b-P(LGA-co-COU) (Babin et al. 2009; Kumar et al. 2012)

(Rifampicin) and an anticancer drug (Paclitaxel) loaded into the PEO-b-P(LGA-co-COU) micelles was further investigated. They found that the two drugs could be released effectively upon NIR light exposure of the micellar solution.

4.3 Upconverting Nanoparticles-Assistant NIR Responsive Drug Nanocarriers

As mentioned above, most of the NIR-responsive nanocarriers are triggered by two-photon absorption of NIR light. However, the photoreactions activated by two-photon absorption of NIR light are generally slow and inefficient due to the low two-photon absorption cross sections of the chromophores. Moreover, the simultaneous absorption of two photons necessitates high laser power density and therefore requires a femtosecond pulse laser. Recently, upconverting nanoparticles (UCNPs) have emerged as an appealing candidate for the application of NIR light (Yang et al. 2012; Gu et al. 2013). Because of the unique ladder-like energy level structures of lanthanide ions (such as Tm^{3+}, Er^{3+}, and Ho^{3+}), UCNPs are able to absorb NIR light and convert it into high-energy photons in a very broad range from the UV to the NIR region. In contrast to two-photon absorption, the excitation of UCNPs by NIR light occurs via sequential, multiple absorptions with real energy levels, which requires much lower power density so that a continuous-wave

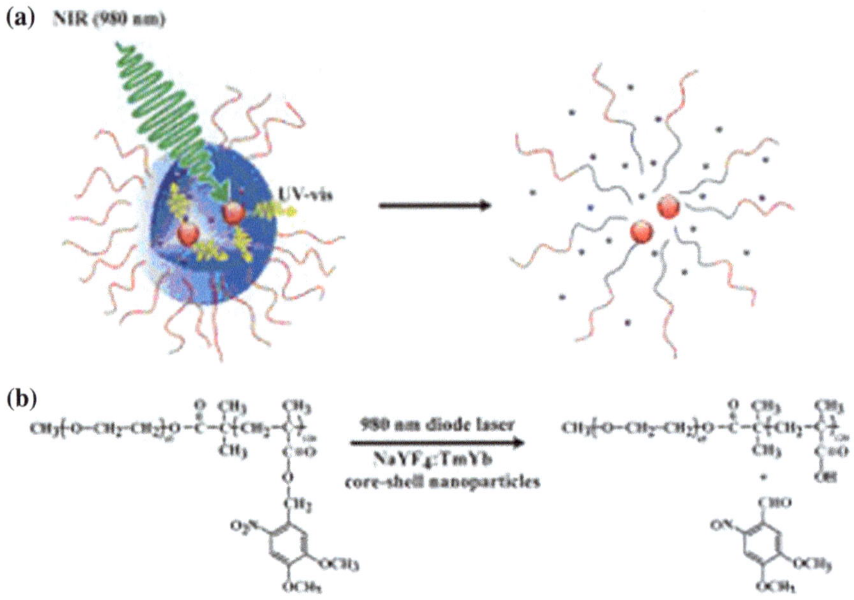

Fig. 17 **a** Schematic illustration of using NIR light excitation of UCNPs to trigger dissociation of micelles. **b** NIR light-triggered photoreaction with the used micelles of PEO-b-PNBMA and UCNPs (Yan et al. 2011). Reproduced by permission of the American Chemical Society

diode NIR laser can be sufficient as the excitation source. Using UCNPs, some UV responsive systems can be converted to NIR responsive systems. Liu and co-workers prepared silica coated UCNPs and then functionalized with a positively charged alkyl amine by photo-responsive o-nitrobenzyl linkers through covalent bonding (Yang et al. 2013). The positively charged nanoparticles can effectively adsorb anionic siRNA through electrostatic attractions and were easily internalized by living cells. Upon NIR light irradiation (980 nm), UCNPs can absorb NIR light and converted it into 365 nm UV light. UV light of 365 nm can cleave the o-nitrobenzyl linkers and converted the positively charged alkyl amine to negatively charged carboxyl group. By this way, the intracellular release of the siRNA can be realized. Zhao and co-workers physically encapsulated UCNPs and Nile Red inside micelles of o-nitrobenzyl containing poly(ethylene oxide)-block-poly(4,5-dimethoxy-2-nitrobenzyl methacrylate) (PEO-b-PNBMA) (Yan et al. 2011). When exposing the micellar solution to 980 nm NIR light, photons in the UV region (365 nm) were emitted by the UCNPs, which in turn were absorbed by o-nitrobenzyl groups on the micelle core-forming block and activated the photocleavage reaction. Finally, the micelles were dissociated and the co-loaded Nile Red was released (Fig. 17). This strategy of using UCNPs as an internal UV or visible light source upon NIR light excitation represents a general and efficient way to circumvent the need of UV or visible light excitation that is a common drawback for photo-responsive systems in biomedical applications.

5 Conclusion and Outlook

In this chapter, we highlighted the recent progresses on photo-responsive polymeric nanocarriers that might realize spatiotemporal and on-demand drug delivery via photo irradiation. A wide variety of photo-responsive moieties as well as synthetic routes were introduced to the drug nanocarriers. The nanocarriers exhibited photo responsiveness and showed great potential in clinical medicine and oncology. Despite these advantages, there are still many challenges and issues that need to be addressed. The main limitations include: (1) The NIR-responsive groups are limited. We should discover more NIR responsive groups with a large absorption cross-section (>1GM); (2) The biodegradability and biocompatibility of the drug nanocarriers should be paid great attention to; (3) The cytotoxicity of the by-products resulting from the photo-induced reactions should be concerned. There is still much room in the field of photo-responsive nanocarriers for further development and the development and applications of photo-responsive polymeric nanocarriers will continue in the future.

References

Almstead J, Urwyler B, Wirz J (1994) Flash photolysis of alpha-diazonaphthoquinones in aqueous solution: determination of rates and equilibria for keto-enol tautomerization of 1-indene-3-carboxylic acid. J Am Chem Soc 116:954–960

Babin J, Pelletier M, Lepage M et al (2009) A new two-photon-sensitive block copolymer nanocarrier. Angew Chem Int Ed 48:3329–3332

Barltrop JA, Plant PJ, Schofield P (1966) Photosensitive protective groups. Chem Commun (London) 22:822–823

Blasco E, Barrio J, Somolinos C et al (2013) Light induced molecular release from vesicles based on amphiphilic linear-dendritic block copolymers. Polym Chem 4:2246–2254

Cabral H, Nishiyama N, Kataoka K (2011) Supramolecular nanodevices: from design validation to theranostic nanomedicine. Acc Chem Res 44:999–1008

Chen CJ, Liu GY, Shi YT et al (2011a) Biocompatible micelles based on comb-like peg derivates: formation, characterization, and photo-responsiveness. Macromol Rapid Commun 32:1077–1081

Chen CJ, Liu GY, Liu XS et al (2011b) Photo-responsive, biocompatible polymeric micelles self-assembled from hyperbranched polyphosphate-based polymers. Polym Chem 2:1389–1397

Chen CJ, Jin Q, Liu GY et al (2012) Reversibly light-responsive micelles constructed via a simple modification of hyperbranched polymers with chromophores. Polymer 53:3695–3703

Christie RJ, Miyata K, Matsumoto Y (2011) Effect of polymer structure on micelles formed between siRNA and cationic block copolymer comprising thiols and amidines. Biomacromolecules 12:3174–3185

Du JZ, Du XJ, Mao CQ et al (2011) Tailor-made dual pH-sensitive polymer-doxorubicin nanoparticles for efficient anticancer drug delivery. J Am Chem Soc 133:17560–17563

Elsabahy M, Wooley KL (2012) Design of polymeric nanoparticles for biomedical delivery applications. Chem Soc Rev 41:2545–2561

Fomina N, Sankaranarayanan J, Almutairi A (2012) Photochemical mechanisms of light-triggered release from nanocarriers. Adv Drug Delivery Rev 64:1005–1020

Gil ES, Hudson SM (2004) Stimuli-reponsive polymers and their bioconjugates. Prog Polym Sci 29:1173–1222
Gohy J, Zhao Y (2013) Photo-responsive block copolymer micelles: design and behavior. Chem Soc Rev 42:7117–7129
Goldburt E, Shvartsman F, Fishman S (1984) Intramolecular interactions in photochromic spiropyran-merocyanine polymers. Macromolecules 17:1225–1230
Goodwin AP, Mynar JL, Ma YZ et al (2005) Synthetic micelle sensitive to ir light via a two-photon process. J Am Chem Soc 127:9952–9953
Gu ZJ, Yan L, Tian G et al (2013) Recent advances in design and fabrication of upconversion nanoparticles and their safe theranostic applications. Adv Mater 25:3758–3779
Habault D, Zhang HJ, Zhao Y (2013) Light-triggered self-healing and shape-memory polymers. Chem Soc Rev 42:7244–7256
Han DH, Tong X, Zhao Y (2011) Fast photodegradable block copolymer micelles for burst release. Macromolecules 44:437–439
Han D, Tong X, Zhao Y (2012) Block copolymer micelles with a dual-stimuli-responsive core for fast or slow degradation. Langmuir 28:2327–2331
Jiang J, Tong X, Zhao Y (2005) A new design for light-breakable polymer micelles. J Am Chem Soc 127:8290–8291
Jiang J, Tong X, Morris D et al (2006) Toward photocontrolled release using light-dissociable block copolymer micelles. Macromolecules 39:4633–4640
Jin Q, Liu GY, Ji J (2010a) Micelles and reverse micelles with a photo and thermo double-responsive block copolymer. J Polym Sci, Part A: Polym Chem 48:2855–2861
Jin Q, Liu GY, Liu XS (2010b) Photo-responsive supramolecular self-assembly and disassembly of an azobenzene-containing block copolymer. Soft Matter 6:5589–5595
Johnson JA, Lu YY, Burts AO et al (2011) Core-clickable PEG-branch-azide bivalent-bottle-brush polymers by ROMP: grafting-through and clicking-to. J Am Chem Soc 133:559–566
Kumar S, Allard JF, Morris D et al (2012) Near-infrared light sensitive polypeptide block copolymer micelles for drug delivery. J Mater Chem 22:7252–7257
Lee H, Wu W, Oh JK et al (2007) Light-induced reversible formation of polymeric micelles. Angew Chem Int Ed 46:2453–2457
Li YM, Qian YF, Liu T et al (2012a) Light-triggered concomitant enhancement of magnetic resonance imaging contrast performance and drug release rate of functionalized amphiphilic diblock copolymer micelles. Biomacromolecules 13:3877–3886
Li WY, Wang YX, Chen LN et al (2012b) Light-regulated host–guest interaction as a new strategy for intracellular PEG-detachable polyplexes to facilitate nuclear entry. Chem Commun 48:10126–10128
Li Y, Gao GH, Lee DS (2013) Stimulus-Sensitive polymeric nanoparticles and their applications as drug and gene carriers. Adv Healthcare Mater 2:388–417
Liu GY, Chen CJ, Li DD et al (2012) Near-infrared light-sensitive micelles for enhanced intracellular drug delivery. J Mater Chem 22:16865–16871
Liu G, Liu W, Dong CM (2013) UV- and NIR-responsive polymeric nanomedicines for on-demand drug delivery. Polym Chem 4:3431–3443
Meng LL, Huang W, Wang DL et al (2013) Chitosan-based nanocarriers with pH and light dual response for anticancer drug delivery. Biomacromolecules 14:2601–2610
Mynar JL, Goodwin AP, Cohen JA et al (2007) Two-photon degradable supramolecular assemblies of linear-dendritic copolymers. Chem Commun 43:2081–2082
Pasparakis G, Manouras T, Argitis P et al (2012) Photodegradable polymers for biotechnological applications. Macromol Rapid Commun 33:183–198
Rapoport N (2007) Physical stimuli-responsive polymeric micelles for anti-cancer drug delivery. Prog Polym Sci 32:962–990
Sailor MJ, Park JH (2012) Hybrid nanoparticles for detection and treatment of cancer. Adv Mater 24:3779–3802
Srinivas G, Harikrishna D, Aliasgar S (2008) A review of stimuli-responsive nanocarriers for drug and gene delivery. J Control Release 126:187–204

Sun L, Yang Y, Dong CM et al (2011) Two-photon-sensitive and sugar-targeted nanocarriers from degradable and dendritic amphiphiles. Small 7:401–406

Sun L, Ma XF, Dong CM et al (2012) NIR-responsive and lectin-binding doxorubicin-loaded nanomedicine from janus-type dendritic PAMAM amphiphiles. Biomacromolecules 13:3581–3591

Tamai N, Miyasaka H (2000) Ultrafast dynamics of photochromic systems. Chem Rev 100:1875–1890

Trenor SR, Shultz AR, Love BJ et al (2004) Coumarins in polymers: from light harvesting to photo-cross-linkable tissue scaffolds. Chem Rev 104:3059–3078

Wan XJ, Liu T, Hu JM et al (2013) Photo-degradable, protein-polyelectrolyte complex-coated, mesoporous silica nanoparticles for controlled co-release of protein and model drugs. Macromol Rapid Commun 34:341–347

Wang YP, Ma N, Wang ZQ et al (2007) Photocontrolled Reversible Supramolecular Assemblies of an Azobenzene-Containing Surfactant with α-Cyclodextrin. Angew Chem Int Ed 46:2823–2826

Wei H, Zhuo RX, Zhang XZ (2013) Design and development of polymeric micelles with cleavable links for intracellular drug delivery. Prog Polym Sci 38:503–535

Weissleder R (2001) A clearer vision for in vivo imaging. Nature Biotechnol 19:316–317

Xiao W, Chen WH, Xu XD et al (2011) Design of a cellular-uptake-shielding "plug and play" template for photo controllable drug release. Adv Mater 23:3526–3530

Yan B, Boyer JC, Branda NR et al (2011) Near-infrared light-triggered dissociation of block copolymer micelles using upconverting nanoparticles. J Am Chem Soc 133:19714–19717

Yang YM, Shao Q, Deng RR et al (2012) In vitro and in vivo uncaging and bioluminescence imaging by using photocaged upconversion nanoparticles. Angew Chem Int Ed 51:3152–3159

Yang YM, Liu F, Liu XG et al (2013) NIR light controlled photorelease of siRNA and its targeted intracellular delivery based on upconversion nanoparticles. Nanoscale 5:231–238

Yesilyurt V, Ramireddy R, Thayumanava S (2011) Photoregulated release of noncovalent guests from dendritic amphiphilic nanocontainers. Angew Chem Int Ed 50:3038–3042

Part II
Nanocarrier Characterization and Function

Uptake and Intracellular Trafficking of Nanocarriers

Helene Andersen, Ladan Parhamifar and S. Moein Moghimi

Abstract Nanocarriers are widely used for delivery of therapeutic and modulatory agents to eukaryotic cells and specific intracellular compartments. Nanocarrier internalization proceeds via different routes and predominantly via clathrin-coated pits, lipid rafts/caveolae endocytosis and macropinocytosis/phagocytosis, depending on the cell type as well as the physicochemical properties of the nanocarrier. The intracellular fate of the nanocarrier is not only dependent on the mode of entry, but may also be modulated by prior surface modification of nanocarriers with organelle-specific localization ligands. This chapter discusses important methodological aspects for studying cellular uptake and intracellular trafficking of nanocarriers.

Keywords Nanocarriers · Endocytosis · Intracellular trafficking

Abbreviation

CME	Clathrin-mediated endocytosis
ER	Endoplasmic reticulum
GFP	Green fluorescence protein
MOC	Manders overlap coefficient
PALM	Photoactivated localization microscopy
PCC	Pearson co-localization coefficient
PEI	Polyethylenimine
STORM	Stochastic optical reconstruction microscopy
TEM	Transmission electron microscopy

H. Andersen · L. Parhamifar · S. Moein Moghimi (✉)
Nanomedicine Research Group, Centre for Pharmaceutical Nanotechnology and Nanotoxicology, Department of Pharmacy, Faculty of Health and Medical Sciences, University of Copenhagen, Universitetsparken 2, DK-2100 Copenhagen, Denmark
e-mail: moien.moghimi@sund.ku.dk

A. Prokop et al. (eds.), *Intracellular Delivery II*,
Fundamental Biomedical Technologies 7, DOI: 10.1007/978-94-017-8896-0_6,

1 Introduction

There have been numerous attempts to develop nanocarriers, which can not only improve drug solubilisation and delivery to different eukarocytic, but also target the desired intracellular compartments (Treuel et al. 2013). Several cellular barriers, however, need to be crossed before nanoparticles can reach their designated intracellular targets. Accordingly, a carrier may be designed to only enter specific cells in relevant tissues and to enter different cells through a specific endocytic pathway (Wang 2012). Likewise, intracellular transport mechanism of various nanocarriers have also received considerable attention (Treuel et al. 2013; Sakhrani and Padh 2013).

Various methods have been developed to study the mechanisms by which nanocarriers are internalized by cells of different origin and to follow their intracellular trafficking (Vercauteren et al. 2012). Here we describe the most studied endocytic pathways and discuss some of the major barriers for uptake and trafficking of nanocarriers. Advantages and disadvantages of commonly used methods for nanocarrier trafficking studies are also discussed.

2 Endocytosis

For nanocarriers to deliver and release their cargo at an intracellular target site the carriers need to enter the cell by crossing the plasma membrane. Most nanoparticles are believed to be internalized by endocytosis (Canton et al. 2012). Endocytosis is an energy-dependent process where particles are internalized in small vesicles. The mostly studied endocytic pathways are clathrin-mediated endocytosis, caveolae-mediated endocytosis and macropinocytosis, but more pathways have been identified that includes clathrin- and caveolae-independent endocytosis and phagocytosis (Doherty and McMahon 2009) (Fig. 1).

2.1 Clathrin-Mediated Endocytosis

Clathrin-mediated endocytosis (CME) is initiated at clathrin-coated pits in plasma membrane (Maxfield et al. 2004). Once the vesicle is formed in a dynamin-dependent manner (Hinshaw 2000), it is uncoated and followed by fusion with or maturation into early or recycling endosomes (Lemmon 2001; Ma et al. 2002; Maxfield et al. 2004). Clathrin is important for the initiation of invaginations at the membrane level and for the formation of the endocytic vesicles. Other assembly proteins are also involved in the formation of invaginations (Brodsky et al. 2001; Kirchhausen 1999).

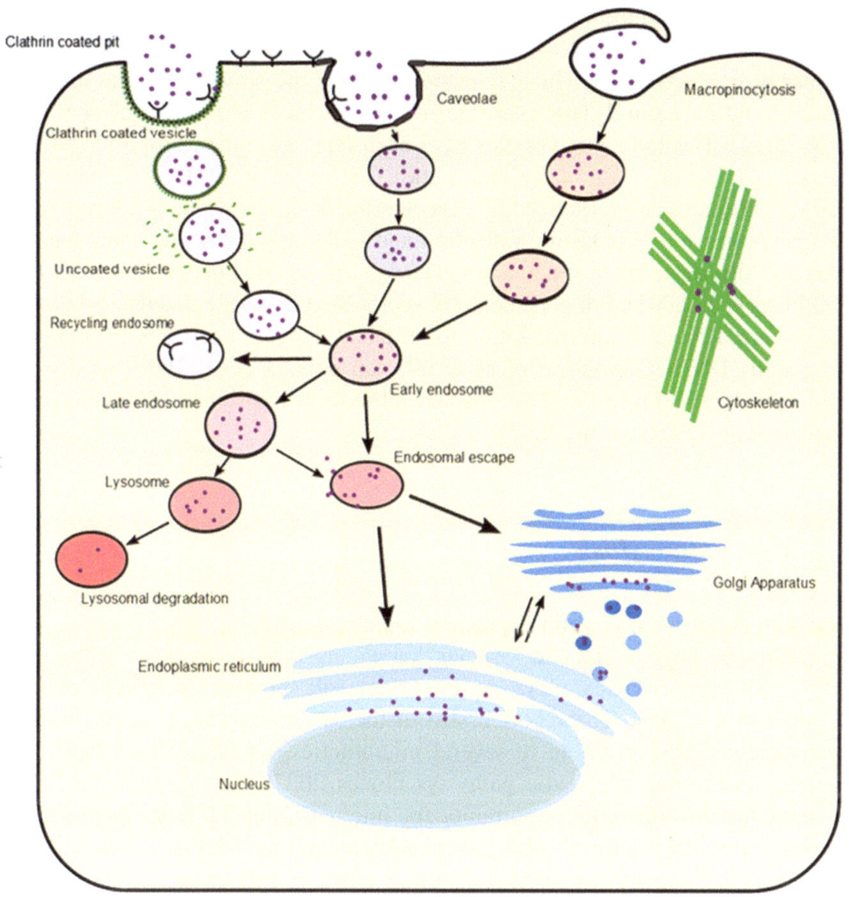

Fig. 1 Schematic overview of intracellular trafficking pathways

The clathrin coated vesicles have been demonstrated to be of various sizes ranging from 50 to 300 nm. Particles reported to be internalized by CME are generally up to 200 nm in diameter (Ehrlich et al. 2004). However, some studies have indicated that larger particles may also be taken up by CME (Moreno-Ruiz et al. 2009; Veiga et al. 2005). CME, however, is important when it comes to cellular uptake and sorting of nutrients, plasma membrane proteins and lipids (Conner et al. 2003).

2.2 Caveolae-Mediated Endocytosis

Caveolae-mediated endocytosis initiates from cholesterol-rich areas on the plasma membrane (Khalil et al. 2006a). Formation of the invaginations is dynamin- and actin-dependent and the vesicles formed are reported to be in 60–80 nm size range (Canton et al. 2012; Hinshaw 2000; Parton et al. 2007). Theses vesicles are coated primarily by caveolin-1, which is responsible for the structure of the endocytic vesicle, and caveolin-2. However, the role of the latter is still not clear (Canton et al. 2012). The endocytic vesicles have been reported to be transported either to the caveosome or to the early endosomes (Parton et al. 2003; Pelkmans et al. 2001). Nanocarriers internalized by caveolae-mediated endocytosis are reported to be transported to endoplasmic reticulum and Golgi apparatus and some nanocarriers are also detected in the nucleus (Harris et al. 2002; Luetterforst et al. 1999; Pelkmans et al. 2001). For gene delivery vectors, caveolae-mediated endocytosis is the preferred route of internalization for efficient delivery and transcription of the exogenous DNA (Gabrielson et al. 2009; van der Aa et al. 2007) as this may substantially avoid lysosomal routing and subsequent lysosomal degradation of the nucleic acids (Harris et al. 2002).

2.3 Macropinocytosis

Macropinocytosis is an actin-dependent membrane ruffling, which results in the formation of large endocytic vesicles known as macropinosomes (Kerr et al. 2009). Macropinocytosis plays an important role in cellular uptake of fluids (Conner et al. 2003). The reported size of the macropinosomes varies, but have been demonstrated to be up to several micrometres, which is much larger than vesicles from other endocytic pathways (Jones 2007). The size of the formed vesicles further allows an opportunity for internalization of larger particles that cannot enter via clathrin- and caveolae-mediated endocytosis. Indeed, some pathogens use macropinocytosis to facilitate entry to different cells (Mercer et al. 2009).

2.4 Factors Influencing the Endocytic Pathway

Several factors have been reported to influence nanoparticle endocytosis. These include nanoparticle physicochemical properties such as size, shape, surface charge and ligand coating (Canton et al. 2012; Zhao et al. 2011).

A study by Rejman et al. (2004) demonstrated that while uptake of particles up to 200 nm was dependent on microtubule, their corresponding larger particles (500 nm) were not. This study further indicated that smaller particles where mainly internalized by CME, whereas the larger particles entered cells through cholesterol-dependent endocytosis and to a lesser extent were routed to the endo-lysosomal pathway (Rejman et al. 2004). However, the literature is not consistent in correlating nanoparticle size with internalization pathway. For instance, some reports have indicated that particles up to 500 nm can be internalized by caveolae-

mediated endocytosis (Georgieva et al. 2011; Rejman et al. 2004), while others report that only particles up to 100 nm can be taken up by this route (Wang et al. 2009). These discrepancies may be due to different experimental setups, nanoparticle surface properties and the cell types examined.
Although the size of the nanoparticles can influence the endocytic pathway, many particles are not uniform in size and have a broad size distribution, which might render them to enter cells by different endocytic pathways, thus complicating the analysis of internalization routes. Others have reported that the shape of the nanoparticles determines the endocytic pathway for internalization (Gratton et al. 2008).

The uptake of particulate carriers of various sizes is also dependent on the cell type (Massignani et al. 2009; Nakai et al. 2003). Zauner et al. (2001) presented a study where they investigated the cellular uptake of microsphere of various sizes by different cell lines. Not all the cell lines were capable of internalizing particles above 1 μm, while all the tested cell lines could take up particles of 20 nm (Zauner et al. 2001). Furthermore, the net surface charge of the particles has also been suggested to influence the internalization pathway. Charged polystyrene or gold particles are reported to internalize to a higher degree than electrically neutral particles (Thorek et al. 2008; Villanueva et al. 2009). Also positively charged particles have been reported to be internalized easier than their corresponding anionic counterparts (Chen et al. 2011). This has been suggested to be due to ionic interaction between the positively charged particle and the negatively charged plasma membrane. In addition to the extent of the particle uptake, the surface charge of the particles may further regulate the pathway of internalisation (Harush-Frenkel et al. 2008; Zuidam et al. 2000). It has been demonstrated that dendrimers coupled to different functional groups enter the cell by different pathways dependent on the functional group. Dendrimers with amine or hydroxyl functional groups enter cells through both CME and caveolae-independent pathways, while dendrimers with functional carboxyl group mainly enter the cells via caveolae-mediated endocytosis (Perumal et al. 2008).

Georgieva et al. (2011) reported that 500 nm nanoparticles coated with either PEI or prion proteins were internalized through different endocytic pathways compared with uncoated particles. The coating of the particles also had an effect on the extent of particles that were co-localizing with recycling endosomes (Georgieva et al. 2011). The percentage of particles removed from cells by exocytosis may also dependent on the size of the particles. For example, the fraction of small 14 nm gold nanoparticles exocytosed was higher than gold particles with a diameter of 100 nm. Even though the percentage of particles exocytosed was reported to be cell-type dependent, the trend is the same and a higher fraction of the small particles being exocytosed (Chithrani et al. 2007). Not only the coating, but also the density of the molecules used for coating can play an important role in determining the endocytic pathway. Coating of liposomes with high density of the cell penetrating peptide R8, shifted the endocytic pathway towards macropinocytosis compared with liposomes coated with low density of the peptide, which were taken up by CME (Khalil et al. 2006b). A study by Li et al. (2011) indicated that PEI/DNA polyplexes are internalized by both clathrin- and caveolae-mediated endocytosis. However, when PEI was chemically modified with the natural polymer cyclodextran, the uptake mechanism was shifted toward caveolae-mediated endocytosis in HEK293T cells, thus demonstrating the role of charge and composition of the carrier in modulating the internalization route (Li et al. 2011).

3 Methods for Identifying Endocytic Pathways

Two methods are commonly used to study the pathways of nanocarriers internalization by cells. These include exclusion methodologies as well as co-localization studies, where overlap between fluorescently labelled particles and fluorescently labelled proteins are investigated by qualitative fluorescent microscopy.

3.1 Exclusion Methodologies

For exclusion determination, the nanocarrier should be fluorescently labelled so the uptake can be measured by flow cytometry or quantitative microscopy (Vercauteren et al. 2012). Several methods to exclude endocytic pathways have been demonstrated. For instance, siRNA can be used to decrease the expression of proteins required for the specific endocytic pathway in question (Zaki and Tirelli 2010). The use of siRNA may induce less cytotoxicity and be more specific than alternative methods used for the exclusion assays (Spoden et al. 2008). Caveolin-1 and dynamin-2 are among the successful proteins that have been down-regulated with siRNA (Huang et al. 2004). However, knockdown of proteins required for certain endocytic pathways by siRNA technology, may lead to up-regulation of some of the other endocytic pathways to compensate (Damke et al. 1995).

The most commonly used form of exclusion assays entail chemically inhibiting specific endocytic pathways. Table 1 lists some of the most commonly used inhibitors. The endocytosis inhibitors do not show exclusive specificity and many of the inhibitors of CME have been reported to cause reorganization of the actin skeleton (Ivanov 2008). Changes in the cytoskeleton might also affect other endocytic pathways, which are reported to be actin-dependent. Especially hypertonic sucrose as inhibitor for CME has been reported to be non-specific (Ivanov 2008). Also one of the mostly used inhibitors for caveolae-mediated endocytosis, methyl-β-cyclodextrin, which causes acute depletion of cholesterol, has been reported to significantly affect the cytoskeleton and also to inhibit CME and macropinocytosis (Kanzaki et al. 2002; Rodal et al. 1999). Inhibitors of macropinocytosis are also non-specific (Ivanov 2008). Even though, it is difficult to inhibit one endocytic pathway without affecting other pathways, the inhibitors are widely useful when studying internalization of nanocarriers.

Table 1 Chemical inhibitors commonly used for exclusion assays

Inhibitor	Pathway	Mechanism	References
Chlorpromazin	CME	Inhibit clathrin coated pit formation	Wang et al. (1993)
Potassium depletion	CME	Translocation of clathrin from plasma membrane to cytosol	Hansen et al. (1993)
Weak acids	CME	Inhibiting the budding of clathrin coated pits	Sandvig et al. (1987)
Nystatin	Raft/caveolae	Flattens the caveolae	Ros-Baro et al. (2001)
Lovastatin	Raft/caveolae	Inhibits HMG-CoA reductase	Liao and Laufs (2005)
Simvastatin	Raft/caveolae	Inhibits HMG-CoA reductase	Liao and Laufs (2005)
Pravastatin	Raft/caveolae	Inhibits HMG-CoA reductase	Liao and Laufs (2005)
Filipin	Raft/caveolae	Changes the structure of caveolae	Ros-Baro et al. (2001)
Okadaic acid	Raft/cacaolae	Removes caveolae from cell surface	Parton et al. (1994)
Amiloride and amiloride analogs	Macropinocytosis and phagocytosis (CME)	Inhibit sodium-proton exchange	Nakase et al. (2004), West et al. (1989)
Genistein	Caveolae	Inhibits tyrosine kinase and thereby phosphorylation of caveolin	Akiyama et al. (1987), Parton et al. (1994)
Nocodazol	Macropinocytosis, phagocytosis	Destabilize microtubuli	Mettlen et al. (2006)
Taxol	Macropinocytosis, phagocytosis	Microtubule stabilizing	de et al. (1981), Roberts et al. (1982)
Latrunculin	Macropinocytosis, phagocytosis—maybe unselective inhibitor	Blocks F-actin polymerization by binding to monomeric F-actin	Mettlen et al. (2006)
Cytochalasin	Macropinocytosis, phagocytosis—maybe unselective inhibitor	Blocks F-action polymerization by binding to the end of fast growing actin filament	Brett et al. (1984)
Wortmanin	Macropinocytosis and phagocytosis	Inhibits PI3K	Mettlen et al. (2006), Salh et al. (1998)
LY294002	Macropinocytosis and phagocytosis	Inhibits PI3K	(Salh et al. 1998)
Methyl- β -cyclodextrin	Raft/caveolae	Removes cholesterol from plasma membrane, flatten and induce mislocalization of caveolin	Kilsdonk et al. (1995), Westermann et al. (2005)

(continued)

Table 1 (continued)

Inhibitor	Pathway	Mechanism	References
Phenylarsine oxide	CME	Reduces energy stores inhibits protein synthesis	Bradley et al. (1993), Visser et al. (2004), Yumoto et al. (2006)
Cholesterol oxidase	Raft/caveolae	Removes caveolin from plasma membrane	Smart et al. (1994)
Hypertonic sucrose	CME	Prevents internalization of receptor bound peptide by inhibiting endosome formation	Daukas et al. (1985)
NH_4Cl	CME	Inhibits budding of clathrin coated pits	Sandvig et al. (1987)
Dansylcadaverine	Phagocytosis	Inhibits formation of phosohatidylcholine	Bradley et al. (1993), Garcia et al. (1982)
Rimantadine	Phagocytosis	Inhibits formation of phosohatidylcholine	Garcia et al. (1982)
Putrescine			Bradley et al. (1993)
N-ethylamaleimide	Caveolae/transcytosis		Schnitzer et al. (1995)

Some of the chemical inhibitors have been reported to induce cell type-dependent cytotoxicity. This can result in incorrect interpretation. Accordingly, appropriate concentration of inhibitors must be selected to optimize the conditions prior to testing (Vercauteren et al. 2010). When analyzing exclusion assays by flow cytometry, positive controls such as transferrin and folic acid, which are known to be internalized by clathrin- and caveolae-mediated endocytosis, respectively, should be included (Rothberg et al. 1990; Schmid 1997). However, flow cytometry may not necessarily distinguish between cells with internalized particles and those with surface bound particles (Ogris et al. 2000; Salvati et al. 2011).

In addition, genetically-modified cell lines have been used to exclude endocytic pathways (Ilina et al. 2012; Rejman et al. 2004). For example, Ilina et al. (2012) reported particle uptake studies with genetically-modified cells where dynamin-dependent and clathrin-mediated endocytic pathways were not operative. They further reported that the use of chemical inhibitors and the genetic blockage do not result in the same level of decrease in transfection efficiency (Ilina et al. 2012). These results further indicate that the interpretation of the inhibition assays can be difficult, and several methods should be considered before reaching a conclusion about the uptake mechanism of particles in specific cell lines. One of the problems with inhibition of endocytic pathways by mutant cell lines is the possibility of another pathway becoming increasingly active, which may aid the uptake of nanocarriers (Damke et al. 1995).

When identifying the endocytic pathway important for efficient delivery of pDNA by PEI, many contradicting reports have been published. Von Gersdorff et al. (2006) reported that clathrin-mediated endocytosis is required for efficient transfection, whereas Hufnagel et al. (2009) reported that fluid phase uptake plays an important role for efficient transfection with PEI/DNA polyplexes. However, it has also been demonstrated that blocking of caveolae-mediated endocytosis can inhibit expression of exogenous DNA (Gabrielson et al. 2009).

Uptake of fluorescently-labeled nanoparticles is usually measured by flow cytometry at various time points after addition of particles. Not all fluorophores are suitable for use in investigating cellular uptake through endocytosis. An example is fluorescein which has been reported to show decreased brightness when pH is below 9 and almost fully quenched at pH 6, which could result in inaccurate interpretation of the results when the particles have reached the acidic environment of endosomes and lysosomes (Geisow 1984).

3.2 Co-localization Microscopy Studies for Uptake Mechanism

Co-localization studies with fluorescence microscopy depend on the availability of fluorescently labeled nanocarriers and fluorescently labeled markers for the specific endocytic pathways that are being investigated (Vercauteren et al. 2012).

A commonly used method is to add the nanocarrier to living cells, which is followed by fixation at various time points after addition. After fixation, immunostaining of the specific endocytic marker is performed to detect possible co-localization between nanocarrier and the marker for the endocytic pathway. Several fixative agents have been reported to cause artifacts, which need to be considered when analyzing the data (Lundberg et al. 2001; Pearson 2007). To avoid such artifacts live-cell imaging may be applied.

To study co-localization by live-cell imaging, the marker protein has to be tagged with an appropriate fluorescent marker. This may include stably transfected cell lines where the marker proteins are coupled to a fluorescent tag such as the Green Fluorescence Protein (GFP) or the Red Fluorescent Protein. However, it is important to control the expression of the fluorescently-labeled marker proteins, since their overexpression might increase the activity of an endocytic pathway, which does not have the same activity in the parental cell line. Furthermore, co-localization between two molecules acquisition, pre-processing of the image and also sample preparation are very important steps to overcome possible artifacts (Abraham et al. 2010; Bolte et al. 2006).

3.3 Other Methods to Study Internalization

Other methods have also been used to analyze the internalization of particles in the nanometer range. Scanning electron microscopy, transmission electron microscopy and atomic force microscopy have been applied to detect reorganization of the plasma membrane (Georgieva et al. 2011; Leroueil et al. 2007). The re-organization of the membrane when endocytosis is initiated varies between the different pathways and the electron density at the invagination sites can be distinguished by electron microscopy thus making it possible to distinguish between endocytic pathways (Doherty and McMahon 2009). Dynamic surface enhanced Raman spectroscopy has also been used to investigate nanoparticle (e.g., gold nanoparticles and carbon nanotubes) internalization and intracellular transport pathways (Ando et al. 2011).

4 Intracellular Transport

From the point when the endocytic vesicles are detached from the membrane the carrier has to be transported to the targeted cellular compartment. Several mechanisms of how the cargo is transported from early endosomes through late endosomes to lysosomes or recycling endosomes have been suggested (Gruenberg et al. 2004; Luzio et al. 2007; Saftig et al. 2009). Both the possibility of gradual maturation and fusions of compartments have been suggested (Stoorvogel et al. 1991). It has further been demonstrated that endosomes and lysosomes can communicate via small vesicles in cell free systems, which indicate that it is not only gradual maturation that takes place (Vida et al. 1999).

Nanoparticles have been reported to accumulate in the peri-nuclear region shortly after internalization (Bae et al. 2012; Bregar et al. 2013; Fichter et al. 2013; Kim et al. 2012; Suh et al. 2003). The movement of carriers has in several cases been demonstrated to be dependent on polymerization of either actin or microtubule (Suh et al. 2003). To investigate the role of microtubule or actin in the cytosolic transport of nanocarriers chemical inhibitors of polymerization of tubulin or actin have been reported to interrupt the transport. PEI/DNA polyplexes have been reported to be dependent on actin during cellular uptake, whilst the intracellular trafficking is suggested to be microtubule dependent. In addition, a lactosylated derivate of PEI was reported to be transported along microtubule (Grosse et al. 2007). Similarly, liposomes have been reported to be transported along microtubule during cellular transport. However, it has been demonstrated that the transfection efficiency is enhanced both when disrupting or stabilizing microtubule, because liposomes no longer are transported to the lysosomes. R8-coated liposomes only reached the periphery of the nuclear when microtubule was not disrupted by nocodazole (Hasegawa et al. 2001; Khalil et al. 2008). On the other hand, PEI-mediated transfection was almost eliminated when disrupting microtubule with nocodazole treatment in COS-7 cells (Grosse et al. 2007). Also free plasmid DNA has been reported to be dependent of microtubule-mediated transport on the route to the nucleus (Vaughan and Dean 2006). Suh et al. (2003) reported that PEI/DNA polyplexes were transported along microtubule. The transport was directed towards the nucleus and carried out by motor protein along microtubule (Suh et al. 2003). However, it should be considered that when destabilizing the microtubule and actin network the cell morphology is highly affected. This can lead to changes in both cell uptake and intracellular transport of nanoparticles (Dos et al. 2011).

Caveolae-mediated endocytosis has in many cases been reported to be an endocytic pathway where delivery to endo-lysosomal pathway can be avoided and trafficking occurs from the caveosome to endoplasmic reticulum and Golgi apparatus (Badizadegan et al. 2000; Lencer et al. 1999). Currently, this transport mechanism has not been completely elucidated. Particles can also be transported between ER and Golgi apparatus. This event has been demonstrated to take advantage of the normal cellular trafficking between the two organelles. Glycofect particles has been demonstrated to be transported in COP I coated vesicle from Golgi to ER (Fichter et al. 2013).

Liposomes internalized by macropinocytosis have also been reported to avoid lysosomal degradation and to be present in the cytosol after endocytosis. When the liposomes where internalized by macropinocytosis the transfection efficiency was significantly higher than when internalized through CME (Khalil et al. 2006b).

4.1 Endosomal Escape

To avoid degradation in the acidic lysosome environment nanoparticles have to escape from early endosomes. One suggested mechanism is the proton-sponge hypothesis and applicable to cationic species. This hypothesis was first suggested in 1997 by Behr and colleagues. Here, polycationic species were suggested to induce influx of negatively charged ions (e.g., chlorides) into the endosomes. This influx is followed by water due to change in osmotic pressure. This may result in endosome swelling and eventual rupture, resulting in cytosolic release of the polycation (Behr 1997). A recent work, however, reported that the lysosomal pH does not change after treating cells with PEI, thus suggesting that the proton sponge hypothesis may be an inadequate explanation for endolysosomal release (Benjaminsen et al. 2013).

Polycations have previously been described to be able to cause perturbations of lipid bilayers, which may be an alternative mechanism by which nanocarriers may escape endo-lysosomes (Bieber et al. 2002). Addition of cationic lipids to lipid carriers has resulted in endosomal escape probably due to destabilization of the lipid bilayer of endosomes (Wasungu et al. 2006). Akita et al. (2010) reported that a pH-sensitive fusogenic peptide could be modified for endosomal escape. One example is the GALA peptide, which can destabilize the lipid bilayer resulting in successful endosomal escape (Akita et al. 2010).

4.2 Cytosolic Transport

After release from the endosomes nanoparticles must reach their designated intracellular targets. For example, if pDNA is delivered the release should happen in the periphery of the nucleus so the DNA is not degraded before it reaches the nucleoplasm (Lechardeur et al. 2005). For cytoplasmic dissociation biodegradable carriers have been designed. The biodegradable carriers will be gradually degraded in the cytoplasm resulting in release of the cargo (Gosselin et al. 2001).

To reach the cellular compartment where the cargo is going to be released an increasing amount of research has been put into targeting the organelle of interest (Sakhrani and Padh 2013). For instance, the ER retention signal consisting of four amino acids (KDEL) has been used to target ER. By targeting ER the endo-lysosomal pathway might be bypassed and lysosomal degradation can be avoided (Acharya et al. 2013; Wang et al. 2013). Mitochondria play an important role in initiation of apoptosis, regulating calcium homeostasis, removal of reactive oxygen species and in ATP synthesis. Hence, targeting of mitochondria through mitochondrial targeting signal peptide can be very important in treatment of various conditions (Heller et al. 2012; Sakhrani and Padh 2013).

4.3 Nuclear Entry

If the cargo of the nanocarrier requires transport to the nucleus, passing the nuclear envelope is a major barrier for efficient delivery. On the nuclear envelope, nuclear pore complexes are widely distributed (Grossman et al. 2012). Molecules that enter the nucleus have to pass through the channel of this complex. Small molecules up to 9 nm in diameter can freely diffuse through the pore channel (Paine et al. 1975), whereas molecules up to 39 nm are transported through the complex by an energy-dependent transport (Pante et al. 2002).

DNA needs to be transported through the nuclear pore complex, but the mechanism behind this still needs further investigation. Some reports suggest that only during mitosis DNA can enter the nucleus or that the transfection efficiency is dependent on the cell cycle, but microinjection of DNA in the cytoplasm has shown that the exogenous DNA can be transcribed in non-dividing cells (Brunner et al. 2000; Dean et al. 2005; Pollard et al. 1998).

It has been reported that a nuclear localization sequence added to the nanocarriers can increase the transport to the nucleus. Also a DNA sequence coding for a binding site for NFkB can increase the transport of DNA to the nucleus resulting in an increase in transfection (Breuzard et al. 2008; Dean et al. 2005).

Entry to the nucleus has been under intense investigation and many contradicting reports have been published. For example, PEI has been suggested to be transported to the nucleus, only to be present in the nucleus after mitosis, whereas other studies have indicated the inability of PEI to reach the nucleus (Brunner et al. 2000; Dowty et al. 1995; Larsen et al. 2012). Not many studies have evaluated the mechanism by which a nanocarrier enters the nucleus, but it has been suggested that polycations might rupture the membrane to access the nucleoplasm (Godbey et al. 1999). The size of the nanocarriers can also be a major obstacle in the transport through the NPC, since only molecules with a diameter of less than 40 nm can be transported through the nuclear pore complex and most nanocarriers have been reported to be above 100 nm (Pante et al. 2002).

5 Techniques to Investigate Cellular Localization of Nanocarriers

Investigations of intracellular trafficking of nanocarriers are in many cases carried out by fluorescence microscopy. However, it is important to interpret the microscopy data with care, since artifacts may be generated and the interpretation of microscopy images can be very subjective (Bolte et al. 2006).

Fixation of cells before staining is widely used, but as previously mentioned the fixation can induce artifacts. Live-cell imaging decreases the amount of artifact that usually can be observed after a fixation procedure. However, to be able to detect which organelles the nanocarrier is transported to it is necessary to use cell permanent organelle specific dyes or to label organelle specific proteins with GFP or similar species (Vercauteren et al. 2012).

There exist several types of fluorescence microscopes. When using confocal microscopes the end product is an image where only light from the focal plane is emitted. For wide-field microscopes also light from the out of focus planes is visible in the final image (Bolte et al. 2006). Images from wide-field microscopes require deconvolution of the images to eliminate the out of focus blur. Especially, if quantifying the co-localization of nanocarriers with organelle markers wide-field images need to be deconvolved but it can also be an advantage to deconvolve confocal images (Bolte et al. 2006; Scriven et al. 2008). Deconvolution requires an accurate point-spread-function, which can be determined with use of various algorithms calculating point-spread-function (PSF) in various ways. The PSF should also be determined for the optical conditions used in experiments (McNally et al. 1999). Co-localization can also be determined visually. An overlap of two channels where the detected light is represented in two different colors will result in color change if there is an overlap between the fluorophores. This, however, is a subjective way of determining overlaps between channels and is only qualitative (Bolte et al. 2006; Dunn et al. 2011). In recent years quantitative algorithms have been developed to calculate the degree of co-localization.

The degree of co-localization can be calculated according to various coefficients. The most commonly used are the Pearson co-localization coefficient (PCC) and the Manders overlap coefficient (MOC). The PCC identify co-localization when a pixel from two different has the same intensity, whereas MOC detect overlap when fluorescence in a pixel from two channels is detected (Bolte et al. 2006; Manders et al. 1992). When using PCC it is important that the same instrument settings are used every time and when imaging nanocarriers their might still be a different in intensity in both the nanocarrier and also in the protein labeled from cell to cell and especially between experiments (Dunn et al. 2011).

For live-cell imaging MOC is the most commonly used coefficient since intensity from organelle markers and the nanocarrier itself can vary between experiments (Dunn et al. 2011). It is important when using MOC that a background threshold is set to avoid false positive read-out, which is a major risk when calculating MOC to analyze the degree of overlap (Bolte et al. 2006; Dunn et al. 2011; Fletcher et al. 2010).

Suh et al. (2003) have performed real-time multiple particle tracking when reporting that intracellular transport of PEI/DNA polyplexes to the periphery of the nuclei is dependent on motor proteins along microtubule (Suh et al. 2003). The same group also reported that polyplexes were actively transported to the early and late endosomes and transport between the endosomes was also observed when performing real-time multiple particle tracking (Suh et al. 2012).

Single-particle tracking is another powerful tool to study the route by which one single nanocarrier is transported through the cell (Ruthardt and Brauchle 2010). This method requires a fast and sensitive camera and the efficient laser excitation. Single-particle tracking is following the trajectory for each separately visible particle. Co-localization of particles are defined as significant correlation between the trajectories (Braeckmans et al. 2010). This method provides the possibility to investigate if the single particles are transported in the same direction and to the same organelles and will provide new insights as how a population of nanoparticles behaves when added to cells (Ruthardt and Brauchle 2010). Other microscopy methods have been applied to study the intracellular distribution of nanocarriers. Among them Raman microscopy and Fluorescence recovery after photo-bleaching have been used to investigate interaction between nanoparticles and mitochondria (Chernenko et al. 2009; Hemmerich et al. 2013; Phizicky et al. 2003).

5.1 High-Resolution Fluorescence Microscopy

Many of the traditional microscopes have a detection limit around 200 nm in the XY-plane and 500 nm in the Z-plane for optimal settings. This gives a problem when imaging nanocarriers with few hundred nanometers in diameter and single particles are not possible to image (Bolte et al. 2006).

An optimized hardware is necessary to perform high-resolution microscopy. Each step from the laser excitation to the camera acquisition has to be optimized compared with conventional microscopy. When performing high resolution microscopy on live cells there is a risk of inducing cell death due to high laser power. Photoactivated localization microscopy (PALM) and stochastic optical reconstruction microscopy (STORM) are two of the most described high resolution microscopy techniques (Henriques et al. 2011). The super resolution techniques create a dataset by excitation of few fluorophores at a time and an image is the reconstructed based on the fluorescent particles. PALM and STORM are two very similar setup but they vary in the fluorophores that are used for the experiments. PALM uses genetically attached fluorophores such as GFP coupled proteins, whereas STORM take advantages of fluorescent dyes (Henriques et al. 2011).

5.2 Transmission Electron Microscopy

Transmission electron microscopy (TEM) is also widely used for co-localization studies. Gold labeled antibodies against a specific antigen are a specific method to investigate the cellular localization of the protein of interest. By TEM it is possible to distinguish between various organelles and cellular compartment and the endocytic vesicles (Georgieva et al. 2011). However the interpretation of electron microscopy can be difficult and may generate artifacts during sample preparation. Sample preparation for electron microscopy includes embedding, slicing and staining of the sample before imaging and this is very time consuming compared to fluorescent imaging (Henriques et al. 2011). Nanocarriers can often be visualized on their own by negative staining.

5.3 Cellular Fractionation

Cellular fractionation has also been used to study the intracellular localization of nanocarriers. Centrifugation steps are carried out to separate the various organelles (He et al. 2013; Shi et al. 2013). For instance, Shi et al. (2013) labeled their polyplexes with [^{3}H]-DNA and [^{14}C]-PEI before adding the polyplexes to the cell growth medium. They separated the lysed cells into nuclear, light mitochondria, heavy mitochondria, microsomes and cytosolic fractions. Quantification of the radioactivity indicated that a large fraction of both PEI and DNA was detected in the nuclei fraction. This method, however, was not sufficient to demonstrate if polyplexes entered the nucleus intact or separately as DNA and/or PEI (Shi et al. 2013). A major disadvantage of this method is the risk of contaminating the different cellular compartments with parts from the other fractions.

6 Conclusion

Uptake and intracellular trafficking of nanocarriers is currently under intense investigation. Many factors can influence the endocytic pathways and need to be taken into consideration when designing assays and analyzing the data. Quantification of co-localization between nanocarriers and the marker protein can be determined in a very subjective manner by looking at the overlay images from two or more channels. However, several algorithms have been developed to give a more objective quantification of the co-localization. Also exclusion assays when studying the uptake of nanocarriers can be difficult to interpret with certainty, since inhibiting the selected pathways can result in unwanted cellular responses such as toxic responses or undesired increased activation of alternative pathways.

Acknowledgment Financial support by the Danish Agency for Science, Technology and Innovation (Det Frie Forskningsrå for Teknologi og Produktion) reference 274-08-0534 is gratefully acknowledged.

References

Abraham T, Allan SE, Levings MK (2010) Deconvolution and chromatic aberration corrections in quantifying colocalization of a transcription factor in three-dimensional cellular space. Micron 41(6):633–640

Acharya S, Hill RA (2013) High efficacy gold-KDEL peptide-siRNA nanoconstruct-mediated transfection in C2C12 myoblasts and myotubes. Nanomedicine doi: 10.1016/j.nano.2013.07.015.doi epub ahead of print

Ando J, Fujita K, Smith NI, Kawata S (2011) Dynamic SERS imaging of cellular transport pathways with endocytosed gold nanoparticles. Nano Lett 11:5344–5348

Akita H, Kogure K, Moriguchi R et al (2010) Nanoparticles for ex vivo siRNA delivery to dendritic cells for cancer vaccines: programmed endosomal escape and dissociation. J Control Rel 143(3):311–317

Akiyama T, Ishida J, Nakagawa S et al (1987) Genistein, a specific inhibitor of tyrosine-specific protein kinases. J Biol Chem 262(12):5592–5595

Badizadegan K, Wolf AA, Rodighiero C et al (2000) Floating cholera toxin into epithelial cells: functional association with caveolae-like detergent-insoluble membrane microdomains. Int J Med Microbiol 290(4–5):403–408

Bae YM, Park YI, Nam SH et al (2012) Endocytosis, intracellular transport, and exocytosis of lanthanide-doped upconverting nanoparticles in single living cells. Biomaterials 33(35):9080–9086

Behr JP (1997) The proton sponge: a trick to enter cells the viruses did not exploit. Chimia 51(1–2):34–36

Benjaminsen RV, Mattebjerg MA, Henriksen JR et al (2013) The possible "proton sponge" effect of polyethylenimine (PEI) does not include change in lysosomal pH. Mol Ther 21(1):149–157

Bieber T, Meissner W, Kostin S et al (2002) Intracellular route and transcriptional competence of polyethylenimine-DNA complexes. J Control Release 82(2–3):441–454

Bolte S, Cordelieres FP (2006) A guided tour into subcellular colocalization analysis in light microscopy. J Microsc 224(Pt 3):213–232

Bradley JR, Johnson DR, Pober JS (1993) Four different classes of inhibitors of receptor-mediated endocytosis decrease tumor necrosis factor-induced gene expression in human endothelial cells. J Immunol 150(12):5544–5555

Braeckmans K, Buyens K, Naeye B et al (2010) Advanced fluorescence microscopy methods illuminate the transfection pathway of nucleic acid nanoparticles. J Control Rel 148(1):69–74

Bregar VB, Lojk J, Sustar V et al (2013) Visualization of internalization of functionalized cobalt ferrite nanoparticles and their intracellular fate. Int J Nanomed 8:919–931

Brett JG, Godman GC (1984) Macrovacuolation induced by cytochalasin: its relation to the cytoskeleton; morphological and cytochemical observations. Tissue Cell 16(3):311–324

Breuzard G, Tertil M, Goncalves C et al (2008) Nuclear delivery of NFkappaB-assisted DNA/polymer complexes: plasmid DNA quantitation by confocal laser scanning microscopy and evidence of nuclear polyplexes by FRET imaging. Nucleic Acids Res 36(12):e71

Brodsky FM, Chen CY, Knuehl C et al (2001) Biological basket weaving: formation and function of clathrin-coated vesicles. Annu Rev Cell Dev Biol 17:517–568

Brunner S, Sauer T, Carotta S et al (2000) Cell cycle dependence of gene transfer by lipoplex, polyplex and recombinant adenovirus. Gene Ther 7(5):401–407

Canton I, Battaglia G (2012) Endocytosis at the nanoscale. Chem Soc Rev 41(7):2718–2739

Chen L, Mccrate JM, Lee JC et al (2011) The role of surface charge on the uptake and biocompatibility of hydroxyapatite nanoparticles with osteoblast cells. Nanotechnology 22(10):105708

Chernenko T, Matthaus C, Milane L et al (2009) Label-free Raman spectral imaging of intracellular delivery and degradation of polymeric nanoparticle systems. ACS Nano 3(11):3552–3559

Chithrani BD, Chan WC (2007) Elucidating the mechanism of cellular uptake and removal of protein-coated gold nanoparticles of different sizes and shapes. Nano Lett 7(6):1542–1550

Conner SD, Schmid SL (2003) Regulated portals of entry into the cell. Nature 422(6927):37–44

Damke H, Baba T, van der Bliek AM et al (1995) Clathrin-independent pinocytosis is induced in cells overexpressing a temperature-sensitive mutant of dynamin. J Cell Biol 131(1):69–80

Daukas G, Zigmond SH (1985) Inhibition of receptor-mediated but not fluid-phase endocytosis in polymorphonuclear leukocytes. J Cell Biol 101(5 Pt 1):1673–1679

Dean DA, Strong DD, Zimmer WE (2005) Nuclear entry of nonviral vectors. Gene Ther 12(11):881–890

Doherty GJ, McMahon HT (2009) Mechanisms of endocytosis. Annu Rev Biochem 78:857–902

Dos ST, Varela J, Lynch I et al (2011) Effects of transport inhibitors on the cellular uptake of carboxylated polystyrene nanoparticles in different cell lines. PLoS ONE 6(9):e24438

Dowty ME, Williams P, Zhang G et al (1995) Plasmid DNA entry into postmitotic nuclei of primary rat myotubes. Proc Natl Acad Sci USA 92(10):4572–4576
Dunn KW, Kamocka MM, McDonald JH (2011) A practical guide to evaluating colocalization in biological microscopy. Am J Physiol Cell Physiol 300(4):C723–C742
Ehrlich M, Boll W, van Oijen A et al (2004) Endocytosis by random initiation and stabilization of clathrin-coated pits. Cell 118(5):591–605
Fichter KM, Ingle NP, McLendon PM et al (2013) Polymeric nucleic acid vehicles exploit active interorganelle trafficking mechanisms. ACS Nano 7(1):347–364
Fletcher PA, Scriven DR, Schulson MN et al (2010) Multi-image colocalization and its statistical significance. Biophys J 99(6):1996–2005
Gabrielson NP, Pack DW (2009) Efficient polyethylenimine-mediated gene delivery proceeds via a caveolar pathway in HeLa cells. J Control Rel 136(1):54–61
Garcia GM, Alonso F, Chiva VA et al (1982) Phospholipid turnover during phagocytosis in human polymorphonuclear leucocytes. Biochem J 206(1):67–72
Geisow MJ (1984) Fluorescein conjugates as indicators of subcellular Ph—a critical-evaluation. Exp Cell Res 150(1):29–35
Georgieva JV, Kalicharan D, Couraud PO et al (2011) Surface characteristics of nanoparticles determine their intracellular fate in and processing by human blood-brain barrier endothelial cells in vitro. Mol Ther 19(2):318–325
Godbey WT, Wu KK, Mikos AG (1999) Tracking the intracellular path of poly(ethylenimine)/DNA complexes for gene delivery. Proc Natl Acad Sci USA 96(9):5177–5181
Gosselin MA, Guo W, Lee RJ (2001) Efficient gene transfer using reversibly cross-linked low molecular weight polyethylenimine. Bioconjug Chem 12(6):989–994
Gratton SE, Ropp PA, Pohlhaus PD et al (2008) The effect of particle design on cellular internalization pathways. Proc Natl Acad Sci USA 105(33):11613–11618
Grosse S, Aron Y, Thevenot G et al (2007) Cytoskeletal involvement in the cellular trafficking of plasmid/PEI derivative complexes. J Control Rel 122(1):111–117
Grossman E, Medalia O, Zwerger M (2012) Functional architecture of the nuclear pore complex. Annu Rev Biophys 41:557–584
Gruenberg J, Stenmark H (2004) The biogenesis of multivesicular endosomes. Nat Rev Mol Cell Biol 5(4):317–323
Hansen SH, Sandvig K, van Deurs B (1993) Clathrin and HA2 adaptors: effects of potassium depletion, hypertonic medium, and cytosol acidification. J Cell Biol 121(1):61–72
Harris J, Werling D, Hope JC et al (2002) Caveolae and caveolin in immune cells: distribution and functions. Trends Immunol 23(3):158–164
Harush-Frenkel O, Rozentur E, Benita S et al (2008) Surface charge of nanoparticles determines their endocytic and transcytotic pathway in polarized MDCK cells. Biomacromolecules 9(2):435–443
Hasegawa S, Hirashima N, Nakanishi M (2001) Microtubule involvement in the intracellular dynamics for gene transfection mediated by cationic liposomes. Gene Ther 8(21):1669–1673
He B, Lin P, Jia Z et al (2013) The transport mechanisms of polymer nanoparticles in Caco-2 epithelial cells. Biomaterials 34(25):6082–6098
Heller A, Brockhoff G, Goepferich A (2012) Targeting drugs to mitochondria. Eur J Pharm Biopharm 82(1):1–18
Hemmerich PH, von Mikecz AH (2013) Defining the subcellular interface of nanoparticles by live-cell imaging. PLoS ONE 8(4):e62018
Henriques R, Griffiths C, Hesper RE et al (2011) PALM and STORM: unlocking live-cell super-resolution. Biopolymers 95(5):322–331
Hinshaw JE (2000) Dynamin and its role in membrane fission. Annu Rev Cell Dev Biol 16:483–519
Huang F, Khvorova A, Marshall W et al (2004) Analysis of clathrin-mediated endocytosis of epidermal growth factor receptor by RNA interference. J Biol Chem 279(16):16657–16661
Hufnagel H, Hakim P, Lima A et al (2009) Fluid phase endocytosis contributes to transfection of DNA by PEI-25. Mol Ther 17(8):1411–1417

Ilina P, Hyvonen Z, Saura M et al (2012) Genetic blockage of endocytic pathways reveals differences in the intracellular processing of non-viral gene delivery systems. J Control Release 163(3):385–395
Ivanov AI (2008) Pharmacological inhibition of endocytic pathways: is it specific enough to be useful? Method Mol Biol 440:15–33
Jones AT (2007) Macropinocytosis: searching for an endocytic identity and role in the uptake of cell penetrating peptides. J Cell Mol Med 11(4):670–684
Kanzaki M, Pessin JE (2002) Caveolin-associated filamentous actin (Cav-actin) defines a novel F-actin structure in adipocytes. J Biol Chem 277(29):25867–25869
Kerr MC, Teasdale RD (2009) Defining macropinocytosis. Traffic 10(4):364–371
Khalil IA, Kogure K, Akita H et al (2006a) Uptake pathways and subsequent intracellular trafficking in nonviral gene delivery. Pharmacol Rev 58(1):32–45
Khalil IA, Kogure K, Futaki S et al (2006b) High density of octaarginine stimulates macropinocytosis leading to efficient intracellular trafficking for gene expression. J Biol Chem 281(6):3544–3551
Khalil IA, Kogure K, Futaki S et al (2008) Octaarginine-modified liposomes: enhanced cellular uptake and controlled intracellular trafficking. Int J Pharm 354(1–2):39–48
Kilsdonk EP, Yancey PG, Stoudt GW et al (1995) Cellular cholesterol efflux mediated by cyclodextrins. J Biol Chem 270(29):17250–17256
Kim AJ, Boylan NJ, Suk JS et al (2012) Non-degradative intracellular trafficking of highly compacted polymeric DNA nanoparticles. J Control Rel 158(1):102–107
Kirchhausen T (1999) Adaptors for clathrin-mediated traffic. Annu Rev Cell Dev Biol 15:705–732
Larsen JD, Ross NL, Sullivan MO (2012) Requirements for the nuclear entry of polyplexes and nanoparticles during mitosis. J Gene Med 14(9–10):580–589
Lechardeur D, Verkman AS, Lukacs GL (2005) Intracellular routing of plasmid DNA during non-viral gene transfer. Adv Drug Deliv Rev 57(5):755–767
Lemmon SK (2001) Clathrin uncoating: Auxilin comes to life. Curr Biol 11(2):R49–R52
Lencer WI, Hirst TR, Holmes RK (1999) Membrane traffic and the cellular uptake of cholera toxin. Biochim Biophys Acta 1450(3):177–190
Leroueil PR, Hong S, Mecke A et al (2007) Nanoparticle interaction with biological membranes: does nanotechnology present a Janus face? Acc Chem Res 40(5):335–342
Li W, Chen L, Huang Z et al (2011) The influence of cyclodextrin modification on cellular uptake and transfection efficiency of polyplexes. Org Biomol Chem 9(22):7799–7806
Liao JK, Laufs U (2005) Pleiotropic effects of statins. Annu Rev Pharmacol Toxicol 45:89–118
Luetterforst R, Stang E, Zorzi N et al (1999) Molecular characterization of caveolin association with the golgi complex: identification of a cis-golgi targeting domain in the caveolin molecule. J Cell Biol 145(7):1443–1459
Lundberg M, Johansson M (2001) Is VP22 nuclear homing an artifact? Nat Biotechnol 19(8):713–714
Luzio JP, Pryor PR, Bright NA (2007) Lysosomes: fusion and function. Nat Rev Mol Cell Biol 8(8):622–632
Ma Y, Greener T, Pacold ME et al (2002) Identification of domain required for catalytic activity of Auxilin in supporting clathrin uncoating by Hsc70. J Biol Chem 277(51):49267–49274
Manders EM, Stap J, Brakenhoff GJ et al (1992) Dynamics of three-dimensional replication patterns during the S-phase, analysed by double labelling of DNA and confocal microscopy. J Cell Sci 103(3):857–862
Massignani M, LoPresti C, Blanazs A et al (2009) Controlling cellular uptake by surface chemistry, size, and surface topology at the nanoscale. Small 5(21):2424–2432
Maxfield FR, McGraw TE (2004) Endocytic recycling. Nat Rev Mol Cell Biol 5(2):121–132
McNally JG, Karpova T, Cooper J et al (1999) Three-dimensional imaging by deconvolution microscopy. Methods 19(3):373–385
Mercer J, Helenius A (2009) Virus entry by macropinocytosis. Nat Cell Biol 11(5):510–520

Mettlen M, Platek A, Van Der Smissen P et al (2006) Src triggers circular ruffling and macropinocytosis at the apical surface of polarized MDCK cells. Traffic 7(5):589–603
Moreno-Ruiz E, Galan-Diez M, Zhu W et al (2009) Candida albicans internalization by host cells is mediated by a clathrin-dependent mechanism. Cell Microbiol 11(8):1179–1189
Nakai T, Kanamori T, Sando S et al (2003) Remarkably size-regulated cell invasion by artificial viruses. Saccharide-dependent self-aggregation of glycoviruses and its consequences in glycoviral gene delivery. J Am Chem Soc 125(28):8465–8475
Nakase I, Niwa M, Takeuchi T et al (2004) Cellular uptake of arginine-rich peptides: roles for macropinocytosis and actin rearrangement. Mol Ther 10(6):1011–1022
Ogris M, Wagner E, Steinlein P (2000) A versatile assay to study cellular uptake of gene transfer complexes by flow cytometry. Biochim Biophys Acta 1474(2):237–243
Paine PL, Moore LC, Horowitz SB (1975) Nuclear envelope permeability. Nature 254(5496):109–114
Pante N, Kann M (2002) Nuclear pore complex is able to transport macromolecules with diameters of about 39 nm. Mol Biol Cell 13(2):425–434
Parton RG, Joggerst B, Simons K (1994) Regulated internalization of caveolae. J Cell Biol 127(5):1199–1215
Parton RG, Richards AA (2003) Lipid rafts and caveolae as portals for endocytosis: new insights and common mechanisms. Traffic 4(11):724–738
Parton RG, Simons K (2007) The multiple faces of caveolae. Nat Rev Mol Cell Biol 8(3):185–194
Pearson H (2007) The good, the bad and the ugly. Nature 447(7141):138–140
Pelkmans L, Kartenbeck J, Helenius A (2001) Caveolar endocytosis of simian virus 40 reveals a new two-step vesicular-transport pathway to the ER. Nat Cell Biol 3(5):473–483
Perumal OP, Inapagolla R, Kannan S et al (2008) The effect of surface functionality on cellular trafficking of dendrimers. Biomaterials 29(24–25):3469–3476
Phizicky E, Bastiaens PI, Zhu H et al (2003) Protein analysis on a proteomic scale. Nature 422(6928):208–215
Pollard H, Remy JS, Loussouarn G et al (1998) Polyethylenimine but not cationic lipids promotes transgene delivery to the nucleus in mammalian cells. J Biol Chem 273(13):7507–7511
Rejman J, Oberle V, Zuhorn IS et al (2004) Size-dependent internalization of particles via the pathways of clathrin- and caveolae-mediated endocytosis. Biochem J 377(Pt 1):159–169
Roberts RL, Nath J, Friedman MM et al (1982) Effects of taxol on human neutrophils. J Immunol 129(5):2134–2141
Rodal SK, Skretting G, Garred O et al (1999) Extraction of cholesterol with methyl-beta-cyclodextrin perturbs formation of clathrin-coated endocytic vesicles. Mol Biol Cell 10(4):961–974
Ros-Baro A, Lopez-Iglesias C, Peiro S et al (2001) Lipid rafts are required for GLUT4 internalization in adipose cells. Proc Natl Acad Sci USA 98(21):12050–12055
Rothberg KG, Ying YS, Kolhouse JF et al (1990) The glycophospholipid-linked folate receptor internalizes folate without entering the clathrin-coated pit endocytic pathway. J Cell Biol 110(3):637–649
Ruthardt N, Brauchle C (2010) Visualizing uptake and intracellular trafficking of gene carriers by single-particle tracking. Top Curr Chem 296:283–304
Saftig P, Klumperman J (2009) Lysosome biogenesis and lysosomal membrane proteins: trafficking meets function. Nat Rev Mol Cell Biol 10(9):623–635
Sakhrani NM, Padh H (2013) Organelle targeting: third level of drug targeting. Drug Des Dev Ther 7:585–599
Salh B, Wagey R, Marotta A et al (1998) Activation of phosphatidylinositol 3-kinase, protein kinase B, and p70 S6 kinases in lipopolysaccharide-stimulated Raw 264.7 cells: differential effects of rapamycin, Ly294002, and wortmannin on nitric oxide production. J Immunol 161(12):6947–6954

Salvati A, Aberg C, dos Santos T et al (2011) Experimental and theoretical comparison of intracellular import of polymeric nanoparticles and small molecules: toward models of uptake kinetics. Nanomedicine 7(6):818–826
Sandvig K, Olsnes S, Petersen OW et al (1987) Acidification of the cytosol inhibits endocytosis from coated pits. J Cell Biol 105(2):679–689
Schmid SL (1997) Clathrin-coated vesicle formation and protein sorting: an integrated process. Annu Rev Biochem 66:511–548
Schnitzer JE, Allard J, Oh P (1995) NEM inhibits transcytosis, endocytosis, and capillary permeability: implication of caveolae fusion in endothelia. Am J Physiol 268(2):H48–H55
Scriven DR, Lynch RM, Moore ED (2008) Image acquisition for colocalization using optical microscopy. Am J Physiol Cell Physiol 294(5):C1119–C1122
Shi J, Chou B, Choi JL et al (2013) Investigation of polyethylenimine/DNA polyplex transfection to cultured cells using radiolabeling and subcellular fractionation methods. Mol Pharm 10(6):2145–2156
Smart EJ, Ying YS, Conrad PA et al (1994) Caveolin moves from caveolae to the golgi apparatus in response to cholesterol oxidation. J Cell Biol 127(5):1185–1197
Spoden G, Freitag K, Husmann M et al (2008) Clathrin- and caveolin-independent entry of human papillomavirus type 16–involvement of tetraspanin-enriched microdomains (TEMs). PLoS ONE 3(10):e3313
Stoorvogel W, Strous GJ, Geuze HJ et al (1991) Late endosomes derive from early endosomes by maturation. Cell 65(3):417–427
Suh J, An Y, Tang BC et al (2012) Real-time gene delivery vector tracking in the endo-lysosomal pathway of live cells. Microsc Res Tech 75(5):691–697
Suh J, Wirtz D, Hanes J (2003) Efficient active transport of gene nanocarriers to the cell nucleus. Proc Natl Acad Sci USA 100(7):3878–3882
Thorek DL, Tsourkas A (2008) Size, charge and concentration dependent uptake of iron oxide particles by non-phagocytic cells. Biomaterials 29(26):3583–3590
Treuel L, Jiang X, Nienhaus GU (2013) New views on cellular uptake and trafficking of manufactured nanoparticles. J Royal Soc Interface 10(82):20120939
van der Aa MA, Huth US, Hafele SY et al (2007) Cellular uptake of cationic polymer-DNA complexes via caveolae plays a pivotal role in gene transfection in COS-7 cells. Pharm Res 24(8):1590–1598
Vaughan EE, Dean DA (2006) Intracellular trafficking of plasmids during transfection is mediated by microtubules. Mol Ther 13(2):422–428
Veiga E, Cossart P (2005) Listeria hijacks the clathrin-dependent endocytic machinery to invade mammalian cells. Nat Cell Biol 7(9):894–900
Vercauteren D, Rejman J, Martens TF et al (2012) On the cellular processing of non-viral nanomedicines for nucleic acid delivery: mechanisms and methods. J Control Rel 161(2):566–581
Vercauteren D, Vandenbroucke RE, Jones AT et al (2010) The use of inhibitors to study endocytic pathways of gene carriers: optimization and pitfalls. Mol Ther 18(3):561–569
Vida T, Gerhardt B (1999) A cell-free assay allows reconstitution of Vps33p-dependent transport to the yeast vacuole/lysosome. J Cell Biol 146(1):85–98
Villanueva A, Canete M, Roca AG et al (2009) The influence of surface functionalization on the enhanced internalization of magnetic nanoparticles in cancer cells. Nanotechnology 20(11):115103
Visser CC, Stevanovic S, Heleen VL et al (2004) Validation of the transferrin receptor for drug targeting to brain capillary endothelial cells in vitro. J Drug Target 12(3):145–150
von Gersdorff K, Sanders NN, Vandenbroucke R et al (2006) The internalization route resulting in successful gene expression depends on both cell line and polyethylenimine polyplex type. Mol Ther 14(5):745–753
Wang AZ (2012) Nanoparticle drug delivery: focusing on the therapeutic cargo. Nanomedicine (Lond.) 7(10):1463–1465

Wang G, Norton AS, Pokharel D et al (2013) KDEL peptide gold nanoconstructs: promising nanoplatforms for drug delivery. Nanomedicine 9(3):366–374
Wang LH, Rothberg KG, Anderson RG (1993) Mis-assembly of clathrin lattices on endosomes reveals a regulatory switch for coated pit formation. J Cell Biol 123(5):1107–1117
Wang Z, Tiruppathi C, Minshall RD et al (2009) Size and dynamics of caveolae studied using nanoparticles in living endothelial cells. ACS Nano 3(12):4110–4116
Wasungu L, Hoekstra D (2006) Cationic lipids, lipoplexes and intracellular delivery of genes. J Control Rel 116(2):255–264
West MA, Bretscher MS, Watts C (1989) Distinct endocytotic pathways in epidermal growth factor-stimulated human carcinoma A431 cells. J Cell Biol 109(1):2731–2739
Westermann M, Steiniger F, Richter W (2005) Belt-like localisation of caveolin in deep caveolae and its re-distribution after cholesterol depletion. Histochem Cell Biol 123(6):613–620
Yumoto R, Nishikawa H, Okamoto M et al (2006) Clathrin-mediated endocytosis of FITC-albumin in alveolar type II epithelial cell line RLE-6TN. Am J Physiol Lung Cell Mol Physiol 290(5):L946–L955
Zaki NM, Tirelli N (2010) Gateways for the intracellular access of nanocarriers: a review of receptor-mediated endocytosis mechanisms and of strategies in receptor targeting. Expert Opin Drug Deliv 7(8):895–913
Zauner W, Farrow NA, Haines AM (2001) In vitro uptake of polystyrene microspheres: effect of particle size, cell line and cell density. J Control Rel 71(1):39–51
Zhao F, Zhao Y, Liu Y et al (2011) Cellular uptake, intracellular trafficking, and cytotoxicity of nanomaterials. Small 7(10):1322–1337
Zuidam NJ, Posthuma G, de Vries ET et al (2000) Effects of physicochemical characteristics of poly(2-(dimethylamino)ethyl methacrylate)-based polyplexes on cellular association and internalization. J Drug Target 8(1):51–66

Mucus as Physiological Barrier to Intracellular Delivery

Eleonore Fröhlich and Eva Roblegg

Abstract Nanoparticle-based delivery systems are versatile tools to improve drug delivery because size, surface charge and surface hydrophobicity can be varied to meet specific physiological requirements. While small size, hydrophobicity, and positive surface charge, in general, improve passage through the plasma membrane and enhance intracellular delivery, these particle parameters may not be optimal if a mucus layer covers the target cells. Many target cells for drug delivery in the gastrointestinal tract but also in the respiratory tract are covered by mucus. The review describes the different compositions of mucus-covered epithelia of the oro-gastrointestinal and the respiratory tract and strategies to reach target cells beyond the mucus layer by the design of appropriate nanocarriers.

Keywords Nanoparticles · Drug delivery · Mucoadhesion · Mucus

Abbreviations

AP-1	Activator protein 1
CREB	cAMP response element binding protein
EGFR	Epidermal growth factor receptor
ERK	Extracellular signal-regulated kinase
IL	Interleukin
JNK	Jun N-terminal kinase
LPS	Lipopolysaccharide
MEK	Mitogen-activated protein kinase kinase
MSK1	Mitogen and stress activated protein kinase
MUC	Mucin

E. Fröhlich (✉)
Center for Medical Research, Medical University of Graz, Graz, Austria
e-mail: eleonore.froehlich@medunigraz.at

E. Roblegg
Institute of Pharmaceutical Sciences, Department of Pharmaceutical Technology, Karl-Franzens-University of Graz, Graz, Austria

A. Prokop et al. (eds.), *Intracellular Delivery II*,
Fundamental Biomedical Technologies 7, DOI: 10.1007/978-94-017-8896-0_7,

MyD88	Myeloid differentiation primary response gene 88
NIK	NFkB-inducing kinase
NFkB	Nuclear factor ‘kappa-light-chain-enhancer’ of activated B-cells
p38	p38 mitogen-activated protein kinase
PTS	proline, serine, and threonine
ROS	Reactive oxygen species
SP1	Specificity protein 1
Src	Rous sarcoma oncogene cellular homolog
STAT	Signal transducer and activator of transcription
TAK1	TGF β-activated kinase

1 Introduction

Non-parenteral, non-invasive routes include ocular, dermal, oral, inhalative, and vaginal applications. Except for the dermal route, all routes need the passage of a mucus layer to reach the epithelium. Out of the mucus-covered routes oro-gastrointestinal and inhalative applications are the most widely used for local and systemic drug delivery.

The rate of drug absorption is a function of ionization, lipophilicity (partition coefficient, i.e., logP), molecular size and (lipid) solubility. For small molecular weight drug compounds logP values can reasonably well predict drug uptake. For nanoparticles (NPs), which have been developed for the delivery of conventional drugs, recombinant proteins, vaccines and nucleotides, mucus may present a considerable hindrance since they are considerably larger than conventional drug compounds.

2 Mucus

The role of mucus is manifold; it provides the interface between external and internal environment, destroys virus and bacteria, traps particles, prevents water loss, humidifies pathways, lubricates movement of materials, and protects surfaces against damaging substances. Under normal conditions, the mucus barrier on epithelia functions as part of the innate immune system and represents an important barrier for the penetration/permeation of NPs. The amounts of mucus production, the structure of the mucus layer, and the rheological properties vary according to the different requirements, mainly mechanical stress and composition of the environment (pH).

2.1 Mucus Producing Cells

In the respiratory tract, intraepithelial mucus cells and submucosal glands produce mucus. Submucosal glands occur in nasal cavity, trachea and bronchi (down to segmental or tertiary bronchi) and are absent in bronchioles. Intraepithelial mucus producing cells are still present in bronchioles and only absent in the last part of the conducting airways, the terminal bronchioles.

In the oral cavity, saliva production by extramucosal glands (submaxillary and sublingual gland, and parotid gland), increases the thickness of the mucus layer, while goblet cells are absent. In the gastrointestinal tract submucosal glands are found in the esophagus and in the duodenum (Brunner glands and small salivary glands distributed in the oral cavity). Intraepithelial mucus-producing cells are located in the stomach mucosa (foveolar cells and surface epithelial cells) and as goblet cells in the mucosa from duodenum to the rectum. Mucus is condensed in intracellular granules and expands tremendously (500 times) and almost instantly (20 ms) upon exocytosis. Inside the granules calcium ions neutralize the abundant negative charges of the mucins. During exocytosis calcium ions diffuse through a membrane pore and the charges are no longer neutralized. The repulsion of polyanionic charges and consequent hydratation leads to expansion of the gel. Mucus secretion in the respiratory and gastrointestinal tract is regulated by acetylcholine, neuropeptides such as vasoactive intestinal peptide, and neurotensin and cytokines (Laboisse et al. 1996; Rogers 2002). In the stomach, also histamine, prostaglandins and gastrin increase mucus secretion. Basal secretion is dependent on cytoskeletal movement of secretory granules, while in stimulated secretion exocytosis is initiated (Specian and Oliver 1991).

2.2 Mucus Composition and Mucins

The basic composition of airway and gastrointestinal mucus is similar. Mucus consists of 97 % water and 3 % solids (mucins, non-mucin proteins, salts, lipids, cellular debris). Out of these solids mucins represent about 30 %. Mucins are classified into the smaller secreted and the larger membrane-associated mucins. While membrane-bound mucins serve for cell adhesion, pathogen binding, and signal transduction, secreted mucins contribute to the viscosity of the mucus. Gel-forming mucins, the main compounds of mucus, are MUC2, MUC5AC, MUC5B, MUC6, and MUC19. MUC7, MUC8, and MUC9 are secreted but not gel-forming. 12 membrane-associated mucins have been identified so far (Finkbeiner et al. 2011). MUC5AC and MUC5B are the most abundant mucins in human airways, while MUC2 predominates in gastrointestinal mucus. Intraepithelial glands and submucosal glands may have different mucin secretion pattern. In the airways, MUC5AC is mainly secreted from goblet cells and MUC5B from submucosal glands (Rogers 2007).

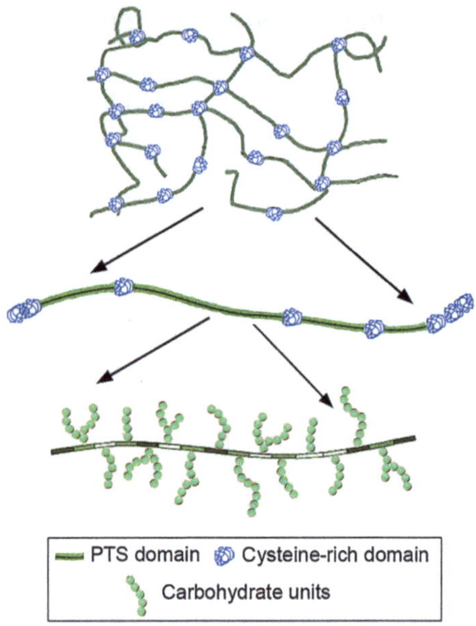

Fig. 1 Architecture of the mucus mesh. Mucin fibres form a network of about 350–500 nm mesh size. The fibres consist of proline, serine and threonine rich regions (PTS domains), covered with carbohydrates, and are linked by cysteine rich domains

Mucins are large glycoproteins, which consist of two domains, regions rich in proline, serine and threonine (PTS domains) and naked hydrophobic regions (Fig. 1). Lipids, present in the mucus to 1–2 %, are associated with the hydrophobic regions of the fibres, while sugar chains are linked through their OH-groups, by so called O-glycosylation, to the PTS domains (Fahy and Dickey 2010). Due to the high glycosylation rate, carbohydrates account for 80 % of dry weight of mucus. Two PTS domains are linked by small cysteine-rich domains, while N-termini are responsible for di- and trimerization of the molecules. Mucin fibres are 3-10 nm in diameter with coverage of typically 20–30 carbohydrates per 100 amino acids. The high content in sialic acid and sulfate creates a strongly negative charge and renders the polymer rigid by charge repulsion (Lai et al. 2009b). Entangled mucins and other mucus constituents with reversible linkage to the polymer, such as lipids and associated proteins (e.g. IgA), primarily form the mucus mesh and also determine the viscosity of mucus (Murty et al. 1984). The entanglement can occur as mesh with pore sizes of about 500 nm and by non-covalent calcium-dependent cross-linking leading to effective mesh spacing of 30–100 nm (Knowles and Boucher 2002).

2.3 Rheology of Mucus

Rheological properties of mucus have been intensely studied by Hanes and coworkers. The response of a fluid to a forced shearing flow is defined by viscosity (resistance to flow) and elasticity (stiffness). Mucus behaves differently under high and low shear: under low shear mucus acts like an elastic solid and regains shape over time, while under high shear mucus behaves like a viscous liquid that eventually deforms irreversibly. After shear, viscosity and elasticity are restored within seconds (Lai et al. 2009b). Hydratation determines viscosity, elastic properties and consistency of the mucus, which for normal airway mucus corresponds to egg white. Hydratation is often about 100 times of mucin weight and determined by the type of the glycan side chains of the mucin. Mucus in water will swell and finally completely dissolve. Salts in percentages of >1 % decrease viscosity, while multivalent cations, such as calcium and magnesium, can collapse the mucus gel and induce cross-links between mucin monomers. Viscoelasticity of mucus is also determined by pH; while the pH of respiratory mucus (nasal, lung) is in the neutral range [7.0–7.4; (Hehar et al. 1999; Jayaraman et al. 2001)], pH of the oro-gastrointestinal tract ranges from 2.0 to 8.0 (Fallingborg 1999). At low pH, mucin fibre production increases and fibres aggregate, creating a denser mucin network in the stomach than in the duodenum. In addition, also proteins, typically IgA, IgM, and lysozyme, further increase viscoelasticity.

2.4 Mucus Architecture and Turnover

In addition to different contributing cells, the architecture of the mucus layer varies between the locations (Fig. 2). The periciliary (or sol) layer of trachea and bronchi contains large membrane-bound glycoproteins as well as tethered mucins and is a part of the glycocalyx (Fig. 2a). Despite its watery consistency, this layer presents a second barrier for particle transport (Randell and Boucher 2006). A more viscous gel layer, where dust and other foreign particles are trapped and removed from the airways, covers this layer. In the stomach and colon, however, one adherent layer, released by mucus cells and impermeable to bacteria, and another more luminal and non-adherent mucin layer, formed by enzymatic processing of mucins, is present (Fig. 2b).

There are also considerable differences in the thickness of the mucin layer. In the nasal cavity the mucus layer is relatively thin [5.5–15 μm (Merkus et al. 1998)] and in the large airways region-specific variations between 7–70 μm have been reported (Krouse 2001). In relation to that, the average thickness of the mucus layer in the oral cavity, varies between 70 and 100 μm (Collins and Dawes 1987; Lagerlof and Dawes 1984). The layer in the gastrointestinal tract is much thicker and can reach up to 1 mm in the colon. The different thickness influences the rate of turnover, which is 20 min for nasal mucus (Ali and Pearson 2007) and 20 h in the gastrointestinal tract (Faure et al. 2002). Mucociliary clearance in the large airways removes foreign particles with a velocity of about 1 mm/min (Salathe 2007).

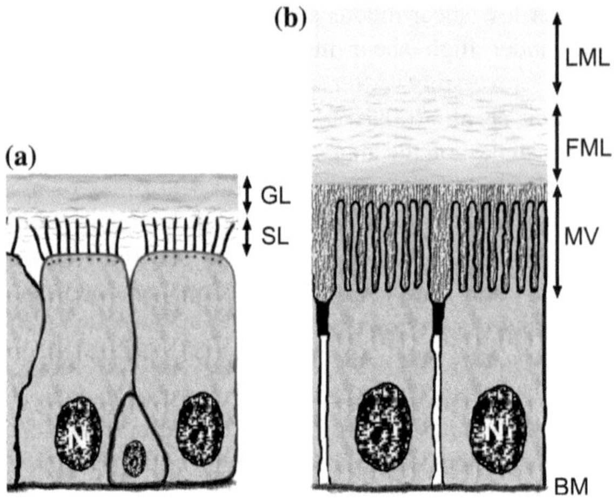

Fig. 2 Composition of bronchial tract (**a**) and gastrointestinal tract (**b**). **a** A simple columnar epithelium is present in the *upper* respiratory tract with mucus composed of a periciliary sol layer (*SL*) with low viscosity, where the kinocilia of the respiratory cells beat, and a gel layer (*GL*) with higher viscosity. **b** The gastrointestinal barrier consists of a simple columnar epithelium and a mucus layer with two strata. Microvilli (*MV*) at the apical surface of the enterocytes are covered by a glycoprotein layer, termed glycocalyx. This layer leads to a firmly attached stratified mucus layer (*FML*) and a loosely attached mucus layer (*LML*). All epithelia reside on a basal lamina (*BM*)

2.5 Disease-Induced Changes of Mucus Properties

Several types of pathologies are accompanied by mucus changes. Lack of MUC2 expression, for instance, indicates poor survival in colon cancer (Elzagheid et al. 2013). For drug delivery changes in mucus amount and properties are relevant. These changes occur predominantly in inflamed respiratory and gastrointestinal epithelia.

Common chronic respiratory diseases, such as chronic obstructive pulmonary disease (COPD), asthma, and cystic fibrosis, cause hyperproduction of mucus by hyperplasia of intraepithelial and submucosal glandular cells. In asthmatic patients, for instance, the number of goblet cells, which make up <5 % in normal bronchial epithelium, increases to 20–25 % of all epithelial cells (Shimura et al. 1996). While in mild asthma only goblet cell number and the amount of stored mucin are increased, in moderate asthma also more mucin is secreted (Ordonez et al. 2001). In addition to hyperproduction, hyperviscosity of mucus is another hallmark of chronic airway diseases. The increased viscosity in asthma is due to both mucus hypersecretion and interaction with plasma exudates in the small airways. According to one hypothesis, plasma proteins break the hydrogen bonds between adjacent mucin molecules, which leads to a greater intertangling between mucin and albumin (Rogers 2007). Another hypothesis is, that the plasma proteins, leading to an increase in solids from 3 to 15 %, may limit hydration and swelling of the mucins. While the consistence of normal airway mucus is similar to egg white, the mucus in asthma patients has a rubber-like quality. Additional disease-related changes regard expression and glycosylation pattern of mucins. Compared to normal epithelia, in allergic mucous metaplasia 40–200 × more MUC5AC and 3–10 × more MUC5B is produced (Fahy and Dickey 2010). The glycosylation pattern of mucins is altered mainly during exacerbation of chronic airway diseases; increased presence of typical tetrasaccharide carbohydrates, sialyl Lewis^x epitopes, and higher degree of sulfatation is typical for exacerbations of COPD and chronic fibrosis (Rose and Voynow 2006).

In a similar way to airway mucus, gastrointestinal diseases can affect secretion of mucus, as well as expression and glycosylation pattern of mucins. Secretion of mucins and viscosity of mucus can be increased or decreased. In general, acute infections of the gastrointestinal tract lead to goblet cell hyperplasia and increased mucus secretion. Helminth parasitic infections cause goblet cells hyperplasia and the worms are trapped within the mucus and motility and feeding capacity reduced (Kim and Khan 2013). Chronic infections, by contrast, lead to depletion of goblet cells with consequent decrease of mucus secretion. Decreased mucus production in combination with decreased viscosity in colitis ulcerosa and other inflammatory colon diseases enhance permeation of pathogens (McGuckin and Hasnain 2013). When comparing the most frequent chronic inflammatory bowel diseases, however, different mucus pattern can be recognized: in Crohn's disease hyperproduction of mucus and abnormal mucin glycosylation is observed, while in colitis ulcerosa decrease mucus production with decreased glycosylation has been reported (Dorofeyev et al. 2013). MUC3, MUC4, and MUC5B levels, for instance, are reduced in patients with Crohn's disease (Shirazi et al. 2000). Helicobacter pylori infection decreased mucin exocytosis and secretion of the gastric mucosa but viscosity of gastric mucus is increased (Markesich et al. 1995). Changes in glycosylation pattern of mucins related to gastrointestinal diseases are less specific.

Disease-related changes in mucus production are mainly caused by regulation of transcription and have been intensely studied in the airway mucus. Bacteria, teichoic acid, peptidoglycan from gram-positive bacteria induce MUC2 expression in the respiratory system (Hauber et al. 2006). They act mainly by activation of the Ras/MEK/ERK pathway (Fig. 3). Similarly, growth factors act mainly by trans-activation of EGF-receptor, using the same pathway (Thorley and Tetley 2007). Cytokines use different signaling pathways. While IL-6 acts via ERK to increase MUC5B levels, TNF-α uses p38MAPK signaling (Li et al. 1998). Activation of JNK/STAT appears to be involved in the action of IL-4, IL-13, IL-9, and IL-6 and leads to MUC5AC over-expression. IL-4 also stimulates MUC2, IL-9 increases MUC5AC, and IL-6 and IL-17 enhance MUC5AC and MUC5B transcription. TNF-α, IL-1, IL-5, IL-6, IL-10, and IL-17 also induce overproduction of mucus. Not all interleukins, however, increase mucin transcription; IL-4 decreases MUC5AC and MUC5B levels. Cholinergic agonists, neuropeptides, microbes and microbial products, toxins, inflammatory cytokines, and reactive oxygen and nitrogen species stimulate exocytosis of mucins (Davis and Dickey 2008).

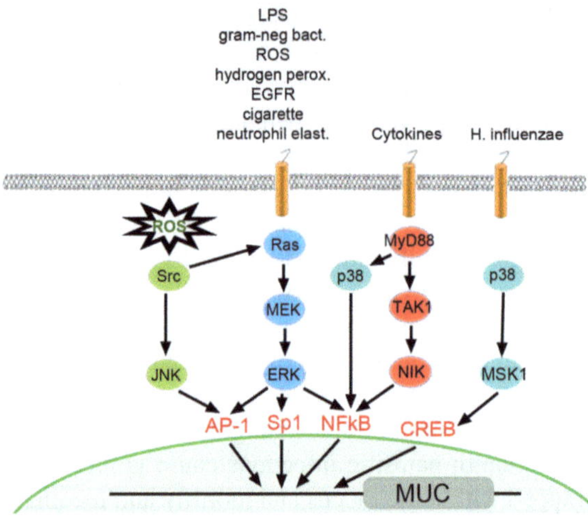

Fig. 3 Regulation of mucin transcription in chronic airway disease. Most pathogens and toxins increase the transcription of mucins through the Ras (raf)/MEK/ERK pathway. Cytokines bind to receptors, which use MyD88/TAK1/NIK signaling. Viruses act by p38/MSK1 signaling and intracellular ROS activate Src/JNK pathway. Transcription factors AP-1, SP1, NFkB, and CREB translocate to the nucleus, bind to the mucin promotor and increase mucin gene transcription

Disease-related mucus changes have a great impact on the passage of particles. Although virus particles ≤55 nm rapidly diffuse through healthy human mucus (Olmsted et al. 2001), 20 nm adeno-associated virus serotype 5 has been trapped in human cystic fibrosis sputum (Hida et al. 2011). Un-coated polystyrene particles are completely trapped into chronic sinusitis mucus (Lai et al. 2011). In their pegylated form, diffusion of the 100 and 500 nm particles is slowed down to a higher degree than diffusion of 200 nm particles. It is proposed that adhesive interactions are more important than steric factors in chronic sinusitis mucus. Permeability order of pegylated polystyrene particles (200 nm > 100 nm> 500 nm) in chronic sinusitis is very similar to cystic fibrosis mucus suggesting that inflammatory processes cause similar effects on particle permeation in airway mucus.

2.6 Role of Particle Size and Surface Properties for Permeation of Macromolecules and Particles Through Mucus

Due to the size-exclusion effect of mucus, size is one decisive parameter for mucus permeation. Only small (<10 nm) uncharged molecules can cross the mucus without hindrance. Due to their hydrophilic surface, also slightly negatively charged polio virus (28 nm) and hepatitis virus rapidly diffuse through intestinal mucus (Frey et al. 1996). Microviscosity for ≤55 nm virus particles is similar to water, 180 nm herpes simplex virus are slowed, and for ≥200 nm particles viscosity is greatly increased in cervicovaginal mucus (Lai et al. 2009a). Diffusion of amidine, carboxyl, carboxylate-modified, and sulfate polystyrene particles through porcine gastric mucus gel decreased markedly at a size of 300 nm, irrespective of surface charge (Norris and Sinko 1997).

Charged surface groups, however, also have an important effect on mucus permeation since they can perform different interactions with the mucin fibres. 50 nm carboxyl polystyrene particles bound to hydrophobic parts of the cervicovaginal mucins and induced the formation of thick cables (Lai et al. 2009a), while 200 nm amine polystyrene particles were retained by binding to mucin (Wang et al. 2011). In this kind of mucus the mobility of small (100 nm) strongly negatively charged particles was lower than that of larger (500 nm) neutral ones (Lai et al. 2007). Neutral NPs are also better able to cross the respiratory mucus layer (Sanders et al. 2000). In the oral cavity, small negatively charged polystyrene particles permeated the mucus layer better than larger ones (Teubl et al. 2013, Roblegg et al. 2012). However, with increasing size, negatively charged particles interacted with the mucus, formed aggregates, and were immobilized in the layer. By changing the surface charge into positive, particles permeated the mucus layer and penetrated the lower regions of the tissue. This appears to be due to the fact that NPs avoid adhesion to mucin fibres and/or circumvent size exclusion effects. Furthermore, neutral spherical NPs permeated the mucus layer and penetrated into the epithelium size- dependent (Teubl et al. 2012). Compared to 25, 50 and 100 nm particles, 200 nm NPs penetrated more rapidly into deeper regions of the mucosa, indicating that the size and also the surface charge strongly determine the uptake behavior. Uncharged particles did not adhere to the mucin fibres indicating that mainly pore size of the mucin fibres influenced their mobility. This phenomenon is still an open question since to our knowledge the detailed structure of the mucus layer in the oral cavity has not been investigated so far.

Data on particle penetration/permeation in gastrointestinal mucus suggest that dense opposing charges on virus particles causing hydrophilic particles with a net negative charge can quickly pass mucus barriers (Aljayyoussi et al. 2012). Positively charged particles are expected to adhere to the mucus and being retarded, while negatively charged particles are electrostatically repelled by the anionic barrier, and uncharged particles hindered by hydrophobic interactions (Aljayyoussi et al. 2012). Also small lipophilic molecules interact with the hydrophobic mucins, they are, however, not able to form polyvalent adhesive bonds and, therefore, are able to cross mucus layers. At similar size, polyacrylic acid/polyallylamine copolymers (neutral) are better transported through porcine mucus than polyacrylic acid (negative charge) and polyallylamine (positive charge) NPs (Laffleur et al. 2013). It appears that, irrespective of the source of the mucus, neutral NPs can cross the mucus layer better than charged ones.

When the surface properties are shielded by polyethylene glycol (PEG), polystyrene particles permeate non-ovulatory cervicovaginal mucus size dependently (Fig. 4). 200–500 nm densely pegylated polymeric NPs transversed cervicovaginal mucus only 4–6 fold slower than water, while 1 μm particles diffused

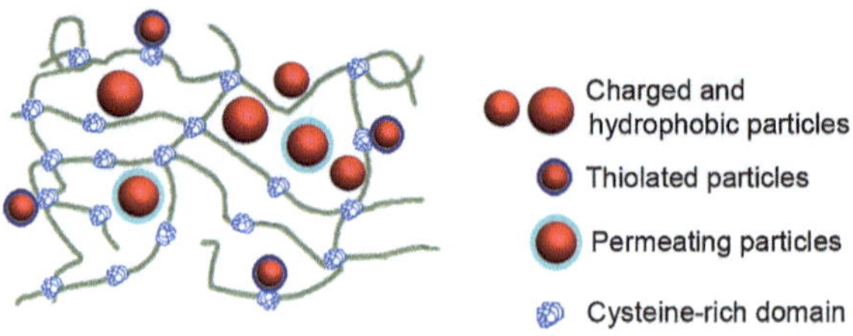

Fig. 4 Interaction of particles with the mucin mesh. Anionic and hydrophobic particles interact with hydrophobic regions outside the PTS domains, while cationic particles interact with carbohydrates at the PTS regions and thiolated particles with cysteine-rich regions. Permeating particles do not interact with mucus fibres to a great extent and mainly mesh size determines their passage

through mucus much slower than 200 and 500 nm particles with the same pegylation. Coating with PEG prevents both ionic and hydrophobic interaction with mucus fibres. The density and the molecular weight of the PEG are important parameters in the increase of particle penetration/permeation through mucus (Wang et al. 2008). Pegylation improved the passage of 500 nm poly(lactic-co-glycolic acid) (PGLA) particles through cervicovaginal mucus 4–6 fold (Lai et al. 2007) and increased mean square displacement (MSD). While efficient diffusivity of pegylated PGLA NPs in mucus is only 8 times lower than in water, it is >12,000 fold lower for the non-pegylated ones (Yu et al. 2012). With PEG coating the nature of the NP appears to play only a small role because pegylated PLGA particles showed similar velocity to pegylated latex beads of the same size. On the other hand, PLGA NPs coated with DNA pass gastric mucus 10 times faster than similarly sized latex particles (Dawson et al. 2004). The use of pegylated solid lipid (stearic acid) NPs as mucus penetrating/permeating delivery systems was demonstrated by Yuan et al. (2013). Bioavailability is increased almost 2 fold and pegylated particles can pass mucosa in the everted gut sac model, while they get trapped without PEG-coating. It appears that pluronic 127 enhances mucus permeation in a similar way to pegylation. Pluronic F127 either integrated or absorbed to liposomes increased mucus permeation 5–7 fold. It is hypothesized that the hydrophilic parts of the Pluronic F127 reduce hydrophobic and electrostatic interactions with the mucin fibres (Li et al. 2011).

Taken together, passage through mucus is the effect of bonding to mucin fibres and of size exclusion or sieving though a mesh. The retention of smaller polystyrene particles can be explained similar to size-exclusion chromatography, where smaller particles diffuse in smaller holes and, thereby, are retained longer. Delivery is also influenced by changes in the mucus mesh size. Addition of non-ionic surfactant decreases the mesh size of cervical mucus from 340 to 130 nm by increasing hydrophobic interactions of the mucin fibres. On the other hand, negatively charged NPs, such as 120 and 24 nm carboxyl polystyrene particles, can reduce the size of mucin aggregates and accelerate expansion of the mucin matrix. This leads to increased hydratation with better diffusion (Lai et al. 2009a).

Notable, also the rigid structure of the mucus (pore) has to be considered. Kirch et al. (2012) demonstrated that in hydroxyethylcellulose gels particles were dispersed in dense gels of small pore sizes and moved by external forces. In respiratory mucus, pore rigidity was higher and thus, confinement in mucus was stronger and mobility of the particles was prevented.

3 Drug Delivery Across the Mucus Layer

3.1 *Mucoadhesive Particles*

Bioadhesion describes a state in which two materials, at least one biological in nature, are held together by interfacial forces for an extended period of time. If the substrate is mucus and not a biological membrane, the term mucoadhesion is used. Materials providing mucoadhesive properties are natural or synthetic polymers. These polymers are being employed in various kinds of delivery systems, such as tablet, powders, films, ointments, gels, and patches. Drugs are integrated into these devices and are being released at the site of action. While local application of the devices by the patient is possible for eye, buccal cavity, vagina, and rectum, mucoadhesive NPs could also improve delivery via the gastrointestinal and respiratory tract, where local application of a mucoadhesive device is not possible.

Mucoadhesion consists of two phases: the contact and the consolidation stage. In the first phase, particles are influenced by repulsive forces (osmotic pressure, electrostatic repulsion) and attractive forces (van der Waals forces, electrostatic attraction) (Smart 2005). When particles have overcome the repulsion, the first phase is finished and moisture allows the mucoadhesive molecules to form weak van der Waals and hydrogen bonds with the mucus. This may occur by interpenetrations of their chains and formation of secondary bonds. Alternatively, particles may dehydrate the mucus until osmotic balance is reached and consolidates the adhesive bond.

The molecular weight (MW) of the polymer has a great influence on the strength of mucoadhesion. For polyethylene glycol of 20,000 MW mucoadhesion is negligible, at 200,000 MW improved, and at 400,000 MW excellent (Roy et al. 2009). Due to a better penetration/permeation into the mucus layer, long chain length causes better adhesion than shorter ones. A reduced flexibility of the chains allows better diffusion of water, improved hydration and entry of particles. Spatial arrangement of the polymer chains is an additional factor in mucoadhesion. Polymer concentration has an application-specific optimum because low concentrations lack mucoadhesive strength, while too high concentrations hinder solvent diffusion. Biological factors, such as pH and ionic strength, determine hydrogen bonding and electrostatic interaction of polymer and mucus. Cationic polymers, chitosan, for instance, act by interaction with glycoproteins of the mucus only at neutral to alkaline pH. Anionic polymers (carbomers, carboxymethylcellulose, alginate) form hydrophobic interactions, hydrogen and van der Waals bonds. Non-ionic polymers (hydroxypropylmethylcellulose, hydroxymethylcellulose, methyl-cellulose), in general, show weak mucoadhesion properties. Thiolated polymers form disulfide bonds with cysteine-rich domains. Lectins bind reversibly to sugar/carbohydrate residues.

Correlation of mucoadhesive properties and longer retention has been demonstrated for oral delivery. Gliadin and sodium alginate NPs were retained up to 25 × better in perfused rat stomach than aqueous solutions (Umamaheshwari et al. 2004; Katayama et al. 1999). Fluorescein isothiocyanate (FTIC) dextran loaded on poly(hydroxyethylmethacrylate) NPs permeated excised stomach mucosa better when coated with thiolated chitosan of different molecular weight (Moghaddam et al. 2009). According to other studies, mucoadhesion was linked to better bioavailability. Lectin-coating, for instance, improved uptake of polystyrene particles by oral gavage in rats (Hunter et al. 2012). The direct correlation of mucoadhesivity and absorption was most clearly seen in the study by Sakuma et al. (1999). The group determined adhesion of particles with different mucoadhesive coatings [poly(N-isopropylacrylamide), poly(vinylamine) and poly(methacrylic acid)] to the gastrointestinal mucosa. In parallel, they determined calcium plasma levels after application of calcitonin together with the particles. Calcium levels increased in parallel to the strength of the mucoadhesive action of the polymers.

It is not clear which mechanisms are involved in increased uptake and bioavailability of mucoadhesive NPs. Some mucoadhesive polymers, such as trimethylchitosan and monocarboxymethylchitosan also act as absorption enhancer (Kato et al. 2003). The role of mucoadhesive polymers in disruption of tight junctions, however, is not clear. While several studies report that both, chitosan solutions and chitosan coatings, cause transient disruption of tight junctions (Rodrigues et al. 2012), other reports state that only protonated soluble chitosan shows this effect (Issa et al. 2005).

3.2 Measurement of Mucoadhesion

There are several in vitro and in vivo techniques to study the interaction of materials with mucus; they are classified into direct and indirect methods (Davidovich-Pinhas and Baianco-Peled 2013). The prior ones measure the force that is necessary to detach the mucoadhesive system from the mucosal surface. The latter ones assess the interactions (e.g., bondings such as van der Waals forces or chemical bondings) between the system and mucins.

3.2.1 Direct Methods

Commonly, direct methods are performed in the tensile mode or in the shear mode. Thus, the applied force acts axially or tangentially and the maximum required force and the total force necessary for detachment are measured (Mortazavi and Smart 1995). Review of the literature confirmed that the tensile strength method is the most known technique in this field, using a (modified) texture analyzer as testing machine (Davidovich-Pinhas and Baianco-Peled 2013; Woertz et al. 2013). In vitro assessment is conducted with excised tissues/substrates from various animals and/or humans (Patlolla et al. 2010). However, since mucus is expected to immediately change the microstructure due to dehydration, results are often not reproducible and/or comparable to humans. As alternative, commercially available mucin powders, as gel, disc or hydrated film are utilized (Tamburic and Craig 1997). The most interesting candidates are mucoadhesive (co-) polymers (e.g., chitosan, polyacrylic acid, thiolated polymers) applied in their dry, semihydrated and hydrated form and formulated into tablets, pellets, films and/or nanocarriers. Furthermore, depending on the anatomical location, testing conditions can be adapted to meet physiological conditions. The sample and the mucosal surface are immersed in liquid and the influence of the pH of the polymer is considered (Grabovaca et al. 2005; Lehr et al. 1990). Apart from tensile mode, dynamic assays under shear forces are of great interest. To measure mucoadhesion under shear, the rotating cylinder method can be used. The sample device is attached to excised mucosa, spun on a cylinder, which is immersed in physiological liquid and agitated under constant speed (Bernkop-Schnürch and Steininger 2000). Additionally, for quantitative assessment, adhesion time (until detachment) is evaluated. Furthermore, assays are performed that mimic shear forces caused by continuous flow to study the binding ability of the polymer and the mucosal surface (Rao and Buri 1989). This method was improved by several groups using a flow-through cell device under temperature-controlled conditions, mimicking laminar and turbulent flow (Blegamwar et al. 2009; Nielsen et al. 1998; Schultz et al. 2000).

To study molecular and surface forces on molecular scale, atomic force microscopy (AFM) as a non-invasive and non-destructive imaging technique with high potential can be performed under physiological conditions. A cantilever mounted tip deflects due to van der Waals repulsive interactions in contact with the sample. A laser beam focused on the free end of the cantilever and reflected into a photodiode detects the cantilever deflection. Adhesion can be measured by force-distance curves and binding of the polymer to the mucosal surface can be examined due to changes in surface roughness. Deacon et al. (2000) studied the interaction between pig gastric mucin and chitosan via AFM. They demonstrated that both materials showed linear filamentous structures, which formed aggregates after mixing. These results were in accordance with earlier studies performed with electron microscopy. Additionally, Patel et al. (2000) showed that AFM is an appropriate and potential technique for imaging and measuring interactions between polymers and buccal mucosal surfaces.

Apart from in vitro and ex vivo testing, also in vivo studies are conducted to evaluate the residence time of the mucoadhesive device. Lehr et al. (1990) investigated polymer-coated microspheres in a rat model by connecting an isolated intestinal rat loop to the rat intestine cavity and measuring the outlet amount. Other possibilities are to insert (invasive) labeled polymers into the gastrointestinal tract and/or to apply the material orally and monitor it via magnetic resonance images (Albrecht et al. 2006).

3.2.2 Indirect Methods

In addition to spectroscopic methods (Fourier transform infrared spectrometer and nuclear magnetic resonance), rheology is an appropriate technique to study mucoadhesive properties of polymers. The later technique is based on the idea that upon mixture of polymer and mucin, the viscosity, and thus, the flow as well as the deformation change (Hassan and Gallo 1990). Strong mucoadhesive polymers result in a more viscous system (compared to the individual viscosity of the polymer and the mucin) due to chain entanglements and various chemical bonds. These changes can be monitored and characteristic interactions/adhesive properties can be determined. Viscosity is described by the loss modulus (G''), indicating the resistance of a gel to flow and elasticity is defined by the storage modulus (G'), which measures the ability of the gel to recover [e.g. (Rossi et al. 2001, Lai et al. 2009b)]. The loss tangent δ ($\tan \delta$) is a simple indicator that describes the viscoelasticity of the sample, calculated from the ratio G'' to G'. Small values reflect a solid like response (gel), compared to larger values, which represent a liquid like response. A large number of data is available in this field, due to the increasing number of (mucoadhesive) polymers, mucin types and measurement devices, such as parallel plates, cones and plates or concentric cylinders (Caramella et al. 1994). Although rheology is a useful method, it should always be used in combination with other methods due to the fact that variations of the used concentrations are like to influence the entire results (Hagerstrom and Edsman 2003).

3.2.3 Applications of Mucoadhesive NPs in Drug Delivery by the Respiratory System

Intranasal delivery is the preferred route to deliver peptides to the brain. The olfactory nerve has a direct connection to the brain and circumvents the tight blood brain barrier. Powders, gels, liquids, sprays and microparticles have been employed to deliver peptides such as desmopressin to treat diabetes insipidus and nocturnal enuresis, oxytoxin to induce lactation, and calcitonin for osteoporosis. Non-peptide drugs, such as sumatriptan and zolmitriptan, in nasal application are used in migraine (Singh et al. 2012).

For the delivery of small peptides, thiolated chitosan and wheat germ agglutinin modified pegylated polylactic acid NPs have been used (Table 1). Also larger peptides, such as insulin, could be successfully delivered to diabetic rats using chitosan, PEG-graft trimethylchitosan, and amine-modified poly(vinyl alcohol) graft poly (L-lactide) NPs (Pires et al. 2009). Another important area is mucosal vaccination, where chitosan NPs crosslinked with polyanionic cyclodextrins

Table 1 Examples for successful delivery of compounds by nanoparticles with mucoadhesive, mucopenetrating and mucolytic properties

Mechanism	Drug compound	Material/composition	Model	Route	References
Adhesive	Low molecular weight heparin	N-trimethyl chitosan	Rat intestine	Oral	Paliwal et al. (2012)
Adhesive	Alendronate	Chitosan-coated liposomes	Caco-2 monolayer	Oral	Han et al. (2012)
Adhesive	Docetaxel	Chitosan coated methyl-b-cyclodextrin(polyisobutylcyanoacrylate)	Rat intestine	oral	Mazzaferro et al. (2012)
Adhesive	Docetaxol	Thiolated chitosan coated methyl methacrylate	Rat intestine	Oral	Saremi et al. (2011)
Adhesive	None	Chitosan -4-thiobutylamidine conjugates	Porcine small intestine	Oral	Sakloetsakun et al. (2010)
Adhesive	Fluorescein isothiocyanate dextran	Thiolated hydroxyethyl cellulose	Porcine small intestine, Caco-2 cells	Oral	Rahmat et al. (2012)
Adhesive	Insulin	Carboxylated chitosan and poly(methyl methacrylate)	Rat	Oral	Cui et al. (2009)
Adhesive	Insulin	Chitosan-insulin polyelectrolyte complexes suspended in oily phase	Rat	Oral	Elsayed et al. (2009)
Adhesive	Insulin	Chitosan-based poloxamer 188	Rat	Oral	Pan et al. (2002)
Adhesive	Insulin	Chitosan/dextran sulfate NPs	Rat intestine	Oral	Sarmento et al. (2007a)
Adhesive	Insulin	Alginate/chitosan	Rat	Oral	Sarmento et al. (2007b)
Adhesive	Fexofenadine	Chitosan-coated liposomes	Rat	Nasal	Qiang et al. (2012)
Adhesive	Plasmid DNA	PLGA: poloxamer and PLGA: poloxamine blend nanostructures	Rat	Nasal	Csaba et al. (2006)
Adhesive	Olanzapine	Poloxamer 188 or 407 nanocubic vesicles	Rat	Nasal	Salama et al. (2012)
Adhesive	Coumarin-6	Wheat germ agglutinin modified PEG-PLA	Rat	Nasal	Liu et al. (2012)
Adhesive	Coumarin-6	Solanun tuberosum lectin-conjugated PLGA	Rat	Nasal	Chen et al. (2012)
Adhesive	Insulin	PEG-grafted chitosan	Rabbit	Nasal	Zhang et al. (2008)

(continued)

Table 1 (continued)

Mechanism	Drug compound	Material/composition	Model	Route	References
Adhesive	Insulin	Chitosan	Rat	Nasal	Wang et al. (2009)
Adhesive	Insulin	Amine-modified poly(vinyl alcohol)-graft-poly(L-lactide)	Rat	Nasal	Simon et al. (2005)
Adhesive	Insulin	3-Arylamidophenylboronic acid-N-acetylglucosamine-copolymer	Rat	Nasal	Zhang et al. (2012)
Adhesive	Leuprolide	Thiolated chitosan	Porcine nasal mucosa	Nasal	Shahnaz et al. (2012)
Adhesive	Vasointestinal active peptide	Wheat germ agglutinin modified PEG-PLA	Rat	Nasal	Gao et al. (2006)
Adhesive	Vaccination	Chitosan phosphatidylcholine-cholesterol-stearyl-amine liposomes	Mouse	Nasal	Figueiredo et al. (2012)
Adhesive	Theophylline	Thiolated chitosan	OVA-challenged mice	Nasal	Lee et al. (2006)
Adhesive	Calcitonin	Surface-modified DL-lactide/glycolide copolymer (PLGA) nanospheres with chitosan	Guinea pigs	Bronchial	Yamamoto et al. (2005)
Adhesive	Low molecular weight heparin	Chitosan/Hyaluronic acid	Rat mast cells	Bronchial	Oyarzun-Ampuero et al. (2009)
Penetrating	DNA	Poly(d,l-lactic-co-glycolic) acid (PLGA) + dimethyldioctadecylammonium bromide	Porcine intestinal mucus	Oral	Dawson et al. (2004)
Penetrating	None	PEGylated solid lipid	Caco-2/HT29, rat intestine	Oral	Yuan et al. (2013)
Penetrating	Coumarin 6	PF127-inlaid liposomes	Caco-2 monolayer	Oral	Li et al. (2011)
Lytic	Fluorescein isothiocyanate dextran	Polyelectrolyte complex of spermidine with polyacrylic acid	Caco-2 monolayer, rat intestine	Oral	Makhlof et al. (2011)

(sulfobutyl-β-cyclodextrin or carbomethyl-β-cyclodextrin) were identified as promising candidates (Csaba et al. 2009). Coating with chitosan increased the efficacy of vaccination according to secreted IgA levels. Despite strong mucoadhesive properties, normal lectins are of limited use due to toxicity and immunogenicity. Wheat germ agglutinin, solanum tuberosum lectin and truncated lectins, however, are successful for delivery (Anand et al. 2012). There is, however, not a general superiority of NP formulations: N-trimethylchitosan/ovalbumin NPs did not perform as well as trimethylchitosan/ovalbumin conjugates according to diffusion through Calu-3 cells and induction of IgA levels in mice (Slutter et al. 2010).

In general, delivery of NPs without encapsulation in microparticular formulations to bronchi or deep lung due to reduced dimensions and, hence, low inertia is impractical. While mucoadhesive microparticles have been evaluated in some studies [e.g. (Sakagami et al. 2001)], only few applications have used NPs. For instance, chitosan/hyaluronic acid NPs loaded with low molecular weight heparin were successfully tested for bronchial delivery ex vivo (Oyarzun-Ampuero et al. 2009) and aerosolized PLGA/chitosan NPs were longer retained in the lungs of guinea pigs (Yamamoto et al. 2005). When loaded with calcitonin, these particles showed a longer and more potent effect on calcium blood levels than chitosan unmodified PLGA NPs.

3.2.4 Applications of Mucoadhesive NPs in Oral Drug Delivery

Applications for mucoadhesive NPs in oral application include systemic treatment of cancer, osteoporosis, pain, etc. Local applications are also possible, for instance, application of thiolated chitosan polymethylacrylic acid NPs loaded with metronidazole benzoate in periodontal disease (Saboktakin et al. 2011).

In vitro and ex-vivo studies showed better adhesion and drug transport for chitosan, thiolated chitosan, chitosan -4-thiobutylamidine conjugates, thiolated hydroxyethyl cellulose particles and better efficacy in vivo of N-trimethyl chitosan, carboxylated chitosan, chitosan-based poloxamer 188, and chitosan/dextran sulfate particles (Table 1). Several types of mucoadhesive NPs improved the delivery of insulin. It is, however, not clear if insulin can cross the intestinal barriers as intact functional molecule because intestinal mucus acts as an enzymatic barrier in addition to a mechanic one (Aoki et al. 2005).

Also other molecules were used to create mucoadhesive NPs for oral delivery: such approaches include thermally hydrocarbonized porous silicon functionalized with self-assembled amphiphilic protein coating consisting of class II hydrophobin from Trichoderma reesei (Sarparanta et al. 2012), disulfide containing poly(cystamine bisacrylamide-4-amino-1-butanol) NPs (Cohen et al. 2012), ethylcellulose (Suwannateep et al. 2011), and pectin liposomal nanocomplexes (Thirawong et al. 2008).

3.3 Alternative Strategies Used for NP-Mediated Drug Delivery

Despite the broad use of mucoadhesive devices, these systems have limitations. Mucin turnover, physiological condition, and presence of irritants can cause inefficient delivery. Mucoadhesion caused considerable disruption of the mucus barrier (McGill and Smyth 2010; Wang et al. 2011). The importance of destruction of the mucus barrier, allowing invasion of pathogens, has not been sufficiently clarified so far. Due to potential problems with barrier disruption, mucolytic particles are rarely employed. Large cationic NPs at high concentrations can create large channels by vigorous electrostatic interaction (Wang et al. 2011). Local disruption of the mucus barrier is caused by <200 nm poly(acrylic acid) and mucolytic enzyme (papain) NPs (Müller et al. 2013). In a similar way, polyelectrolyte complex of spermidine with polyacrylic acid enhanced FITC dextran transport in Caco-2 cells and permeation of calcitonin through the rat intestine (Makhlof et al. 2011).

More recently, another strategy has been developed, where mucus integrity is completely preserved and NPs, due to dense PEG-coating, cross the mucus layer. Efficacy of mucus penetrating particles in vivo, currently, has only been demonstrated for vaginal application, where it was efficient in the delivery of acyclovir to murine vagina (Ensign et al. 2012). Mucus permeating particles were also more efficient than solutions in gene delivery (Yu et al. 2012). While conventional mucoadhesive NPs aggregated in the mucus, protection against herpes simplex virus 2 infection in mice by vaginal patches with Acyclovir-loaded mucus penetrating particles was better (53 vs. 16 %) than with the soluble drug. In the gastrointestinal tract, pegylated solid lipid stearic acid (PSA) NPs and Pluronic F127 either integrated or absorbed to liposomes displayed mucus penetrating properties (Yuan et al. 2013; Li et al. 2011).

The use of mucus penetrating strategies could be suitable in chronic diseases of the respiratory system. Cystic fibrosis mucus has one of the highest viscosities and presents a particular challenge because, due to recurrent infections, these patients have also a high need for proper medication. PEG-PSA NPs penetrated/permeated cervicovaginal mucus and cystic fibrosis sputum with 50-fold higher mean square displacement (MSD) than uncoated latex particles in the same size (Tang et al. 2009). 40 % of the PEG-PSA NPs were transported by diffusion compared to <6 % of the latex particles and 31 % of these particles were able to cross a cystic fibrosis sputum layer of 10 μm compared to 0.6 % of the latex particles. Effective diffusivity of the PEG-PSA NPs was 0.023 $\mu m^2/s$, compared to the diffusion coefficient of un-coated 59-1,000 nm latex particles with 0 $\mu m^2/s$ (Olmsted et al. 2001). PEG-PSA NPs, therefore, might have a greater chance to reach the epithelial layer without being cleared from the organism than mucoadhesive NPs.

4 Concluding Remarks

A direct comparison of different particle delivery systems is not possible because, in general, the NP-based systems are compared to solutions containing the respective drug. Based on the long experience with mucoadhesive devices and data with mucoadhesive NPs, it appears reasonable to conclude that this strategy improves drug delivery across/into mucus-covered epithelia. On the other hand, the differences in mucus structure, mucus turnover at different anatomical sites of the human body and pathological mucus changes suggest that also other types of NPs might be appropriate delivery systems. While mucoadhesive NPs could be suitable for applications in healthy epithelia with low mucus turnover, like in the gastrointestinal tract, mucus penetrating NPs or mucolytic NPs could be more efficient for delivery to diseased airways with strong production of highly viscous mucus.

Acknowledgements Support of the studies by FP6 European integrated project "NanoBio-pharmaceutics", NMP4-CT-2006-026723, the Research and Technology Development in Project Cluster NANO-HEALTH and the Austrian Research Promotion Agency (FFG) project 826136 is gratefully acknowledged.

References

Albrecht K, Greindl M, Kremser C, Wolf C, Debbagec B, Bernkop-Schürch A (2006) Comparative in vivo mucoadhesion studies ofthiomer formulations using magnetic resonance imaging and fluorescence detection. J Control Release 115:78–84

Ali MS, Pearson JP (2007) Upper airway mucin gene expression: a review. Laryngoscope 117:932–938

Aljayyoussi G, Abdulkarim M, Griffiths P, Gumbleton M (2012) Pharmaceutical nanoparticles and the mucin biopolymer barrier. BioImpacts BI 2:173–174

Anand U, Feridooni T, Agu R (2012) Novel mucoadhesive polymers for nasal delivery. In: Seizer A (ed) Recent advances in novel drug carrier systems. InTech

Aoki Y, Morishita M, Takayama K (2005) Role of the mucous/glycocalyx layers in insulin permeation across the rat ileal membrane. Int J Pharm 297:98–109

Bernkop-Schnürch A, Steininger S (2000) Synthesis and characterization of mucoadhesive thiolated polymers. Int J Pharm 194:239–247

Blegamwar V, Shah V, Surana SJ (2009) Formulation and evaluation of oral mucoadhesive multiparticulate sytsem containing metoprolol tartrate: an in vitro-ex vivo characterization. Curr Drug Deliv 6:113–121

Caramella C, Bonferoni M, Rossi S, Ferrari F (1994) Rheological and tensile tests for the assessment of polymer-mucin interactions. Eur J Pharm Biopharm 40:213–217

Chen J, Zhang C, Liu Q, Shao X, Feng C, Shen Y, Zhang Q, Jiang X (2012) Solanum tuberosum lectin-conjugated PLGA nanoparticles for nose-to-brain delivery: in vivo and in vitro evaluations. J Drug Target 20:174–184

Cohen S, Coue G, Beno D, Korenstein R, Engbersen JF (2012) Bioreducible poly(amidoamine)s as carriers for intracellular protein delivery to intestinal cells. Biomaterials 33:614–623

Collins LM, Dawes C (1987) The surface area of the adult human mouth and thickness of the salivary film covering the teeth and oral mucosa. J Dent Res 66:1300–1302

Csaba N, Sanchez A, Alonso MJ (2006) PLGA: poloxamer and PLGA:poloxamine blend nanostructures as carriers for nasal gene delivery. J Control Release 113:164–172
Csaba N, Garcia-Fuentes M, Alonso MJ (2009) Nanoparticles for nasal vaccination. Adv Drug Deliv Rev 61:140–157
Cui F, Qian F, Zhao Z, Yin L, Tang C, Yin C (2009) Preparation, characterization, and oral delivery of insulin loaded carboxylated chitosan grafted poly(methyl methacrylate) nanoparticles. Biomacromolecules 10:1253–1258
Davidovich-Pinhas M, Baianco-Peled H (2013) Mucoadhesion: a review of characterization techniques. Expert Opin 7:259–271
Davis CW, Dickey BF (2008) Regulated airway goblet cell mucin secretion. Annu Rev Physiol 70:487–512
Dawson M, Krauland E, Wirtz D, Hanes J (2004) Transport of polymeric nanoparticle gene carriers in gastric mucus. Biotechnol Prog 20:851–857
Deacon M, McGurk S, Roberts C, Williams P, Tendler S, Davies M, Davies S, Harding S (2000) Atomic force microscopy of gastric mucin and chitosan mucoadhesive systems. Biochem J 348:557–563
Dorofeyev AE, Vasilenko IV, Rassokhina OA, Kondratiuk RB (2013) Mucosal barrier in ulcerative colitis and Crohn's disease. Gastroenterol Res Pract 2013:431231
Elsayed A, Remawi MA, Qinna N, Farouk A, Badwan A (2009) Formulation and characterization of an oily-based system for oral delivery of insulin. Eur J Pharm Biopharm 73:269–279
Elzagheid A, Emaetig F, Buhmeida A, Laato M, El-Faitori O, Syrjanen K, Collan Y, Pyrhonen S (2013) Loss of MUC2 expression predicts disease recurrence and poor outcome in colorectal carcinoma. Tumour Biol J Int Soc Oncodevelop Biol Med 34:621–628
Ensign LM, Tang BC, Wang YY, Tse TA, Hoen T, Cone R, Hanes J (2012) Mucus-penetrating nanoparticles for vaginal drug delivery protect against herpes simplex virus. Sci Trans Med 4:138ra179
Fahy JV, Dickey BF (2010) Airway mucus function and dysfunction. N Engl J Med 363:2233–2247
Fallingborg J (1999) Intraluminal pH of the human gastrointestinal tract. Dan Med Bull 46:183–196
Faure M, Moennoz D, Montigon F, Fay LB, Breuille D, Finot PA, Ballevre O, Boza J (2002) Development of a rapid and convenient method to purify mucins and determine their in vivo synthesis rate in rats. Anal Biochem 307:244–251
Figueiredo L, Cadete A, Goncalves LM, Corvo ML, Almeida AJ (2012) Intranasal immunisation of mice against Streptococcus equi using positively charged nanoparticulate carrier systems. Vaccine 30:6551–6558
Finkbeiner WE, Zlock LT, Morikawa M, Lao AY, Dasari V, Widdicombe JH (2011) Cystic fibrosis and the relationship between mucin and chloride secretion by cultures of human airway gland mucous cells. Am J Physiol Cell Mol Physiol 301:L402–L414
Frey A, Giannasca KT, Weltzin R, Giannasca PJ, Reggio H, Lencer WI, Neutra MR (1996) Role of the glycocalyx in regulating access of microparticles to apical plasma membranes of intestinal epithelial cells: implications for microbial attachment and oral vaccine targeting. J Exp Med 184:1045–1059
Gao X, Tao W, Lu W, Zhang Q, Zhang Y, Jiang X, Fu S (2006) Lectin-conjugated PEG-PLA nanoparticles: preparation and brain delivery after intranasal administration. Biomaterials 27:3482–3490
Grabovaca V, Guggi D, Bernkop-Schnürch A (2005) Comparison of the mucoadhesive properties of various polymers. Adv Drug Deliv Rev 57:1713–1723
Hagerstrom H, Edsman K (2003) Limitations of the rheological mucoadhesion method: the effect of the choice of conditions and the rheological synergism parameter. Eur J Pharm Biopharm 18:349–357
Han HK, Shin HJ, Ha DH (2012) Improved oral bioavailability of alendronate via the mucoadhesive liposomal delivery system. Eur J Pharm Sci 46:500–507

Hassan E, Gallo J (1990) A simple rheological method for the in vitro assessment of mucin-polymer bioadhesive bond strength. Pharm Res 7:491–495

Hauber HP, Foley SC, Hamid Q (2006) Mucin overproduction in chronic inflammatory lung disease. Can Respir J J Can Thorac Soc 13:327–335

Hehar SS, Mason JD, Stephen AB, Washington N, Jones NS, Jackson SJ, Bush D (1999) Twenty-four hour ambulatory nasal pH monitoring. Clin Otolaryngol Allied Sci 24:24–25

Hida K, Lai SK, Suk JS, Won SY, Boyle MP, Hanes J (2011) Common gene therapy viral vectors do not efficiently penetrate sputum from cystic fibrosis patients. PLoS ONE 6:e19919

Hunter AC, Elsom J, Wibroe PP, Moghimi SM (2012) Polymeric particulate technologies for oral drug delivery and targeting: a pathophysiological perspective. Nanomed Nanotechnol Biol Med 8(Suppl 1):S5–S20

Issa M, Köping-Höggård M, Artursson P (2005) Chitosan and the mucosal delivery of biotechnology drugs. Drug Discov TodayTechnol 2:1–6

Jayaraman S, Joo NS, Reitz B, Wine JJ, Verkman AS (2001) Submucosal gland secretions in airways from cystic fibrosis patients have normal [Na(+)] and pH but elevated viscosity. Proc Natl Acad Sci USA 98:8119–8123

Katayama H, Nishimura T, Ochi S, Tsuruta Y, Yamazaki Y, Shibata K, Yoshitomi H (1999) Sustained release liquid preparation using sodium alginate for eradication of Helicobacter pyroli. Biol Pharm Bull 22:55–60

Kato Y, Onishi H, Machida Y (2003) Application of chitin and chitosan derivatives in the pharmaceutical field. Curr Pharm Biotechnol 4:303–309

Kim J, Khan W (2013) Goblet cells and mucins: role in innate defense in enteric infections. Pathogens 2:55–70

Kirch J, Schneider A, Abou B, Hopf A, Schaefer UF, Schneider M, Schall C, Wagner C, Lehr CM (2012) Optical tweezers reveal relationship between microstructure and nanoparticle penetration of pulmonary mucus. Proc Natl Acad Sci USA 109:18355–18360

Knowles MR, Boucher RC (2002) Mucus clearance as a primary innate defense mechanism for mammalian airways. J Clin Invest 109:571–577

Krouse ME (2001) Is cystic fibrosis lung disease caused by abnormal ion composition or abnormal volume? J Gen Physiol 118:219–222

Laboisse C, Jarry A, Branka JE, Merlin D, Bou-Hanna C, Vallette G (1996) Recent aspects of the regulation of intestinal mucus secretion. Proc Nutr Soc 55:259–264

Laffleur F, Hintzen F, Shahnaz G, Rahmat D, Leithner K, Bernkop-Schnurch A (2013) Development and in vitro evaluation of slippery nanoparticles for enhanced diffusion through native mucus. Nanomedicine [Epub ahead of print]

Lagerlof F, Dawes C (1984) The volume of saliva in the mouth before and after swallowing. J Dent Res 63:618–621

Lai SK, O'Hanlon DE, Harrold S, Man ST, Wang YY, Cone R, Hanes J (2007) Rapid transport of large polymeric nanoparticles in fresh undiluted human mucus. Proc Natl Acad Sci USA 104:1482–1487

Lai SK, Wang YY, Hanes J (2009a) Mucus-penetrating nanoparticles for drug and gene delivery to mucosal tissues. Adv Drug Deliv Rev 61:158–171

Lai SK, Wang YY, Wirtz D, Hanes J (2009b) Micro- and macrorheology of mucus. Adv Drug Deliv Rev 61:86–100

Lai SK, Suk JS, Pace A, Wang YY, Yang M, Mert O, Chen J, Kim J, Hanes J (2011) Drug carrier nanoparticles that penetrate human chronic rhinosinusitis mucus. Biomaterials 32:6285–6290

Lee DW, Shirley SA, Lockey RF, Mohapatra SS (2006) Thiolated chitosan nanoparticles enhance anti-inflammatory effects of intranasally delivered theophylline. Respir Res 7:112

Lehr C, Bouwstra J, Tukker J, Junginger H (1990) Intestinal transit of bioadhesive microspheres in an in situ loop in the rat-a comparative study with copolymers and blends based on poly (acrylic acid). J Control Release 13:51–62

Li JD, Feng W, Gallup M, Kim JH, Gum J, Kim Y, Basbaum C (1998) Activation of NF-kappaB via a Src-dependent Ras-MAPK-pp90rsk pathway is required for Pseudomonas aeruginosa-induced mucin overproduction in epithelial cells. Proc Natl Acad Sci USA 95:5718–5723

Li X, Chen D, Le C, Zhu C, Gan Y, Hovgaard L, Yang M (2011) Novel mucus-penetrating liposomes as a potential oral drug delivery system: preparation, in vitro characterization, and enhanced cellular uptake. Int J Nanomed 6:3151–3162

Liu Q, Shen Y, Chen J, Gao X, Feng C, Wang L, Zhang Q, Jiang X (2012) Nose-to-brain transport pathways of wheat germ agglutinin conjugated PEG-PLA nanoparticles. Pharm Res 29:546–558

Makhlof A, Werle M, Tozuka Y, Takeuchi H (2011) A mucoadhesive nanoparticulate system for the simultaneous delivery of macromolecules and permeation enhancers to the intestinal mucosa. J Control Release 149:81–88

Markesich DC, Anand BS, Lew GM, Graham DY (1995) Helicobacter pylori infection does not reduce the viscosity of human gastric mucus gel. Gut 36:327–329

Mazzaferro S, Bouchemal K, Skanji R, Gueutin C, Chacun H, Ponchel G (2012) Intestinal permeation enhancement of docetaxel encapsulated into methyl-beta-cyclodextrin/poly(isobutylcyanoacrylate) nanoparticles coated with thiolated chitosan. J Control Release 162:568–574

McGill SL, Smyth HD (2010) Disruption of the mucus barrier by topically applied exogenous particles. Mol Pharm 7:2280–2288

McGuckin MA, Hasnain SZ (2014) There is a 'uc' in mucus, but is there mucus in UC? Gut 63:216–217

Merkus FW, Verhoef JC, Schipper NG, Marttin E (1998) Nasal mucociliary clearance as a factor in nasal drug delivery. Adv Drug Deliv Rev 29:13–38

Moghaddam FA, Atyabi F, Dinarvand R (2009) Preparation and in vitro evaluation of mucoadhesion and permeation enhancement of thiolated chitosan-pHEMA core-shell nanoparticles. Nanomed Nanotechnol Biol Med 5:208–215

Mortazavi S, Smart J (1995) An investigation of some factors influencing the in vitro assessment of mucoadhesion. Int J Pharm 116:223–230

Müller C, Leithner K, Hauptstein S, Hintzen F, Salvenmoser W, Bernkop-Schnürch A (2013) Preparation and characterization of mucus-penetrating papain/poly(acrylic acid) nanoparticles for oral drug delivery applications. J Nanopart Res 15(1):1–13

Murty VL, Sarosiek J, Slomiany A, Slomiany BL (1984) Effect of lipids and proteins on the viscosity of gastric mucus glycoprotein. Biochem Biophys Res Commun 121:521–529

Nielsen L, Schubert L, Hansen J (1998) Bioadhesive drug delivery systems. I. Characterization of mucoadhesive properties of systems based on glyceryl monooleate and glycerylmonolinoleate. Eur J Pharm Biopharm 6:231–239

Norris D, Sinko P (1997) Effect of size, surface charge, and hydrophobicity on the translocation of polystyrene microspheres through gastrointestinal mucin. J Appl Polym Sci 63:1481–1492

Olmsted SS, Padgett JL, Yudin AI, Whaley KJ, Moench TR, Cone RA (2001) Diffusion of macromolecules and virus-like particles in human cervical mucus. Biophys J 81:1930–1937

Ordonez CL, Khashayar R, Wong HH, Ferrando R, Wu R, Hyde DM, Hotchkiss JA, Zhang Y, Novikov A, Dolganov G, Fahy JV (2001) Mild and moderate asthma is associated with airway goblet cell hyperplasia and abnormalities in mucin gene expression. Am J Respir Crit Care Med 163:517–523

Oyarzun-Ampuero FA, Brea J, Loza MI, Torres D, Alonso MJ (2009) Chitosan-hyaluronic acid nanoparticles loaded with heparin for the treatment of asthma. Int J Pharm 381:122–129

Paliwal R, Paliwal SR, Agrawal GP, Vyas SP (2012) Chitosan nanoconstructs for improved oral delivery of low molecular weight heparin: In vitro and in vivo evaluation. Int J Pharm 422:179–184

Pan Y, Li YJ, Zhao HY, Zheng JM, Xu H, Wei G, Hao JS, Cui FD (2002) Bioadhesive polysaccharide in protein delivery system: chitosan nanoparticles improve the intestinal absorption of insulin in vivo. Int J Pharm 249:139–147

Patel D, Smith J, Smith A, Grist N, Barnett P, Smart J (2000) An atomic force microscopy investigation of bioadhesive polymer adsorption onto human buccal cells. Int J Pharm 200:271–277

Patlolla A, Patlolla B, Tchounwou P (2010) Evaluation of cell viability, DNA damage, and cell death in normal human dermal fibroblast cells induced by functionalized multiwalled carbon nanotube. Mol Cell Biochem 338:225–232

Pires A, Fortuna A, Alves G, Falcao A (2009) Intranasal drug delivery: how, why and what for? J Pharm Pharm Sci 12:288–311

Qiang F, Shin HJ, Lee BJ, Han HK (2012) Enhanced systemic exposure of fexofenadine via the intranasal administration of chitosan-coated liposome. Int J Pharm 430:161–166

Rahmat D, Muller C, Barthelmes J, Shahnaz G, Martien R, Bernkop-Schnurch A (2012) Thiolated hydroxyethyl cellulose: Design and in vitro evaluation of mucoadhesive and permeation enhancing nanoparticles. Eur J Pharm Biopharm 83:149–155

Randell SH, Boucher RC (2006) Effective mucus clearance is essential for respiratory health. Am J Respir Cell Mol Biol 35:20–28

Rao K, Buri P (1989) A novel in situ method to test polymers and coated microparticles for bioadhesion. Int J Pharm 52:265–270

Roblegg E, Fröhlich E, Meindl C, Teubl B, Zaversky M, Zimmer A (2012) Evaluation of a physiological in vitro system to study the transport of nanoparticles through the buccal mucosa. Nanotoxicology 6:399–413

Rodrigues S, Dionísio M, Remuñán López C, Grenha A (2012) Biocompatibility of Chitosan carriers with application in drug delivery. J Funct Biomater 3:615–641

Rogers DF (2002) Pharmacological regulation of the neuronal control of airway mucus secretion. Curr Opin Pharmacol 2:249–255

Rogers DF (2007) Physiology of airway mucus secretion and pathophysiology of hypersecretion. Respir Care, 52:1134–1146 (discussion 1146–1139)

Rose MC, Voynow JA (2006) Respiratory tract mucin genes and mucin glycoproteins in health and disease. Physiol Rev 86:245–278

Rossi S, Ferrari F, Bonferoni MC, Caramella C (2001) Characterization of chitosan hydrochloride–mucin rheological interaction: influence of polymer concentration and polymer: mucin weight ratio. Eur J Pharm Sci 12:479–485

Roy S, Pal K, Anis A, Pramanik K, Prabhakar B (2009) Polymers in mucoadhesive drug delivery system: a brief note. Des Monomers Polym 12:483–495

Saboktakin MR, Tabatabaie RM, Maharramov A, Ramazanov MA (2011) Development and in vitro evaluation of thiolated chitosan–Poly(methacrylic acid) nanoparticles as a local mucoadhesive delivery system. Int J Biol Macromol 48:403–407

Sakagami M, Sakon K, Kinoshita W, Makino Y (2001) Enhanced pulmonary absorption following aerosol administration of mucoadhesive powder microspheres. J Control Release 77:117–129

Sakloetsakun D, Perera G, Hombach J, Millotti G, Bernkop-Schnurch A (2010) The impact of vehicles on the mucoadhesive properties of orally administrated nanoparticles: a case study with chitosan-4-thiobutylamidine conjugate. AAPS PharmSciTech 11:1185–1192

Sakuma S, Sudo R, Suzuki N, Kikuchi H, Akashi M, Hayashi M (1999) Mucoadhesion of polystyrene nanoparticles having surface hydrophilic polymeric chains in the gastrointestinal tract. Int J Pharm 177:161–172

Salama HA, Mahmoud AA, Kamel AO, ABDEL HADY M, AWAD GA (2012) Phospholipid based colloidal poloxamer-nanocubic vesicles for brain targeting via the nasal route. Colloids Surf B Biointerfaces 100:146–154

Salathe M (2007) Regulation of mammalian ciliary beating. Annu Rev Physiol 69:401–422

Sanders NN, de Smedt SC, van Rompaey E, Simoens P, de Baets F, Demeester J (2000) Cystic fibrosis sputum: a barrier to the transport of nanospheres. Am J Respir Crit Care Med 162:1905–1911

Saremi S, Atyabi F, Akhlaghi SP, Ostad SN, Dinarvand R (2011) Thiolated chitosan nanoparticles for enhancing oral absorption of docetaxel: preparation, in vitro and ex vivo evaluation. Int J Nanomed 6:119–128

Sarmento B, Ribeiro A, Veiga F, Ferreira D, Neufeld R (2007a) Oral bioavailability of insulin contained in polysaccharide nanoparticles. Biomacromolecules 8:3054–3060

Sarmento B, Ribeiro A, Veiga F, Sampaio P, Neufeld R, Ferreira D (2007b) Alginate/chitosan nanoparticles are effective for oral insulin delivery. Pharm Res 24:2198–2206
Sarparanta MP, Bimbo LM, Makila EM, Salonen JJ, Laaksonen PH, Helariutta AM, Linder MB, Hirvonen JT, Laaksonen TJ, Santos HA, Airaksinen AJ (2012) The mucoadhesive and gastroretentive properties of hydrophobin-coated porous silicon nanoparticle oral drug delivery systems. Biomaterials 33:3353–3362
Schultz M, Callow M, Callow J (2000) A turbulent channel flow apparatus for the determination of the adhesion strength of microfouling organisms. Biofouling 15:243–251
Shahnaz G, Vetter A, Barthelmes J, Rahmat D, Laffleur F, Iqbal J, Perera G, Schlocker W, Dunnhaput S, Augustijns P, Bernkop-Schnurch A (2012) Thiolated chitosan nanoparticles for the nasal administration of leuprolide: bioavailability and pharmacokinetic characterization. Int J Pharm 428:164–170
Shimura S, Andoh Y, Haraguchi M, Shirato K (1996) Continuity of airway goblet cells and intraluminal mucus in the airways of patients with bronchial asthma. Eur Respir J 9:1395–1401
Shirazi T, Longman RJ, Corfield AP, Probert CS (2000) Mucins and inflammatory bowel disease. Postgrad Med J 76:473–478
Simon M, Wittmar M, Kissel T, Linn T (2005) Insulin containing nanocomplexes formed by self-assembly from biodegradable amine-modified poly(vinyl alcohol)-graft-poly(L-lactide): bioavailability and nasal tolerability in rats. Pharm Res 22:1879–1886
Singh A, Singh A, Satheesh Madhav N (2012) Nasal cavity: a promising transmucosal platform for drug delivery and research approaches from nasal to brain targetting. J Drug Delivery Thera 2:22–33
Slutter B, Bal SM, Que I, Kaijzel E, Lowik C, Bouwstra J, Jiskoot W (2010) Antigen-adjuvant nanoconjugates for nasal vaccination: an improvement over the use of nanoparticles? Mol Pharm 7:2207–2215
Smart JD (2005) The basics and underlying mechanisms of mucoadhesion. Adv Drug Deliv Rev 57:1556–1568
Specian RD, Oliver MG (1991) Functional biology of intestinal goblet cells. Am J Physiol 260:C183–C193
Suwannateep N, Banlunara W, Wanichwecharungruang SP, Chiablaem K, Lirdprapamongkol K, Svasti J (2011) Mucoadhesive curcumin nanospheres: biological activity, adhesion to stomach mucosa and release of curcumin into the circulation. J Control Release 151:176–182
Tamburic S, Craig DQM (1997) A comparison of different in vitro methods for measuring mucoadhesive performance. Eur J Pharm Biopharm 44:159–167
Tang BC, Dawson M, Lai SK, Wang YY, Suk JS, Yang M, Zeitlin P, Boyle MP, Fu J, Hanes J (2009) Biodegradable polymer nanoparticles that rapidly penetrate the human mucus barrier. Proc Natl Acad Sci USA 106:19268–19273
Teubl BJ, Meindl C, Eitzlmayr A, Zimmer A, Frohlich E, Roblegg E (2012) In vitro permeability of neutral polystyrene particles via buccal mucosa. Small 9:457–467
Teubl BJ, Absenger M, Frohlich E, Leitinger G, Zimmer A, Roblegg E (2013) The oral cavity as a biological barrier system: design of an advanced buccal in vitro permeability model. Eur J Pharm Biopharm 84:386–393
Thirawong N, Thongborisute J, Takeuchi H, Sriamornsak P (2008) Improved intestinal absorption of calcitonin by mucoadhesive delivery of novel pectin-liposome nanocomplexes. J Control Release 125:236–245
Thorley AJ, Tetley TD (2007) Pulmonary epithelium, cigarette smoke, and chronic obstructive pulmonary disease. Int J Chronic Obstructive Pulm Dis 2:409–428
Umamaheshwari RB, Ramteke S, Jain NK (2004) Anti-Helicobacter pylori effect of mucoadhesive nanoparticles bearing amoxicillin in experimental gerbils model. AAPS PharmSciTech 5:e32
Wang YY, Lai SK, Suk JS, Pace A, Cone R, Hanes J (2008) Addressing the PEG mucoadhesivity paradox to engineer nanoparticles that "slip" through the human mucus barrier. Angew Chem 47:9726–9729

Wang X, Zheng C, Wu Z, Teng D, Zhang X, Wang Z, Li C (2009) Chitosan-NAC nanoparticles as a vehicle for nasal absorption enhancement of insulin. J Biomed Mater Res Part B Appl Biomater 88:150–161

Wang YY, Lai SK, So C, Schneider C, Cone R, Hanes J (2011) Mucoadhesive nanoparticles may disrupt the protective human mucus barrier by altering its microstructure. PLoS ONE 6:e21547

Woertz C, Preis M, Breitkreutz J, Kleinebudde P (2013) Assessment of test methods evaluating mucoadhesive polymers and dosage forms: an overview. Eur J Pharm Biopharm 85:843–853

Yamamoto H, Kuno Y, Sugimoto S, Takeuchi H, Kawashima Y (2005) Surface-modified PLGA nanosphere with chitosan improved pulmonary delivery of calcitonin by mucoadhesion and opening of the intercellular tight junctions. J Control Release 102:373–381

Yu T, Wang Y, Yang M, Schneider C, Zhong W, Pulicare S, Choi W, Mert O, Lai SK, Hanes J (2012) Biodegradable mucus-penetrating nanoparticles composed of diblock copolymers of polyethylene glycol and poly(lactic-co-glycolic acid). Drug Deliv Trans Res 2:124–128

Yuan H, Chen CY, Chai GH, Du YZ, Hu FQ (2013) Improved transport and absorption through gastrointestinal tract by PEGylated solid lipid nanoparticles. Mol Pharm 10:1865–1873

Zhang X, Zhang H, Wu Z, Wang Z, Niu H, Li C (2008) Nasal absorption enhancement of insulin using PEG-grafted chitosan nanoparticles. Eur J Pharm Biopharm 68:526–534

Zhang X, Wang Y, Zheng C, Li C (2012) Phenylboronic acid-functionalized glycopolymeric nanoparticles for biomacromolecules delivery across nasal respiratory. Eur J Pharm Biopharm 82:76–84

Investigation of Nanoparticles in Biological Objects by Electron Microscopy Techniques

Gabriela Kratošová, Kateřina Dědková, Ivo Vávra and Fedor Čiampor

Abstract Electron microscopy (EM) is the most used technique for materials characterisation. Both methods—scanning electron microscopy (SEM) and transmission electron microscopy (TEM) with high resolution mode have become extensively used especially in nanomaterials research. This chapter provides a summary of modern EM techniques used for the analysis of nanoparticles. Regarding a broad spectrum of various nanomaterials being prepared to date or just naturally occurring in the environment, several case studies focusing on nanoparticles analysis by EM methods are introduced. Modified EM methods such as cryo techniques and environmental SEM (ESEM) are also mentioned because of their importance and great potential in research areas combining nanotechnology and biology.

Keywords Transmission electron microscopy (TEM) · Scanning electron microscopy (SEM) · Nanomaterials characterisation

Abbreviations

AEMs	Analytical electron microscopes
BF	Bright field
BSE	Back-scattered electron
BSA	Bovine serum albumin
CCD	Charge-coupled device

G. Kratošová (✉) · K. Dědková · I. Vávra
Nanotechnology Centre, VSB—Technical University of Ostrava, 17. listopadu 15/2172, 708 33 Ostrava, Poruba, Czech Republic
e-mail: gabriela.kratosova@vsb.cz

I. Vávra
Institute of Electrical Engineering, Slovak Academy of Sciences, Dúbravská cesta 9, 841 04 Bratislava, Slovak Republic

F. Čiampor
Institute of Virology, SAS, Dúbravská cesta 9, 841 04 Bratislava, Slovak Republic

A. Prokop et al. (eds.), *Intracellular Delivery II*,
Fundamental Biomedical Technologies 7, DOI: 10.1007/978-94-017-8896-0_8,

Cryo-ET	Cryogenic electron tomography
Cryo-SEM	Cryogenic scanning electron microscopy
Cryo-TEM	Cryogenic transmission electron microscopy
DF	Dark field
DNA	Deoxyribonucleic acid
EELS	Electron energy-loss spectrometry
EDS	Energy dispersive spectrometer
EDX	Energy dispersive X-ray analysis
EM	Electron microscopy
ET	Electron tomography
ESEM	Environmental scanning electron microscopy
FIB	Focused ion beam
FEG	Field emission gun
HAADF	High-angle angular dark field
MTB	Magnetotactic bacteria
SE	Secondary electrons
SEM	Scanning electron microscopy
SEMs	Scanning electron microscopes
SWNT	Single walled carbon nanotubes
TEM	Transmission electron microscopy
TEMs	Transmission electron microscopes
WDS	Wavelength dispersive spectrometer

1 Introduction

The downsizing and development of engineered nanomaterials have invoked a need for more detailed morphological characterisation, quantitative composition of nanoparticles and the determination of at least the character of chemical bounds. In connection with this requirement, electron microscopy techniques have been considerably developed in the last two decades which has enabled the observation and analysis of the diversity of objects at the atomic level.

The size distribution and localisation of nanoparticles may be determined using conventional SEMs, but for detailed analysis of nanostructured materials conventional TEMs should be utilized. Microscopes, which work in the visible region of the electromagnetic spectra as light and/or confocal microscopy, may reveal some properties of bulk nanostructured materials, but these methods are not able to analyse single nanoparticles.

Recently, some of the most modern TEMs and SEMs may operate at 100,000× magnification and more. Additionally, EM in combination with chemical analysis and other imaging methods such as optical or scanning probe microscopy may help us to comprehensively characterise specific nanomaterial.

In this chapter we will discuss some of the latest upgrades to modern EM and its potential in nanotechnology and nanomaterials characterisation, especially with regard to the investigation of nanoparticles associated on/in cells. As illustration, some case studies and/or examples of the application of EM techniques in nanoparticles research are presented.

2 Specimen-Beam Interactions

In both systems—TEM and SEM, the interaction of accelerated electrons with a specimen is essential for the image construction. The series of interactions with nuclei and electrons of the sample occurs when an electron beam strikes the surface of the sample. The interactions make a group of various secondary products such as X-rays, heat, light and electrons with different energy. Many of these by-products are used to generate the images and give additional data about the sample. The most frequent signals in EM are mediated by secondary (SE) and back-scattered electrons (BSE) (Fleger et al. 1993).

Secondary electrons are produced by an inelastic interaction of the beam with the surface of the specimen. The average energy of secondary electrons is around 3–5 eV and therefore are easily attracted to the detector. The major factors that contribute to the ability to produce a topographical image in SEM are absorption and the escape of secondary electrons from the detector. An image of the highest resolution is provided when the secondary electrons escape from a small volume of the whole volume of specimen-beam interactions. Secondary electrons which escaped and electrons which did not reach the detector do not contribute to the final image and they appear as shadows or darker places in contrast (Fleger et al. 1993).

An elastic interaction produces backscattered electrons, beam electrons which have scattered backwards. Backscattered electrons have energy equal to 60–80 % of the initial energy of the electron beam; therefore, a special detector is required (a semiconductor is common). The average atomic number of the sample strongly affects the production of backscattered electrons; the higher the atomic number, the higher volume of backscattered electrons which makes the image brighter. This is applied in material contrast when elements with a higher atomic number appear brighter than elements with a lower atomic number.

Several other products of inelastic scattering are useful in image formation such as X-rays or Auger electrons—electrons with low-energy. X-ray analysis requires a separate detector and an X-ray analyzer. An X-ray is useful for determining the elemental composition of the specimen with elements in the range of atomic numbers from 8 to 99. For specimens with elements in the range of atomic numbers from 2 to 10, the measurement of the energy level of Auger electrons can be used. This measurement also requires a special detector and a high-quality vacuum system (Fleger et al. 1993).

As sources of electron beams several types of cathodes are used. A thermoemission cathode releases electrons when it is heated up to a specific temperature and this type of cathode is the most commonly used in conventional EM. An autoemission cathode (cold cathode or cold emitter) releases electrons on the basis of interactions with the strong electric field which surrounds it. In the case of the most powerful electron microscopes, a field emission gun (FEG) is used as the source of the electron beam. This source may last several years.

Sources of electrons have to be small, stable in the long-term, and user-friendly. They should also exhibit low noise, have a low energy electron emission range, have high intensity electron emission in a small spatial angle, and low operating costs. These conditions are fulfilled by cold emitters and most recently by Schottky emission sources. Compared to thermoemission sources, these emitters are smaller, their intensity is 100× higher and they have better long-term stability as well. The Schottky source of electrons is the most advanced available source of electrons and is usually used in SEMs. More detailed information about method principles and instrumentation may be found in the appropriate literature (Fleger et al. 1993; Hoppert 2003; Williams and Carter 2009).

3 Conventional Electron Microscopy Techniques

Generally, electron microscope techniques are the essential and most employed methods in both top research and academic areas and microscopes are an integral part of many laboratories. In principle, two systems of electron microscopy are differentiated—scanning electron microscopy (SEM) and transmission electron microscopy (TEM). The operating accelerating voltage of the beam and sample preparation protocols are the main differences between these two techniques.

3.1 Transmission Electron Microscopy

In TEM mode accelerated electrons are transmitted through an ultra-thin specimen. The TEM specimen should be carefully prepared from the bulk sample to be transparent for electrons. More detail information about the ultra-thin specimen preparation may be found in the literature (Williams and Carter 2009).

Transmitted electrons carry information about the specimen structure and the final image is recorded by hitting a fluorescent screen, photographic plate or CCD camera (charge-coupled device). Using the CCD the image may be displayed in real time on a computer monitor.

Recently, TEMs work with an overall resolution in the order of tenths of nanometers, and this resolution is sufficient to observe protein macromolecules. A higher accelerating voltage beam is emitted with a shorter wavelength and a lower value of resolution is achieved. Regarding this fact, older TEMs were enormously high and special buildings had to be constructed (the microscope tubus length was over two floors). Nowadays, more and more precise corrections for electromagnetic lenses have been developed (Cs corrections), so that better resolutions may be acquired with a standard sized TEMs.

In the interface of specimen-beam interactions there is a strong influence of a crystal structure, average proton number and the specific weight (mass density) of the observed material on the final contrast. Biological samples are composed of light elements, which do not induce sufficient scattering of the primary electrons during transmission to the specimen. This is a limiting problem in the imaging of such samples. Consequently, these samples may be treated using additional chemicals containing heavy metal ions (Hoppert 2003). However, contrasting or shadowing using additional heavy metal ions is not suitable in the case of nanomaterials due to structural distortion.

As was mentioned before, the accelerating voltage value influences the final resolution. More particular atoms may be observed when using an accelerating voltage of about 200 kV. Nevertheless, the quality of the electron optics is also essential. Atomic resolution may also be achieved together with minor specimen damage when applying Cs correction and an accelerating voltage of about 80 kV. One of the most effective methods used to acquire atomic resolution is called the Z-contrast method. This method can be applied when a microscope is equipped with an autoemission cathode, a scanning transmission electron mode and a high-angle angular dark field detector (HAADF). At present, it is possible to use autoemission emitters to achieve 0.15 nm beam size and current around 1 nA, which is enough for atomic resolution in metals. The ability of TEM to resolve particular atoms is being exploited mainly to study the growth and internal structure of nanomaterials—nanoparticles, nanowires, nanotubes, etc., where this method can provide valuable information.

In biology TEM is used to obtain high resolution images of the internal cellular structures of animal and plant tissues, microbes, viruses, phages and DNA. TEM may also be used for the localization of elements, enzymes, and proteins, for the study of membrane interfaces and the structure of macromolecules. The so-called immuno-electron microscope is used for protein detection using labeled antibodies with colloidal gold or feritin markers (Hoppert 2003).

Material engineers mainly apply the TEM method in the investigation of composite interfaces and the examination of dislocations in metals, polymer structures, thin films, clay mineral layers and the crystal lattice structure of catalytic materials. It may also be used to examine element localisation (Fleger et al. 1993).

3.1.1 TEM Applications in Nanotechnology

In this section we would like to present some examples of TEM application in the investigation of nanoparticles, which are in some way associated with biomass or cells. These examples demonstrate the important role of TEM in nanotechnology research.

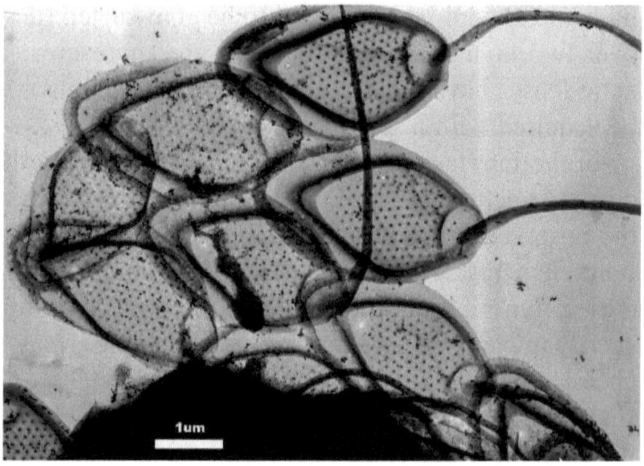

Fig. 1 TEM micrograph of sample containing parts of *Mallomonas kalinae* cells (*Chrysophyceae*) with associated gold nanoparticles, which were predominantly situated in the debris of biomolecules (scale bar 1 μm)

Various organisms have been used to date to produce metal nanoparticles (intra- or extracellular) (Schröfel et al. 2011; Kratošová et al. 2014). The preparation of bio-nanoparticles is based on metal–cell interactions, and the final biosynthesized nanoparticles are mixed with a complex system of cell debris containing metabolite products and different biomolecules, which protect the nanoparticles against aggregation. As an example, the biosynthesis of gold nanoparticles using *Mallomonas kalinae* (Fig. 1) or *Diadesmis gallica* (Fig. 2) is presented here in the form of a TEM microphotograph. Bio-nanoparticles may be used in a colloid solution or as a bio-nanocomposite. Biomass with the associated particles may be dried and stored in the form of powder and directly applied to heterogeneous catalysis (Schröfel and Kratošová 2011).

In TEM there are two main imaging modes—bright-field (BF) imaging and dark-field (DF) imaging. To produce a standard BF image, the electron beam must be able to penetrate the sample with many electrons being transmitted through it. A DF image can be created using only the widely (elastically) scattered electrons from the specimen. This type of image has a high contrast and some details on the specimen may appear more clearly. The difference between BF and DF imaging is illustrated in Figs. 2, 3, 4.

These TEM imaging modes (BF and DF) may be used in nanotechnology to characterize the structures of nanoparticles in nanofluids (see Fig. 3); and in particular to study the penetration of nanoparticles into cells or their localisation in the cell interfaces (see Fig. 4).

A nanofluid is a fluid containing nanoparticles. These fluids are engineered colloidal suspensions of nanoparticles in a base fluid. The nanoparticles used in nanofluids are typically made of metals, oxides, carbides, or carbon nanotubes. Common base fluids include water, ethyleneglycol, kerosene, mineral oil, etc. Nanofluids have distinct properties that make them potentially useful in many applications, e.g. in heat transfer, microelectronics, fuel cells, pharmaceutical processes, etc. They exhibit enhanced thermal conductivity and a convective heat transfer coefficient compared to the base fluid.

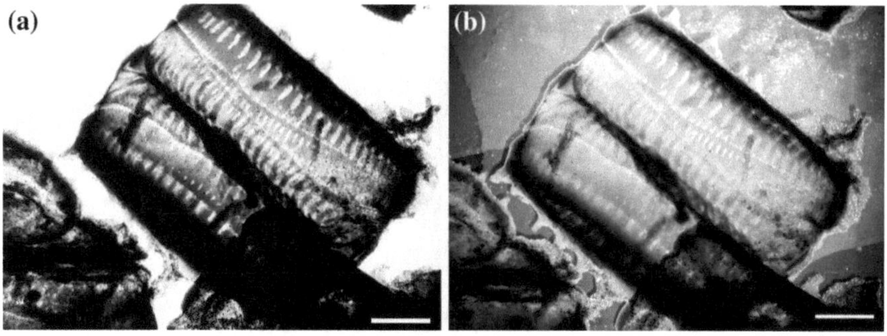

Fig. 2 TEM micrographs of diatom cell (*Diadesmis gallica*) used for biosynthesis of gold nanoparticles. Gold nanoparticles are being associated on the silica frustules, stabilized by secreted biomolecules and form so called bionanocomposite material with promising application potential (scale bar 1 μm). The image **a** is acquired in the BF mode, **b** in the DF mode with more visible gold nanoparticles in the debris round the cell (compare imaging of the same sample with cryo-SEM mode in the Sect. 4.2)

Nanofluids containing magnetic nanoparticles are called magnetic fluids. Magnetite nanoparticles are used as a magnetic moment carrier. The properties of magnetic fluids can be easily influenced by external magnetic fields and thus they are widely applicable in industrial practice as well as biomedicine, especially in the targeted transport of drugs in cancer treatment, cardiovascular disease treatment and radiodiagnostics (Závišová et al. 2011; Šafařík et al. 2012).

Magnetotactic bacteria (MTB) are extremely interesting microorganisms containing magnetite nanocrystals which form so-called magnetosomes in their cells (see Fig. 5). MTB use these magnetosomes to migrate along the magnetic field. Magnetosomes (organelles with chains of magnetite nanocrystals) are composed of real nanocrystals, which indicate extraordinary crystalline perfection. Magnetite nanocrystals may also be prepared in laboratory conditions; however, their quality and variability are much lower.

Bacterial magnetic particles may be exploited in various biotechnological or biomedical applications and have a great potential in nanotechnology. Isolated (exfoliated) magnetosomes can be incorporated in polymer. If the polymerization is performed in an external magnetic field, we can obtain a composite material with strong magnetic anisotropy (Prozorov et al. 2013).

3.2 Scanning Electron Microscopy

SEM produces images of the specimen surface with strong 3D plasticity and a large depth of field, which allows the spatial arrangements of all structures to be viewed. SEM is used when information about the surface or near-surface region is required. The element analysis features connected to SEM are the most important

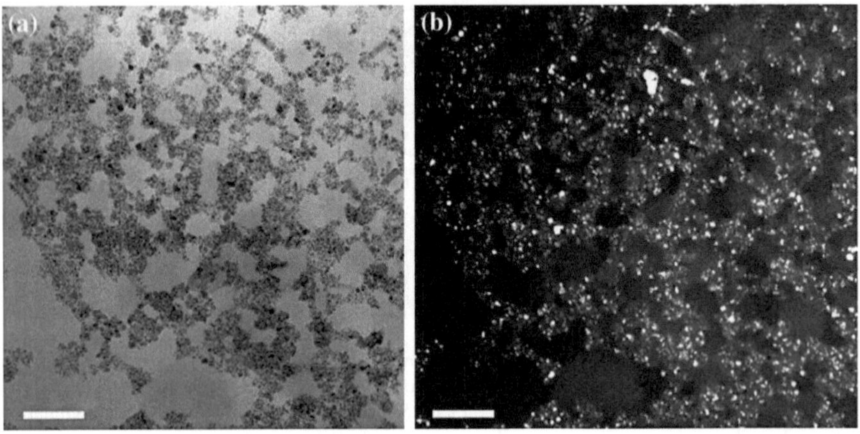

Fig. 3 Bright (**a**) and dark field (**b**) TEM micrographs of 12 nm magnetic nanoparticles in water (scale bar 200 nm). The dark field mode is used for the evidence of the nanoparticles crystallinity and easier detection of nanoparticles in biological specimens. The TEM specimen was prepared by drying of small drop on thin amorphous carbon foil

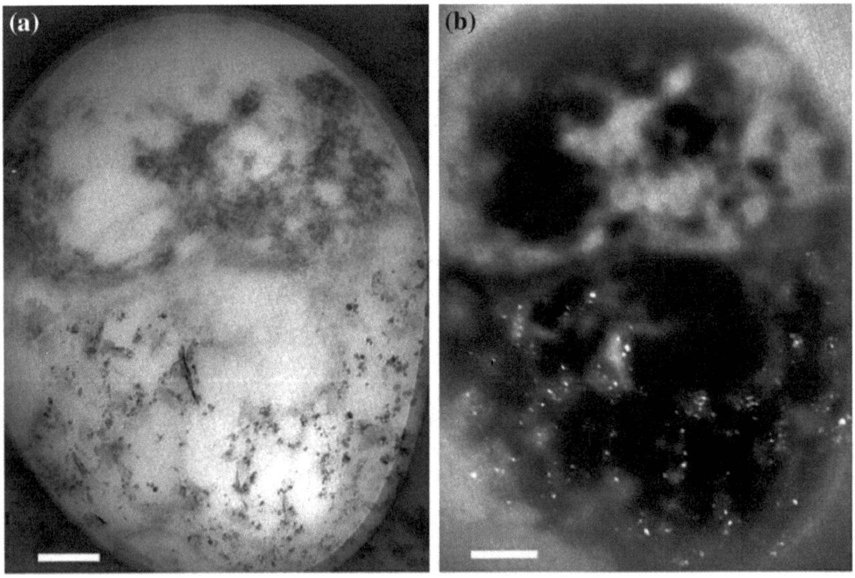

Fig. 4 The BF (**a**) and DF (**b**) micrographs of the region near nucleus membrane (scale bar 200 nm). The human marrow bone cells were in vitro contaminated by magnetite nanoparticles. It is clearly seen that nanoparticles do not penetrate into nucleus (the upper part of micrographs)

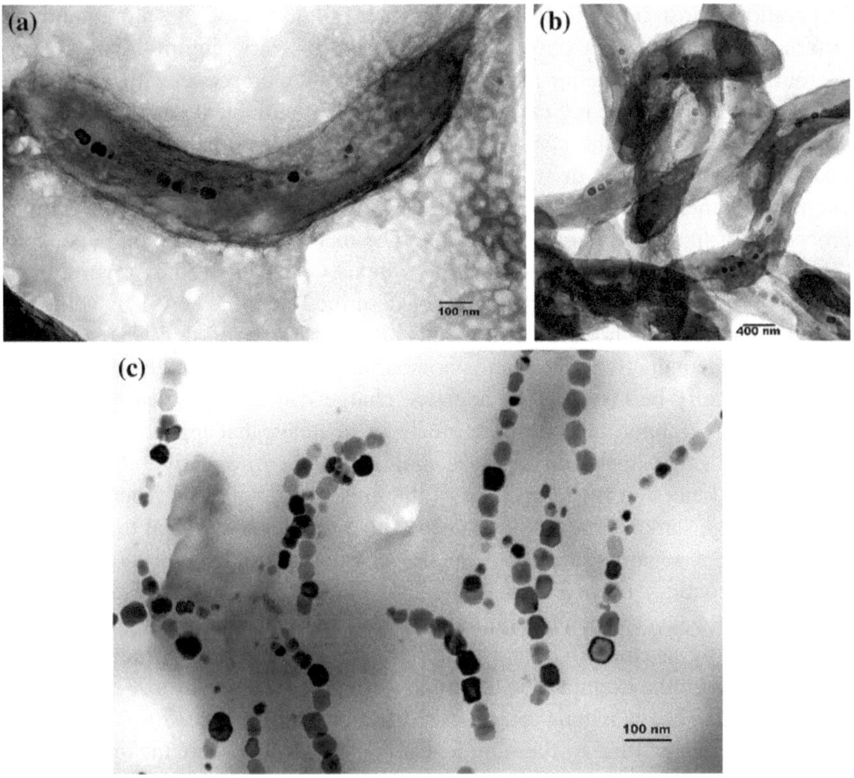

Fig. 5 Magnetotactic bacteria *Magnetospirillum magneticum*. **a** TEM micrograph of single bacterium. **b** TEM micrograph of the magnetotactic bacteria colony. **c** TEM micrograph of the exfoliated bacteria and magnetosomes

for material science research. This method is widespread in many fields of science, technology and industry.

SEM operates using a range of magnification up to 100,000 fold with a resolution limit below 1 nm. The ratio of the size of the displayed image to the size of the scanned area on the specimen gives the magnification. The reduction of the scanned area increases the magnification. The diameter of the focused beam on the surface of the specimen determines the image resolution. The source of the electron beam has an influence on the resolution of image as well (Fleger et al. 1993).

In the case of SEM, the size of a sample is limited by the specimen chamber only and the sample preparation is much easier and less complex than for TEM. There are three main criteria which are required. The sample must be (1) devoid of water or other solvents which can evaporate and contaminate the vacuum system, (2) firmly mounted and (3) electrically conductive. On this account, conductive materials (metals, semiconductors) may be observed without any special preparation. Most biological samples do not fulfill any of the aforementioned criteria and therefore they have to be treated before positioning in the microscope chamber. Allowances being made for that fact, a microscope with an optional vacuum or low voltage regime is used. Regarding biological samples, it is also necessary to consider the character of the tissues. Bones, hair, teeth, insect cuticulas, diatom frustules or woods only need to be metal-coated to ensure conductivity of their surface. On the contrary, soft tissues require a more complex procedure of fixation and/or dehydration. The method of preparation depends on the type of sample and the information which is needed. Details regarding the preparation of biological samples are not described here because it is somewhat complicated and it is not the point of this chapter (Fleger et al. 1993).

As the nonconductive specimen is bombarded by the electron beam, a negative charge gradually builds up. This negative charge can then produce lines on the screen or image, abnormal contrast or breaks in the image that appear as if the image has been split. SE operation mode is more sensitive to negative charging than backscattered electron mode because of the high energy of BSE. Therefore, most nonconductive samples have to be coated with a thin layer of metal to prevent the formation of a negative charge. A sputter coating of gold is the most common method of coating; however, platinum or palladium coating is also used. The thickness of the layer which is required varies somewhat according to the texture of sample, usually around 10–30 nm. Carbon coating can be used instead of heavy metal coating for the examination of samples using BSE mode or for X-ray analysis. This is because heavy metals can obscure the atomic differences for BSE and will absorb the X-ray emitted by the sample. Carbon does not obscure the differences in atomic number and has a low absorption factor for X-rays (Fleger et al. 1993).

3.2.1 SEM Applications in Nanotechnology

The morphological characterization of nanomaterials is important both for developmental and nanotoxicological studies. The shape of observed nanoparticles can be visualized by SEM in an impressive manner. The study of morphological properties is important because it can affect their physical and biological effect. Spherical, rod-like structures, cuboids and many other structures have been observed and identified. Nanoparticles in dispersion have the tendency to agglomerate. These processes can be investigated using SEM as well. Aggregation tendencies can be reduced using stabilizers such as bovine serum or human serum albumin. The positive effect of stabilizers on zinc oxide and titanium dioxide nanoparticles in aqueous suspensions has been documented by SEM using a working protocol for fixing single isolated nanoparticles on poly-L-lysine coated substrates (Tantra et al. 2011).

The shape of nanoparticles may play an important role in biomedical applications because it influences the interaction of nanoparticles with target sites. SEM is employed for visualization of the particles morphologies. Recent studies have used SEM to characterize nanocomposite films as matrices for silver nanoparticles (Srivastava et al. 2011).

Electron microscopy techniques play an important role in the field of characterisation of drug delivery systems. Use of these methods allows the retrieval of information regarding the microstructure and elemental composition of several drug delivery systems such as nanoemulsions, liposomes, lipid particles, liquid particles, nanofibers etc. (Klang et al. 2013). The surface morphology of polymeric triclosan nanoparticles was investigated using SEM and TEM. It is difficult to interpret the information acquired from SEM about nanoparticle fusion and film formation. The shape of nanoparticles and the aggregation phenomena were investigated via TEM. These nanoparticles are interesting for researchers because they could be employed as drug carrier systems for the treatment of various skin diseases (Dominguez-Delgado et al. 2011).

SEM with EDS has served a role in the investigation of interactions between tungsten nanofibers and epithelial cells in vivo. *Porcellio scaber* (as a model organism) ingested tungsten nanofibers and its digestive gland epithelium was observed under SEM. EDS confirmed the presence of nano-WO_x. These particles were found inserted into the digestive gland cells, but no conventional toxicity markers such as mortality or weight changes were observed (Millaku et al. 2013).

SEM in connection with focused ion beam (FIB) cutting has enabled researchers to understand the evolution of pore volume, pore-solid interfacial area, pore shape, pore connectivity and pore number in the course of sintering nanoparticles of zinc oxide. FIB-SEM 3D reconstruction is a method that provides the opportunity for the observation of pore evolution and to understand the sintering process in nanoscale (Chen et al. 2013).

The presence of various micro- and nano-sized metallic particles in human tonsils has also been studied using SEM (Zeleník et al. 2013). These particles may potentially cause an inflammatory response as well as neoplastic changes in human tonsils. The aim of the study was to evaluate the palatine nosils of patients with chronic tonsillitis and spinocellular carcinoma to determine the presence of nano-sized particles (Fig. 6).

4 Microscopy at Cryo-Temperatures

One of the factors which supports the development of new research areas such as biomimetics, nanobiology, bio-nanotechnology, etc. is the rapid expansion of cryo-microscopic imaging techniques. These techniques, due to their high-resolution, shift the research more closely to studying the living organism's ultrastructures.

Using present-day EM techniques, not only engineered materials at nanometer scale may be observed; but also the secrets of plant and animal worlds can be discovered. Nature proves to be a specialist in nanotechnologies even though we have not been able to evaluate its sophistication until now due to the emergence of electron microscopes with high-resolution abilities. There are many examples

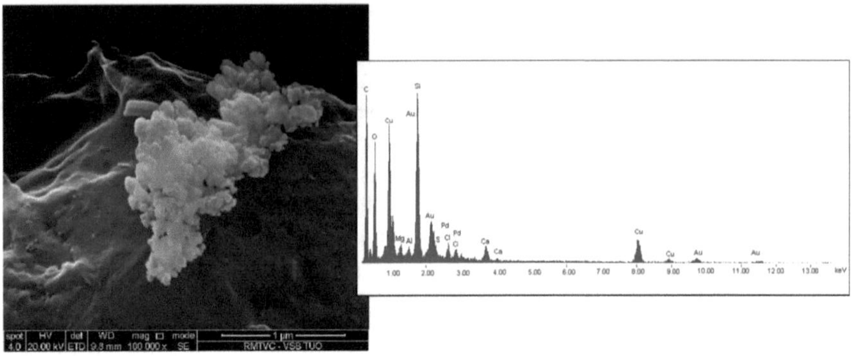

Fig. 6 SEM micrograph of nanoparticles agglomeration observed in human tonsils. According to the EDS analysis it was confirmed that particles are composed of some cupreous salt (sample was placed on the glass, Au and Pd is originated of evaporative coating of the sample)

from nature which are inspiring the current top scientists who are focused on material research and biomimetics.

The currently available microscopic techniques are able to divide the studied object directly in the microscope chamber, cut the sample, etch its surface layers as well as model the shape of the sample using an ion beam (see Sect. 5.2). Because of these advances a lot of new and unique information is being discovered.

It can be said that we are still at the beginning of the investigation of the biological micro-world in all its complexity. It will probably take a long time to understand (if ever) all the details of living organisms with all their functions. However, electron microscopy may help us considerably to understand many exceedingly complicated and complex structures and processes of living organisms.

4.1 Cryogenic Transmission Electron Microscopy

Cryogenic transmission electron microscopy (cryo-TEM) is an analytical method for discovering the structure of soft nanostructured materials such as proteins, liposomes, lipid emulsions, nucleic acids etc. This method is based on the examination of a sample prepared by ultra-fast cooling and the conversion of a liquid sample to a vitrified specimen in TEM. Cryo-TEM reveals global supramolecular structures and local aggregate-specific details at the nanometer resolution. This is an advantage in multiple structure systems with different morphologies, size or internal order. Cryo-TEM is also used to study time-dependent processes because of its ability to take down metastable and short-lived intermediates which is important to understanding and proving self-assembly mechanisms or structures (Ziserman 2011).

The sample is visualized in a frozen hydrated state in cryo-TEM. Some additional features to basic TEM are required. A specialized holder for the microscopic grid holding the specimen is the most important part of the cryo equipment. The holder contains a small dewar as a cooling agent at its end. Cooling of the tip of the holder is performed by a thermally conductive connection between the dewar and tip. The grid with vitrified film is placed and fixed into the cooled tip (Williams and Carter 2009). Thermal contact between holder tip and grid is crucial to ensure appropriate cooling of the sample. If cooling of the sample is inappropriate, then a freeze drying effect can damage the sample. A shutter or gliding shield is incorporated into the holder and can protect the specimen from contamination by cubic ice during transport from the preparation or transfer unit to the TEM. An external temperature monitoring system observes the temperature during the procedure. Tilting of the sample during observation is also possible with many holders. Tilting provides information about the three-dimensional structure. An additional protection system is needed during the transportation of the holder between the sample preparation unit and TEM to avoid any formation of ice on the shutter or holder. Freezing can damage the shutter system and bring moisture into the vacuum (Jiang and Chiu 2007).

Samples containing water and other volatiles have to be chemically or physically fixed to avoid evaporation and supramolecular motion which may lead to blurred images. A drop of sample is adsorbed onto a carbon film, residue solution is blotted with filtration paper, and some chemical compound is added to the chemical fixation. Heavy metal salts (mainly uranyl acetate) are frequently used. After this, the sample is air-dried and examined in TEM at room temperature. This method is called negative staining and is cheap, easy to perform, and provides high contrast (De Carlo and Harris 2011). Negative staining is useful for macromolecular complexes and stable assemblies. However, stain-sample interaction can lead to the stain limiting the resolution or pH changes because of the stain or drying. Negative-staining is unusable for systems with complex phase behavior such as nanostructured liquids because uncontrolled phase and structural changes are caused by evaporation or changes in temperature. Change of state of the matter, from fluid to glassy is included in cryo-TEM. Ultra fast cooling is the only way to convert a thin film of the fluid suspension to a vitrified, low vapor specimen. Nanostructures are protected against solvent crystallization and associated optical artifacts. They are also protected against mechanical damage or redistribution of solutes using high cooling rates of hundreds of degrees in milliseconds which lead to the maximization of the surface area-to-volume ratio of the specimen. Proper cryogens are low-temperature liquids with high thermal conductivities far from their boiling point such as liquid nitrogen or ethane which are commonly used.

Plunger devices are used to vitrificate specimens. They are closed chambers which enable sample preparation under controlled temperatures, typically around 0 and −75 °C. Open plungers cannot be used for studying nanostructured fluids with complex phase behavior. The most important steps of specimen preparation are blotting and plunging. A small drop is placed onto a perforated carbon film-supported grid held by tweezers and pre-equilibrated in the plunger chamber at the chosen temperature and close to saturation. The rest of the solution is removed by blotting the filtration paper which produces a thin liquid film suspended across the holes of the grid. The grid is then placed into the cryogen to make a vitrified sample. If the specimen is 100–200 nm thick, we can say it was successfully prepared. Nanoparticles are embedded in a clean layer of glassy ice. Vitrified ice indicates that the water molecules didn't rearrange into crystalline ice during the process of fixation thus it may be said the nanoparticles preserved their bulk state (Danino 2012).

The most critical step of sample preparation is blotting because it determines the thickness of the film and its quality. In some specimen preparation devices the sample is blotted and plunged manually but most modern devices e.g. Vitrobot (FEI), CP3 (Gatan) or Leica EM GP enable automated blotting and plunging functions which makes the specimen preparation simpler and enables many research laboratories to use cryo-TEM successfully.

4.1.1 Imaging of Vitrified Specimens

Imaging of vitreous specimens is challenging because of low inherent contrast and high sensitivity to electron beam radiation damage. Low molecular weight elements such as hydrogen, carbon, oxygen or nitrogen are the structural elements of many nanostructured fluids. Their contrast is usually very low which means that they are almost invisible. The most contrast-enhancing mechanism used in cryo-TEM is phase contrast which is based on converting wave phase differences into amplitude differences. It is achieved by focusing the objective lenses which make regions of the different inner electron potentials relative to the vitreous ice visible. The electron contrast transfer function can also cause loss of resolution.

Improvement of specimen preparation devices and digital imaging have led to the use of cryo-TEM as a common characterisation tool for soft molecular assemblies and nanostructured liquid systems when it helps to understand the micellization processes and micellar structure. Cryo-TEM also has huge potential for the investigation of the structures of colloidal drug delivery systems e.g. liposomes, lipid emulsions and suspensions or nanoparticles based on liquid crystalline phases (Kuntsche et al. 2011).

Single walled carbon nanotubes (SWNT) dispersed in bovine serum albumin (BSA) have been imaged without a staining agent. SWNT with a diameter around 1.4 nm are easily distinguishable in cryo-TEM. A study of the pH effect on SWNT dispersed in BSA was performed; and it was found that if the electrostatic charge of BSA is minimized, a thick layer of BSA with a high spatial density is formed on the SWNT. When the pH was around 10, a thinner BSA layer was formed on the SWNT (Edri and Regev 2010).

Electron tomography (ET) is a three-dimensional structural investigation using TEM. This method enables the characterisation of structures at a nanometer-scale resolution. The use of cryo-electron tomography has allowed for the determination of the 3D architecture of intermediate steps of the transport of hydrophilic silica nanoparticles into unilamelar neutral liposomes through an internalization process. Nanoparticles which have a strong interaction with lipid membranes can be transported through the membrane by an invagination without the need for energy. Investigation of this process is important for understanding the cytotoxicity of nanoparticles (Le Bihan et al. 2009). Cryo-ET is widely used in biology to map in 3D the biological macromolecules at near-atomic resolution to get more information about their spatial organization, function and dynamics (Zhang et al. 2010).

4.2 Cryogenic Scanning Electron Microscopy

Cryogenic scanning electron microscopy (cryo-SEM) combines the high performance imaging of SEM with cryogenic sample preparation which enables the investigation of structures and materials in their native, hydrated state with nanometer image resolution. Cryo-SEM is becoming used routinely in laboratories with broad applications in the life and material sciences.

Coating is very important in cryo-SEM. Not just because it provides a conductive pathway, but because coating can also enhance the surface detail observable in the SEM image. Biological samples are usually made of relatively light elements such as carbon, hydrogen and oxygen. The presence of heavy metal coating can reduce beam penetration and enhance surface detail in images of samples composed of light elements (Greiser 2009; Klang et al. 2013).

Contamination is the primary source of artifacts and interferences in cryo-SEM because a very cool sample surface can easily collect contaminants from the environment. Contaminants can be from the microscope (vacuum system) or the sample can produce them itself such as water molecules sublimated from the sample surface. Anti-contamination devices or plasma cleaning systems in the sample chamber can reduce the presence of contaminants (Greiser 2009; Klang et al. 2013).

Samples rich in water-containing structures on their surface need to be submitted to plunge-freezing with liquid nitrogen. These types of samples are usually left for 2 min in the preparation chamber at low temperature for removing of natural surface ice. Afterwards they are sputter-coated for conductivity and then transferred to the specimen chamber in SEM (Greiser 2009; Klang et al. 2013).

The cryo-fracture technique enables the observation of the internal structures of the specimen at high-resolution. In the first step, the specimen is cooled to a specified temperature and then it is fractured within the cryo chamber. The crucial step is the control of the sample temperature which affects the nature of the fracture. The exposed surface is typically sputter-coated before the observation (Greiser 2009; Klang et al. 2013).

The cryo sublimation technique reveals even greater detail of the specimen. This method is based on raising the temperature of a hydrated free-fracture specimen to a point at which water begins to sublimate at a controlled rate (-90 to -100 °C). Lowering the temperature again can stop this etching process at any time. This process can be controlled in the specimen chamber or in the cryo chamber (Greiser 2009; Klang et al. 2013).

Internal dehydration of the specimen can occur when the sample is left long enough at a slightly higher temperature relative to the anti-contaminator. A completely dried sample is the result. If the dried sample is removed from the SEM and the sputter-coating is sloughed off in a water bath, a metal-sputtered replica of the original surface can be retrieved. A supported grid collects this replica sample and it can be examined with SEM for high-resolution investigation (Greiser 2009; Klang et al. 2013).

Cryo-SEM has revealed microstructural evolution in dry cast cellulose acetate membranes. The authors have used time-section cryogenic SEM to visualize the phase inversion process when rapidly freezing specimens. When drying starts, the specimens are ostensibly identical and are observed at various intermediate stages of their controlled dry cast processing. The frozen samples are then fractured with control to expose cross-sections, sublimated to reveal phase-separated topography, sputter-coated with platinum layer and finally imaged in the cryo-SEM under low accelerating voltages and probe currents. A series of images of time-sections document the stages of evolution as a function of processing time and space (Parkash et al. 2006).

Cryo-SEM seems to be a reliable technique for micro-morphological studies of microbial sediments and microbial soil crusts. Clay-amended sand has been characterised via cryo-SEM as well. Cryo-SEM was applied to the study of wet-state model soil colloids. The samples were fixed by plunge-freezing with nitrogen slush—a highly thermal conductive liquid. The comparison of the wet-state conventional SEM images with freeze-dried cryo-SEM images revealed that the drying process affects the microstructure of the samples by promoting the formation of aggregates. Cryo-SEM observation can also give morphological interpretations of the particular behavior of some natural colloids such as the high stability of humic-allophanes (Négre et al. 2004).

Ultrastructural information of the internal structure of hydrated samples has been obtained via cryo-SEM. Observation of cryo-fractured calcium pectinate beads appears to be excellent for studying the detailed morphology of structural development and it is much better than conventional SEM because of freeze-drying the sample preparation (Sriamorsnak et al. 2008).

A bicontinuous cubic liquid crystalline system formed by phytantriol and water were investigated under cryo-SEM with a field emission gun as the source of the electron beam. This method examined the three-dimensional and surface structure of the bulk cubic phase and cubosomes with the observed structural features. The results gained from this investigation coincide well with mathematical models (Rizwan et al. 2007).

The cryo-SEM technique was also used to observe bio-gold nanoparticles synthesized by use of brown algae. Due to material contrast the nanogold particles are clearly visible on the diatom frustules (Fig. 7a), and they could be easily differentiated from the particles of ferrofluid which was used to magnetically modify the bio-nanocomposite (Fig. 7b).

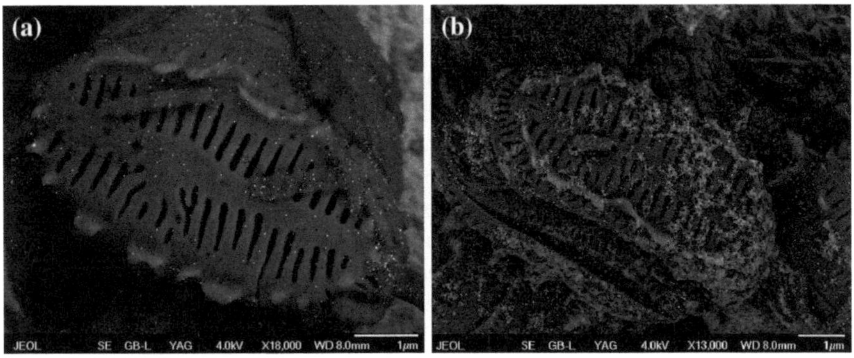

Fig. 7 Cryo-SEM micrographs of **a** Silica frustules of *Diadesmis gallica* and biosynthesized nanogold (*bright dots*). Gold nanoparticles remain captured in the extracellular space, or are directly associated with the diatom frustule structures. **b** Silica frustules with nanogold magnetically modified with gray ferrofluid veil (scale bar 1 μm)

5 Microscopy at Subatmospheric Pressure

5.1 Environmental SEM

As it has been said before, samples containing water and other volatile molecules cannot be examined in SEM, thus biological samples must be dried and many other samples such as food, gels or semisolid materials cannot be observed. Cryo-SEM can solve this problem in the case of some vapor-producing samples. Nevertheless, even cryo-SEM is not able to examine the drying process of adhesives, the curing of cement, the melting of alloys or the process of crystallization. All of these disadvantages have been overcome applying environmental SEM (ESEM). The main difference between SEM and ESEM is in the use of a near-atmospheric environment in the sample chamber which is more conducive to the examination of wet samples than by a high vacuum in SEM. Low vacuum ranges (from 1.3 to 2.7 Pa) enable the examination of many biological and food samples without the need for drying. Using a specimen heater allows for the observation of the melting or solidification of metal alloys. The drying process can be observed when the vacuum level increases (Fleger et al. 1993).

The basic principle of producing the image is the same with conventional SEM. However, ESEM deviates substantially from SEM in several aspects. ESEM is equipped with special electron detectors (Everhart-Thornley) and a different pumping system to allow the transfer of the electron beam from the source (which is surrounded by a high vacuum) to the specimen chamber (at nearly atmospheric conditions). Another deviation from conventional SEM, with enormous advantages, is the resolution of a problem with the accumulation of charge on the surface of non-conductive samples. The gas in the ESEM is electrically conductive. The conductivity of the gas is due to the ionization that occurs with the incident electron beam and the ionizing SE and BSE signals (Danilatos 1993). Every specimen transfer requires the venting of the specimen chamber to the ambient pressure (100 kPa). This requires that a large volume of gas must be pumped out and replaced with the gas of interest which is usually water vapor. During the transfer, there is a decrease of the relative humidity below 100 % which could be a problem in the case of experiments requiring uninterrupted 100 % relative humidity (Cameron and Donald 1994).

ESEM has new and unique possibilities such as the opportunity to examine hydrated specimens, since any pressure greater than 609 Pa allows them to be maintained in their liquid phase at temperatures above 0 °C. Non-conductive samples do not have to be sputtered-coated with noble metals. The gas itself is used as a detection medium which results in novel imaging possibilities, as opposed to vacuum SEM detectors. Plain plastic uncoated scintillating BSE detectors can function without charging; therefore, the highest possible signal-to-noise ratio at the lowest possible accelerating voltage is produced by these detectors. This is due to the fact that the BSE do not disperse energy to the metal coating used for the vacuum SEM. Consequently, specimens can be examined easily and more quickly without modification of the natural surface or the creation of artifacts during the preparation process. Phase interactions such as gas/liquid/solid can be studied in situ and in real time or recorded for subsequent utilizations. Biological specimens can be kept fresh and viable. Thus, ESEM represents a radical departure from conventional electron microscopy, where vacuum conditions hinder the universal adoption of electron beam imaging.

Some representative applications are in the following areas: biology, medicine and medical applications (pharmaceutical systems), industry, in situ studies and material science.

5.2 Focused Ion Beam Technique

In recent years, other beams of charged particles have also been added to the scanning electron microscope. The focused ion beam (FIB) microscope is constructed as an environmental microscope. It is possible to lower the vacuum in the microscope chamber and remove the electrical charge from the specimen surface.

Ion beams enable not only a specific sample imaging, but due to their destructive character they may also be used for in situ modification and treatment of the observed sample directly in the microscope chamber. In addition to inorganic samples, the inner physique of microorganisms or small animals may also be studied in detail, as wee can see in Fig. 8.

6 Electron Microscopy and Chemical Analysis

Most electron microscopes are equipped with some analytical spectrometer to ensure the chemical analysis of the sample. These microscopes are considered to be more powerful instruments called analytical electron microscopes (AEMs). The

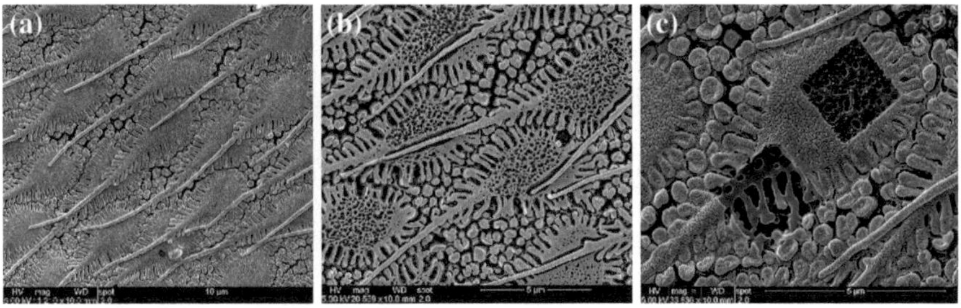

Fig. 8 The SEM (FIB) micrographs of the water beetle surface (genus *Graphelmis*). **a** Original surface. **b** Surface after ion beam removing by Ga ion beam in dual beam microscope. **c** Detail of local ion beam etching of the surface. It is clear that the morphology of the beetle surface is different in the depth of approx. 1 μm. Pictures document the fact, that living organisms are structured in sub-micron dimensions

fundamental principle of this analytical technique is that each element has a unique atomic structure; and after the specimen interacts with electrons, atoms from the sample emit characteristic X-rays, which can then be further evaluated. In this way qualitative and quantitative analysis, distribution of elements (mapping) or line analysis may be performed. The optimum accelerating voltage is determined by the elements present in the studied sample. To obtain correct intensities, the accelerating voltage should not be less than twice the highest excitation energy of any element present (Williams and Carter 2009; Fleger et al. 1993).

Existing detectors may evaluate characteristic X-rays according to the energy (Energy Dispersive Spectrometer—EDS or Energy Dispersive X-Ray Analysis—EDX) or the wavelength (Wavelength Dispersive Spectrometer—WDS). The first method (EDS) is more common because of price and the rate of spectra acquisition, the second (WDS) is more precise.

It is not a problem to analyse the chemical composition of the agglomeration of some nanoparticles (as demonstrated in Fig. 6). In this case the spectra is acquired from a selected area of the agglomeration. In the case of individual nanoparticles, the application of EDX analysis is rather problematic due to the extremely small dimensions of the particle (in the order of 10^{-3} μm^3). Other methods should be preferred for more precise analysis of individual nanoparticles (e.g. Electron Energy-Loss Spectrometry—EELS), due to the dominant influence of statistical errors, decreasing concentration, and uncertainties in background corrections.

6.1 Electron Energy-Loss Spectrometry

Electron Energy-Loss Spectrometry (EELS) is the analytical method which evaluates the energy distribution of electrons that have interacted inelastically with the specimen. These inelastic collisions display the electronic structure of the

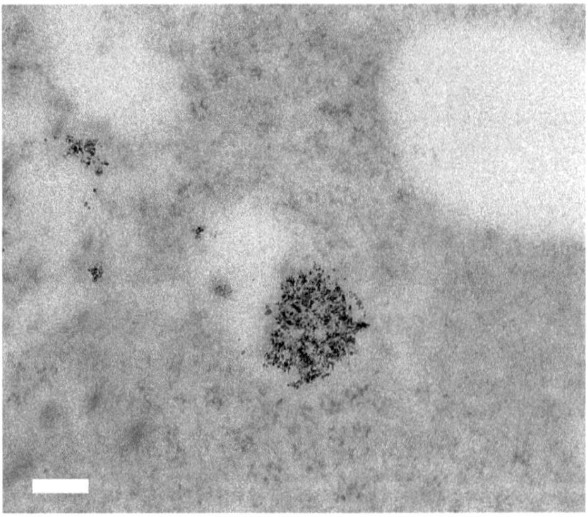

Fig. 9 Bright field (EELS) micrograph (0-loss) showing the distribution of magnetite nanoparticles in human cell cytoplasm. Nanoparticles are assembled predominantly in vacuolas (scale bar 200 nm)

specimen atoms, which in turn reveals the details of the nature of these atoms - their bonding and nearest-neighbor distributions, and their dielectric response. We can apply the EELS method to any material—amorphous or crystalline (Williams and Carter 2009). Using TEM optics together with the EELS spectrometer, it is possible to construct so-called filtered imaging which corresponds to electron spectroscopic images which reveal the elemental distribution of particular elements.

Additionally, thickness information is present in the electron energy-loss spectrum, since the intensity of inelastically scattered electrons increases with the specimen thickness. EELS is applicable over a wide range of thicknesses, and with parallel collection spectrometers it is also possible to obtain thickness maps of thin foils or films (Williams and Carter 2009).

Due to its unique resolution, this method is extraordinarily suited for chemical analysis, localization and distribution assessment of individual nanoparticles in various environments. As seen in Figs. 9 and 10, EELS detected the penetration of magnetite nanoparticles into the cytoplazma of human cells and the localization of nanoparticles predominantly in vacuolas.

7 Conclusions and Perspectives

Nanoparticles in biological objects represent a type of composite material of organic/inorganic substances. The problem of structural analysis by electron microscopy is complicated by a significant difference in the specific mass of both components. Consequently, the biological component has a lower radiation resistivity compared to the inorganic nanoparticle. Moreover, nanoparticles are

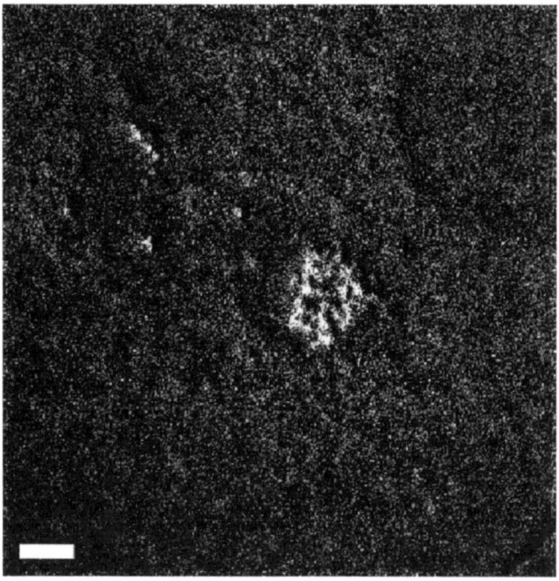

Fig. 10 EELS elemental map of Fe distribution reveals that magnetite nanoparticles are also uniformly distributed in cytoplasma (scale bar 200 nm)

often embedded in the organic (usually amorphous) which causes a higher dissipation of the electron beam. Additionally, the nanoparticles have small dimensions and therefore the microscope must have a good resolution.

In the last two decades, new techniques of sample preparation and new methods of electron microscopy have been developed. The implementation of the CCD technique has allowed for the substantial reduction of radiation damage to samples. Together with new analytical methods (e.g. EELS), such investigations as were presented have become possible.

Acknowledgments This paper has been elaborated from the framework of the *Nanotechnology the basis for international cooperation Project, reg. no. CZ.1.07/2.3.00/20.0074* supported by the Operational Programme "Education for competitiveness" funded by the Structural Funds of the European Union and the state budget of the Czech Republic. The authors gratefully acknowledge support from the SGS project No. SP2013/48. The authors also would like to thank Jay Davis for valuable language corrections.

References

Cameron RE, Donald AM (1994) Minimizing sample evaporation in the environmental scanning electron microscope. J Microsc 173:227–237

Chen B, Xia Z, Lu K (2013) Understanding sintering characteristics of ZnO nanoparticles by FIB-SEM three dimensional analysis. J Eur Ceram Soc 33:2499–2507

Danilatos GD (1993) Environmental scanning electron microscope—some critical issues. Scan Microsc Suppl 7:57–80

Danino D (2012) Cryo-TEM of soft molecular assemblies. Curr Opin Colloids Interface Sci 17:316–329

De Carlo S, Harris JR (2011) Negative staining and cryo-negative staining of macromolecules and viruses for TEM. Micron 42(2):117–131

Dominguez-Delgado CL, Rodriguez-Cruz LM, Escobar-Chaves JJ, Calderon-Lojero LO, Quintanar-Guerrerro D, Ganem A (2011) Preparation and characterization of triclosan nanoparticles intended to be used for the treatment of acne. Eur J Pharm Biopharm 79:102–107

Edri E, Regev O (2010) Cryo-staining techniques in Cryo-TEM studies of dispersed nanotubes. Ultramicroscopy 110:751–757

Fleger SL, Heckman JW Jr, Klomparens KK (1993) Scanning and transmission electron microscopy: an introduction. Oxford University Press, New York. ISBN 978-0-19-510751-7

Greiser J (2009) Advances in Cryo-SEM: from micrometers to nanometers. The New American Laboratory Website—the Ultimate Online Resource of Laboratory Scientist. http://www.americanlaboratory.com/914-Application-Notes/577-Advances-in-Cryo-SEM-From-Micrometers-to-Nanometers/

Hoppert M (2003) Microscopic techniques in biotechnology. Wiley, Weinheim

Jiang W, Chiu W (2007) Cryoelectron microscopy of icosahedral virus particles. In: Kuo J (ed) Electron microscopy: methods and protocols, vol 369, 2nd edn., Methods in Molecular Biology, Humana Press, Totowa, pp 345–363

Klang V, Valenta C, Matsko NB (2013) Electron microscopy of pharmaceutical systems. Micron 44:45–74

Kratošová G, Vávra I, Horská K, Životský O, Němcová Y, Bohunická M, Slabotinský J, Rosenbergová K, Schröfel A (2014) Synthesis of metallic nanoparticles by silica based algae—outline, prospect and applications. In: Rai M, Posten C (eds) Sustainable green synthesis of nanoparticles. Springer, Berlin

Kuntsche J, Horst JC, Bunjes H (2011) Cryogenic transmission electron microscopy (Cryo-TEM) for studying the morphology of colloidal drug delivery systems. Int J Pharm 417:120–137

Le Bihan O, Bonnafous P, Marak L, Bickel T, Trepout S, Mornet S, De Haas F, Talbot H, Taveau J-C, Lambert O (2009) Cryo-electron tomography of nanoparticle transmigration into liposome. J Struct Biol 168:419–425

Millaku A, Drobne D, Tokar M, Novak S, Remškar M, Pipan-Tkalec Ž (2013) Use of scanning electron microscopy to monitor nanofibre/cell interaction in digestive epithelial cells. J Hazard Mater 260:47–52

Négre M, Leone P, Trichet J, Défrage C, Boero V, Gennari M (2004) Characterization of model soil colloids by cryo-scanning electron microscopy. Geoderma 121:1–16

Parkash SS, Francis LF, Scriven LE (2006) Microstructure evolution in dry cast cellulose acetate membranes by cryo-SEM. J Membr Sci 283:328–338

Prozorov T, Bazylinski DA, Mallaparagada SK, Prozorov R (2013) Novel magnetic nanomaterials inspired by magnetic bacteria. Mater Sci Eng R74:133–172

Rizwan SB, Dong Y-D, Hook BJ, Rades T, Hook S (2007) Characterization of bicontinuous cubic liquid crystalline systems of phytantriol and water using cryo field emission scanning electron microscopy (cryo FESEM). Micron 38:478–485

Šafařík I, Pospíšková K, Horská K, Šafaříková M (2012) Potential of magnetically responsive (nano)biocomposites. Soft Matter 8:5407–5413

Schröfel A, Kratošová G (2011) Biosynthesis of metallic nanoparticles and their application. In: Prokop A (ed) Intracellular delivery: fundamentals and applications. Springer, Berlin, pp 373–409

Schröfel A, Kratošová G, Krautová M, Dobročka E, Vávra I (2011) Biosynthesis of gold nanoparticles using diatoms—silica-gold and EPS-gold bionanocomposite formation. J Nanopart Res 13(8):3207–3216

Sriamornsnak P, Thirawong N, Cheewatanakornkool K, Burapapadh K, Sae-Ngow W (2008) Cryo-scanning electron microscopy (cryo-SEM) as a tool for studying the ultrastructure during bead formation by iontropic gelation of calcium pectinate. Int J Pharm 352:115–122

Srivastava R, Tiwari DK, Dutta PK (2011) 4-(Ethoxycarbonyl)phenyl-1-aminooxobutanoic acid-chitosan complex as a new matrix for silver nanocomposite film: preparation, characterization and antibacterial activity. Int J Biol Macromol 49:863–870

Tantra R, Tompkins J, Quincey P (2011) Characterization of the de-agglomeration effects of bovine serum albumin on nanoparticles in aqueous suspensions. Colloids Surf, B 75:275–281

Williams DB, Carter CB (2009) Transmission electron microscopy. A textbook for materials science, 2nd edn. Springer, New York

Závišová V, Koneracká M, Múčková M, Lazová J, Juríková A, Lancy G, Tomašovičová N, Timko M, Kováč J, Vávra I, Fabián M, Feoktzsov A, Garamus V, Avdeev MV, Kopčanský P (2011) Magnetic fluid poly(ethylene glycol) with moderate anticancer activity. J Magn Magn Mater 323:1408–1412

Zeleník K, Kukutschová J, Dvořáčková J, Bielniková H, Peikertová P, Čábalová L, Komínek P (2013) Possible role of nano-sized particles in chronic tonsillitis and tonsillar carcinoma: a pilot study. Eur Arch Oto-Rhino-Laryngol: Head Neck 270(2):705–709

Zhang X, Jin L, Fang Q, Hui WH, Zhou ZH (2010) 3.3 Å Cryo-EM structure of a nonenveloped virus revers a primic mechanism for cell entry. Cell 141(3):472–482

Ziserman L, Lee HY, Raghavan SR, Mor A, Danino D (2011) Unraveling the mechanism of nanotube formation by chiral self-assembly of amphiphiles. J Am Chem Soc 133(8): 2511–2517

Gold Nanoparticles for High Resolution Imaging in Modern Immunocytochemistry

Adam Schröfel, Dušan Cmarko, Eva Bártová and Ivan Raška

Abstract This perspective presents the possibilities for utilization of gold nanoparticles in immunocytochemistry. The unique combination of gold nanoparticle properties is just beginning to be fully recognized for the range of diagnostic and therapeutic applications. The current findings from bionanotechnology and emerging cryo-electron microscopy techniques can lead to novel perspectives for targeted delivery of gold conjugates into cells and tissues. The chapter offers a survey of modern electron microscopy methods with a special focus on immunolabeling. We also discuss the impact of gold nanoparticles on cellular systems and the potential possibilities for their targeted delivery into cells and tissues.

Abbreviations

CEMOVIS	Cryo-electron microscopy of the vitrified sections
CLEM	Correlation light-electron microscopy
EM	Electron microscopy
FM	Fluorescence microscopy
FS	Freeze substitution
GNP	Gold nanoparticle
HPF	High pressure freezing
IEM	Immuno-electron microscopy
LM	Light microscopy

A. Schröfel (✉) · D. Cmarko · I. Raška
Institute of Cellular Biology and Pathology, 1st Faculty of Medicine,
Charles University in Prague, Prague, Czech Republic
e-mail: adam.schrofel@gmail.com

E. Bártová
Institute of Biophysics, Academy of Sciences of the Czech Republic, v.v.i.,
Brno, Czech Republic

A. Prokop et al. (eds.), *Intracellular Delivery II*,
Fundamental Biomedical Technologies 7, DOI: 10.1007/978-94-017-8896-0_9,

NP	Nanoparticle
RME	Receptor mediated endocytosis
ROS	Reactive oxygen species

1 Introduction

Colloidal gold has a long history of utilization starting probably in ancient China (stain for porcelain), moving further to Europe in 17th century (coloring of glass with so-called "purple of Cassius") (Roth 1996). When Michael Faraday observed that colloidal gold solutions have properties that differ from bulk material more than 150 years ago, the modern era of gold colloid synthesis began. Since this important invent, synthesis of various gold conjugates has enabled a vast array of successful applications, including imaging methods (Horisberger 1981), bioelectronics (Xiao et al. 2003), and detection methods (Rosi and Mirkin 2005). Especially its utilization as an electron dense label has been exploited for decades and creates an inseparable part of modern immunocytochemical research.

Straightforwardness of "seeing what you actually study" makes microscopy-based research very effective and provides vast of information. The direct visualization of reality is very powerful in comparison to other indirect techniques such as x-ray crystallography, and should create an inseparable part of almost all papers focusing on fields ranging from cell biology to material science.

Microscopy methods are therefore widely used not only for the structural physical description but also for functional information. Unlike light microscopy (LM) which has already reached its physical limits [although currently breaks them with super-resolution techniques (Toomre and Bewersdorf 2010)], electron microscopy (EM) techniques very likely entering an era of atomic resolution boosted by introduction of cryogenic methods, direct electron detectors (Knott and Genoud 2013), and better electron-optical performance and stability (Zhang and Hong Zhou 2011). Moreover, electron microscopy has still very broad possibilities for further progress.

The rationale of this chapter is to describe the current perspectives of colloidal gold utilization in transmission EM with special emphasis on cell biology and immunocytochemistry. New findings from gold nano(bio)technology together with progressive (cryo) EM methods create unique opportunities for further development of scientific approaches by connecting the most recent findings from both fields. Moreover, better understanding of the actual situation inside the cell, more detailed description of the complex system and relations between the cells content, gold particles, antibodies and other factors is crucial for the correct description of studied systems. On the other hand—immunolabeling with traditional gold conjugates needs certain improvement due to arguable compatibility with cryo-EM and cryo-EM tomography. The possibility of gold/antibody conjugate delivery to the area of interest would dramatically change the situation. It follows that electron microscopists should ask questions such as:

1. How to get antibody/gold conjugates into the cells?
2. What is the fate of gold particles/gold conjugates inside the cell/on the section?
3. Is gold toxic? How and in what concentration? How do the gold particles influence their surroundings?
4. Is it possible to guide the gold particles to the target areas inside the cell? Can be gold/antibody conjugates be further modified by ligands to achieve this?
5. What type of information can we get from such conjugates?
6. What can we learn from cellular uptake and drug delivery studies? What do they have in common with immunolabeling?

We will try to address these questions or discuss them in the following text. The future of EM should include not only cutting-edge methods, but also easily-available, higher-throughput, and more universally applicable techniques.

2 Brief history of Biological EM and Where Did "Magical" Gold Nanoparticles Appear?

Biological ultrastructural research was born more than 60 years ago. Electron microscopy became an important branch of cellular biology with the new specimen preparation techniques containing the first embedding media (Newman et al. 1949), and sharp glass (Latta and Hartmann 1950) and diamond knives (Fernandéz-Morán 1953) respectively. Together with the major modifications of the then ordinary microtomes and the launch of dedicated ultramicrotomes (Sjöstrand 1953; Porter and Blum 1953) were researchers able to cut appropriate ultrathin sections. Major improvement in chemical fixation was done by Novikoff (1959) and Sabatini et al. (1963) by the introduction of aldehydes. The almost total lack of contrast in fixed sections was further solved by the section staining method using uranyl and lead salts developed by Watson (1958).

Although the first successful attempt to localize a certain enzyme in the intact cell was acid phosphatase splitting of phosphate-containing substrates causing electron dense lead salt precipitates (Sheldon et al. 1955), ferritin became the first marker for immunocytochemistry utilization at the ultrastructural level (Singer 1959).

The real breakthrough to immuno-electron microscopy (IEM) was the study by Faulk and Taylor (1971) which brought the method of antibody conjugation with colloidal gold for direct electron microscopy visualization of the surface antigens of *salmonella*. The use of primary antibodies conjugated with gold particles allows ultrastructural localization of various antigen epitopes within the specimen. Since this "revolution in immuno-chemistry", there have been many studies dedicated to the application of functionalized nanoparticles—biomolecule recognizing conjugates (antibodies, lectins, enzymes, aptamers, etc.) (Dykman and Khlebtsov 2011).

3 Gold Nanoparticle Properties and Their Functionalization

The general rule in nanotechnology is that nanomaterials exhibit a number of special properties relative to bulk material. It is also well known that gold nanoparticles (GNP) exhibit properties which differ even more from most of the other nanoparticles (Dreaden et al. 2011). Their photochemical and photophysical properties, a crucial property being the plasmon absorption and scattering, allow their utilization in diagnostic assays and photothermal therapy, both suitable e.g. for cancer treatment (Ferrari 2005). Gold nanoparticles have an interesting surface chemistry which leads to tunable affinity for binding ligands on their surfaces. Multivalent GNP can stabilize and shield unstable drugs and assist with their efficient delivery to the region of interest (Thamphiwatana et al. 2013). Due to their tunable sizes and shapes and dimensions comparable to proteins, GNP can participate in the cellular processes or modify them [review by Dreaden et al. (2012)] (Fig. 1).

Researchers are now emphasizing drug vectorization, combining the benefits of gold and enabling simultaneous diagnostic, targeting, and therapeutic functionality. The current goal is, however (1) to suppress side effects due to nanotoxicity; (2) to improve therapeutic efficiency; (3) to increase the biodistribution and delivery possibilities, and (4) to overcome the problems of solubility, stability and pharmacokinetics of drugs (Boisselier and Astruc 2009).

Gold nanoparticles can be synthesized in many different ways, sizes and shapes. There is a host of available literature describing the utilization of mostly chemical, but also physical and biological methods (Zhao et al. 2013). However, it should be mentioned that the process of synthesis (especially its chemistry) directly influence the chemical behavior of resulting GNP. In particular, capping agents or surfactants, which are used for the protection and stabilization of the nanoparticle, can dramatically change the nanoparticle properties such as charge or hydrophobicity (Fong et al. 2013). The surfactant cetyltrimethylammonium bromide (CTAB) which is crucial in many preparations of gold nanorods and other shapes is cytotoxic at micromolar concentrations (Alkilany et al. 2009).

Chemical modification of the nanoparticle surface is also necessary to provide biological compatibility and specificity to GNPs as well as adequate stability in environments with high ionic strength which can occur in cells. Functional groups which can be bound to GNPs and utilized for the attachment of biomolecules include thiolate, dithiolate, dithiocarbamate, amine, carboxylate, selenide, isothiocyanate, or phosphine groups (Dreaden et al. 2012). The choice of conjugation chemistry usually depends on desired stability and chemistry of the resulting biomolecule.

Gold nanoparticles can be conjugated with a vast array of biomolecules using simple physical methods such as hydrophobic–hydrophobic interactions. Adsorption of these macromolecules to GNPs is therefore not based on covalent bonds but on complex electrochemical interactions. Bound macromolecules

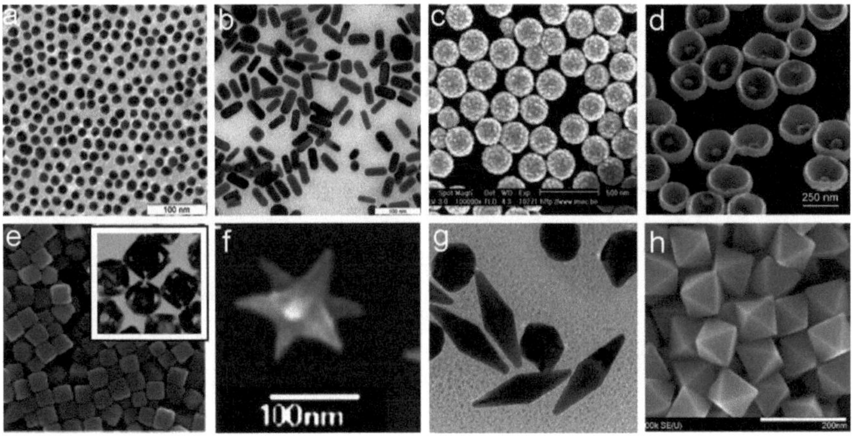

Fig. 1 TEM images of 15 nm colloidal gold (**a**), 15 × 50 nm gold nanorods (**b**), 160 (core)/17 (shell) nm silica/gold nanoshells (**c**, SEM), 250 nm Au nanobowls with 55 Au seed inside (**d**), silver cubes and gold nanocages (insert) (**e**), nanostars (**f**), bipyramids (**g**), and octahedral (**h**) (Reprinted from Khlebtsov and Dykman (2010), with permission from Elsevier)

typically retain their biological activity which is essential for the subsequent application. Until now, colloidal gold has been conjugated with a wide variety of biomolecules including proteins A and G, immunoglobulins, lectins, glycoproteins, dextran, enzymes, streptavidin, and hormones (Dykman and Khlebtsov 2011).

4 Immuno-electron Microscopy Techniques

Since cellular events create a very dynamic environment, conventional EM structural studies may not be sufficient for their sufficient description. IEM techniques may serve as a linker fulfilling lack of information between biochemistry, molecular biology, and ultrastructural studies. Utilization of antibodies, as well as protein A, conjugated with an electron dense gold particle helps to describe cellular processes by placing macromolecular functions within a structural context of the cell (Yamashita 2010). Although IEM is considered to be one of the best methods for localizing proteins in cells, its successful application strongly depends on the preservation of the epitope antigenicity and the specificity of antigen-primary antibody interaction. The most limiting factors are the damage of fragile suitable antigen epitopes during the sample processing (fixation, embedding) and the lack of specific antibodies. IEM methods are therefore still far from being ultimate and universal for any cell type, any tissue, and any target protein (Fig. 2).

Generally, the antigens can be affected during the fixation, by solvents (ethanol, acetone), resins and by heat during polymerization—the antigenicity is therefore

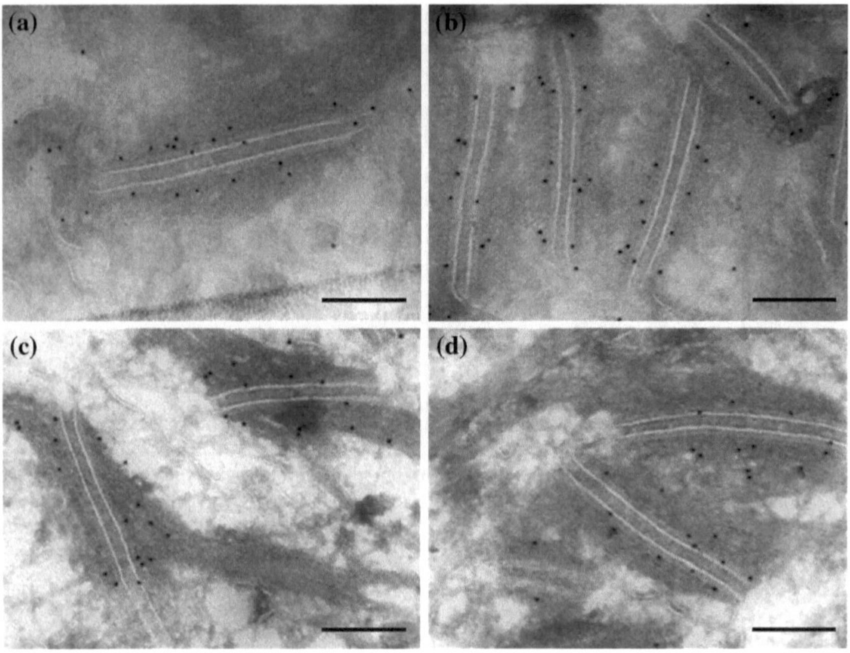

Fig. 2 Cryopreserved desmosomes of human skin with immunogold labeling. **a**, **b** Labeled with Mab 33-3D against the cytoplasmic domain of Dsg2. **c**, **d** Labeled with Mab 11-F against the (**c**) terminus of DP. Note in (**c**) the keratin intermediate filaments can be clearly seen. Scale bars = 230 nm (Reprinted from Scothern and Garrod (2008), with permission from Elsevier)

usually compromised. On the other hand, each of these steps can be either avoided or modified by implementing different IEM protocols. For more detailed information about the traditional IEM techniques see the review by De Paul et al. (2012). One of the most sensitive methods for EM immunolabeling is of ultrathin thawed cryo-sections, according to the method of Tokuyasu (1973). The main asset is that this method keeps the hydrated environment prior to the immunolabeling. Antigens remain hydrated and therefore more accessible in comparison with resin section labelling. Although the ultrastructure is of poorer quality than in the resin embedding method due to chemical fixation, high pressure freezing (HPF) methods (McDonald 1999) brought new insights to this relatively traditional technique. HPF at least partially solves the concern that the penetration time of aldehyde fixatives is too long to immediately stop the entire cellular processes which leads to structural and distributional artifacts. Fixation by HPF, instead of conventional chemical fixation, followed by freeze-substitution (FS)/chemical fixation, rehydration and further processing for Tokuyasu cryo-sectioning leads to an improved preservation of both ultrastructure and antigenicity (Ripper et al. 2008; van Donselaar et al. 2007) (Fig. 3). A novel approach to the preparation of cryo-immobilized material for IEM called VIS2FIX was recently introduced by

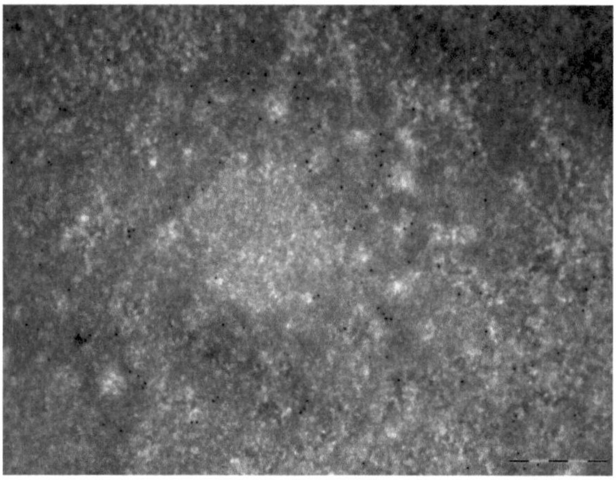

Fig. 3 Tokuyasu cryosection of HeLa cell nucleolus with incorporated fluorouridine. Labeled with anti-BrDU, 12 nm gold

Karreman et al. (2011). Vitreous sections of the sample cryo-immobilized by HPF are fixed, brought to room temperature, and immuno-labeled by means of FS. Thanks to the small volume of each section, fixation by this method is much faster than existing methods and allows investigators to test different fixation conditions.

5 Some Remarks About Antibodies and Their Validation

For electron microscopists gold particles and antibodies belong together. Antibodies conjugated with the gold particles allow high resolution detection and localization of a multiplicity of antigens, both on and within the cells. Antibodies are among the most frequently used biomolecules in basic and clinical research and the range of the commercially available antibodies grows every year. Their usage is very specific, especially for electron microscopy where the success rate of the labeling is usually much lower than in immunohistochemistry assays or even light microscopy. The presence of primary and secondary antibodies makes the situation even more difficult. Since there is a vast of literature about the use antibodies in electron microscopy (Webster et al. 2008), here we will limit to some important and maybe surprising remarks.

Unfortunately, there are still no universally accepted standardized methods for the validity of available antibodies. Starting with the editorials from the editor-in-chief of The Journal of Comparative Neurology (Saper 2005; Saper and Sawchenko 2003), there has been an increasing concern about the practices of biotechnology companies. The inadequate antibody performance may lead not only to wasted time and money but also to misleading or contradictory results. Couchman (2009) very aptly expressed the frustration of many researchers who are using antibodies in their everyday research and the situation between the commercial suppliers and scientist was also nicely described by Kalyuzhny

(2009). Just as an example—Berglund et al. (2008) tested 5,436 commercial antibodies from 51 different antibody providers as part of their project and found only 49 % working in their immunohistochemistry screens. For some suppliers the success rate was 100 % and for some 0 %. Authors conclude that "results enforce the need for publicly available validation results for antibodies because half of the commercially available antibodies did not pass our quality assurance." Another example directly from our lab is temperature-dependent performance of commercial monoclonal anti-actin antibody (clone AC-40, Sigma, A4700) used for immunofluorescence detection of actin in human diploid fibroblast cells (Fidlerova et al. 2005).

Nevertheless, the precise and standardized validation process for any obtained antibody can be performed in any immunohistochemistry lab and should precede any research activity with the reagent. The validation process should lead to proof of specificity, selectivity, and reproducibility in the context in which is the antibody to be used (Bordeaux et al. 2010). Besides the presentation of common pitfalls when working with antibodies and levels of commercial antibody validation of several vendors, authors of the aforementioned review also share their own procedure for the antibody validation.

On the other hand, besides the difficulties with some commercial antibody suppliers, there are still a number of good reasons why various "good" antibodies may deliver variable or even incorrect results. For the instance, microscopists should be aware of problems caused by the fixation. Epitopes exposed in the native proteins may not be exposed in the fixed tissue and vice versa. Therefore, an antibody will recognize one epitope in a native tissue and will recognize another epitope when applied to fixed tissue (Saper and Sawchenko 2003).

6 Current trends in Transmission Electron Microscopy and Why We Still Need Gold/Antibody Conjugates

The field of transmission electron microscopy of biological samples dramatically changed with the specimen vitrification (Adrian et al. 1984), cooled microscope stages and modern detectors which need only a small dose of electrons. This allows high resolution imaging of truly native (frozen) structures. Starting with the imaging of phages, vitrified in the thin layer of ice by plunge freezing, we can now visualize frozen thin sections of cell culture or tissue and use high pressure freezers and cryo-ultramicrotomes (Maimon et al. 2012; Maeshima et al. 2010; Al-Amoudi et al. 2004).

Strategies for electron signal filtering, tomographic specimen holders, faster camera chips, image averaging and development of reconstruction techniques for both symmetric and non-symmetric structures have greatly improved the image quality of low-contrast and fragile samples and enabled single particle analysis methods (Ruprecht and Nield 2001; Zhou 2008). This approach can reveal the true

3D structure of sub-cellular complexes, proteins and other biomolecules starting with the prokaryote ribosome from *Escherichia coli* at 40 Å resolution (Frank et al. 1991). Recently, Bai et al. (2013) reported a ribosomal reconstruction at ~4 Å resolution. Such projects show how cryo-electron techniques now challenge the resolution of crystallography. Moreover, many proteins and protein complexes are difficult to impossible to crystalize.

Vitrification of entire cells and tissues is key for the preservation of molecular structures in their native environment. Electron microscopy of frozen hydrated sections of vitreous ice and vitrified biological samples (cryo-electron microscopy of vitreous sections = CEMOVIS) made a significant turning point in structural biology (McDowall et al. 1983). Rather than purified proteins in a frozen buffer, we can visualize the specific molecules directly within cells. Al-Amoudi et al. (2007) demonstrated the methods abilities by unveiling the molecular architecture of cadherins in native epidermal desmosomes. Additionally, the authors showed correlation with structural data from X-ray crystallography, offering a better understanding of the molecular arrangements. Furthermore, a vast number of proteins cannot be crystallized and cryo-EM is only suitable method for the structural analysis (Kang et al. 2013).

Unfortunately, all these progressive EM methods are not compatible with immunogold labelling and provide only overall structural information without more specific or functional detail. The "all frozen" workflow from the fixation step to the observation in the microscope restricts the usage of any additional treatments. The only possibility for the structure examination is to incorporate markers into living cells prior to their vitrification, because currently used gold/antibody conjugates are too big to penetrate into the cell.

One option is the utilization of the fluorescence signal taken before the freezing in the approach called correlative light-electron microscopy (CLEM) (Lidke and Lidke 2012; Sjollema et al. 2012; Hübner et al. 2013). The ability to perform both fluorescence and electron microscopy on the same specimen makes it possible to identify where a fluorescently labeled proteins are located in the ultrastructures visualized by cryo-EM. However, fluorescence imaging lacks the fine structural information; a fluorescent dot can represent an entire organelle or aggregates of proteins or membranes. The spectrum of fluorescent tags which can be incorporated is relatively wide and contains e.g. photoswitchable organic dyes (Heymann et al. 2009), genetically encoded fluorescent tags (Patterson et al. 2010), or smaller quantum dots which can be conjugated to antibodies or ligands (Dertinger et al. 2009). Although there are few reports of biomarkers which can be visualized both by fluorescence and electron microscopy (Gaietta et al. 2002; Ruiz-González et al. 2013; Shu et al. 2011) their application is not yet widespread.

It follows that there is still a lack of universal labeling strategies for cryo-EM methods. The revolution in structural research caused by the fast vitrification of water has still relatively limited impact in matching the structure information with the particular function or process. This has these important implications: (1) gold/antibody conjugates are still of great importance, they are the most versatile tool; (2) we can combine the methods containing HPF with the conventional IEM methods in order to get better ultrastructure preservation together with high antigenicity; (3) there is much interest in the development of biomarkers for CEMOVIS methods and for CLEM.

7 GNP Interaction with Cells and Their Toxicity

When considering the potential use of gold nanoparticles in cellular biology or biomedicine, one should always take into account their cellular impact. Cells exposed to the nanoparticles may behave and respond differently from unaffected ones (Boisselier and Astruc 2009). Cell response can be measured in many ways; some of them are commonly used generally for the assessment of drug impact. Membrane disruption can be monitored by lactate dehydrogenase release (LDH assay), metabolic activity is measured by observation of the enzymatic activity of mitochondria (MTT assay) (Marquis et al. 2009). Other, usually commercially available solutions are, for example, assays monitoring oxidative stress by measuring the level of ROS (reactive oxygen species), or micro-arrays based on the polymerase chain reaction which analyze expression levels of genes related to cell stress (Alkilany and Murphy 2010). Regardless of the exact numbers describing cellular toxicity of gold, which may differ due to the method of nanoparticle preparation and cell line (this topic was recently reviewed e.g. by Khlebtsov and Dykman (2011)), GNP definitely have a certain impact on the cells (Van Lehn et al. 2013).

Besides the examination of cell response, the crucial task is to monitor the cell-nanoparticle interactions: (1) where are the nanoparticle localized inside the cell; (2) what is the uptake mechanism; (3) what happens to the nanoparticles over time. The better understanding of the GNP uptake mechanism can play a significant role in applications such as intracellular drug and gene delivery (Verma et al. 2008; Van Lehn et al. 2013), or targeted pre-embedding immunolabeling in TEM.

8 In Vitro Cellular Uptake of Gold Nanoparticles

Cells are generally equipped with different tools for the internalization of macromolecules and particles, including phagocytosis, micropinocytosis, and receptor-mediated endocytosis (RME) pathways (Conner and Schmid 2003). These pathways use different receptors, signaling pathways (Dobrovolskaia and McNeil 2007) and the uptake depends on the particle size and type. For example, particles bigger than 500 nm are usually internalized by phagocytosis, while the receptor-mediated pathway is activated for the uptake of smaller molecules (Hess and Tseng 2007).

Gold nanoparticles commonly used in biology have dimensions up to 100 nm and the primary mechanism of cell entry is RME (Shukla et al. 2005). The uptake of different shapes of transferrin coated GNP by a three types of cultured cell lines was studied by Chithrani and Chan (2007). The RME mechanism was proved by the temperature dependence of cellular uptake (70 % decrease between 4 and 37 °C).

Chithrani and Chan (2007) also reported a study on the influence of GNP sizes and revealed that it plays a very critical role both for the uptake rate and its extent. 50 nm transferrin-coated GNP exhibited higher uptake rates and extents compared to smaller or larger sizes in the range of 10–100 nm. This "optimal" uptake size can be explained by the processes and interactions which are occurring in the cellular membrane. This so called "wrapping effect" describes the nanoparticle encapsulation by the membrane and is influenced the free energy resulting from the ligand-receptor interaction and diffusion kinetics of the receptor (Alkilany and Murphy 2010). Based on previous considerations, Chaudhuri et al. (2011) presented a mathematical model explaining the size-dependent endocytosis of nanoparticles and the size-dependent uptake mechanism.

Besides the ligand-mediated uptake mechanism, specially modified nanoparticles can enter the cells also by means of direct penetration. Verma et al. (2008) reported non-destructive endocytosis-independent entry of gold nanospheres ($\sim$5 nm) decorated with two capping molecules (anionic and hydrophobic). Similarly, Van Lehn et al. (2013) presented the synthesis of anionic, monolayer-protected GNP which have been shown to nondisruptively penetrate cellular membranes and also showed that a critical first step in the penetration process is the fusion of such GNPs with the membrane lipid bilayer.

When considering the intracellular localization of GNP, the general conclusion is that the gold nanoparticles are trapped in vesicles (such as lysosomes) and that they are also not able to enter the nucleus (Chithrani and Chan 2007). However, there are reports indicating the possibility for the nuclear penetration using both functionalized (Ojea-Jiménez et al. 2012; Gu et al. 2009) and non-functionalized GNP (Tsoli et al. 2005) and even their binding to DNA (Tsoli et al. 2005).

9 Immunolabeling of CEMOVIS Cryosections: How to Get the Particles Inside the Cell?

The ultimate goal for the in vitro immunolabeling would be a universal electron dense particle carrying the primary antibody which will be able to penetrate the cell and bind to the area (antigen) of interest. Furthermore, everything should be done without influencing of natural cellular processes. Unfortunately, such markers for the pre-embedding prior to HPF are not yet available. The most frequent method used for most fluorescent microscopy and some electron microscopy techniques in order to detect intracellular antigens is cell permeabilization. Permeabilization provides access to the intracellular or intraorganellar antigens and uses basically two general types of reagents: (1) organic solvents which dissolve lipids from cell membranes and make them permeable and (2) detergents, such as saponin, interacts with membrane cholesterol, leaving holes

Fig. 4 The typical example of CEMOVIS micrograph—*Drosophyla meganoster* embryo

in the membrane (Jamur and Oliver 2010). The drawback of this method is obvious; it drastically disrupt the cellular environment, start diverse processes and pathways and even change the cells physically. Although one can still get some reasonable results and good immunolabeling, there is no reason to undergo complicated specimen preparation techniques when the structural preservation is already damaged (Fig. 4).

Other approach utilize sharp objects, which temporarily rupture the cell membrane and the chosen compound is either injected into the cell (Zhang and Yu 2008) or passively taken up by diffusion (Liang et al. 2007) The microinjection of single cell is well developed and widely used technique. It is believed that the membrane holes heel up immediately and affected cells can recover without any serious consequences. An advantage of this technique is also the possibility to target individual cell organelles such as nuclei. The drawbacks are smaller capacity and fact that one can inject only one cell at a time. Scassellati et al. (2010) illustrated the potential of microinjection in a study describing intracellular sphingomyelin distribution and function. Gold nanoparticles coupled with enzyme sphingomyelinase were injected into hepatoma tissue culture cells. The gold particles were localized in the nucleoplasm in an individual or clustered labelling pattern. This indicates the successful delivery of the enzyme to its target. Interestingly, the control with free gold particles showed that the majority of GNP remained accumulated close to the place of injection, regardless of the incubation time (Fig. 5).

The so-called "cell scratching method" is much less defined, but has larger capacity. The idea is very simple—scratching the surface of adhered cells with a sharp tiny tip, causing their rupture alongside its trajectory. Most of the cells in culture stay unaffected, and some might die, but a certain population will recover and be temporarily accessible for the compound of choice (Liang et al. 2007) (Fig. 6).

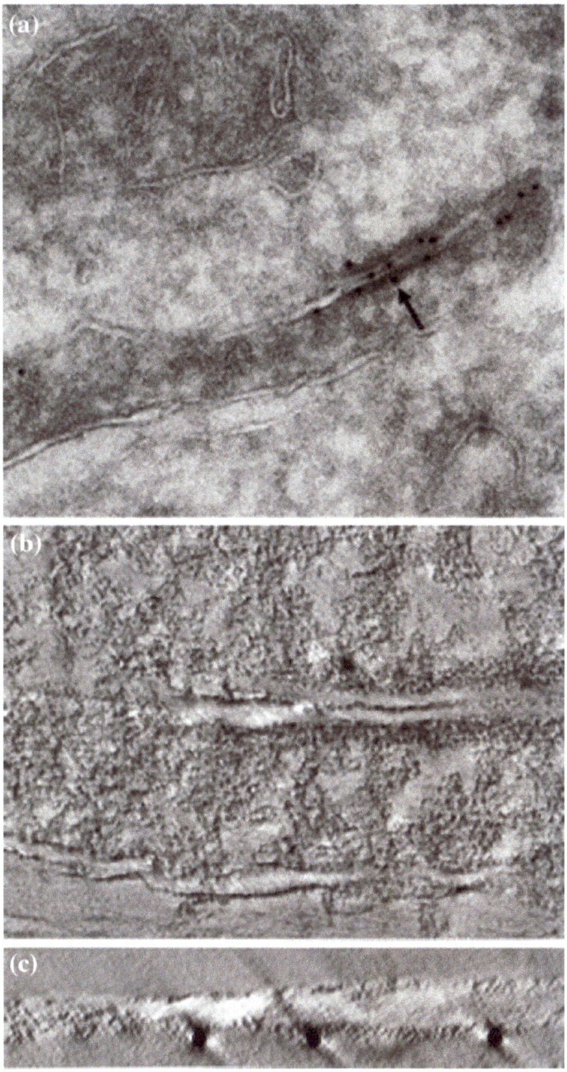

Fig. 5 Tomography of a cryosection immunolabeled with a connexin 43. **a** A slice from the original tilt series acquired with an Tecnai F20. The *black arrow* indicates 10 nm gold that is localized to a gap junction within an A543 cell (Similar to the area shown in Fig. 4.) Bar ¼ 100 nm. **b** A slice from the middle of a reconstructed tomogram. **c** A slice from the tomogram that has been rotated perpendicular to the section plane. Note the gold label remains on the surface of the section. The tilt series was acquired using a Tecnai F20 (FEI, Eindhoven) and reconstructed using the IMOD software. Bar ¼ 50 nm (Reprinted from Peters and Pierson (2008), with permission from Elsevier)

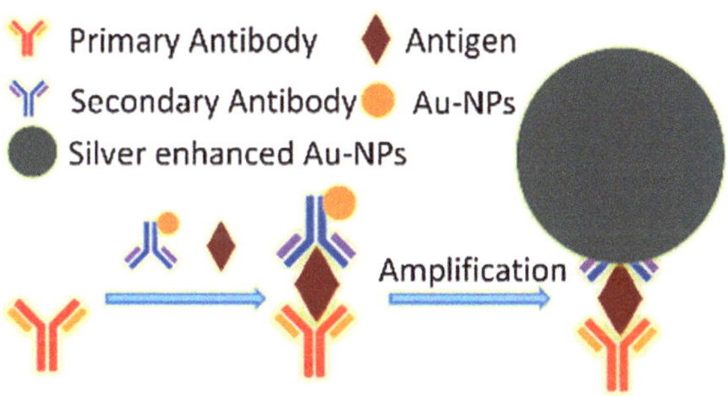

Fig. 6 Schematic diagram of the sandwich immunoassay and silver amplification of AuNP tags (Reprinted from Liu et al. (2011), with permission from American Chemical Society)

The last possible (but still theoretical) method is the utilization of facts about the GNP uptake by the cells, which was described above. The idea of a gold/antibody conjugate with some signaling ligand is very similar to popular drug delivery approaches (Peer et al. 2007).

10 Conclusion and Future Aspects

Gold nanoparticles are extremely important for multiple biological fields due to their unique properties. The potential for their utilization for targeted cellular delivery is raising with more and more new nanoparticle systems, which have shown promising properties. On the other hand, there is still a need to perform careful studies of their impacts on cells and tissues. We have to diminish the cellular response to circulating gold nanoparticles, and to increase their targeting selectivity.

Even today, the use of different electron microscopy techniques and approaches is still a tradeoff between nicely preserved ultrastructure and antigenicity. Therefore, there is still a great potential for novel labelling strategies and nanogold will very likely play an important role. The potential connection of cryo-EM techniques and novel nanoparticles for labelling would create a tool for very precise detection of specific molecules in their native state without chemical fixative and contrasting agents.

Acknowledgements This book chapter was supported by the Grant Agency of the Czech Republic (grant number P302/12/G157) and by the Charles University in Prague (projects UNCE 204022 and Prvouk/1LF/1). This publication is also supported by the project “BIOCEV—Biotechnology and Biomedicine Centre of the Academy of Sciences and Charles University” (CZ.1.05/1.1.00/02.0109), from the European Regional Development Fund and project CZ.2.16/3.1.00/24010 from the European Regional Development Fund. Many thanks to Guy Hagen for the language corrections.

References

Adrian M, Dubochet J, Lepault J, McDowall AW (1984) Cryo-electron microscopy of viruses. Nature 308(5954):32–36. doi:10.1038/308032a0

Al-Amoudi A, Chang JJ, Leforestier A, McDowall A, Salamin LM, Norlen LPO, Richter K, Blanc NS, Studer D, Dubochet J (2004) Cryo-electron microscopy of vitreous sections. EMBO J 23(18):3583–3588. doi:10.1038/sj.emboj.7600366

Al-Amoudi A, Diez DC, Betts MJ, Frangakis AS (2007) The molecular architecture of cadherins in native epidermal desmosomes. Nature 450(7171):832–837. doi:10.1038/nature05994

Alkilany AM, Murphy CJ (2010) Toxicity and cellular uptake of gold nanoparticles: what we have learned so far? J Nanopart Res 12(7):2313–2333. doi:10.1007/s11051-010-9911-8

Alkilany AM, Nagaria PK, Hexel CR, Shaw TJ, Murphy CJ, Wyatt MD (2009) Cellular uptake and cytotoxicity of gold nanorods: molecular origin of cytotoxicity and surface effects. Small 5(6):701–708. doi:10.1002/smll.200801546

Bai X-C, Fernandez IS, McMullan G, Scheres SH, Kühlbrandt W (2013) Ribosome structures to near-atomic resolution from thirty thousand cryo-EM particles. eLife 2. doi:10.7554/eLife.00461

Berglund L, Bjoerling E, Oksvold P, Fagerberg L, Asplund A, Szigyarto CAK, Persson A, Ottosson J, Wernerus H, Nilsson P, Lundberg E, Sivertsson A, Navani S, Wester K, Kampf C, Hober S, Ponten F, Uhlen M (2008) A genecentric Human Protein Atlas for expression profiles based on antibodies. Mol Cell Proteomics 7(10):2019–2027. doi:10.1074/mcp.R800013-MCP200

Boisselier E, Astruc D (2009) Gold nanoparticles in nanomedicine: preparations, imaging, diagnostics, therapies and toxicity. Chem Soc Rev 38(6):1759–1782. doi:10.1039/B806051G

Bordeaux J, Welsh AW, Agarwal S, Killiam E, Baquero MT, Hanna JA, Anagnostou VK, Rimm DL (2010) Antibody validation. Biotechniques 48(3):197–209. doi:10.2144/000113382

Chaudhuri A, Battaglia G, Golestanian R (2011) The effect of interactions on the cellular uptake of nanoparticles. Phys Biol 8(4):9. doi:10.1088/1478-3975/8/4/046002

Chithrani BD, Chan WCW (2007) Elucidating the mechanism of cellular uptake and removal of protein-coated gold nanoparticles of different sizes and shapes. Nano Lett 7(6):1542–1550. doi:10.1021/nl070363y

Conner SD, Schmid SL (2003) Regulated portals of entry into the cell. Nature 422(6927):37–44. doi:10.1038/nature01451

Couchman JR (2009) Commercial antibodies: the good, bad, and really ugly. J Histochem Cytochem 57(1):7–8. doi:10.1369/jhc.2008.952820

De Paul AL, Mukdsi JH, Petiti JP, Gutiérrez S, Quintar AA, Maldonado CA, Torres AI (2012) Immunoelectron microscopy: a reliable tool for the analysis of cellular processes, applications of immunocytochemistry. In: Dehghani H (ed) Applications of immunocytochemistry. In Tech, pp 66–96. doi:10.5772/33108

Dertinger T, Colyer R, Iyer G, Weiss S, Enderlein J (2009) Fast, background-free, 3D super-resolution optical fluctuation imaging (SOFI). Proc Natl Acad Sci. doi:10.1073/pnas.0907866106

Dobrovolskaia MA, McNeil SE (2007) Immunological properties of engineered nanomaterials. Nat Nanotechnol 2(8):469–478. doi:10.1038/nnano.2007.223

Dreaden EC, Mackey MA, Huang XH, Kang B, El-Sayed MA (2011) Beating cancer in multiple ways using nanogold. Chem Soc Rev 40(7):3391–3404. doi:10.1039/c0cs00180e

Dreaden EC, Alkilany AM, Huang X, Murphy CJ, El-Sayed MA (2012) The golden age: gold nanoparticles for biomedicine. Chem Soc Rev 41(7):2740–2779. doi:10.1039/C1CS15237H

Dykman LA, Khlebtsov NG (2011) Gold nanoparticles in biology and medicine: recent advances and prospects. Acta Naturae 3(2):34–55

Faulk WP, Taylor GM (1971) An immunocolloid method for the electron microscope. Immunochemistry 8(11):1081–1083

Fernandéz-Morán H (1953) A diamond knife for ultrathin sectioning. Exp Cell Res 5(1):255–256. doi:10.1016/0014-4827(53)90112-8

Ferrari M (2005) Cancer nanotechnology: opportunities and challenges. Nat Rev Cancer 5(3):161–171. doi:10.1038/nrc1566

Fidlerova H, Masata M, Malinsky A, Fialova M, Cvackova Z, Louzecka A, Koberna K, Berezney R, Raska I (2005) Replication-coupled modulation of early replicating chromatin domains detected by anti-actin antibody. J Cell Biochem 94(5):899–916. doi:10.1002/jcb.20374

Fong Y-Y, Gascooke JR, Visser BR, Harris HH, Cowie BCC, Thomsen L, Metha GF, Buntine MA (2013) Influence of cationic surfactants on the formation and surface oxidation states of gold nanoparticles produced via laser ablation. Langmuir 29(40):12452–12462. doi:10.1021/la402234k

Frank J, Penczek P, Grassucci R, Srivastava S (1991) 3-Dimensional reconstruction of the 70s *Escherichia-coli* ribosome in ice—the distribution of ribosomal-RNA. J Cell Biol 115(3):597–605. doi:10.1083/jcb.115.3.597

Gaietta G, Deerinck TJ, Adams SR, Bouwer J, Tour O, Laird DW, Sosinsky GE, Tsien RY, Ellisman MH (2002) Multicolor and electron microscopic imaging of connexin trafficking. Science 296(5567):503–507. doi:10.1126/science.1068793

Gu Y-J, Cheng J, Lin C-C, Lam YW, Cheng SH, Wong W-T (2009) Nuclear penetration of surface functionalized gold nanoparticles. Toxicol Appl Pharmacol 237(2):196–204. doi:10.1016/j.taap.2009.03.009

Hess H, Tseng Y (2007) Active intracellular transport of nanopartides: opportunity or threat? ACS Nano 1(5):390–392. doi:10.1021/nn700407v

Heymann JAW, Shi D, Kim S, Bliss D, Milne JLS, Subramaniam S (2009) 3D Imaging of mammalian cells with ion-abrasion scanning electron microscopy. J Struct Biol 166(1):1–7. doi:10.1016/j.jsb.2008.11.005

Horisberger M (1981) Colloidal gold—a cytochemical marker for light and fluorescent microscopy and for transmission and scanning electron-microscopy. Scan Electron Microsci 9–32

Hübner B, Cremer T, Neumann J (2013) Correlative microscopy of individual cells: sequential application of microscopic systems with increasing resolution to study the nuclear landscape. In: Shav-Tal Y (ed) Imaging gene expression, vol 1042. Methods in molecular biology. Humana Press, pp 299–336. doi:10.1007/978-1-62703-526-2_21

Jamur M, Oliver C (2010) Permeabilization of cell membranes. In: Oliver C, Jamur MC (eds) Immunocytochemical methods and protocols, vol 588. Methods in molecular biology. Humana Press, pp 63-66. doi:10.1007/978-1-59745-324-0_9

Kalyuzhny AE (2009) The dark side of the immunohistochemical moon: industry. J Histochem Cytochem 57(12):1099–1101. doi:10.1369/jhc.2009.954867

Kang HJ, Lee C, Drew D (2013) Breaking the barriers in membrane protein crystallography. Int J Biochem Cell Biol 45(3):636–644. doi:10.1016/j.biocel.2012.12.018

Karreman MA, van Donselaar EG, Gerritsen HC, Verrips CT, Verkleij AJ (2011) VIS2FIX: a high-speed fixation method for immuno-electron microscopy. Traffic 12(7):806–814. doi:10.1111/j.1600-0854.2011.01199.x

Khlebtsov NG, Dykman LA (2010) optical properties and biomedical applications of plasmonic nanoparticles. J Quant Spectrosc Radiat Transfer 1–35

Khlebtsov N, Dykman L (2011) Biodistribution and toxicity of engineered gold nanoparticles: a review of in vitro and in vivo studies. Chem Soc Rev 40(3):1647–1671. doi:10.1039/C0CS00018C

Knott G, Genoud C (2013) Is EM dead? J Cell Sci 126(20):4545–4552. doi:10.1242/jcs.124123

Latta H, Hartmann F (1950) Use of a glass edge in thin sectioning for electron microscopy. Proc Soc Exp Biol Med 74(2):436–439

Liang C-C, Park AY, Guan J-L (2007) In vitro scratch assay: a convenient and inexpensive method for analysis of cell migration in vitro. Nat Protoc 2(2):329–333. doi:10.1038/nprot.2007.30

Lidke DS, Lidke KA (2012) Advances in high-resolution imaging—techniques for three-dimensional imaging of cellular structures. J Cell Sci 125(11):2571–2580. doi:10.1242/jcs.090027

Liu R, Liu X, Tang Y, Wu L, Hou X, Lv Y (2011) Highly sensitive immunoassay based on immunogold—silver amplification and inductively coupled plasma mass spectrometric detection. Anal Chem 83(6):2330–2336

Maeshima K, Hihara S, Eltsov M (2010) Chromatin structure: does the 30-nm fibre exist in vivo? Curr Opin Cell Biol 22(3):291–297. doi:10.1016/j.ceb.2010.03.001

Maimon T, Elad N, Dahan I, Medalia O (2012) The human nuclear pore complex as revealed by cryo-electron tomography. Structure 20(6):998–1006. doi:10.1016/j.str.2012.03.025

Marquis BJ, Love SA, Braun KL, Haynes CL (2009) Analytical methods to assess nanoparticle toxicity. Analyst 134(3):425–439. doi:10.1039/B818082B

McDonald K (1999) High-pressure freezing for preservation of high resolution fine structure and antigenicity for immunolabeling. In: Nasser Hajibagheri MA (ed) Electron microscopy

methods and protocols, vol 117. Methods in molecular biology™. Humana Press, pp 77–97. doi:10.1385/1-59259-201-5:77
McDowall AW, Chang JJ, Freeman R, Lepault J, Walter CA, Dubochet J (1983) Electron microscopy of frozen hydrated sections of vitreous ice and vitrified biological samples. J Microsc 131(1):1–9. doi:10.1111/j.1365-2818.1983.tb04225.x
Newman SB, Borysko E, Swerdlow M (1949) Ultramicrotomy by a new method. Anal Chem 21(11):1435
Novikoff AB (1959) The proximal tubule cell in experimental hydronephrosis. J Biophys Biochem Cytol 6(1):136–138. doi:10.1083/jcb.6.1.136
Ojea-Jiménez I, García-Fernández L, Lorenzo J, Puntes VF (2012) Facile preparation of cationic gold nanoparticle-bioconjugates for cell penetration and nuclear targeting. ACS Nano 6(9):7692–7702. doi:10.1021/nn3012042
Patterson G, Davidson M, Manley S, Lippincott-Schwartz J (2010) Superresolution imaging using single-molecule localization. Annu Rev Phys Chem 61(1):345–367. doi:10.1146/annurev.physchem.012809.103444
Peer D, Karp JM, Hong S, FaroKhzad OC, Margalit R, Langer R (2007) Nanocarriers as an emerging platform for cancer therapy. Nat Nanotechnol 2(12):751–760. doi:10.1038/nnano.2007.387
Peters PJ, Pierson J (2008) Immunogold labeling of thawed cryosections. Methods Cell Biol 88:131–149
Porter KR, Blum J (1953) A study in microtomy for electron microscopy. Anatomical Rec 117(4):685–709. doi:10.1002/ar.1091170403
Ripper D, Schwarz H, Stierhof Y-D (2008) Cryo-section immunolabelling of difficult to preserve specimens: advantages of cryofixation, freeze-substitution and rehydration. Biol Cell 100(2):109–123. doi:10.1042/BC20070106
Rosi NL, Mirkin CA (2005) Nanostructures in biodiagnostics. Chem Rev 105(4):1547–1562. doi:10.1021/cr030067f
Roth J (1996) The silver anniversary of gold: 25 years of the colloidal gold marker system for immunocytochemistry and histochemistry. Histochem Cell Biol 106(1):1–8
Ruiz-González R, Cortajarena AL, Mejias SH, Agut M, Nonell S, Flors C (2013) Singlet oxygen generation by the genetically encoded tag miniSOG. J Am Chem Soc 135(26):9564–9567. doi:10.1021/ja4020524
Ruprecht J, Nield J (2001) Determining the structure of biological macromolecules by transmission electron microscopy, single particle analysis and 3D reconstruction. Prog Biophys Mol Biol 75(3):121–164. doi:10.1016/s0079-6107(01)00004-9
Sabatini DD, Bensch K, Barrnett RJ (1963) Cytochemistry and electron microscopy: the preservation of cellular ultrastructure and enzymatic activity by aldehyde fixation. J Cell Biol 17(1):19–58. doi:10.1083/jcb.17.1.19
Saper CB (2005) An open letter to our readers on the use of antibodies. J Comp Neurol 493(4):477–478. doi:10.1002/cne.20839
Saper CB, Sawchenko PE (2003) Magic peptides, magic antibodies: guidelines for appropriate controls for immunohistochemistry. J Comp Neurol 465(2):161–163. doi:10.1002/cne.10858
Scassellati C, Albi E, Cmarko D, Tiberi C, Cmarkova J, Bouchet-Marquis C, Verschure PJ, van Driel R, Magni MV, Fakan S (2010) Intranuclear sphingomyelin is associated with transcriptionally active chromatin and plays a role in nuclear integrity. Biol Cell 102(6):361–375. doi:10.1042/BC20090139
Scothern A, Garrod D (2008) Visualization of desmosomes in the electron microscope (Chap. 18). Methods Cell Biol 88:347–366
Sheldon H, Zetterqvist H, Brandes D (1955) Histochemical reactions for electron microscopy: acid phosphatase. Exp Cell Res 9(3):592–596. doi:10.1016/0014-4827(55)90092-6
Shu X, Lev-Ram V, Deerinck TJ, Qi Y, Ramko EB, Davidson MW, Jin Y, Ellisman MH, Tsien RY (2011) A genetically encoded tag for correlated light and electron microscopy of intact cells, tissues, and organisms. PLoS Biol 9(4):e1001041. doi:10.1371/journal.pbio.1001041

Shukla R, Bansal V, Chaudhary M, Basu A, Bhonde RR, Sastry M (2005) Biocompatibility of gold nanoparticles and their endocytotic fate inside the cellular compartment: a microscopic overview. Langmuir 21(23):10644–10654. doi:10.1021/la0513712

Singer SJ (1959) Preparation of an electron-dense antibody conjugate. Nature 183(4674):1523–1524. doi:10.1038/1831523a0

Sjollema KA, Schnell U, Kuipers J, Kalicharan R, Giepmans BNG (2012) Correlated light microscopy and electron microscopy. In: MullerReichert T, Verkade P (eds) Correlative light and electron microscopy, vol 111. Methods in cell biology. Elsevier Academic Press Inc, San Diego, pp 157–173. doi:10.1016/b978-0-12-416026-2.00009-1

Sjöstrand FS (1953) A new microtome for ultrathin sectioning for high resolution electron microscopy. Experientia 9(3):114–115. doi:10.1007/BF02178346

Thamphiwatana S, Fu V, Zhu J, Lu D, Gao W, Zhang L (2013) Nanoparticle-stabilized liposomes for pH-responsive gastric drug delivery. Langmuir 29(39):12228–12233. doi:10.1021/la402695c

Tokuyasu KT (1973) A technique for ultracryotomy of cell suspensions and tissues. J Cell Biol 57(2):551–565. doi:10.1083/jcb.57.2.551

Toomre D, Bewersdorf J (2010) A new wave of cellular imaging. In: Schekman R, Goldstein L, Lehmann R (eds) Annual review of cell and developmental biology, vol 26. Annual review of cell and developmental biology. Annual Reviews, Palo Alto, pp 285–314. doi:10.1146/annurev-cellbio-100109-104048

Tsoli M, Kuhn H, Brandau W, Esche H, Schmid G (2005) Cellular uptake and toxicity of Au55 clusters. Small 1(8–9):841–844. doi:10.1002/smll.200500104

van Donselaar E, Posthuma G, Zeuschner D, Humbel BM, Slot JW (2007) Immunogold labeling of cryosections from high-pressure frozen cells. Traffic 8(5):471–485. doi:10.1111/j.1600-0854.2007.00552.x

Van Lehn RC, Atukorale PU, Carney RP, Yang Y-S, Stellacci F, Irvine DJ, Alexander-Katz A (2013) Effect of particle diameter and surface composition on the spontaneous fusion of monolayer-protected gold nanoparticles with lipid bilayers. Nano Lett 13(9):4060–4067. doi:10.1021/nl401365n

Verma A, Uzun O, Hu YH, Hu Y, Han HS, Watson N, Chen SL, Irvine DJ, Stellacci F (2008) Surface-structure-regulated cell-membrane penetration by monolayer-protected nanoparticles. Nat Mater 7(7):588–595. doi:10.1038/nmat2202

Watson ML (1958) Staining of tissue sections for electron microscopy with heavy metals. J Biophys Biochem Cytol 4(4):475–478. doi:10.1083/jcb.4.4.475

Webster P, Schwarz H, Griffiths G (2008) Preparation of cells and tissues for immuno EM. In: Allen TD (ed) Introduction to electron microscopy for biologists, vol 88. Methods in cell biology. pp 45–58. doi:10.1016/s0091-679x(08)00403-2

Xiao Y, Patolsky F, Katz E, Hainfeld JF, Willner I (2003) "Plugging into enzymes": nanowiring of redox enzymes by a gold nanoparticle. Science 299(5614):1877–1881. doi:10.1126/science.1080664

Yamashita S (2010) The Post-embedding method for immunoelectron microscopy of mammalian tissues: a standardized procedure based on heat-induced antigen retrieval. In: T immunoelectron microscopy, vol 657. Methods in molecular biology, pp 237–248

Zhang X, Hong Zhou Z (2011) Limiting factors in atomic resolution cryo electron microscopy: no simple tricks. J Struct Biol 175(3):253–263. doi:10.1016/j.jsb.2011.05.004

Zhang Y, Yu L-C (2008) Single-cell microinjection technology in cell biology. Bioessays 30(6):606–610. doi:10.1002/bies.20759

Zhao PX, Li N, Astruc D (2013) State of the art in gold nanoparticle synthesis. Coord Chem Rev 257(3–4):638–665. doi:10.1016/j.ccr.2012.09.002

Zhou ZH (2008) Towards atomic resolution structural determination by single-particle cryo-electron microscopy. Curr Opin Struct Biol 18(2):218–228. doi:10.1016/j.sbi.2008.03.004

Design of Functional Polymers for Intracellular Nucleic Acids Delivery

Hiroyasu Takemoto and Nobuhiro Nishiyama

Abstract For intracellular nucleic acid delivery, the fulfillment of certain requirements is crucial: protection of nucleic acids from enzymatic degradation in the cell exterior, efficient cellular internalization of nucleic acids, and control over intracellular distribution of nucleic acids. Polyion complex (PIC), composed of nucleic acids and polycations through electrostatic interaction, is one of the most developed delivery carriers to meet such requirements. This chapter describes rational strategies in the design of polymers for PIC-based nucleic acid delivery.

Keywords Nucleic acids delivery · Polycation · Polyion complex (PIC) · Endosomal escape · Biodegradability · Nucleic acids modification

Abbreviations

DET	Diethylenetriamine
GSH	Glutathione
pDNA	Plasmid deoxyribonucleic acid
PEI	Polyethyleneimine
PIC	Polyion complex
siRNA	Small interfering ribonucleic acid
SNA	Small nucleic acid

H. Takemoto · N. Nishiyama (✉)
Polymer Chemistry Division, Chemical Resources Laboratory, Tokyo Institute of Technology, R1-11, 4259 Nagatsuta, Midori-ku, Yokohama 226-8503, Japan
e-mail: nishiyama@res.titech.ac.jp

A. Prokop et al. (eds.), *Intracellular Delivery II*,
Fundamental Biomedical Technologies 7, DOI: 10.1007/978-94-017-8896-0_10,

1 Introduction

Nucleic acids, such as plasmid DNA (pDNA) and small interfering RNA (siRNA), have gained growing attention as therapeutic drug, due to their ability to control gene expression in the cell; pDNA produces its encoded protein, and siRNA silences mRNA translation into protein in a sequence specific manner (Chen and Okayama 1987; Elbashir et al. 2001). However, although nucleic acids need to be delivered into the cell interior in order to exert their biological activity, they are hardly uptaken by the cell because of their anionic charges (Scholz and Wagner 2012; Kanasty et al. 2013). Cellular membrane is densely coated with anionic glycocalyx, such as glycosaminoglycan, glycolipid and glycoprotein, and thus anionic nucleic acids give rise to electrostatic repulsion with glycocalyx, leading to inefficient cellular uptake of nucleic acids (Palte and Raines 2012). Moreover, enzymes at the cell exterior easily digest nucleic acids, resulting in decrease of their biological activity (Scholz and Wagner 2012; Kanasty et al. 2013). Therefore, nucleic acids delivery system that allows efficient cellular uptake and protection from enzymes is necessary to be developed.

There are several types of nucleic acids delivery system, e.g., virus-based system, lipid-based system, and inorganic particle-based system (Scholz and Wagner 2012; Kanasty et al. 2013; Airenne et al. 2013; Ramishetti and Huang 2012; Zhang and Kataoka 2009). Among them, polymer-based system has also been explored due to the following advantages: low immunogenicity derived from non-viral structure, feasibility of quantity synthesis and modifiability for multiple functionality (Kataoka et al. 2001; Kang et al. 2012; Miyata et al. 2012). Generally, cationic polymer is used for polymer-based nucleic acids delivery. Cationic polymer electrostatically interacts with nucleic acids and forms polyion complex (PIC). PIC formation neutralizes the anionic charges of nucleic acids and facilitates the cellular uptake of them, and protects nucleic acids from enzymatic degradation. In this chapter, we describe polymer-based nucleic acids delivery strategy.

2 Polycation Structure for PIC-Based Nucleic Acids Delivery

It is well demonstrated that PIC formation facilitates the cellular uptake of nucleic acids and protects them from enzymatic degradation (Kataoka et al. 2001; Kang et al. 2012; Miyata et al. 2012). However, PIC is uptaken mainly through endocytosis toward lysosomal digestion. Considering that nucleic acids need to be intact in order to exert their biological activity, rational design of the functional polymer is highly required to be developed for the nucleic acids transfer from the endosome into the cytosol, i.e., endosomal escape. In addition, molecular weight of polycations should be also considered. Although high-molecular-weight polycations are needed for stable PIC formation, they can hardly release nucleic acids at the cell interior towards efficient biological activity. And polycations potentially induce cytotoxicity with their molecular weight increasing (Fischer et al. 2003). Therefore, after the nucleic acids delivery into the cell, polycations need to be degraded for low cytotoxicity as well as unstable PIC formation leading to nucleic acids release. In this section, we describe the polycation design for endosomal escape and biodegradability.

2.1 Polycation Design for Endosomal Escape

Polycations need to selectively affect endosomal membrane for efficient endosomal escape with low cytotoxicity; polycations should exert minimal interaction with extracellular membrane for negligible adverse effects. In this regard, it is known that the endosomal pH decreases to ~5 during endosomal maturation, and therefore acidic pH-responsive polycations have been developed in order to facilitate endosomal escape with low cytotoxicity. "Proton sponge effect" is a well-known mechanism to develop the polycations for facilitated endosomal escape (Boussif et al. 1995; Thomas and Klibanov 2002). Proton sponge effect is induced by the compounds that have highly buffering capacity at pH 5.0–7.4. One of the examples is polyethyleneimine (PEI) (Fig. 1), which is the gold standard for nucleic acids delivery (Demeneix and Behr 2005; Jager et al. 2012). The protonation ratio of amino groups in PEI is ~20 % at physiological pH 7.4, whereas ~40 % at late-endosomal pH 5.0 (Suh et al. 1994). Buffering effect induces extensive influx of proton as well as chloride ion during endosomal maturation, followed by influx of water due to the increase of osmotic pressure in endosome. Excess influx of water into endosome leads to endosome rupture, and further leads to the release of entrapped components from endosome into the cytosol (Fig. 2). Actually, it has been observed that the decelerated pH decrease and increased chloride ion accumulation in endosome were associated with 140 % increase in the relative endosomal volume (Sonawane et al. 2003). Meanwhile, the direct interaction between the polycation and endosomal membrane is also widely accepted mechanism for facilitated endosomal escape (Hong et al. 2006; Miyata et al. 2008; Uchida et al. 2011). For example, polyaspartic acid derivative bearing diaminoethane pendant groups, PAsp(DET) (Fig. 3), has two pKa's of 6.2 and 8.9; the amino groups in each side chain of PAsp(DET) are mono-protonated at physiological pH 7.4, whereas di-protonated at late-endosomal pH 5.0 (Miyata et al. 2008; Kanayama et al. 2006) (Fig. 4). The di-protonated PAsp(DET) side chains can strongly interact with membrane constituents due to the increased cationic charge density, leading to the endosomal membrane destabilization (Fig. 5). The enhanced membrane destabilization activity at acidic pH has been demonstrated by hemolysis assay (Miyata et al. 2008). The hemolysis activity of PAsp(DET) at pH 5.5 was 10 times stronger than at pH 7.4; PAsp(DET) destabilizes the membrane selectively at late-endosomal pH and facilitates the release of entrapped

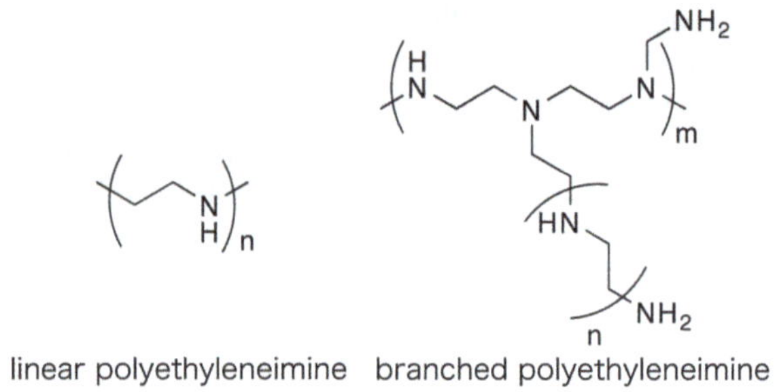

Fig. 1 Chemical structure of polyethyleneimine

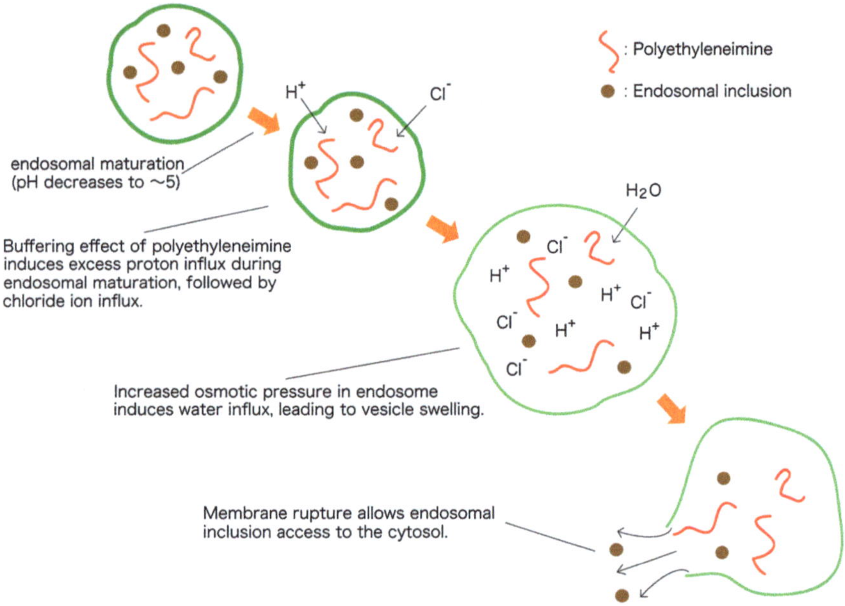

Fig. 2 Proposed mechanism of proton sponge effect

components, suggesting the strong potential for endosomal escape with low cytotoxicity. The both of the two distinct mechanisms, i.e., proton sponge effect and direct interaction between the polycation and endosomal membrane, are considered to have a synergistic effect toward facilitated endosomal escape. Facilitated endosomal escape by the functional polycations, e.g., PEI and PAsp(DET), is reported to enable 100 times increase in gene expression of pDNA payload for cultured cells (Boussif et al. 1995; Miyata et al. 2008).

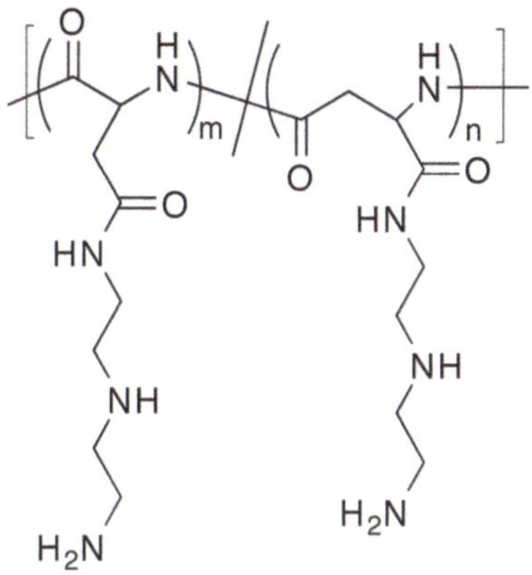

Fig. 3 Chemical structure of PAsp(DET)

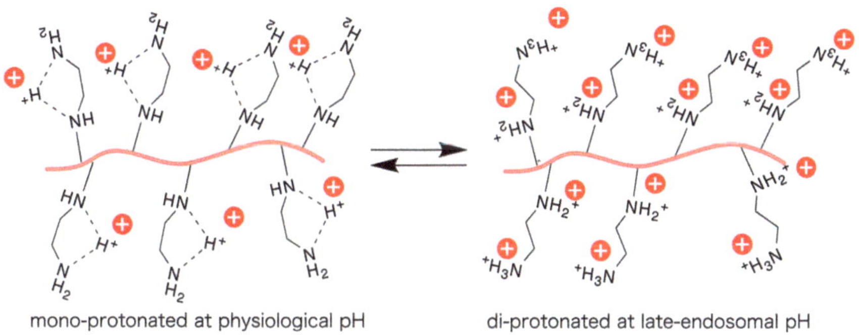

Fig. 4 pH-responsive protonation of PAsp(DET) side chain

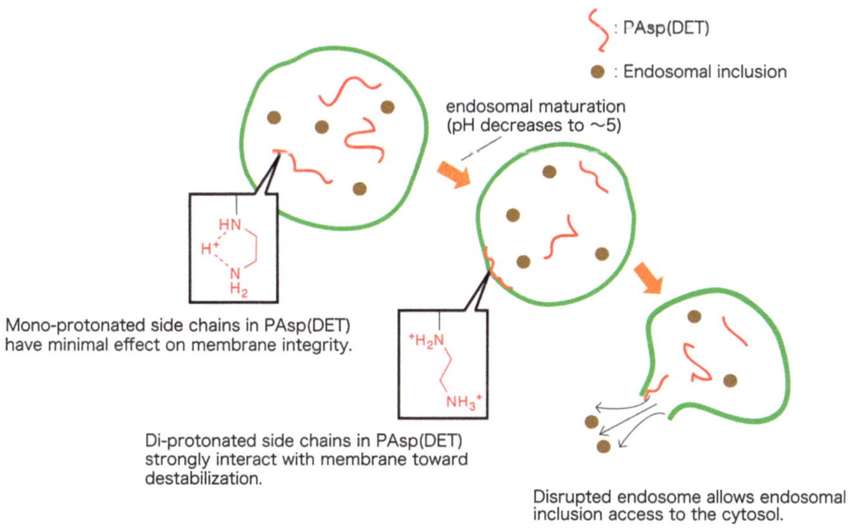

Fig. 5 Proposed mechanism of endosomal membrane disruption

2.2 Polycation Design for Biodegradability

The biodegradability of polycations is also an important property for PIC-based nucleic acids delivery. High-molecular-weight polycations tend to form stable PIC due to lots of cationic charges toward strong electrostatic interaction with nucleic acids, but also induce compromised releasability of nucleic acids in the cytosol as well as high cytotoxicity (Fischer et al. 2003; Kunath et al. 2003). On the other hand, low-molecular-weight polycations do not have enough cationic charges for stable PIC formation, and consequently they do not deliver nucleic acids into the cell interior efficiently. Therefore, biodegradable property needs to be installed into polycation for the stable PIC formation at the cell exterior and the efficient release of nucleic acids payload at the cell interior as well as low cytotoxicity. For example, PEI has great potential for nucleic acids delivery, but its biodegradable property is limited. Therefore, several modification strategies have been challenged for PEI in order to degrade PEI selectively at the cell interior (Gosselin et al. 2001; Carlisle et al. 2004; Kim et al. 2005; Knorr et al. 2008). Disulfide linkage is one of the promising strategies to endow polymers with biodegradability (Gosselin et al. 2001; Carlisle et al. 2004). Glutathione (GSH), which is a tripeptide composed of glutamic acid, cysteine, and glycine, is a biological reductant, and exists at the concentration of $\sim 10\ \mu M$ at the cell exterior, but ~ 10 mM at the cell interior (Meister and Anderson 1983; Saito et al. 2003). This concentration difference induces disulfide cleavage selectively at the cell interior. Also, acid-cleavable linkages, such as imine, acetal, and ketal, are developed for PEI degradation responsive to late-endosomal pH (Kim et al. 2005; Knorr et al. 2008). Small-molecular-weight PEIs conjugated with biodegradable linkage allow enhanced pDNA gene expression as well as low cytotoxicity. Meanwhile, PAsp(DET) is reported to perform unique behavior of biodegradation. The poly(amino acid) main chain of PAsp(DET) can moderately degrade through self-catalytic reaction at 37 °C. Auto-degradability of asparagine-containing peptide can explain the mechanism of self-catalytic reaction in PAsp(DET); the amide nitrogen at asparagine residue makes a nucleophilic attack on carbonyl group at peptide bond, and induces peptide bond cleavage via succinimide formation (Geiger and Clarke 1987). Consequently, PAsp(DET) has comparable pDNA delivery ability to PEI without compromising cell homeostasis, e.g., maintained house-keeping gene expression system after transfection (Itaka et al. 2010).

3 Chemical Modification of Nucleic Acids for Efficient Delivery

Small nucleic acids (SNAs), such as siRNA and antisense oligomeric nucleic acid, are chemically synthesized based on solid-phase polymerization, and thus SNA production with on-demand sequence is commercially available. Moreover, reactive groups can be introduced into 5′ and/or 3′ ends of SNAs; primary amine, thiol, azide, carboxylic acid, and biotin are available as an end group of commercial SNAs. Therefore, SNAs can be modified with functional molecules not only for improvement in their existing ability but also for further functionalization with entirely new properties. Actually, poly(ethylene glycol) and ligand modification lead to enhanced stability against enzyme and facilitated cellular uptake, respectively (Lee et al. 2011; Alam et al. 2011). In this section, we introduce several methodologies for SNA modification.

3.1 SNA Modification for PIC Stabilization

Small molecular weight and small number of anionic charges of SNAs are apt to result in unstable PIC formation with polycation, due to weak electrostatic interaction (Takemoto et al. 2010; Suma et al. 2012). Also, in case of double stranded SNAs, such as siRNA, the length is shorter than persistence length, e.g., siRNA is ~10 times shorter than the persistence length of double stranded RNA. This short structure of double stranded SNAs appears as their rigid property, possibly leading to inefficient entanglement with polycations and associated unstable PIC formation. Therefore, multimerized SNA structures have been developed for increased anionic charges and entanglement with polycation towards stable PIC formation. Sticky siRNA is the first multimerized SNA; the extended overhang nucleotide of a monomeric siRNA moderately hybridizes with that of another monomeric siRNA, and linearly conjugated monomeric siRNA is produced (Bolcato-Bellemin et al. 2007) (Fig. 6). Sticky siRNA has been reported to form stable PIC with PEI and induce strong gene silencing in vitro and in vivo. Covalently conjugated monomeric siRNA via disulfide linkage (multi-siRNA) has been also reported (Mok et al. 2010) (Fig. 7). In this report, when siRNA PIC was formed with PEI, multi-siRNA PIC also showed strong gene silencing effect in vitro and in vivo. Interestingly, PIC prepared from multi-siRNA, which is composed of only two monomeric siRNA molecules, was demonstrated to achieve significant increase in gene silencing effect compared to monomeric siRNA PIC.

3.2 SNA Modification for Facilitated Endosomal Escape

The endosomal escape is one of the critical keys for efficient siRNA delivery into the cytosol. Thus, we have reported the siRNA conjugate with PAsp(DET) for facilitated endosomal escape as well as stable PIC formation (Takemoto et al. 2013). In design, several siRNA molecules were covalently introduced into PAsp(DET) side chains via maleic acid amide (siRNA-releasable/endosome-disrupting conjugate, REC) (Fig. 8). Maleic acid amide is stable at physiological neutral pH to link siRNA and mask PAsp(DET) into biologically inert polyanion,

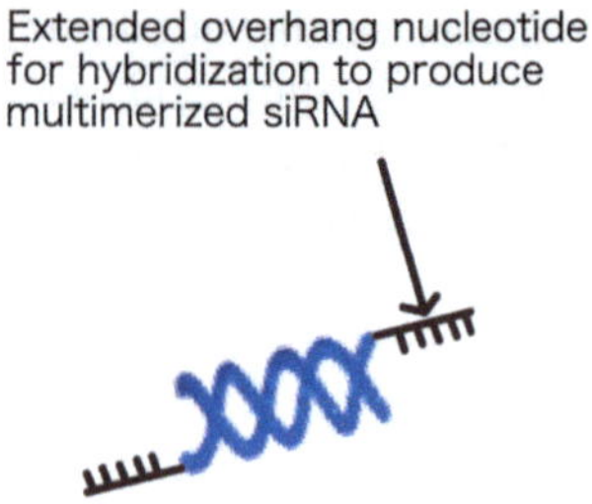

Fig. 6 Illustration of the monomer structure of sticky siRNA

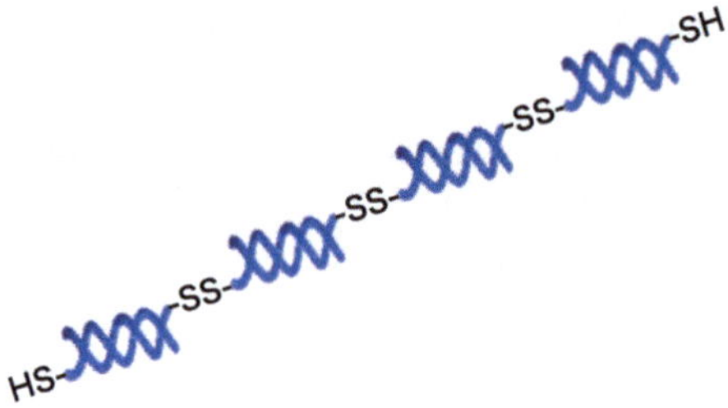

Fig. 7 Illustration of the structure of multi-siRNA

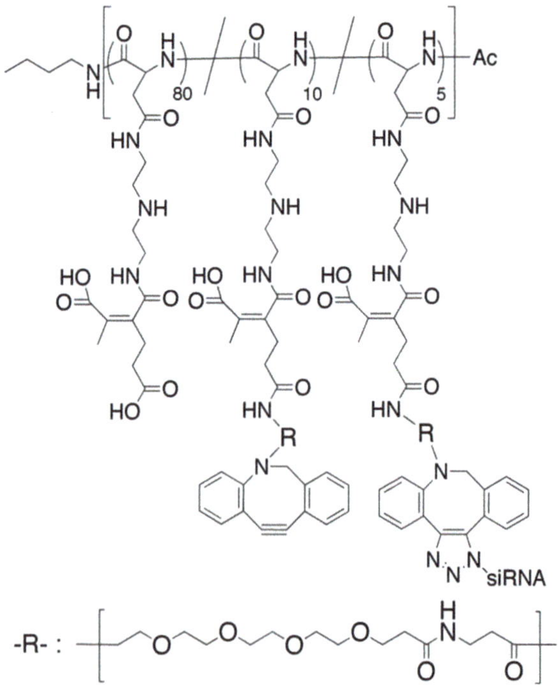

Fig. 8 Chemical structure of REC

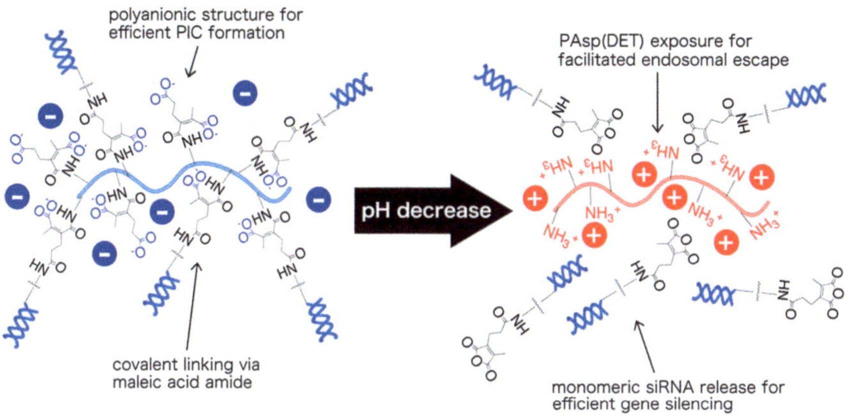

Fig. 9 Illustration of the strategy of REC [Ref. Takemoto et al. (2013)]

but is hydrolyzed at late-endosomal acidic pH to release siRNA and expose PAsp(DET). Increased anionic charges in REC, compared to monomeric siRNA, allow stable PIC formation at the cell exterior, and after cellular uptake, exposed PAsp(DET) and released siRNA enable facilitated endosomal escape and ready recruitment into gene silencing pathway, respectively (Fig. 9). Consequently, when siRNA PIC was formed with PAsp(DET), REC PIC treatments for cultured cells enabled significant enhancement in endosomal escape and subsequent gene silencing effect, compared to monomeric siRNA PIC treatment.

4 Conclusion and Future Perspective

Nucleic acids have great potential as a therapeutic drug, and many researchers have made remarkable efforts on the development of the nucleic acids-based therapeutics, e.g., more than 10,000 reports regarding siRNA have been published per one year. However, in order to realize the nucleic acids-based therapeutics, delivery strategy should satisfy several requirements: protection from enzymatic degradation and facilitated cellular uptake. Furthermore, after cellular uptake, the intracellular fate of nucleic acids needs to be controlled, and intact nucleic acids should be recruited into the expected biological process to exert their therapeutic activity. Polymer-based nucleic acids delivery carrier has been demonstrated to induce the biological activity of nucleic acids with low cytotoxicity. Polymer-based carrier has low immunogenicity, compared to virus-based carrier, and is able to be synthesized at industrial scale, leading to low hurdle for commercial production. We hope that polymer-based delivery carrier will be further developed toward the success in nucleic acids-based therapeutics.

References

Airenne KJ, Hu YC, Kost TA, Smith RH, Kotin RM, Ono C, Matsuura Y, Wang S, Yla-Herttuala S (2013) Baculovirus: an insect-derived vector for diverse gene transfer applications. Mol Ther 21:739–749

Alam MR, Ming X, Fisher M, Lackey JG, Rajeev KG, Manoharan M, Juliano RL (2011) Multivalent cyclic RGD conjugates for targeted delivery of small interfering RNA. Bioconjug Chem 22:1673–1681

Bolcato-Bellemin AL, Bonnet ME, Creusat G, Erbacher P, Behr JP (2007) Sticky overhangs enhance siRNA-mediated gene silencing. Proc Natl Acad Sci USA 104:16050–16055

Boussif O, Lezoualc'h F, Zanta MA, Mergny MD, Scherman D, Demeneix B, Behr JP (1995) A versatile vector for gene and oligonucleotide transfer into cells in culture and in vivo: polyethylenimine. Proc Natl Acad Sci USA 92:7297–7301

Carlisle RC, Etrych T, Briggs SS, Preece JA, Ulbrich K, Seymour LW (2004) Polymer-coated polyethylenimine/DNA complexes designed for triggered activation by intracellular reduction. J Gene Med 6:337–344

Chen C, Okayama H (1987) High-efficiency transformation of mammalian cells by plasmid DNA. Mol Cell Biol 7:2745–2752

Demeneix B, Behr JP (2005) Polyethylenimine (PEI). Adv Genet 53:215–230

Elbashir SM, Harborth J, Lendeckel W, Yalcin A, Weber K, Tuschl T (2001) Duplexes of 21-nucleotide RNAs mediate RNA interference in cultured mammalian cells. Nature 411:494–498

Fischer D, Li Y, Ahlemeyer B, Krieglstein J, Kissel T (2003) In vitro cytotoxicity testing of polycations: influence of polymer structure on cell viability and hemolysis. Biomaterials 24:1121–1131

Geiger T, Clarke S (1987) Deamidation, isomerization, and racemization at asparaginyl and aspartyl residues in peptides. Succinimide-linked reactions that contribute to protein degradation. J Biol Chem 262:785–794

Gosselin MA, Guo W, Lee RJ (2001) Efficient gene transfer using reversibly cross-linked low molecular weight polyethylenimine. Bioconjug Chem 12:989–994

Hong S, Leroueil PR, Janus EK, Peters JL, Kober MM, Islam MT, Orr BG, Baker JR Jr, Banaszak Holl MM (2006) Interaction of polycationic polymers with supported lipid bilayers and cells: nanoscale hole formation and enhanced membrane permeability. Bioconjugate Chem 17:728–734

Itaka K, Ishii T, Hasegawa Y, Kataoka K (2010) Biodegradable polyamino acid-based polycations as safe and effective gene carrier minimizing cumulative toxicity. Biomaterials 31:3707–3714

Jager M, Schubert S, Ochrimenko S, Fischer D, Schubert US (2012) Branched and linear poly(ethylene imine)-based conjugates: synthetic modification, characterization, and application. Chem Soc Rev 41:4755–4767

Kanasty R, Dorkin JR, Vegas A, Anderson D (2013) Delivery materials for siRNA therapeutics. Nat Mater 12:967–977

Kanayama N, Fukushima S, Nishiyama N, Itaka K, Jang WD, Miyata K, Yamasaki Y, Chung UI, Kataoka K (2006) A PEG-based biocompatible block catiomer with high buffering capacity for the construction of polyplex micelles showing efficient gene transfer toward primary cells. ChemMedChem 1:439–444

Kang HC, Huh KM, Bae YH (2012) Polymeric nucleic acid carriers: current issues and novel design approaches. J Control Release 164:256–264

Kataoka K, Harada A, Nagasaki Y (2001) Block copolymer micelles for drug delivery: design, characterization and biological significance. Adv Drug Deliv Rev 47:113–131

Kim YH, Park JH, Lee M, Kim YH, Park TG, Kim SW (2005) Polyethylenimine with acid-labile linkages as a biodegradable gene carrier. J Control Release 103:209–219

Knorr V, Ogris M, Wagner E (2008) An acid sensitive ketal-based polyethylene glycol-oligoethylenimine copolymer mediates improved transfection efficiency at reduced toxicity. Pharm Res 25:2937–2945

Kunath K, von Harpe A, Fischer D, Petersen H, Bickel U, Voigt K, Kissel T (2003) Low-molecular-weight polyethylenimine as a non-viral vector for DNA delivery: comparison of physicochemical properties, transfection efficiency and in vivo distribution with high-molecular-weight polyethylenimine. J Control Release 89:113–125

Lee SH, Mok H, Park TG (2011) Di- and triblock siRNA-PEG copolymers: PEG density effect of polyelectrolyte complexes on cellular uptake and gene silencing efficiency. Macromol Biosci 11:410–418

Meister A, Anderson ME (1983) Glutathione. Annu Rev Biochem 52:711–760

Miyata K, Oba M, Nakanishi M, Fukushima S, Yamasaki Y, Koyama H, Nishiyama N, Kataoka K (2008) Polyplexes from poly(aspartamide) bearing 1,2-diaminoethane side chains induce pH-selective, endosomal membrane destabilization with amplified transfection and negligible cytotoxicity. J Am Chem Soc 130:16287–16294

Miyata K, Nishiyama N, Kataoka K (2012) Rational design of smart supramolecular assemblies for gene delivery: chemical challenges in the creation of artificial viruses. Chem Soc Rev 41:2562–2574

Mok H, Lee SH, Park JW, Park TG (2010) Multimeric small interfering ribonucleic acid for highly efficient sequence-specific gene silencing. Nat Mater

Palte MJ, Raines RT (2012) Interaction of nucleic acids with the glycocalyx. J Am Chem Soc 134:6218–6223

Ramishetti S, Huang L (2012) Intelligent design of multifunctional lipid-coated nanoparticle platforms for cancer therapy. Ther Deliv 3:1429–1445

Saito G, Swanson JA, Lee KD (2003) Drug delivery strategy utilizing conjugation via reversible disulfide linkages: role and site of cellular reducing activities. Adv Drug Deliv Rev 55:199–215

Scholz C, Wagner E (2012) Therapeutic plasmid DNA versus siRNA delivery: common and different tasks for synthetic carriers. J Controlled Release 161:554–565

Sonawane ND, Szoka FC Jr, Verkman AS (2003) Chloride accumulation and swelling in endosomes enhances DNA transfer by polyamine-DNA polyplexes. J Biol Chem 278:44826–44831

Suh J, Paik HJ, Hwang BK (1994) Ionization of poly(ethylenimine) and poly(allylamine) at various pH's. Bioorg Chem 22:318–327

Suma T, Miyata K, Ishii T, Uchida S, Uchida H, Itaka K, Nishiyama N, Kataoka K (2012) Enhanced stability and gene silencing ability of siRNA-loaded polyion complexes formulated from polyaspartamide derivatives with a repetitive array of amino groups in the side chain. Biomaterials 33:2770–2779

Takemoto H, Ishii A, Miyata K, Nakanishi M, Oba M, Ishii T, Yamasaki Y, Nishiyama N, Kataoka K (2010) Polyion complex stability and gene silencing efficiency with a siRNA-grafted polymer delivery system. Biomaterials 31:8097–8105

Takemoto H, Miyata K, Hattori S, Ishii T, Suma T, Uchida S, Nishiyama N, Kataoka K (2013) Acidic pH-responsive siRNA conjugate for reversible carrier stability and accelerated endosomal escape with reduced IFNα-associated immune response. Angew Chem Int Ed 52:6218–6221

Thomas M, Klibanov AM (2002) Enhancing polyethylenimine's delivery of plasmid DNA into mammalian cells. Proc Natl Acad Sci USA 99:14640–14645

Uchida H, Miyata K, Oba M, Ishii T, Suma T, Itaka K, Nishiyama N, Kataoka K (2011) Odd-even effect of repeating aminoethylene units in the side chain of N-substituted polyaspartamides on gene transfection profiles. J Am Chem Soc 133:15524–15532

Zhang M, Kataoka K (2009) Nano-structured composites based on calcium phosphate for cellular delivery of therapeutic and diagnostic agents. Nano Today 4:508–517

Membrane-Domain-Selective Drug Targeting Based on Lipid Modification

Takeshi Mori and Yoshiki Katayama

Abstract Cellular lipid bilayers are critical platforms for the myriad of functions performed by membrane proteins. Drugs modified with lipid membrane anchors can selectively target membrane proteins. In this review, we discuss preferences of representative lipidated molecules for specific cellular membrane domains and the recent progress in lipidated drug delivery.

Keywords Cell membrane · Raft · Inner leaflet · Outer leaflet · Membrane protein · Lipid

Abbreviations

ER	Endoplasmic reticulum
GPI	Glycosylphosphatidylinositol
PEG	Polyethylene glycol
CLIC	Clathrin-independent carrier
GPCR	G protein-coupled receptor
APP	Amyloid precursor protein
HIV	Human immunodeficiency virus
HBV	Hepatitis B virus
PAR1	Protease-activated receptor 1
CXCR4	CXC-type receptor 4
SMO	Smoothened

T. Mori (✉) · Y. Katayama
Department of Applied Chemistry, Faculty of Engineering, Kyushu University, 744 Motooka, Nishi-ku, Fukuoka 819-0395, Japan
e-mail: mori.takeshi.880@m.kyushu-u.ac.jp

Y. Katayama
e-mail: ykatatcm@mail.cstm.kyushu-u.ac.jp

T. Mori · Y. Katayama
Center For Future Chemistry, Kyushu University, 744 Motooka, Nishi-ku, Fukuoka 819-0395, Japan

A. Prokop et al. (eds.), *Intracellular Delivery II*,
Fundamental Biomedical Technologies 7, DOI: 10.1007/978-94-017-8896-0_11,

IGF1R	Insulin-like growth factor 1 receptor
STAT3	Signal transducer and activator of transcription 3
SH2	Src homology 2
TRPV1	Transient receptor potential vanilloid 1
PAT	Palmitoyl acyl transferase

1 Introduction

Phospholipids, cholesterol, and fatty acids are the main components of cell membranes. Because of the high affinity of these lipids to bilayer membranes, lipid modification can be used to anchor molecules to the membranes. The lipid structure determines where the lipid-modified molecules localize with respect to plasma, endosomal, lysosomal, endoplasmic reticulum (ER), and Golgi membranes. The structure also governs the partitioning of the lipid-modified molecules in raft or non-raft domains of plasma membranes. Thus, lipids can selectively deliver molecules, such as fluorescent probes, to specific membrane domains. In this way, lipid modification has been used for subcellular targeting of drugs to specific kinds and domains of membranes where therapeutic target proteins reside. The considerable number of membrane-penetrating proteins, membrane bound proteins, and membrane-anchored proteins are potential targets of lipid-modified drugs. In each case, the activity of inhibitors of membrane proteins can be significantly improved by lipid modification (Ingallinella et al. 2009).

In this review, we first systematically categorize preferences of lipid groups. Then we summarize recent progress in intracellular drug delivery that uses lipid modification.

2 Cellular Localization of Lipid-Modified Molecules

First we summarized the representative lipid-modified molecules and their preference to the kinds and domain of the membranes. We categorized these lipid-modified molecules into three groups; small molecules, proteins (peptides), and polymers. The molecular structures of these molecules are shown in Fig. 1.

2.1 Lipid-Modified Small Molecules

Fluorescent dyes modified with alkyl chains are commercially available as cell tracking reagents (e.g., CellTracker, PKH, and CellBrite). Being hydrophobic, they are less soluble in aqueous media, especially at high ionic strength. Thus modification

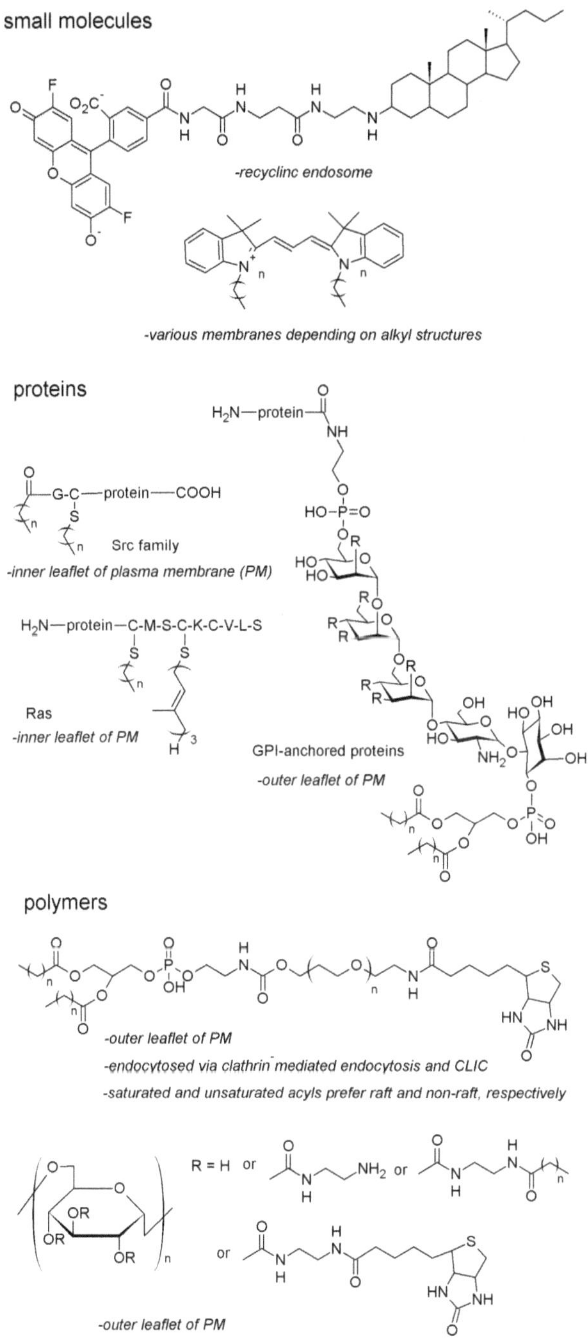

Fig. 1 Representative lipid-modified molecules

of the dyes is usually performed in an isotonic aqueous medium with low ionic strength or by using bovine serum albumin as a transfer reagent of the dyes to the membrane. Hydrophobic lipid-modified dyes can permeate plasma membranes, which leads to distribution of the dyes not only to the membrane inner leaflet, but also nonspecifically to endosomal, lysosomal, ER, Golgi and nuclear membranes. The length and saturation of the lipid alkyl chains significantly affect the affinity. Dyes with relatively long, saturated alkyl chains partition to mature endosomes, while those with short or unsaturated alkyl chains partition to recycling endosomes (Mukherjee et al. 1999). If the dyes include hydrophilic head groups or linkers, their membrane permeability is reduced and they localize predominantly on the outer leaflet of the plasma membrane. Therefore, they require endocytosis for transfer to endosomal and lysosomal membranes. Peterson's group reported that for some internalized dyes that were modified with cholesterol with relatively hydrophilic linkers, an appreciable fraction returns to the plasma membrane via recycling endosomes (Boonyarattanakalin et al. 2004).

2.2 Lipid-Modified Proteins and Peptides

Acyl modification of proteins is an important post-transcriptional modification by specific enzymes such as N-myristoyltransferase, palmitoyl acyltransferase, Rasp, farnesyltransferase, and geranylgeranyltransferase that takes place at the C- or N-termini (Resh 2006). It controls intracellular localization of proteins. For example, S-palmitoylation of H-Ras increases partitioning into the raft domains of the inner leaflets of plasma membranes. Src-family tyrosine kinases are myristoylated at an N-terminal glycine residue and then palmitoylated at a nearby cysteine residue.

Another class of lipid-modified proteins is glycosylphosphatidylinositol (GPI)-anchored proteins. GPI modification occurs in the ER membrane on the C-terminus of proteins and eventually the GPI-anchored proteins are transferred to the outer leaflet of plasma membrane.

Lipid-modified peptides have been originally synthesized to raise the membrane permeability of the inhibitory peptides (Bogoyevitch et al. 2005). Hydrophilic peptides which are inhibitors of intracellular enzymes such as protein kinases can be taken up by cells by modifying a lipid such as palmitoyl and stearoyl group to their termini to raise their inhibitory efficacy.

Silvius's group reported that short peptides taken from the substrate of the above-mentioned enzymes for acyl modification could also be substrates of these enzymes. The dually acylated peptides distributed preferentially to the raft domains of cell membranes (Wang et al. 2001; Schroeder et al. 1996). In addition, Silvius's group studied the stability and raft distribution of model peptides of acylated regions of C- or N-terminus proteins in liposomes (Silvius 2002).

2.3 *Lipid-Modified Polymers*

The intracellular localization of polyethylene glycol (PEG)-modified phosphatidyl ethanolamine was also reported by Silvius's group. Saturated dipalmitoyl conjugates were substantially partitioned into the raft domain, while polyunsaturated dilinoleoyl conjugates were excluded (Wang et al. 2005). These saturated and unsaturated conjugates are taken up both by clathrin-mediated endocytosis and by clathrin-independent carriers (CLIC). The latter is typically observed for the endocytosis of GPI-anchored proteins. The contribution of CLIC becomes dominant with the increased size of PEG domains (Bhagatji et al. 2009).

We reported that dextran modified with multiple palmitoyl groups anchors stably to a cell surface and is rapidly endocytosed, eventually locating to lysosomes (Tobinaga et al. 2014). This technique enables intracellular delivery of antigen proteins within dendritic cells for cancer immunotherapy.

3 Membrane-Targeted Lipidated Drugs

Covic and Kuliopulos's team initially proposed the idea of lipidated drugs to target specific membrane domains for inhibition of G protein-coupled receptors (GPCRs) (Covic et al. 2002). After this pioneering work, many groups have designed drugs that modulate the activity of membrane-bound proteins. These drugs are peptide-based with sequences taken from domains of proteins that interact with target membrane proteins. Here, we categorize these drugs according to the targeted membrane domains, i.e., outer or inner leaflet of plasma and/or endosomal membranes. Typical lipidated drugs are summarized in Table 1.

3.1 *Outer Leaflet-Targeted Lipidated Drugs*

3.1.1 Inhibition of β-Secretase for Treatment of Alzheimer's Disease

Simon's group reported on an inhibitor of β-secretase based on a lipidated peptide that can work in vivo, via intra-brain injection (Rajendran et al. 2008). β-secretase is known to generate amyloid β-peptides from amyloid precursor protein (APP), which is involved in Alzheimer's disease. Amyloid genesis of APP by β-secretase occurs in lipid raft domains in endosomes. They modified dihydroxycholesterol, which is known to partition to lipid rafts, on the C-terminus of the inhibitory peptides with linkers of various lengths until they found the optimal length for effective inhibition (Schieb et al. 2010).

Table 1 Lipidated peptides drugs

Location of targeting	Sequence	Application	Origin of peptide	Target protein	Ref.
outer leaflet of plasma membrane	Ac-WMEWDREINNYTSLIHSLIEESQNQQEKNEQ ELLGSGC(Chol)-NH_2	inhibition of HIV infection	gp41	gp41	Ingallinella at al. 2009
	C(Chol)KKKKGSGIEPHDWTKNITDKIDQIIHDFVDK	inhibition of Ebola virus infection	GP1	GP1	Higgins et al. 2013
	Chol-VNKKIEEIDKKIEELNKKLEELEKKLEEV NKK-H	inhibition of Newcastle disease virus and infectious bronchitis virus infection	Fusogenic glycoprotein	Fusogenic glycoprotein	Li et al. 2013
	Myr-GQNLSTSNPLGFFPDHQLDPAFRANTANP DWDFNPNKDTWPDANKVG	inhibition of HBV infection	preS1	unknown	Petersen et al. 2008
inner leaflet of plasma membrane	Pal-KKSRALF-NH_2	inhibition of platelet activation, anti-cancer	PAR1	unknown	Covic et al. 2002; Yang et al. 2009
	Pal-RCLSSSAVANRS-NH_2	inhibition of platelet activation	PAR1	unknown	Covic et al. 2002
	Pal-RSSAMDENSEKKRKSAIK-NH_2	anti-inflammation	PAR2	unknown	Sevigny et al. 2011
	Pal-SGRRYGHALR-NH_2	inhibition of platelet activation	PAR4	unknown	Covic et al. 2002
	Pal-MGYQKKLRSMTDKYRL-NH_2	HSPC mobilization	CXCR4	Gi	Tchernychev et al. 2010
	Pal-RGVMTLFSIKSNHPGLLS-HH_2	anti-cancer	SMO	unknown	Remsberg et al. 2007
	Myr-lLARRPTKGIHEY	inhibition of platelet activation	$G_{\alpha 13}$	PAR1, RhoA Integrin $\alpha_{\pi b}\beta_3$	Huang et al. 2007; Gong et al. 2010
	Pal-MGETVNKIAQES-NH_2	Ca_{2+} channel inhibition	TRPV1 channel	TRPV1 channel	Valente et al. 2011
	Pal-QLNLKEYNLV-NH_2	Na^+ channel inhibition	$G_{\alpha q/11}$	M (Kv7) channel	Robbins et al. 2006
	Pal-KVGFFKR	platelet activation	Integrin α	unknown	Bernard et al. 2009
	Pal-KVGAAKR	inhibition of platelet activation	Integrin α	unknown	Bernard et al. 2009
	Pal-RNNSRLGNGVLY-NH_2	inhibition of proliferation	IGF1R	IGF1R	Johannessen et al. 2011
	Pal-RRERRDLfTE-NH_2	inhibition of platelet activation	CD226	CD226	Edwards et al. 2007
	DTKYLEQLHKLYKKK(Pal)	anti-cancer	STAT3	STAT3	Timofeeva et al. 2007
	Chol-GpYLPQTV-NH_2	anti-cancer	STAT3	STAT3	Avadisian et al. 2011
	Myr-GCK(NBD)-NH_2	probing PAT activity	Src family	type-2 PAT	Schroeder et al. 1996
	NBD-CLC(Farn)-OMe	probing PAT activity	Ras	type-1 PAT	Schroeder et al. 1996
inside of endosomal membrane	H-EVNStaVAEF-3GI-4GI-3GI-4GI-Asp(ODhc)-NH_2	treating Alzheimer's disease	amyloid precursor protein	β-secretase	Rajendran et al. 2008

Pal: palmitoyl, Myr: myristoyl, Sta: (3S,4S)-4-amino-3-hydroxy-6-methylheptanoic acid, 3GI: 12-amino-4, 7, 10-trioxadodecanoyl, 4GI: 15-amino-4, 7, 10, 13-tetraoxapentadecanoyl, D(ODhc): O-dihydrocholesteroyl-L-asparatyl, NBD: 4-nitrobenzo-2-oxa-l, 3 diazole, Farn: farnesyl

3.1.2 Inhibition of Virus Infection

The lipid modification of the inhibitor peptide was then applied to the development of inhibitors of virus infection. For inhibitors of human immunodeficiency virus (HIV), Pessi's group added a cholesterol group to the C-terminus of the inhibitory peptide C34 (enfuvirtide) that was taken from gp41 of HIV fusogenic protein (Ingallinella et al. 2009). Half-maximal inhibitory concentration values of the cholesteryl peptide exhibited a 400-fold improvement relative to unmodified C34. The same design concept was applied to the inhibitory peptides of Ebola virus (Higgins et al. 2013), as well as to Newcastle disease virus and infectious bronchitis virus (Li et al. 2013).

Urban's group developed an inhibitor of hepatitis B virus (HBV) infection by using the original peptides taken from an envelope L-protein of HBV without further modification. The myristoylated preS1 peptide exists in the N-terminus region of L-protein and plays a critical role in HBV infection. It successfully inhibited HBV infection in vivo by preventing virus entry (Petersen et al. 2008).

3.2 Inner Leaflet-Targeted Lipidated Drugs

3.2.1 Non-enzyme-Mediated Targeting

Modulation of GPCR Signaling

Pepducins which originate from peptide sequences of GPCRs were the first lipidated drugs that targeted the inner leaflet of the plasma membrane (Covic et al. 2002). They were taken from the intracellular loop regions of protease-activated receptor 1 (PAR1) and PAR4 GPCRs to inhibit in vivo the signal transduction leading to thrombin-mediated platelet aggregation. Within seconds to minutes, the lipidated peptides flip into the inner leaflet; the mechanism is not clear (Wielders et al. 2007). Several reviews of pepducins have been published (O'Callaghan et al. 2012; Carlson et al. 2012; Tressel et al. 2011). So far, they have been designed to target nine GPCRs including PAR2 (Sevigny et al. 2011), CXC-type receptor 4 (CXCR4) (Quoyer et al. 2013; Tchernychev et al. 2010), and smoothened (SMO) (Remsberg et al. 2007; Johannessen et al. 2011). The pepducins for SMO and PAR1 (Yang et al. 2009) inhibited the growth of cancer cells, while pepducins for PAR2 and CXCR4 suppressed inflammation and mobilization of bone marrow hematopoietic cells, respectively. Because of the promising variety of pepducin functions, Anchor Therapeutics, Inc. was established to develop them commercially.

GPCR signaling can also be modulated with lipidated peptides originating from G-proteins. Brown's group reported that palmitoylated peptides taken from G proteins inhibit in vitro the M-type (Kv7) potassium channel in neurons (Robbins et al. 2006). Le Breton's team found that a myristoylated C-terminal peptide fragment of $G_{\alpha 13}$ could inhibit signaling to $G_{\alpha 13}$ from the platelet receptor integrin

$\alpha_{IIb}\beta_3$ (Gong et al. 2010). This lipidated peptide was originally designed to inhibit PAR1-mediated platelet activation via $G_{\alpha 13}$ (Huang et al. 2007).

Inhibition of Receptor and Non-receptor Tyrosine Kinase Signaling

Tarasova's group reported the inhibition of signal transduction from the IGF1R receptor tyrosine kinase with a peptide taken from the juxtamembrane domain of insulin-like growth factor 1 receptor (IGF1R) (Johannessen et al. 2011). They also reported that a peptide from the N-domain of signal transducer and activator of transcription 3 (STAT3) inhibits the signal transduction that induces apoptosis (Timofeeva et al. 2007). Gradinaru and Gunning's team used a peptide taken from the Src homology 2 (SH2) domain of STAT3 to inhibit STAT3-mediated signal transduction (Avadisian et al. 2011).

Inhibition of Adhesion Proteins

To find a peptide that inhibits platelet activation, Kenny and Shields relied on a bioinformatics approach to find 47 highly expressed transmembrane proteins, and chose 52 peptides from the proteins and 26 paralogous peptides to screen in vitro activity (Edwards et al. 2007). A highly effective peptide originated from CD226, which is a platelet and an endothelial adhesion factor. Moran's group reported that a highly conserved sequence of the juxtamembrane region of integrin α inhibited the platelet aggregation, while the replacement of two amino acids of the peptide inversely inhibited the activation (Bernard et al. 2009).

Inhibition of Channel Protein

Ferrer-Montiel's group reported that myristoylated peptide taken from the TRP domain of the transient receptor potential vanilloid 1 (TRPV1) channel inhibits channel activity both in vitro and in vivo (Valente et al. 2011).

3.2.2 Enzyme-Mediated Targeting

Silvius's group reported that very short N-myristoylated peptides are substrates of type 2 palmitoyl acyl transferase (PAT), which catalyzes palmitoylation of the Src family for translocation into the inner leaflets of plasma membranes (Schroeder et al. 1996). These peptides are fluorescent probes for the detection of intracellular PAT activity (Schroeder et al. 1996; Creaser et al. 2002; Varner et al. 2003). Silvius et al. designed other substrate peptides for type 1 PAT that were taken from the C-terminus of H- and N-Ras (Schroeder et al. 1996). These peptides are used for cell-based screening of PAT inhibitors (Ducker et al. 2006).

These peptides have not been used for drugs, but will be useful as inhibitors that actively change location. Hamachi and Tsukiji's team recently reported on the translocation of folates by using a substrate peptide for type 2 PAT to localize the folate-binding protein to the inner plasma membrane leaflet (Ishida et al. 2013).

4 Conclusion and Perspective

We have reviewed recent progress in lipidated drugs for membrane-domain-selective targeting of membrane proteins. Formulation will be an issue for practical usage. Pepducins circulate in the blood stream for long periods, with an elimination half-life of 4.7 h (Carlson et al. 2012). This is probably because of their interactions with serum albumin, which is a carrier protein of fatty acids. To further improve the circulation in the blood stream and to avoid degradation by endogenous protease, liposomes could be used as carriers because they can load lipidated drugs effectively (Rubas et al. 1986).

References

Avadisian M, Fletcher S, Liu B et al (2011) Artificially induced protein-membrane anchorage with cholesterol-based recognition agents as a new therapeutic concept. Angew Chem Int Ed 50:6248–6253

Bernard E, Parthasarathi L, Cho MK et al (2009) Ligand switching in cell-permeable peptides: manipulation of the α-integrin signature motif. ACS Chem Biol 4:457–471

Bhagatji P, Leventis R, Comeau J et al (2009) Steric and not structure-specific factors dictate the endocytic mechanism of glycosylphosphatidylinositol-anchored proteins. J Cell Biol 186:615–628

Bogoyevitch MA, Barr RK, Ketterman AJ (2005) Peptide inhibitors of protein kinases-discovery, characterization and use. Biochim Biophys Acta 1754:79–99

Boonyarattanakalin S, Martin SE, Dykstra SA et al (2004) Synthetic mimics of small mammalian cell surface receptors. J Am Chem Soc 126:16379–16386

Carlson KE, McMurry TJ, Hnt SW III (2012) Pepducins: lipopeptide allosteric modulators of GPCR signaling. Drug Deliv Today Technol 9:e33–e39

Covic L, Misra M, Badar J et al (2002) Pepducin-based intervention of thrombin-receptor signaling and systemic platelet activation. Nat Med 8:1161–1165

Creaser SP, Peterson BR (2002) Sensitive and rapid analysis of protein palmitoylation with a synthetic cell-permeable mimic of Src oncoproteins. J Chem Soc 124:2444–2445

Ducker CE, Griffel LK, Smith RA et al (2006) Discovery and characterization of inhibitors of human palmitoyl acyltransferases. Mol Cancer Ther 5:1647–1659

Edwards RJ, moran N, Devocelle M et al (2007) Bioinformatic discovery of novel bioactive peptides. Nat Chem Biol 3:108–112

Gong H, Shen B, Flevaris P et al (2010) G protein subunit $G_{\alpha13}$ binds to integirn $\alpha_{IIb}\beta_3$ aand mediates integrin "outside-in" signaling. Science 327:340–343

Higgins CD, Koellhoffer JF, Chandran K et al (2013) C-peptide inhibitors of Ebola virus glycoprotein-mediated cell entry: effects of conjugation to cholesterol and side chain-side chain crosslinking. Bioorg Med Chem Lett 23:5356–5360

Huang JS, Dong L, Kozasa T et al (2007) Signaling through Gα_{13} switch region I is essential for protease-activated receptor 1-mediated human platelet shape change, aggregation, and secretion. J Biol Chem 282:10210–10222

Ingallinella P, Bianchi E, Ladwa NA et al (2009) Addition of a cholesterol group to an HIV-1 peptide fusion inhibitor dramatically increases its antiviral potency. Pro Natl Acad Sci USA 106:5801–5806

Ishida M, Watanabe H, Takigawa K et al (2013) Synthetic self-localizing ligands that control the spatial location of proteins in living cells. J Am Chem Soc 135:12684–12689

Johannessen L, Remsberg J, Gaponenko V et al (2011) Peptide structure stabilization by membrane anchoring and its general applicability to the development of potent cell-permeable inhibitors. Chem Bio Chem 12:914–921

Li CG, Tang W, Chi XJ et al (2013) A cholesterol tag at the N terminus of the relatively broad-spectrum fusion inhibitory peptide targets an earlier stage of fusion glycoprotein activation and increases the peptide's antiviral potency in vivo. J Virol 87:9223–9232

Mukherjee S, Soe TT, Maxfield FR (1999) Endocytic sorting of lipid analogues differing solely in the chemistry of their hydrophobic tails. J Cell Biol 144:1271–1284

O'Callaghan K, Kuliopulos A, Covic L (2012) Turning receptors on and off with intracellular pepducins: new insights into G-protein-coupled receptor drug development. J Biol Chem 287:12787–12796

Petersen J, Dandri M, Mier W et al (2008) Prevention of hepatitis B virus infection in vivo by entry inhibitors derived from the large envelope protein. Nat Biotechnol 26:335–341

Quoyer J, Janz JM, Luo J et al (2013) Pepducin targeting the C-X-C chemokine receptor type 4 acts as a biased agonist favoring activation of the inhibitory G protein. Pro Natl. Acad Sci USA 110:E5088–E5097

Rajendran L, Schneider A, Schlechtingen G et al (2008) Efficient inhibition of the Alzheimer's disease β-secretase by membrane targeting. Science 320:520–523

Remsberg JR, Lou H, Tarasov SG et al (2007) Structural analogues of smoothened intracellular loops as potent inhibitors of hedgehog pathway and cancer cell growth. J Med Chem 50:4534–4538

Resh MD (2006) Trafficking and signaling by fatty-acylated and prenylated proteins. Nat Chem Biol 2:584–590

Robbins J, March SJ, Brown DA (2006) Probing the regulation of M (Kv7) potassium channels in intact neurons with membrane-targeted peptides. J Neurosci 26:7950–7961

Rubas W, Superasxo A, Weder HG et al (1986) Treatment of murine L1210 lymphoid leukemia and melanoma B16 with lipophilic cytosine arabinoside prodrugs incorporated into unilamellar liposomes. Int J Cancer 37:149–154

Schieb H, Weidlich S, Schlechtingen G et al (2010) Structural design, solid-phase synthesis and activity of membrane-anchored b-secretase inhibitors on Ab generation from wild-type and Swedish mutant APP. Chem Eur J 16:14412–14423

Schroeder H, Leventis R, Shahinian S et al (1996) Lipid-modified, cysteinyl-containing peptides of diverse structures are efficiently S-acylated at the plasma membrane of mammalian cells. J Cell Sci 134:647–660

Sevigny LM, Zhang P, Bohm A et al (2011) Interdicting protease-activated receptor-2-driven inflammation with cell-penetrating pepducins. Pro Natl Acad Sci USA 108:8491–8496

Silvius JR (2002) Lipidated peptides as tools for understanding the membrane interactions of lipid-modified proteins. Curr Topics Membr 52:371–395

Tchernychev B, Ren Y, Sachdev P et al (2010) Discovery of a CXCR4 agonist pepducin that mobilizes bone marrow hematopoietic cells. Pro Natl Acad Sci USA 107:22255–22259

Timofeeva OA, Gaponenko V, Kockett SJ et al (2007) Rationally designed inhibitors identify STAT3 N-domain as a promising anticancer drug target. ACS Chem Biol 2:799–809

Tobinaga K, Li C, Takeo M et al (2014) Rapid and serum-insensitive endocytotic delivery of proteins using biotinylated polymers attached via multivalent hydrophobic anchors. J Controlled Release (in press)

Tressel SL, Koukos G, Tchernychev B et al (2011) Pharmacology, biodistribution, and efficacy of GPCR-based pepducins in disease models. Method Mol Biol 683:259–275

Valente P, Fernandez-Carvajal A, Camprubi-Robles M et al (2011) Membrane-tethered peptides patterned after the TRP domain (TRPducins) selectively inhibit TRPV1 channel activity. FASEB J 25:1628–1640

Varner AS, Ducker CE, Xia Z et al (2003) Characterization of human palmitoyl-acyl transferase activity using peptides that mimic distinct palmitoylation motifs. Biochem J 373:91–99

Wang TY, Leventis R, Silvius JR (2001) Partitioning of lipidated peptide sequences into liquid-ordered lipid domains in model and biological membranes. Biochemistry 40:13031–13040

Wang TY, Leventis R, Silvius JR (2005) Artificially lipid-anchored proteins can elicit clustering-induced intracellular signaling events in Jurkat T-lymphocytes independent of lipid raft association. J Biol Chem 280:22839–22846

Wielders SJ, Bennaghmouch A, Reutelingsperger CPM et al (2007) Anticoagulant and antithrombotic properties of intracellular protease activated receptor antagonists. J Thromb Haemost 5:571–576

Yang E, Boire A, Agrwal A et al (2009) Blockade of PAR1 signaling with cell-penetrating pepducins inhibits Akt survival pathways in breast cancer cells and suppresses tumor survival and metastasis. Cancer Res 69:6223–6231

Multifunctional Protein-Based Nanoparticles for Cancer Theranosis

Luca Vannucci, Elisabetta Falvo and Pierpaolo Ceci

Abstract Selective delivery of therapeutic and/or diagnostic (theranostic) agents to diseased sites represents a major challenge to improve the outcome of current therapy and our ability to detect cancer cells at early stages or in the spread sites. A promising route to reach this goal is the design and engineering of functionalized nanoparticle (NP)-based carriers for targeted delivery of drug or diagnostic agents. Protein-based nanocarriers are attracting growing interest due to their exceptional characteristics, namely biodegradability, solubility, functionalization versatility and extraordinary binding capacity of various drugs. We highlight the use of these cage-shaped protein-based materials, with special emphasis on ferritin, as smart building blocks for the development of multifunctional NPs for cancer theranosis.

Keywords Theranosis · Nanoparticles · Protein-based nanocarriers · Drug-delivery · Functionalization · Targeted therapy · Biocompatibility

Abbreviations

NP	Nanoparticle
Hsp	Heat shock proteins
Ft	Ferritin
EPR	Enhanced permeability and retention
mAbs	Monoclonal antibodies
MRI	Magnetic resonance imaging
Fe_3O_4	Magnetite
PEG	Polyethylene glycol

L. Vannucci (✉)
Institute of Microbiology, Academy of Sciences of the Czech Republic (ASCR), v.v.i, Prague, Czech Republic
e-mail: vannucci@biomed.cas.cz

E. Falvo · P. Ceci
CNR—National Research Council of Italy, Institute of Molecular Biology and Pathology, Rome, Italy

A. Prokop et al. (eds.), *Intracellular Delivery II*,
Fundamental Biomedical Technologies 7, DOI: 10.1007/978-94-017-8896-0_12,

CTP	Carboxyl terminal peptide
VLP	Viral-like particles
CPMV	Cowpea mosaic virus
CCMV	Cowpea Chlorotic Mottle Virus
sHsp	Small heat shock proteins
Dps	DNA-binding proteins from starved cells
HFt	Human Ft
USPIO	Ultrasmall Superparamagnetic Iron Oxide
TMB	3,3′, 5,5′-Tetramethylbenzidine
DAB	3,3′-Diaminobenzidine
PET	Positron emission tomography
NIRF	Near infrared fluorophores
MB	Methylene blue
MFH	Magnetic fluid hyperthermia
RGD	Arginine-glycine-aspartate
α-MSH	A-melanocyte-stimulating hormone
EGF	Epidermal growth factor
FcBP	Fc-binding peptide
IgGs	Immunoglobulins G
TfR1	Transferrin receptor 1

1 Introduction

Selective delivery of therapeutic and/or diagnostic (theranostic) agents to diseased sites represents a major challenge to improve the outcome of current therapy and our ability to detect cancer cells at early stages or in the spread sites. In fact, despite rapid advances in diagnostic procedures and treatments, the overall survival rate from cancer has not improved substantially over the past 30 years (Howlader et al. 2010). The World Health Organization estimates that 84 million people will die of cancer between 2005 and 2015 (www.cdc.gov). Accordingly, there is a clear need for the development of novel approaches for the accurate detection of the early-stages of cancer and for targeted therapies.

A promising route to reach this goal is the design and engineering of functionalized nanoparticle (NP)-based carriers for targeted delivery of drug or diagnostic agents. The targeted delivery of nanomaterials can overcome difficulties associated with conventional free anticancer drugs, including rapid clearance, insolubility under aqueous conditions and a lack of selectivity, all resulting in non-specific toxicity toward normal cells and low therapeutic indices (Minten et al. 2009).

In this framework, several drug-delivery systems have been designed for a number of drug-carrier platforms including synthetic (gels, silica, polymers) and natural (lipids, proteins, oligosaccharides) (Kateb et al. 2011).

Protein-based nanocarriers are attracting growing interest due to their exceptional characteristics, namely biodegradability, solubility, functionalization versatility and extraordinary binding capacity to various drugs. A variety of proteins have been used and characterized for drug-delivery including virus-derived capsids (Minten et al. 2009, 2011), heat shock proteins (Hsp) (Flenniken et al. 2006; Choi et al. 2011) and ferritin (Ft) (Blazkova et al. 2013; Vannucci et al. 2012; Kitagawa et al. 2012). Viral capsids, Hsp and Ft are reported as protein cages, based on their particular shape. These proteins are characterized by a quaternary structure consisting in an assembly of multiple subunits endowed with the same fold. These assemblies enclose hollow spaces that can be used as ideal templates for the encapsulation of nano-material cargos. Indeed, the uniformity of the quaternary structure guarantees the attainment of NPs that are highly homogeneous in both size and shape, and the interior of the cage provides an isolated environment, shielded from bulk solution, where chemical reactions can take place. Furthermore, the protein surface comprises of diverse chemical groups (i.e., primary amines, carboxylates, thiols) that can be genetically and/or chemically manipulated in order to confer specific functionalities to the nano-cage. Additional advantages of protein cages include their generally remarkable stability, which can match and even be higher than that of nonprotein-based molecules, as well as high solubility in water. Another important aspect is the possibility to be produced at low cost as a recombinant protein in industrial bacterial strains on a large-scale (grams or even kilograms).

In this chapter, we highlight the use of these cage-shaped protein-based materials, with special emphasis to the ferritin ones, as smart building blocks for the development of multifunctional NPs for cancer theranosis.

2 Targeting Strategies

In general, nanoparticle-based materials can be delivered to tumors by passive and/or active targeting (Fig. 1).

2.1 *Passive Targeting*

Passive targeting can be ascribed to the enhanced permeability and retention (EPR) effect, determined by both extravasation of macromolecules through the leaky and poorly differentiated neo-vascular tumor system and lack of functional lymphatics, which result in the accumulation of extravasated nano-materials at the tumor site (Danhier et al. 2010; Matsumura and Maeda 1986). In fact, the ability of vascular endothelium to present open fenestrations was described for the liver sinusoidal endothelium (Oda et al. 2003), as well as the vascular endothelium in inflamed tissues, in hypoxic areas of infarcted myocardium (McDonald et al. 1999;

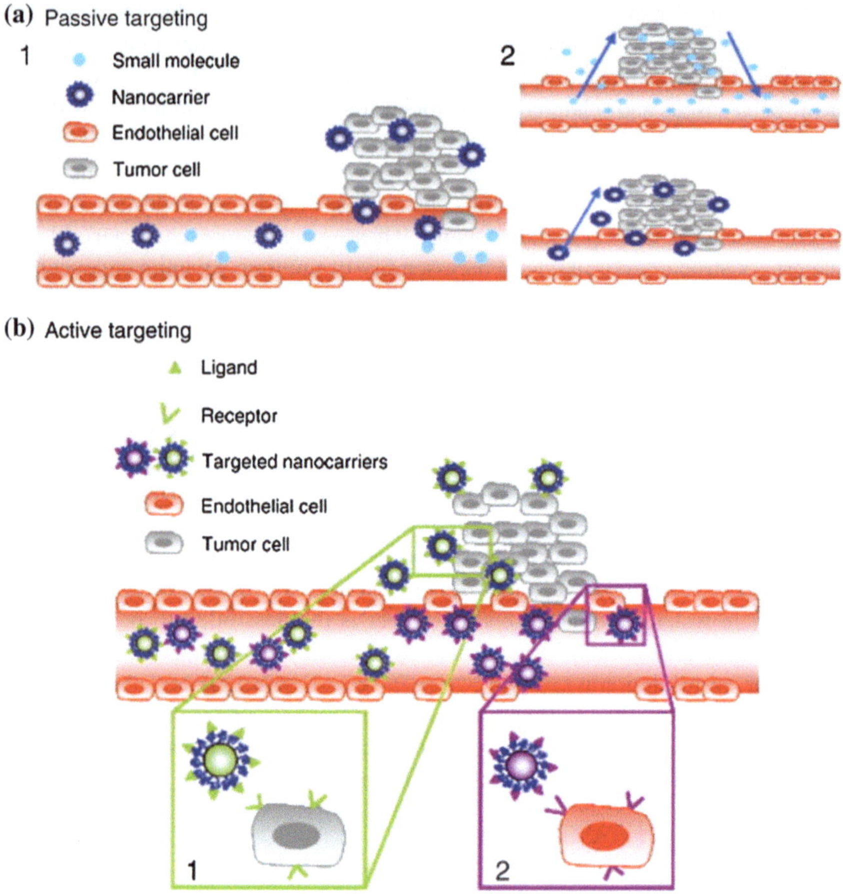

Fig. 1 Drug delivery of nanocarriers in cancer: passive and active targeting. **a** Nanocarriers can accumulate passively in solid tumor tissue exploiting the enhanced permeability and retention effect. NPs that enter into the tumor microenvironment are not removed efficiently and are thus accumulated and retained inside the tumor. In contrast, drugs alone diffuse freely in and out the tumor blood vessels because of their small size. This often can determine low drug accumulation. **b** Active targeting is mediated by NP-conjugated ligands that are able to bind with high affinity and selectivity to target molecules over-expressed by tumor cells as compared to healthy tissues. From (Danhier et al. 2010) To exploit the tumor microenvironment: Passive and active tumor targeting of nanocarriers for anti-cancer drug delivery Danhier et al. (2010). Reproduced with permission from Elsevier B.V., all reserved rights

Galaup et al. 2012) or in tumors (Nagy et al. 2012). In particular, tumor blood vessels are generally characterized by abnormalities such as a high proportion of proliferating endothelial cells and aberrant basement membrane formation leading to an enhanced vascular permeability. Particles, such as nanocarriers (in the size range of 10–150 nm), can extravasate and accumulate inside the interstitial space. Moreover, lymphatic vessels are insufficiently represented, abnormally structured and non-functional in tumors. This condition further contributes to an inefficient drainage from the tumor tissue, due to an increased interstitial pressure and a reduced intra- extravascular gradient (Baronzio et al. 2012). NPs that enter into the tumor microenvironment are not removed efficiently and are thus accumulated and retained inside the tumor. Using this mechanism, a very high local concentration of drug-loaded nanocarriers can be delivered to the tumor site at, for instance, a 5–50-fold higher rate than in normal tissue within few days (Iyer et al. 2006).

However, passive targeting approaches also suffer from several limitations. Targeting cancer cells using the EPR effect is not feasible in all tumors because the degree of tumor vascularization, porosity of tumor vessels and interstitial pressure can vary with the tumor type. For example, hypovascular tumors such as prostate and pancreatic cancers are very difficult to reach. Moreover, even within a single tumor, huge differences with regard to vascular permeability can be found, with parts in which macromolecules as large as 200 nm are able to extravasate and penetrate, whereas in other parts, even molecules of 10 nm are unable to enter the interstitium. Another limitation can be the presence of necrotic areas, especially in larger neoplasms (Rossin et al. 2005). A possible advantage for the use of small nanoparticles (around 10 nm) could be the possibility of passive transport even inside this difficult tissue once in the interstitial fluid circulation (Huang et al. 2012).

Finally, although tumor targeting consists in passive targeting and active targeting, the active targeting process cannot be separated from the passive because it occurs only after passive accumulation in tumors.

2.2 Active Targeting

To enhance intracellular delivery, active targeting approaches have been developed. Active targeting is mediated by NP-conjugated ligands that are able to bind with high affinity and selectivity to target molecules over-expressed by tumor cells as compared to healthy tissues (Danhier et al. 2010; Friedman et al. 2013). Active drug targeting is generally implemented to improve target cell recognition and cell uptake. Targeting ligands which have been explored to date include peptides, small organic molecules, oligosaccharides and monoclonal antibodies (mAbs) (Friedman et al. 2013). The mAbs have been widely used as tumor-homing molecules for the targeted delivery of NPs; however, several limitations including large size, difficulty in conjugation to NPs and high manufacturing costs have hampered their use. Thus, looking to obtaining similar selectivity toward the target, NPs decorated with smaller-sized ligands (including peptides) have attracted greater attention these days (Talekar et al. 2011).

In general, two cellular targets can be distinguished in the active targeting strategy: (a) the targeting of cancer cell and (b) the targeting of tumor vasculature. In the latter strategy, the nanocarriers have more chances to reach the target as it does not depend on extravasation and penetration across tumor interstitium since they encounter their target receptors much more frequently than do cancer cell-targeted NPs (Sakhrani and Padh 2013; Bedi et al. 2013; Wang et al. 2011).

A broad spectrum of chemical approaches has been used to conjugate targeting moieties to NP surfaces. These methods can be categorized as conventional bio-conjugation strategies (direct conjugation, linker chemistry, physical interactions), click chemistry or hybridization methods (Aubin-Tam 2013; Yu et al. 2012). Furthermore, only nanosystems based on proteins allow an additional and more powerful conjugation method, i.e. the genetic engineering approach. Genetic engineering represents a much easier and more reproducible method of generating protein-based NPs with exactly the same architecture as those obtained with chemical modifications. In fact, peptide sequences can be genetically inserted into the amino acid protein sequence to build a homogenous and ready-to use construct for selective cellular recognition and thus avoid repeated and expensive chemical reactions (Vannucci et al. 2012; de la Rica and Matsui 2010).

3 Multifunctional Nanoparticles for Theranosis

Multifunctional integrated systems based on NPs that combine differing properties such as tumor targeting, therapy, and imaging in an all-in-one system are providing more useful multimodal approaches in the battle against cancer. These systems have been intensively studied with the aim to overcome limitations associated with conventional cancer diagnosis and therapy, such as rapid clearance, insolubility under aqueous conditions and lack of selectivity (Rossin et al. 2005; Bao et al. 2013; Howell et al. 2013; Chen et al. 2013a).

An ideal multifunctional nanocarrier would allow for the simultaneously loading of therapeutics, ligands for cell specific targeting and fluorescent materials for ease of detection. Moreover, the loaded cancer drugs should be locally released in a controlled fashion by a defined stimulus in the environment (such as pH, temperature, specific proteases, etc.). The addition of magnetic properties, such as magnetite (Fe_3O_4) clusters, would further extend the functionality of the nano-carrier. In this way, the localization of nanocarriers to tumors by magnetic resonance imaging (MRI) can be also accomplished by using magnetic fields to concentrate the carrier directly in tumors while magnetic fluid hyperthermia can be used to directly kill cancer cells or as a trigger for controlled release. Also, the presence of both fluorescent and magnetic tracers could also render the nanocarriers a useful bimodal imaging agent (Wadajkar et al. 2013; Hayashi et al. 2013; Wilhelm et al. 2013; Chen et al. 2013b; Xiao et al. 2013; Chatterjee et al. 2011; Yoon et al. 2012; Tietze et al. 2012).

Nanocarriers are frequently functionalized to protect them from the reticulo–endothelial system and to increase the biocompatibility, solubility and stability in the bloodstream. The common method to do this consists of coating the surface of the particles with polyethylene glycol (PEG), a procedure called PEGylation. The coating of PEG chains to the surface of NPs results in an increase in the blood circulation half-life by several orders of magnitude (10–100) (Ferrari et al. 2013; Essa et al. 2011). In fact, by creating a hydrophilic protective layer around the NPs, steric repulsion forces repel the absorption of opsonin proteins, thereby blocking and delaying the opsonization process (Sant et al. 2008; Romberg et al. 2008; Xie et al. 2007).

Several additional approaches have been devised in recent years to extend the life span of NPs by slowing their clearance from the body, such as polysaccharide dextran decoration, PASylation and CTP (carboxyl terminal peptide) conjugation (Kotagiri et al. 2013; Schlapschy et al. 2013; Fares et al. 2007).

In particular, the last two approaches are based on recently developed technologies which involve genetic fusion or chemical conjugation with polypeptide sequences composed of the amino acids Pro, Ala, and Ser (for PASylation) or derived from the carboxyl terminal of human chorionic gonadotropin β subunit (for CTP). These sequences can readily be attached to a wide array of existing proteins and material, stabilizing them in the bloodstream and greatly extending its life span without additional toxicity or loss of desired biological activity (Fares et al. 2007; Schlapschy et al. 2013). Moreover, in the case of protein-based materials, PAS or CTP-modified proteins can be manufactured using established recombinant DNA techniques in widely used protein expression systems. Hence, the benefits of these technologies are substantial decreasing manufacturing complexity or high costs.

4 Protein-Based Systems

A protein-based nanomedicine platform utilizes natural or synthetic protein as a template for the production of a different multifunctional nanosystem (Lee and Wang 2006). A variety of proteins have been used and characterized for drug-delivery including viral capsids, heat shock proteins (Hsp) and ferritin (Ft) (Fig. 2).

4.1 Virus-Based Capsids

The viral capsid is the structural protein shell of a virus. Virus particles typically consist of several hundreds to thousands of protein molecules, which self-assemble to form a hollow scaffold packaging the viral nucleic acid. Their shape can be icosahedral or helical (Fig. 2). Their sizes are diverse due to the diversity of the viruses that have diameters at the nanometer level, which makes the viral capsids form nanometer-sized platforms (Fig. 2).

Viral-like particles (VLP) are exceptionally robust, they can be produced in large quantities in short time and they present programmable scaffolds (Snijder et al. 2012; Minten et al. 2009). VLPs offer advantages over synthetic nanomaterials,

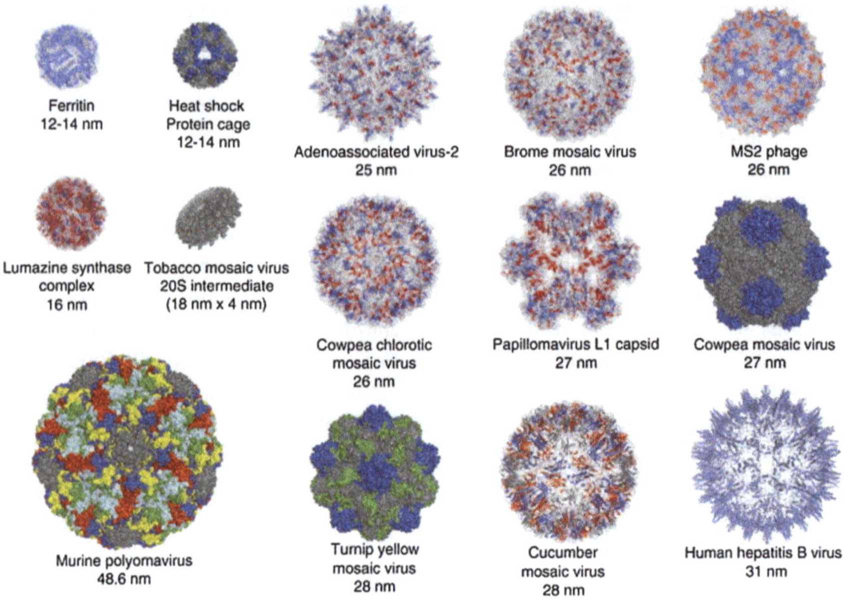

Fig. 2 Types of protein cage structures currently being developed for applications in bionanotechnology applications. Images generated using PyMol (http://www.pymol.org). From Lee and Wang (2006) adaptations of nanoscale viruses and other protein cages for medical applications Lee and Wang (2006). Reproduced with permission from Elsevier B.V., all reserved rights

primarily because they are biocompatible and biodegradable. VLPs derived from plant viruses and bacteriophages are particularly advantageous, because they are less likely to be pathogenic in humans and therefore less likely to induce undesirable side effects.

There are numerous reports in the literature for the use of virus capsids for the synthesis of biomaterial and their uses in nanotechnology and therapeutic applications (Manchester and Singh 2006; Franzen and Lommel 2009; Wen et al. 2012; Singh and Kostarelos 2009). The most widely used and understood viral capsids at this moment include Cowpea mosaic virus, CPMV (Steinmetz et al. 2009; Yildiz et al. 2013), Cowpea Chlorotic Mottle Virus, CCMV (Allen et al. 2005; Kaiser et al. 2007), and MS2 bacteriophage (Ashley et al. 2011; Anderson et al. 2006). Viral capsids have been widely used in drug and gene delivery (Galaway and Stockley 2013; Larocca et al. 2002; Arcangeli et al. 2013; Medina-Kauwe 2013; Tegerstedt et al. 2005), with antibodies (Gleiter and Lilie 2001; Stubenrauch et al. 2001; Frolova et al. 2010), for diagnostic imaging (Cormode et al. 2010; Chen et al. 2012; Carrico et al. 2012), with carbohydrates (Yin et al. 2012; Miermont et al. 2008), oligonucleotides (Tong et al. 2009; Lau et al. 2011), proteins (Pokorski et al. 2011; Banerjee et al. 2010), for fluorescent labeling (Rhee et al. 2011; Lu et al. 2012; Jung and Anvari 2013), and metallic nanoparticles (Chorny et al. 2013; Hwang et al. 2011).

However, before testing VLPs for actual nanomedicinal applications, a thorough evaluation of their toxicities and biodistributions in vivo is necessary. VLPs derived from bacteriophages and plant viruses are considered to be much safer because humans are not natural hosts for the parent viruses, although there have been few studies describing the characterization of such VLP platforms in vivo. Moreover, innate immunity response to the viral components has been found and can challenge the efficiency and efficacy of the VLPs as well as producing unwanted effects. Biodistribution and toxicity is sometimes discussed in the studies on the various types of nanovectors, but systemic studies on the argument are still lacking. We suggest the following references as a general orientation while specific observations can be found in the original papers cited in this chapter (Singh et al. 2007; Sakurai et al. 2008; Campos and Barry 2007; Thacker et al. 2009; Sharma et al. 2012; Ai et al. 2011; Ding and Wu 2012).

4.2 Heat Shock Proteins

Small heat shock proteins (sHsp) are generally composed of 24 subunits that autoassemble to form a cage with an exterior diameter of 12 nm and an interior of 6.5 nm. Hsp are produced at high levels in response to cellular stress and assist in the correct folding of proteins. Like other protein cages, its architecture is made up of exterior, interior, and interface surfaces. The external surface is characterized by the presence of large-size pore, about 3 nm, which can be used for free cargo exchange between the bulk solution (exterior) and the interior.

Like other viral capsids and ferritin proteins, sHsp were also used as a reaction vessel for biomimetic mineralization reactions. Transition metals, iron oxide and alloy nanoparticles were entrapped and synthesized within the sHsp protein cage (Varpness et al. 2005; Bode et al. 2011). Additionally, through site-directed mutation on the interior surface, organic drug molecules were also attached to the interior surface of the sHsp (Flenniken et al. 2003; Choi et al. 2011). Recently, ligand-conjugated Hsp cages via genetic or chemical modification were also reported to bind specifically to melanoma cells or human hepatocellular carcinoma in vitro or to SCC7 squamous carcinoma in vivo (Flenniken et al. 2006; Toita et al. 2012; Choi et al. 2011).

However, as discussed for viral capsids, before testing VLPs for real nanomedicinal applications, a thorough evaluation of their toxicities and biodistributions in vivo is necessary. In fact, most of the sHsp so far used are of bacterial origin (i.e. *Methanococcus jannaschii*) and therefore will need accurate evaluation of their toxic and immunological effects. A preliminary study on the in vivo biodistribution was reported by Douglas and co-workers (Kaiser et al. 2007).

4.3 Ferritins

Members of the ferritin family, including ferritins (Fts) and mini-ferritins (DNA-binding proteins from starved cells, Dps), are among the most widely studied cage-like proteins. These proteins are ubiquitous in nature and are involved in iron homeostasis and storage. The unique template structures of these proteins have been used for the synthesis of a variety of new, non-physiological mineral cores (Galvez et al. 2005, 2010; Kasyutich et al. 2010; Prastaro et al. 2009; Suzuki et al. 2009; Uchida et al. 2009, 2010; Fan et al. 2010; Jeong et al. 2005; Wong et al. 1998; Inoue et al. 2011; Kostiainen et al. 2011, 2013; Okuda et al. 2005) and/or organic molecules (Blazkova et al. 2013; Ma-Ham et al. 2011; Yan et al. 2010; Simsek and Kilic 2005) within the protein shell.

Apoferritin is a highly symmetrical multimeric protein consisting of 24 subunits that self-assemble into a shell-like molecule enclosing a hollow cavity with external and internal diameters of 12 nm and 8 nm, respectively (Harrison and Arosio 1996).

Dps's are formed by 12 subunits with an outer diameter of 8.5 nm and an inner diameter of 5 nm (Chiancone and Ceci 2010). Dps are a member of the ferritin superfamily and are found exclusively in prokaryotes. The Dps protein cage was also utilized as a size and shape constrained nanoreactor, similar to ferritin, but due to its bacterial nature, they have not been used for medical applications.

In contrast, for biomedical applications, NPs based on the human Ft (HFt) present a number of favorable properties with respect to other systems (Dominguez-Vera et al. 2010). HFt is a physiological protein that has high solubility and stability in water, blood and buffers, as well as low toxicity, all of which are desirable features for in vivo applications in human. Furthermore, HFt can be easily functionalized through genetic engineering and/or chemical reactions involving one of the many chemical groups exposed to the exterior (primary amines, carboxylates, thiols), granting for the rational design of new nanometric tools for delivery, imaging or therapy. Moreover, since their biological function is iron sequestration and storage, they are natively tailored for metal uptake and NP incorporation which may preclude the occasional release of toxic metal ions during blood circulation. The exceptional stability of the Ft cage structure over a wide range of temperatures (up to 80–100° C) and pH (3–10) makes large scale production at a low cost through recombination techniques possible.

The dimension of the HFt is an important feature of this protein cage. In fact, the HFt small size ($\leq$15 nm) increases the chances of passing human body barriers and reaching specific targets. Indeed, the dimensions of the NPs, which must be small enough to penetrate capillary fenestrations and large enough to avoid rapid clearance through the kidney (the ideal diameter being lower than 30 nm and greater than 6–8 nm, respectively), are one of the key prerequisites for efficient targeted delivery, together with a long-circulating capability of the carrier and high specificity of the selector towards the target receptor.

4.3.1 Theranostic Use of Ferritin-Based NPs

Imaging agents are among the compounds that can be successfully loaded within the interior cavity of ferritin proteins. The superparamagnetism of magnetite (Fe_3O_4)-enclosed ferritin NPs (named magnetoferritin) makes them ideal contrast agents for MRI for tumor diagnosis. MRI can provide high spatial resolution and functional anatomic and physiological information with simultaneous non-invasive imaging. Indeed, Douglas and co-workers were the first group that have been able to synthesize magnetite NPs encapsulated within the internal cavity of recombinant HFt, which possessed T2* MRI properties comparing favorably with known iron oxide MRI contrast agents (e.g., Ultrasmall Superparamagnetic Iron Oxide, USPIO; (Uchida et al. 2006, 2008). The same group has hypothesized that ferritins, which often accumulate in human atherosclerotic plaques, may serve as an intrinsic vehicle for targeting plaque macrophages (Uchida et al. 2009) and has demonstrated that modified ferritin cages can be used as fluorescence or MRI agents for in vivo detection of vascular macrophages (Terashima et al. 2011).

Recently, Fan et al. (2012) demonstrated the use of magnetoferritin as a dual-functional reagent allowing simultaneous targeting and visualization of tumors. In brief, they found that magnetoferritin NPs have intrinsic peroxidase-like activity that can be exploited to produce the same color reaction as peroxidase enzymes in the presence of peroxidase substrates. For instance, magnetoferritin NPs were able to react with 3,3′,5,5′—Tetramethylbenzidine (TMB) to produce a blue color and 3,3′—Diaminobenzidine (DAB) to produce a brown precipitate. This ability is based on the mineral cores consisting of magnetite or maghemite. In this way, Fan and co-workers established a new method to detect tumor tissues. The ability of ferritin to target tumor cells and tissues will be discussed below in this chapter.

In addition, other three groups from different laboratories, namely the groups of Aime, Domınguez-Vera and Kimura, have prepared water-soluble gadolinium NPs with NMR longitudinal and transverse relaxivities higher than the ones of clinically approved paramagnetic Gd-chelates, thus indicating the great appeal of these novel classes of MRI contrast agents (Crich et al. 2006; Makino et al. 2011; Sanchez et al. 2009).

A different approach, in terms of the nature of tracer to be used for medical imaging, has been recently reported (Lin et al. 2011). HFt has been loaded with radioactive metal ions (^{64}Cu). Such a ferritin nanotracer possessed positron emission tomography (PET) functionalities for high sensitive tumor imaging. In addition, when conjugated with near infrared fluorophores (NIRF) on the protein surface, these constructs can be used for multimodal imaging such as PET and NIRF imaging (Lin et al. 2011).

The loading of drugs, metal-based or not, inside the ferritin cavity or conjugating them on the external surface are other appealing opportunities for future tumor therapies.

Generally, the method used for drug-encapsulation is based simply on the addition of various metal ions or compounds to apoferritin under specific conditions allowing for the diffusion of the ions or compounds inside the protein that concentrate inside the cavity. This method exploits the presence of pores and channels (diameter of ~0.4 nm) traversing the protein shell that generate an electrostatic gradient favoring the concentration of cations. For metal particles and other compounds unable to pass through the ferritin channels for size or charge reasons, an alternative approach from the general method described above has been used. In this method, the protein cage has been reversibly disassembled in extremely acidic conditions (~pH 2.0) and the desired compounds have been passively encapsulated within the cavity by raising the pH to neutral values (Blazkova et al. 2013; Ma-Ham et al. 2011; Yan et al. 2010; Simsek and Kilic 2005; Yang et al. 2007).

Additionally, HFt can be easily functionalized through genetic engineering and/or chemical conjugation involving one of the many chemical groups naturally exposed to the protein exterior (primary amines, carboxylates, thiols). In this way, the exterior surface is another platform to be used for drugs loading, such as chemotherapeutics, toxins and cytotoxic peptides.

However, at present a few studies have begun to design ferritin NPs as a carrier to deliver drugs for the purpose of therapy. For example, Xing et al. (2009) successfully encapsulated platinum-based anticancer drugs in the cavity of horse spleen ferritin. Each ferritin molecule was able to encapsulate about 50 cisplatin molecules and these constructs were able to efficiently induce apoptosis on rat pheochromocytoma cells in vitro (Xing et al. 2009).

In another example, apoferritin was used to encapsulate doxorubicin inside the internal cavity using Cu(II) as a helper agent. These doxorubicin-loaded ferritin nanocages showed a higher tumor uptake, better tumor growth inhibition and less cardiotoxicity than free doxorubicin (Zhen et al. 2013). Overall, such a technology holds great potential in clinical translation.

Recently, Yan and co-workers reported that photosensitizers can also be encapsulated inside the ferritin cage (Yan et al. 2010). As a model, they demonstrated the successful encapsulation of methylene blue (MB) in apoferritin via a dissociation–reassembly process controlled by pH. The resulting MB-containing apoferritin nanocages showed a positive effect on singlet oxygen production and cytotoxic effects on MCF-7 human breast adenocarcinoma cells when irradiated at the appropriate wavelength (i.e. 633 nm).

An approach for using outer surface of apoferritin nanoparticles as a possible site for linking therapeutic molecules was proposed by Kwon and collaborators in 2011 (Kwon et al. 2012). They genetically modified the human heavy chain ferritin to have only one cysteine exhibited per subunit on the surface. To each cysteine, a β-cyclodextrin (24 per cage) was attached through thiol-maleimide Michael-type addition followed by copper(I) catalyzed azide-alkyne cycloaddition. This allowed formation of inclusion complexes with fluorescein isothiocyanate-conjugated adamantane and to slowly release it reversibly in a buffer solution. Such an approach is proposed as a new possibility for non-covalently capturing hydrophobic molecules, such as insoluble drugs, and for a limited period allowing them to be delivered to the desired cells or/and tissues (Kwon et al. 2012).

Another potential application of ferritin in the biomedical field has been proposed by Babincová et al. (2000), who suggested exploiting the magnetic properties of the ferritin iron core for magnetic fluid hyperthermia (MFH). MFH is a promising new cancer treatment aimed at killing tumor cells inducing their apoptosis/necrosis by increasing the temperature over the physiological tolerance threshold of the target. The procedure has been successfully used in glioma, prostate, liver, and breast tumors. Magnetic NPs should be applied directly to the tumor or injected into the body intravenously and diffuse selectively into cancerous tissues. Adding a safe high-frequency magnetic field (100–400 kHz), leads the particles to heat up, raising the temperature of the tumor cells without damaging the normal ones. However, at present, data showing the heating capacity of super-paramagnetic cores encapsulated in ferritins remains lacking.

4.3.2 Tumor Targeting Using Ferritin-Based NPs

For all the applications described so far, the development of molecules endowed with the ability to specifically direct NPs to selected cells and tissues would be of great value. In this direction, the exterior surface of the ferritin assembly possesses all the features necessary to operate as an appropriate platform for specific cell targeting/delivery. As mentioned above, modification of the protein exterior surface can be achieved either chemically or genetically. For instance, short peptide sequences and full length antibodies or their fragments, able to recognize specific cell receptors, can be genetically conjugated with the N-terminal region of HFt.

In previous reports, HFt has been genetically conjugated with an arginine-glycine-aspartate (RGD)-containing tumor targeting peptide which is recognized by a large number of integrin molecules expressed by melanoma, glioma, and other tumors as well as by normal cells (Li et al. 2012a; Kitagawa et al. 2012). Different groups found that the RGD-modified magnetoferritin could bind to many types of tumor cells, including amelanotic melanoma, glioblastoma, and lung adenocarcinoma cells (Zhen et al. 2013; Uchida et al. 2006). However, in some cases, it seemed that these ferritin NPs have a quick washout from the tumor and that RGD-integrin targeting would not increase accumulation of NPs in tumor. Therefore, it is not clear about the in vivo behavior of targeted ferritin-based NPs and whether active targeting would effectively guide NPs to tumor sites.

Recently, our group has genetically linked HFt to α-melanocyte-stimulating hormone (α-MSH) (Vannucci et al. 2012). The MSH peptide binds to melanocortin receptors that are overexpressed by melanoma cells and metastases, and are expressed to a lesser extent only by melanocytes (Miao and Quinn 2008). As a consequence, it is expected to be significantly more selective than the RGD peptide linked to HFt in previous studies, in that the RGD moiety is recognized by a large number of integrin molecules which are expressed by both normal and tumor cells. Furthermore, to increase the circulation time of ferritin, as well as to minimize nonspecific binding versus other human cells, we chemically conjugated the HFt surface with PEG molecules (Fig. 3). Targeted HFt-NPs, loaded with MRI

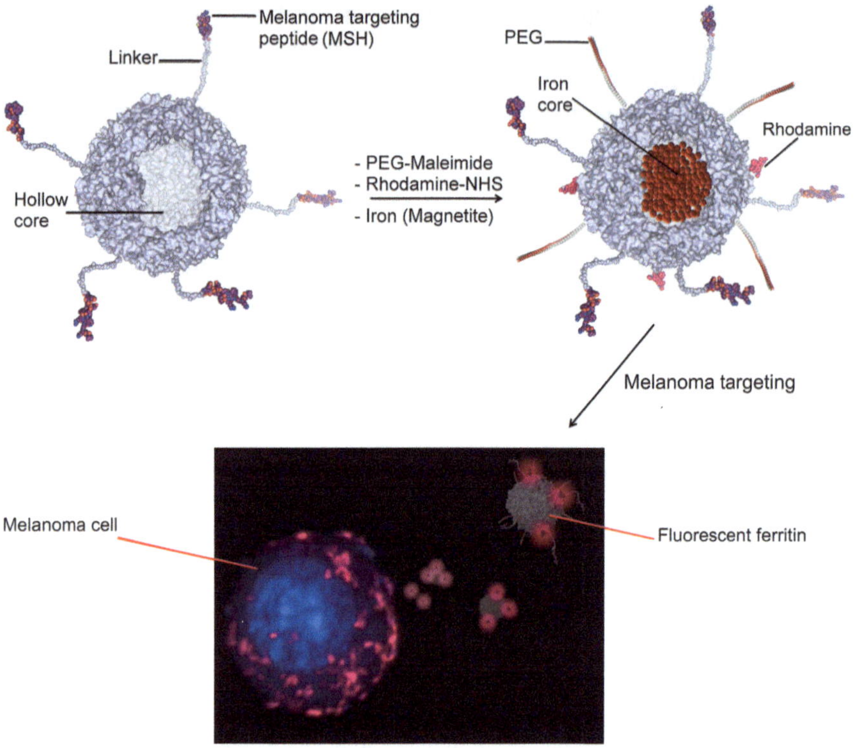

Fig. 3 *Top*. Example of a HFt-based nanoplatform for melanoma targeting. HFt protein was derivatized with a melanoma targeting peptide (MSH) that has been genetically joined to the N-terminus of each of the 24 subunits by a linker peptide (only 5 of the 24 derivatized N-termini are shown, for clarity). Additionally, novel functionalities (PEG, rhodamine, magnetic iron) have been added to this nanoconstruct by chemical derivatization. *Bottom*. Confocal laser scanning microscopy of a melanoma cell. The shown confocal image reveals an evident binding and uptake of HFt-based NPs after *in vitro* incubation. (*red*: NP fluorescence; *blue*: nucleus stained with DAPI. From International Journal of Nanomedicine by DOVE Medical Press. Reproduced with permission of DOVE Medical Press in the format Book via Copyright Clearance Center, see Ref. Vannucci et al. 2012)

or fluorescent tracers, were shown to accumulate significantly at the melanoma level for at least one week after systemic administration with high selectivity in comparison to both other tissues and a different type of tumor (adenocarcinoma). Untargeted HFt-NPs also showed a melanoma localization, but this was less pronounced and disappeared more rapidly than with the targeted counterparts, indicating that active targeting provides a relevant contribution to melanoma localization (Vannucci et al. 2014). Also, the HFt-MSH NPs were also able to target melanoma metastases that represent the actual cause of melanoma mortality.

Other genetic ferritin-based constructs have been developed recently in different laboratories. In these cases, larger targeting moieties were used, such as a full polypeptide or a portion of antibody. Li et al. (2012b) generated an epidermal growth factor (EGF)-HFt chimeric protein. The HFt NPs bearing EGF on the surface were able to accumulate in breast tumors in a mouse xenograft model. Dehal et al. (2010) produced and assessed magnetized fusion proteins consisting of the antigen-binding portion of an antibody (single chain variable fraction) fused to the ferritin protein. The resultant fusion protein was shown to be magnetizable and capable of binding target antigens. However, it should be noted that both procedures described above generated fusion proteins as insoluble products. This meant additional purification steps, lower yields, and increased costs that could preclude possible translation into the clinic.

Another elegant strategy was recently developed by Kang et al. that genetically introduced a Fc-binding peptide (FcBP) into the sequence of ferritin (Kang et al. 2012). FcBP-presenting ferritin (FcBP-ferritin) formed very stable non-covalent complexes with both human and rabbit immunoglobulins G (IgGs) through the simple molecular recognition between the Fc region of the antibodies and the Fc-binding peptide clusters inserted onto the surface of FcBP-ferritin. This approach has the great advantage of producing precisely-oriented antibodies on the surfaces of the protein cage simply by mixing without any complicated chemical conjugation.

All the targeting strategies described above pointed out to the joint of the ferritin surface with selective molecules, endowed with the ability to specifically direct NPs to selected cells and tissues. However, HFt *per se* has the ability to effectively bind cancer cells. A recent breakthrough finding was the identification of a ferritin receptor, transferrin receptor 1 (TfR1) (Li et al. 2010). In fact, the expression of TfR1 in proliferating cells, such as cancer cells, may be up to 100-fold higher than in normal cells. Fan et al. (2012, 2013) have exploited this finding and they successfully targeted TfR1 on different tumor tissues (using the visualization technique described above) to distinguish between tumors and normal tissues. In particular, they screened 474 clinical samples and found that H-ferritin specifically binds to the nine most common solid tumors, including liver, lung, colon, cervical, ovarian, prostate, breast, and thymus cancers. However, although this approach is very useful and powerful for ex vivo analyses, it should be improved and re-calibrated for its use in vivo. In fact, it is well known that when injected systemically, naked ferritin NPs possess short half-life in the bloodstream, thus limiting their desirable large and effective accumulation at the tumor sites. Therefore, future research should be directed towards fully utilizing the intrinsic properties of the ferritin system for in vivo applications.

Acknowledgements P.C. thanks the Italian Association for Cancer Research (AIRC), Milan (IT) for funding under grant agreement No. MFAG10545, and the Italian Ministry of Economy and Finance for funding the Project "FaReBio di Qualità". L.V. acknowledges RVO 61388971 (CZ), Fondazione Anna Villa e Felice Rusconi Fund (IT), ENI Czech Republic s.r.o. (CZ), Manghi Czech Republic s.r.o. Fund (CZ), Paul's Bohemia Trading s.r.o. (CZ) and Torino-Praga Invest s.r.o (CZ).

References

Ai J, Biazar E, Jafarpour M, Montazeri M, Majdi A, Aminifard S, Zafari M, Akbari HR, Rad HG (2011) Nanotoxicology and nanoparticle safety in biomedical designs. Int J Nanomed 6:1117–1127

Allen M, Bulte JWM, Liepold L, Basu G, Zywicke HA, Frank JA, Young M, Douglas T (2005) Paramagnetic viral nanoparticles as potential high-relaxivity magnetic resonance contrast agents. Magnet Reson Med 54(4):807–812 10.1002/Mrm.20614

Anderson EA, Isaacman S, Peabody DS, Wang EY, Canary JW, Kirshenbaum K (2006) Viral nanoparticles donning a paramagnetic coat: conjugation of MRI contrast agents to the MS2 capsid. Nano Lett 6(6):1160–1164.doi:10.1021/nl060378g

Arcangeli C, Circelli P, Donini M, Aljabali AA, Benvenuto E, Lomonossoff GP, Marusic C (2013) Structure-based design and experimental engineering of a plant virus nanoparticle for the presentation of immunogenic epitopes and as a drug carrier. J Biomol Struct Dyn. doi:10.1080/07391102.2013.785920

Ashley CE, Carnes EC, Phillips GK, Durfee PN, Buley MD, Lino CA, Padilla DP, Phillips B, Carter MB, Willman CL, Brinker CJ, Caldeira Jdo C, Chackerian B, Wharton W, Peabody DS (2011) Cell-specific delivery of diverse cargos by bacteriophage MS2 virus-like particles. ACS Nano 5(7):5729–5745. doi:10.1021/nn201397z

Aubin-Tam ME (2013) Conjugation of nanoparticles to proteins. Methods Mol Biol 1025:19–27. doi:10.1007/978-1-62703-462-3_3

Babincova M, Leszczynska D, Sourivong P, Babinec P (2000) Selective treatment of neoplastic cells using ferritin-mediated electromagnetic hyperthermia. Med Hypotheses 54(2):177–179. doi:10.1054/mehy.1999.0011

Banerjee D, Liu AP, Voss NR, Schmid SL, Finn MG (2010) Multivalent display and receptor-mediated endocytosis of transferrin on virus-like particles. ChemBioChem 11(9):1273–1279. doi:10.1002/cbic.201000125

Bao G, Mitragotri S, Tong S (2013) Multifunctional nanoparticles for drug delivery and molecular imaging. Annu Rev Biomed Eng 15:253–282. doi:10.1146/annurev-bioeng-071812-152409

Baronzio G, Schwartz L, Kiselevsky M, Guais A, Sanders E, Milanesi G, Baronzio M, Freitas I (2012) Tumor interstitial fluid as modulator of cancer inflammation, thrombosis Immunity and Angiogenesis. Ant-icancer Res 32(2):405–414

Bedi D, Gillespie JW, Petrenko VA, Ebner A, Leitner M, Hinterdorfer P, Petrenko VA (2013) Targeted delivery of siRNA into breast cancer cells via phage fusion proteins. Mol Pharmaceut 10(2):551–559. doi:10.1021/Mp3006006

Blazkova I, Nguyen HV, Dostalova S, Kopel P, Stanisavljevic M, Vaculovicova M, Stiborova M, Eckschlager T, Kizek R, Adam V (2013) Apoferritin modified magnetic particles as Doxorubicin carriers for anticancer drug delivery. Int J Mol Sci 14(7):13391–13402. doi: ijms140713391 [pii] 10.3390/ijms140713391

Bode SA, Minten IJ, Nolte RJ, Cornelissen JJ (2011) Reactions inside nanoscale protein cages. Nanoscale 3(6):2376–2389. doi:10.1039/c0nr01013h

Campos SK, Barry MA (2007) Current advances and future challenges in adenoviral vector biology and targeting. Curr Gene Ther 7(3):189–204. doi:10.2174/156652307780859062

Carrico ZM, Farkas ME, Zhou Y, Hsiao SC, Marks JD, Chokhawala H, Clark DS, Francis MB (2012) N-Terminal labeling of filamentous phage to create cancer marker imaging agents. ACS Nano 6(8):6675–6680. doi:10.1021/nn301134z

Chatterjee DK, Diagaradjane P, Krishnan S (2011) Nanoparticle-mediated hyperthermia in cancer therapy. Ther Deliv 2(8):1001–1014

Chen J, Shao R, Zhang XD, Chen C (2013a) Applications of nanotechnology for melanoma treatment, diagnosis, and theranostics. Int J Nanomed 8:2677–2688. doi: 10.2147/IJN.S45429 ijn-8-2677 [pii]

Chen W, Cao Y, Liu M, Zhao Q, Huang J, Zhang H, Deng Z, Dai J, Williams DF, Zhang Z (2012) Rotavirus capsid surface protein VP4-coated Fe(3)O(4) nanoparticles as a theranostic platform for cellular imaging and drug delivery. Biomaterials 33(31):7895–7902. doi: S0142-9612(12)00785-5 [pii] 10.1016/j.biomaterials.2012.07.016

Chen WH, Xu XD, Jia HZ, Lei Q, Luo GF, Cheng SX, Zhuo RX, Zhang XZ (2013b) Therapeutic nanomedicine based on dual-intelligent functionalized gold nanoparticles for cancer imaging and therapy in vivo. Biomaterials. doi: S0142-9612(13)00898-3 [pii] 10.1016/j.biomaterials.2013.07.084

Chiancone E, Ceci P (2010) The multifaceted capacity of Dps proteins to combat bacterial stress conditions: Detoxification of iron and hydrogen peroxide and DNA binding. Biochimica et Biophysica Acta (BBA) —General Subjects 1800(8):798–805

Choi SH, Kwon IC, Hwang KY, Kim IS, Ahn HJ (2011) Small heat shock protein as a multifunctional scaffold: integrated tumor targeting and caspase imaging within a single cage. Biomacromolecules 12(8):3099–3106. doi:10.1021/Bm200743g

Chorny M, Fishbein I, Tengood JE, Adamo RF, Alferiev IS, Levy RJ (2013) Site-specific gene delivery to stented arteries using magnetically guided zinc oleate-based nanoparticles loaded with adenoviral vectors. Faseb J 27(6):2198–2206. doi:10.1096/Fj.12-224659

Cormode DP, Jarzyna PA, Mulder WJM, Fayad ZA (2010) Modified natural nanoparticles as contrast agents for medical imaging. Adv Drug Deliver Rev 62(3):329–338. doi:10.1016/j.addr.2009.11.005

Crich SG, Bussolati B, Tei L, Grange C, Esposito G, Lanzardo S, Camussi G, Aime S (2006) Magnetic resonance visualization of tumor angiogenesis by targeting neural cell adhesion molecules with the highly sensitive gadolinium-loaded apoferritin probe. Cancer Res 66(18):9196–9201. doi:10.1158/0008-5472.Can-06-1728

Danhier F, Feron O, Preat V (2010) To exploit the tumor microenvironment: Passive and active tumor targeting of nanocarriers for anti-cancer drug delivery. J Control Release 148(2):135–146. doi: S0168-3659(10)00710-8 [pii] 10.1016/j.jconrel.2010.08.027

de la Rica R, Matsui H (2010) Applications of peptide and protein-based materials in bionanotechnology. Chem Soc Rev 39(9):3499–3509. doi:10.1039/B917574c

Dehal PK, Livingston CF, Dunn CG, Buick R, Luxton R, Pritchard DJ (2010) Magnetizable antibody-like proteins. Biotechnol J 5(6):596–604. doi:10.1002/biot.200900273

Ding H, Wu F (2012) Image guided biodistribution and pharmacokinetic studies of theranostics. Theranostics 2(11):1040–1053. doi: thnov02p1040 [pii] 10.7150/thno.4652

Dominguez-Vera JM, Fernandez B, Galvez N (2010) Native and synthetic ferritins for nanobiomedical applications: recent advances and new perspectives. Future Med Chem 2(4):609–618. doi:10.4155/fmc.09.171

Essa S, Rabanel JM, Hildgen P (2011) Characterization of rhodamine loaded PEG-g-PLA nanoparticles (NPs): effect of poly(ethylene glycol) grafting density. Int J Pharm 411(1–2):178–187. doi: S0378-5173(11)00174-8 [pii] 10.1016/j.ijpharm.2011.02.039

Fan K, Cao C, Pan Y, Lu D, Yang D, Feng J, Song L, Liang M, Yan X (2012) Magnetoferritin nanoparticles for targeting and visualizing tumour tissues. Nat Nanotechnol 7(7):459–464. doi: nnano.2012.90 [pii] 10.1038/nnano.2012.90

Fan K, Gao L, Yan X (2013) Human ferritin for tumor detection and therapy. Wiley Interdiscip Rev Nanomed Nanobiotechnol 5(4):287–298. doi:10.1002/wnan.1221

Fan R, Chew SW, Cheong VV, Orner BP (2010) Fabrication of gold nanoparticles inside unmodified horse spleen apoferritin. Small 6(14):1483–1487. doi:10.1002/smll.201000457

Fares F, Ganem S, Hajouj T, Agai E (2007) Development of a long-acting erythropoietin by fusing the carboxyl-terminal peptide of human chorionic gonadotropin beta-subunit to the coding sequence of human erythropoietin. Endocrinology 148(10):5081–5087. doi:10.1210/en.2007-0026

Ferrari R, Colombo C, Casali C, Lupi M, Ubezio P, Falcetta F, D'Incalci M, Morbidelli M, Moscatelli D (2013) Synthesis of surfactant free PCL-PEG brushed nanoparticles with tunable degradation kinetics. Int J Pharm 453(2):551–559. doi:10.1016/j.ijpharm.2013.06.020

Flenniken ML, Willits DA, Brumfield S, Young MJ, Douglas T (2003) The small heat shock protein cage from Methanococcus jannaschii is a versatile nanoscale platform for genetic and chemical modification. Nano Lett 3(11):1573–1576. doi:10.1021/Nl0347861

Flenniken ML, Willits DA, Harmsen AL, Liepold LO, Harmsen AG, Young MJ, Douglas T (2006) Melanoma and lymphocyte cell-specific targeting incorporated into a heat shock protein cage architecture. Chem Biol 13(2):161–170. doi:10.1016/j.chembiol.2005.11.007

Franzen S, Lommel SA (2009) Targeting cancer with 'smart bombs': equipping plant virus nanoparticles for a 'seek and destroy' mission. Nanomedicine (Lond) 4(5):575–588. doi:10.2217/nnm.09.23

Friedman AD, Claypool SE, Liu R (2013) The Smart Targeting of Nanoparticles. Curr Pharm Des. doi: CPD-EPUB-20130304-8 [pii]

Frolova OY, Petrunia IV, Komarova TV, Kosorukov VS, Sheval EV, Gleba YY, Dorokhov YL (2010) Trastuzumab-binding peptide display by Tobacco mosaic virus. Virology 407(1):7–13. doi: S0042-6822(10)00522-2 [pii] 10.1016/j.virol.2010.08.005

Galaup A, Gomez E, Souktani R, Durand M, Cazes A, Monnot C, Teillon J, Le Jan S, Bouleti C, Briois G, Philippe J, Pons S, Martin V, Assaly R, Bonnin P, Ratajczak P, Janin A, Thurston G, Valenzuela DM, Murphy AJ, Yancopoulos GD, Tissier R, Berdeaux A, Ghaleh B, Germain S (2012) Protection against myocardial infarction and no-reflow through preservation of vascular integrity by Angiopoietin-like 4. Circulation 125(1):140–U557. doi:10.1161/Circulationaha.111.049072

Galaway FA, Stockley PG (2013) MS2 viruslike particles: a robust, semisynthetic targeted drug delivery platform. Mol Pharm 10(1):59–68. doi:10.1021/mp3003368

Galvez N, Sanchez P, Dominguez-Vera JM (2005) Preparation of Cu and CuFe prussian blue derivative nanoparticles using the apoferritin cavity as nanoreactor. Dalton T 15:2492–2494. doi:10.1039/B506290j

Galvez N, Valero E, Dominguez-Vera JM, Masciocchi N, Guagliardi A, Clemente-Leon M, Coronado E (2010) Structural and magnetic characterization of Pd nanoparticles encapsulated in apoferritin. Nanotechnology 21(27):274017. doi:10.1088/0957-4484/21/27/274017

Gleiter S, Lilie H (2001) Coupling of antibodies via protein Z on modified polyoma virus-like particles. Protein Sci 10(2):434–444. doi:10.1110/ps.31101

Harrison PM, Arosio P (1996) The ferritins: molecular properties, iron storage function and cellular regulation. Biochim Biophys Act 1275:161–203

Hayashi K, Nakamura M, Sakamoto W, Yogo T, Miki H, Ozaki S, Abe M, Matsumoto T, Ishimura K (2013) Superparamagnetic nanoparticle clusters for cancer theranostics combining magnetic resonance imaging and hyperthermia treatment. Theranostics 3(6):366–376. doi: thnov03p0366 [pii] 10.7150/thno.5860

Howell M, Wang C, Mahmoud A, Hellermann G, Mohapatra SS, Mohapatra S (2013) Dual-function theranostic nanoparticles for drug delivery and medical imaging contrast: perspectives and challenges for use in lung diseases. Drug Deliv Transl Res 3(4):352–363. doi:10.1007/s13346-013-0132-4

Howlader N, Ries LA, Mariotto AB, Reichman ME, Ruhl J, Cronin KA (2010) Improved estimates of cancer-specific survival rates from population-based data. J Natl Cancer Inst 102(20):1584–1598. doi: djq366 [pii] 10.1093/jnci/djq366

Huang K, Ma H, Liu J, Huo S, Kumar A, Wei T, Zhang X, Jin S, Gan Y, Wang PC, He S, Liang XJ (2012) Size-dependent localization and penetration of ultrasmall gold nanoparticles in cancer cells, multicellular spheroids, and tumors in vivo. ACS Nano 6(5):4483–4493. doi:10.1021/nn301282m

Hwang JH, Lee S, Kim E, Kim JS, Lee CH, Ahn IS, Jang JH (2011) Heparin-coated superparamagnetic nanoparticle-mediated adeno-associated virus delivery for enhancing cellular transduction. Int J Pharmaceut 421(2):397–404. doi:10.1016/j.ijpharm.2011.10.019

Inoue I, Zheng B, Watanabe K, Ishikawa Y, Shiba K, Yasueda H, Uraoka Y, Yamashita I (2011) A novel bifunctional protein supramolecule for construction of carbon nanotube-titanium hybrid material. Chem Commun (Camb) 47(47):12649–12651. doi:10.1039/c1cc15221a

Iyer AK, Khaled G, Fang J, Maeda H (2006) Exploiting the enhanced permeability and retention effect for tumor targeting. Drug Discov Today 11(17–18):812–818. doi:10.1016/j.drudis.2006.07.005

Jeong GH, Yamazaki A, Suzuki S, Yoshimura H, Kobayashi Y, Homma Y (2005) Cobalt-filled apoferritin for suspended single-walled carbon nanotube growth with narrow diameter distribution. J Am Chem Soc 127(23):8238–8239. doi:10.1021/ja0505144

Jung B, Anvari B (2013) Virus-mimicking optical nanomaterials: near infrared absorption and fluorescence characteristics and physical stability in biological environments. ACS Appl Mater Interfaces 5(15):7492–7500. doi:10.1021/am401800w

Kaiser CR, Flenniken ML, Gillitzer E, Harmsen AL, Harmsen AG, Jutila MA, Douglas T, Young MJ (2007) Biodistribution studies of protein cage nanoparticles demonstrate broad tissue distribution and rapid clearance in vivo. Int J Nanomed 2(4):715–733

Kang HJ, Kang YJ, Lee YM, Shin HH, Chung SJ, Kang S (2012) Developing an antibody-binding protein cage as a molecular recognition drug modular nanoplatform. Biomaterials 33(21):5423–5430. doi:10.1016/j.biomaterials.2012.03.055

Kasyutich O, Ilari A, Fiorillo A, Tatchev D, Hoell A, Ceci P (2010) Silver ion incorporation and nanoparticle formation inside the cavity of pyrococcus furiosus ferritin: structural and size-distribution analyses. J Am Chem Soc 132(10):3621–3627. doi:10.1021/Ja910918b

Kateb B, Chiu K, Black KL, Yamamoto V, Khalsa B, Ljubimova JY, Ding H, Patil R, Portilla-Arias JA, Modo M, Moore DF, Farahani K, Okun MS, Prakash N, Neman J, Ahdoot D, Grundfest W, Nikzad S, Heiss JD (2011) Nanoplatforms for constructing new approaches to cancer treatment, imaging, and drug delivery: what should be the policy? Neuroimage 54:S106–S124. doi:10.1016/j.neuroimage.2010.01.105

Kitagawa T, Kosuge H, Uchida M, Dua MM, Iida Y, Dalman RL, Douglas T, McConnell MV (2012) RGD-conjugated human ferritin nanoparticles for imaging vascular inflammation and angiogenesis in experimental carotid and aortic disease. Mol Imaging Biol 14(3):315–324. doi:10.1007/s11307-011-0495-1

Kostiainen MA, Ceci P, Fornara M, Hiekkataipale P, Kasyutich O, Nolte RJ, Cornelissen JJ, Desautels RD, van Lierop J (2011) Hierarchical self-assembly and optical disassembly for controlled switching of magnetoferritin nanoparticle magnetism. ACS Nano 5(8):6394–6402. doi:10.1021/nn201571y

Kostiainen MA, Hiekkataipale P, Laiho A, Lemieux V, Seitsonen J, Ruokolainen J, Ceci P (2013) Electrostatic assembly of binary nanoparticle superlattices using protein cages. Nat Nanotechnol 8(1):52–56. doi: nnano.2012.220 [pii] 10.1038/nnano.2012.220

Kotagiri N, Lee JS, Kim JW (2013) Selective pathogen targeting and macrophage evading carbon nanotubes through dextran sulfate coating and pegylation for photothermal theranostics. J Biomed Nanotechnol 9(6):1008–1016. doi:10.1166/jbn.2013.1531

Kwon C, Kang YJ, Jeon S, Jung S, Hong SY, Kang S (2012) Development of protein-cage-based delivery nanoplatforms by polyvalently displaying beta-cyclodextrins on the surface of ferritins through copper(I)-catalyzed azide/alkyne cycloaddition. Macromol Biosci 12(11):1452–1458. doi:10.1002/mabi.201200178

Larocca D, Burg MA, Jensen-Pergakes K, Ravey EP, Gonzalez AM, Baird A (2002) Evolving phage vectors for cell targeted gene delivery. Curr Pharm Biotechnol 3(1):45–57

Lau JL, Baksh MM, Fiedler JD, Brown SD, Kussrow A, Bornhop DJ, Ordoukhanian P, Finn MG (2011) Evolution and protein packaging of small-molecule RNA aptamers. ACS Nano 5(10):7722–7729. doi:10.1021/Nn2006927

Lee LA, Wang Q (2006) Adaptations of nanoscale viruses and other protein cages for medical applications. Nanomedicine-UK 2(3):137–149. doi: S1549-9634(06)00107-9 [pii] 10.1016/j.nano.2006.07.009

Li K, Zhang ZP, Luo M, Yu X, Han Y, Wei HP, Cui ZQ, Zhang XE (2012a) Multifunctional ferritin cage nanostructures for fluorescence and MR imaging of tumor cells. Nanoscale 4(1):188–193. doi:10.1039/C1nr11132a

Li L, Fang CJ, Ryan JC, Niemi EC, Lebron JA, Bjorkman PJ, Arase H, Torti FM, Torti SV, Nakamura MC, Seaman WE (2010) Binding and uptake of H-ferritin are mediated by human transferrin receptor-1. Proc Natl Acad Sci USA 107(8):3505–3510. doi:10.1073/pnas.0913192107 [pii]

Li X, Qiu LH, Zhu P, Tao XY, Imanaka T, Zhao J, Huang YG, Tu YP, Cao XN (2012b) Epidermal growth factor-ferritin H-chain protein nanoparticles for tumor active targeting. Small 8(16):2505–2514. doi:10.1002/smll.201200066

Lin X, Xie J, Niu G, Zhang F, Gao H, Yang M, Quan Q, Aronova MA, Zhang G, Lee S, Leapman R, Chen X (2011) Chimeric ferritin nanocages for multiple function loading and multimodal imaging. Nano Lett 11(2):814–819. doi:10.1021/nl104141g

Lu X, Thompson JR, Perry KL (2012) Encapsidation of DNA, a protein and a fluorophore into virus-like particles by the capsid protein of cucumber mosaic virus. J Gen Virol 93(Pt 5):1120–1126. doi:10.1099/vir.0.040170-0 vir.0.040170-0[pii]

Ma-Ham AH, Wu H, Wang J, Kang XH, Zhang YY, Lin YH (2011) Apoferritin-based nanomedicine platform for drug delivery: equilibrium binding study of daunomycin with DNA. J Mater Chem 21(24):8700–8708. doi:10.1039/C0jm04321d

Makino A, Harada H, Okada T, Kimura H, Amano H, Saji H, Hiraoka M, Kimura S (2011) Effective encapsulation of a new cationic gadolinium chelate into apoferritin and its evaluation as an MRI contrast agent. Nanomed-Nanotechnol 7(5):638–646. doi:10.1016/j.nano.2011.01.015

Manchester M, Singh P (2006) Virus-based nanoparticles (VNPs): platform technologies for diagnostic imaging. Adv Drug Deliv Rev 58(14):1505–1522. doi:10.1016/j.addr.2006.09.014 S0169-409X(06)00176-1[pii]

Matsumura Y, Maeda H (1986) A new concept for macromolecular therapeutics in cancer chemotherapy: mechanism of tumoritropic accumulation of proteins and the antitumor agent smancs. Cancer Res 46(12 Pt 1):6387–6392

McDonald DM, Thurston G, Baluk P (1999) Endothelial gaps as sites for plasma leakage in inflammation. Microcirculation 6(1):7–22. doi:10.1038/sj.mn.7300053

Medina-Kauwe LK (2013) Development of adenovirus capsid proteins for targeted therapeutic delivery. Ther Deliv 4(2):267–277. doi:10.4155/tde.12.155

Miao Y, Quinn TP (2008) Peptide-targeted radionuclide therapy for melanoma. Crit Rev Oncol Hematol 67(3):213–228. doi:10.1016/j.critrevonc.2008.02.006 S1040-8428(08)00046-2[pii]

Miermont A, Barnhill H, Strable E, Lu X, Wall KA, Wang Q, Finn MG, Huang X (2008) Cowpea mosaic virus capsid: a promising carrier for the development of carbohydrate based antitumor vaccines. Chemistry 14(16):4939–4947. doi:10.1002/chem.200800203

Minten IJ, Hendriks LJ, Nolte RJ, Cornelissen JJ (2009) Controlled encapsulation of multiple proteins in virus capsids. J Am Chem Soc 131(49):17771–17773. doi:10.1021/ja907843s

Minten IJ, Wilke KDM, Hendriks LJA, van Hest JCM, Nolte RJM, Cornelissen JJLM (2011) Metal-ion-induced formation and stabilization of protein cages based on the cowpea chlorotic mottle epidermal growth factor-ferritin h-chain protein nanoparticles for tumor active targetingvirus. Small 7(7):911–919. doi:10.1002/smll.201001777

Nagy JA, Dvorak AM, Dvorak HF (2012) Vascular hyperpermeability, angiogenesis, and stroma generation. Cold Spring Harb Perspect Med 2(2):a006544. doi:10.1101/cshperspect.a006544 a006544[pii]

Oda M, Yokomori H, Han JY (2003) Regulatory mechanisms of hepatic microcirculation. Clin Hemorheol Microcirc 29(3–4):167–182

Okuda M, Kobayashi Y, Suzuki K, Sonoda K, Kondoh T, Wagawa A, Kondo A, Yoshimura H (2005) Self-organized inorganic nanoparticle arrays on protein lattices. Nano Lett 5(5):991–993. doi:10.1021/nl050556q

Pokorski JK, Hovlid ML, Finn MG (2011) Cell targeting with hybrid Qbeta virus-like particles displaying epidermal growth factor. Chembiochem 12(16):2441–2447. doi:10.1002/cbic.201100469

Prastaro A, Ceci P, Chiancone E, Boffi A, Cirilli R, Colone M, Fabrizi G, Stringaro A, Cacchi S (2009) Suzuki-Miyaura cross-coupling catalyzed by protein-stabilized palladium nanoparticles under aerobic conditions in water: application to a one-pot chemoenzymatic enantioselective synthesis of chiral biaryl alcohols. Green Chem 11(12):1929–1932. doi:10.1039/B915184b

Rhee JK, Hovlid M, Fiedler JD, Brown SD, Manzenrieder F, Kitagishi H, Nycholat C, Paulson JC, Finn MG (2011) Colorful virus-like particles: fluorescent protein packaging by the Q beta capsid. Biomacromolecules 12(11):3977–3981. doi:10.1021/Bm200983k

Romberg B, Hennink WE, Storm G (2008) Sheddable coatings for long-circulating nanoparticles. Pharm Res 25(1):55–71. doi:10.1007/s11095-007-9348-7

Rossin R, Pan D, Qi K, Turner JL, Sun X, Wooley KL, Welch MJ (2005) 64Cu-labeled folate-conjugated shell cross-linked nanoparticles for tumor imaging and radiotherapy: synthesis, radiolabeling, and biologic evaluation. J Nucl Med 46(7):1210–1218 46/7/1210[pii]

Sakhrani NM, Padh H (2013) Organelle targeting: third level of drug targeting. Drug Des Devel Ther 7:585–599. doi: S45614 dddt-7-585 [pii] 10.2147/DDDT

Sakurai H, Kawabata K, Sakurai F, Nakagawa S, Mizuguchi H (2008) Innate immune response induced by gene delivery vectors. Int J Pharmaceut 354(1–2):9–15. doi:10.1016/j.ijphann.2007.06.012

Sanchez P, Valero E, Galvez N, Dominguez-Vera JM, Marinone M, Poletti G, Corti M, Lascialfari A (2009) MRI relaxation properties of water-soluble apoferritin-encapsulated gadolinium oxide-hydroxide nanoparticles. Dalton Trans 5:800–804. doi:10.1039/b809645g

Sant S, Poulin S, Hildgen P (2008) Effect of polymer architecture on surface properties, plasma protein adsorption, and cellular interactions of pegylated nanoparticles. J Biomed Mater Res A 87(4):885–895. doi:10.1002/jbm.a.31800

Schlapschy M, Binder U, Borger C, Theobald I, Wachinger K, Kisling S, Haller D, Skerra A (2013) PASylation: a biological alternative to PEGylation for extending the plasma half-life of pharmaceutically active proteins. Protein Eng Des Sel 26(8):489–501. doi: 10.1093/protein/gzt023 gzt023[pii]

Sharma A, Madhunapantula SV, Robertson GP (2012) Toxicological considerations when creating nanoparticle-based drugs and drug delivery systems. Expert Opin Drug Met 8(1):47–69. doi:10.1517/17425255.2012.637916

Simsek E, Kilic MA (2005) Magic ferritin: a novel chemotherapeutic encapsulation bullet. J Magn Magn Mater 293(1):509–513. doi:10.1016/j.jmmm.2005.01.066

Singh P, Prasuhn D, Yeh RM, Destito G, Rae CS, Osborn K, Finn MG, Manchester M (2007) Bio-distribution, toxicity and pathology of cowpea mosaic virus nanoparticles in vivo. J Controlled Release 120(1–2):41–50. doi:10.1016/j.jconrel.2007.04.003

Singh R, Kostarelos K (2009) Designer adenoviruses for nanomedicine and nanodiagnostics. Trends Biotechnol 27(4):220–229. doi:10.1016/j.tibtech.2009.01.003

Snijder J, Ivanovska IL, Baclayon M, Roos WH, Wuite GJ (2012) Probing the impact of loading rate on the mechanical properties of viral nanoparticles. Micron 43(12):1343–1350. doi: S0968 4328(12)00134-5 [pii] 10.1016/j.micron.2012.04.011

Steinmetz NF, Lin T, Lomonossoff GP, Johnson JE (2009) Structure-based engineering of an icosahedral virus for nanomedicine and nanotechnology. Curr Top Microbiol Immunol 327:23–58

Stubenrauch K, Gleiter S, Brinkmann U, Rudolph R, Lilie H (2001) Conjugation of an antibody Fv fragment to a virus coat protein: cell-specific targeting of recombinant polyoma-virus-like particles. Biochem J 356(Pt 3):867–873

Suzuki M, Abe M, Ueno T, Abe S, Goto T, Toda Y, Akita T, Yamada Y, Watanabe Y (2009) Preparation and catalytic reaction of Au/Pd bimetallic nanoparticles in Apo-ferritin. Chem Commun 32:4871–4873

Talekar M, Kendall J, Denny W, Garg S (2011) Targeting of nanoparticles in cancer: drug delivery and diagnostics. Anticancer Drugs 22(10):949–962. doi:10.1097/CAD.0b013e32834a4554 00001813-201111000-00001 [pii]

Tegerstedt K, Franzen AV, Andreasson K, Joneberg J, Heidari S, Ramqvist T, Dalianis T (2005) Murine polyomavirus virus-like particles (VLPs) as vectors for gene and immune therapy and vaccines against viral infections and cancer. Anticancer Res 25(4):2601–2608

Terashima M, Uchida M, Kosuge H, Tsao PS, Young MJ, Conolly SM, Douglas T, McConnell MV (2011) Human ferritin cages for imaging vascular macrophages. Biomaterials 32(5):1430–1437. doi: S0142-9612(10)01184-1 [pii] 10.1016/j.biomaterials.2010.09.029

Thacker EE, Timares L, Matthews QL (2009) Strategies to overcome host immunity to adenovirus vectors in vaccine development. Expert Rev Vaccines 8(6):761–777. doi:10.1586/Erv.09.29

Tietze R, Lyer S, Durr S, Alexiou C (2012) Nanoparticles for cancer therapy using magnetic forces. Nanomedicine-UK 7(3):447–457. doi:10.2217/Nnm.12.10

Toita R, Murata M, Tabata S, Abe K, Narahara S, Piao JS, Kang JH, Hashizume M (2012) Development of human hepatocellular carcinoma cell-targeted protein cages. Bioconjugate Chem 23(7):1494–1501. doi:10.1021/Bc300015f

Tong GJ, Hsiao SC, Carrico ZM, Francis MB (2009) Viral capsid DNA aptamer conjugates as multivalent cell-targeting vehicles. J Am Chem Soc 131(31):11174–11178. doi:10.1021/Ja903857f

Uchida M, Flenniken ML, Allen M, Willits DA, Crowley BE, Brumfield S, Willis AF, Jackiw L, Jutila M, Young MJ, Douglas T (2006) Targeting of cancer cells with ferrimagnetic ferritin cage nanoparticles. J Am Chem Soc 128(51):16626–16633. doi:10.1021/ja0655690

Uchida M, Kang S, Reichhardt C, Harlen K, Douglas T (2010) The ferritin superfamily: supramolecular templates for materials synthesis. Biochim Biophys Acta 8:834–845. doi: S0304-4165(09)00340-7 [pii] 10.1016/j.bbagen.2009.12.005

Uchida M, Terashima M, Cunningham CH, Suzuki Y, Willits DA, Willis AF, Yang PC, Tsao PS, McConnell MV, Young MJ, Douglas T (2008) A human ferritin iron oxide nano-composite magnetic resonance contrast agent. Magnet Reson Med 60(5):1073–1081. doi:10.1002/Mrm.21761

Uchida M, Willits DA, Muller K, Willis AF, Jackiw L, Jutila M, Young MJ, Porter AE, Douglas T (2009) Intracellular distribution of macrophage targeting ferritin-iron oxide nanocomposite. Adv Mater 21(4):458. doi:10.1002/adma.200801209

Vannucci L, Falvo E, Fornara M, Di Micco P, Benada O, Krizan J, Svoboda J, Hulikova-Capkova K, Morea V, Boffi A, Ceci P (2012) Selective targeting of melanoma by PEG-masked protein-based multifunctional nanoparticles. Int J Nanomed 7:1489–1509. doi:10.2147/IJN.S28242 ijn-7-1489[pii]

Vannucci L, Falvo E, Failla CM, Carbo M, Fornara M, Canese R, Cecchetti S, Rajsiglova L, Stakheev D, Krizan J, Boffi A, Carpinelli G, Morea V, Ceci P (2014) In vivo targeting of cutaneous melanoma using an MSH-engineered human protein cage bearing fluorophore and MRI tracers. J Biomed Nanotechnol (in press)

Varpness Z, Peters JW, Young M, Douglas T (2005) Biomimetic synthesis of a H-2 catalyst using a protein cage architecture. Nano Lett 5(11):2306–2309. doi:10.1021/Nl0517619

Wadajkar AS, Menon JU, Kadapure T, Tran RT, Yang J, Nguyen KT (2013) Design and application of magnetic-based theranostic nanoparticle systems. Recent Pat Biomed Eng 6(1):47–57. doi:10.2174/1874764711306010007

Wang Z, Chui WK, Ho PC (2011) Nanoparticulate delivery system targeted to tumor neovasculature for combined anticancer and antiangiogenesis therapy. Pharm Res-Dordr 28(3):585–596. doi:10.1007/s11095-010-0308-2

Wen AM, Lee KL, Yildiz I, Bruckman MA, Shukla S, Steinmetz NF (2012) Viral nanoparticles for in vivo tumor imaging. J Vis Exp 69:e4352. doi:10.3791/4352 4352[pii]

Wilhelm S, Hirsch T, Patterson WM, Scheucher E, Mayr T, Wolfbeis OS (2013) Multicolor upconversion nanoparticles for protein conjugation. Theranostics 3(4):239–248. doi: thnov03p0239 [pii] 10.7150/thno.5113

Wong KKW, Douglas T, Gider S, Awschalom DD, Mann S (1998) Biomimetic synthesis and characterization of magnetic proteins (magnetoferritin). Chem Mater 10(1):279–285. doi:10.1021/cm970421o

Xiao Q, Zheng X, Bu W, Ge W, Zhang S, Chen F, Xing H, Ren Q, Fan W, Zhao K, Hua Y, Shi J (2013) A core/satellite multifunctional nanotheranostic for in vivo imaging and tumor eradication by radiation/photothermal synergistic therapy. J Am Chem Soc. doi:10.1021/ja404985w

Xie J, Xu C, Kohler N, Hou Y, Sun S (2007) Controlled PEGylation of monodisperse Fe_3O_4 nanoparticles for reduced non-specific uptake by macrophage cells. Adv Mater 19(20):3163. doi:10.1002/adma.200701975

Xing R, Wang X, Zhang C, Zhang Y, Wang Q, Yang Z, Guo Z (2009) Characterization and cellular uptake of platinum anticancer drugs encapsulated in apoferritin. J Inorg Biochem 103(7):1039–1044. doi: S0162-0134(09)00095-6 [pii] 10.1016/j.jinorgbio.2009.05.001

Yan F, Zhang Y, Kim KS, Yuan HK, Vo-Dinh T (2010) Cellular uptake and photodynamic activity of protein nanocages containing methylene blue photosensitizing drug. Photochem Photobiol 86(3):662–666. doi:10.1111/j.1751-1097.2009.00696.x

Yang Z, Wang X, Diao H, Zhang J, Li H, Sun H, Guo Z (2007) Encapsulation of platinum anticancer drugs by apoferritin. Chem Commun (Camb) 33:3453–3455. doi:10.1039/b705326f

Yildiz I, Lee KL, Chen K, Shukla S, Steinmetz NF (2013) Infusion of imaging and therapeutic molecules into the plant virus-based carrier cowpea mosaic virus: cargo-loading and delivery. J Control Release 172:568. doi:10.1016/j.jconrel.2013.04.023 S0168-3659(13)00242-3[pii]

Yin ZJ, Nguyen HG, Chowdhury S, Bentley P, Bruckman MA, Miermont A, Gildersleeve JC, Wang Q, Huang XF (2012) Tobacco mosaic virus as a new carrier for tumor associated carbohydrate antigens. Bioconjugate Chem 23(8):1694–1703. doi:10.1021/Bc300244a

Yoon HY, Saravanakumar G, Heo R, Choi SH, Song IC, Han MH, Kim K, Park JH, Choi K, Kwon IC, Park K (2012) Hydrotropic magnetic micelles for combined magnetic resonance imaging and cancer therapy. J Control Release 160(3):692–698. doi: S0168-3659(12)00253-2 [pii] 10.1016/j.jconrel.2012.04.012

Yu MK, Park J, Jon S (2012) Targeting strategies for multifunctional nanoparticles in cancer imaging and therapy. Theranostics 2(1):3–44. doi:10.7150/thno.3463 thnov02p0003[pii]

Zhen Z, Tang W, Chen H, Lin X, Todd T, Wang G, Cowger T, Chen X, Xie J (2013) RGD-modified apoferritin nanoparticles for efficient drug delivery to tumors. ACS Nano 7(6):4830–4837. doi:10.1021/nn305791q

Cytocompatible Phospholipid Polymers for Non-invasive Nanodevices

Tomohiro Konno

Abstract The design of cytocompatible polymer materials without any cytotoxicity is an essential and important part in the preparation of nanodevices for the delivery of bioactive molecules. The phospholipid polymer 2-methacryloyloxyethyl phosphorylcholine (MPC) is a strong candidate to provide such nanodevices because of its excellent cytocompatibility. Water-soluble MPC polymers bearing hydrophobic monomer units can form a stable polymer aggregate structure in biological milieus because of their amphiphilic nature. The obtained polymer aggregate can solubilize poorly soluble molecules, including bioactive molecules. In addition, MPC polymers containing active ester units to immobilize bioactive molecules are useful to investigate the bioactivity of immobilized molecules because non-specific interactions with biomolecules are reduced. The concept of a "cell-shuttle" that can penetrate the cell membrane without showing any cytotoxicity will be described.

Keywords Phospholipid polymer · Polymer aggregate · Solubilization · Polymer nanoparticles · Biointerfaces · Bioconjugate · Cell membrane penetration · Paclitaxel · Cell-shuttle

Abbreviations

MPC	2-Methacryloyloxyethyl phosphorylcholine
BMA	*n*-Butyl methacrylate
MEONP	*p*-Nitrophenyloxycarbonyl poly(oxyethylene) methacrylate
PLA	Poly(L-lactic acid)
R8	Octaarginine peptide
PTX	Paclitaxel

T. Konno (✉)
Department of Bioengineering, School of Engineering, The University of Tokyo, 7-3-1, Hongo, Bunkyo-ku, Tokyo 113-8656, Japan
e-mail: konno@bioeng.t.u-tokyo.ac.jp

A. Prokop et al. (eds.), *Intracellular Delivery II*,
Fundamental Biomedical Technologies 7, DOI: 10.1007/978-94-017-8896-0_13,

1 Amphiphilic Phospholipid Polymers as Colloid Biomaterials

Molecular assemblies consisting of low-molecular-weight phospholipids such as liposomes and lipid microspheres have the potential to deliver hydrophobic bioactive molecules, including anti-cancer drugs, because of their excellent cytocompatibility. Although much effort has been made to improve their effectiveness in drug delivery systems, most attempts have failed because of the lack of mechanical and chemical stability under physiological conditions. Thus, the stability of these phospholipid assemblies must be improved.

The cell membrane is a sophisticated, nanostructured barrier in living organisms (Singer and Nicholson 1972). The cell membrane is mainly composed of phospholipid molecules, which play an important role in various bioreactions. The structure of the cell membrane is the most attractive candidate for the fabrication of nanostructured biomaterials. Phospholipid molecules are the fundamental unit in the construction of the cell membrane. In particular, phosphorylcholine, an electrically neutral, zwitterionic head group, is one of the major phospholipid polar groups on the cell membrane. In 1990, Ishihara et al. reported the significant functions of the phospholipid polymer 2-methacryloyloxyethyl phosphorylcholine (MPC) (Ishihara et al. 1990). MPC is a methacrylate derivative and can copolymerize with any other vinyl monomers. In particular, MPC polymers containing an *n*-butyl methacrylate unit are widely used as coating materials for various medical devices (Moro et al. 2004; Snyder et al. 2007). Even when blood is applied to the surface of MPC polymers, protein adsorption and platelet adhesion are suppressed, confirming that MPC polymers have excellent cytocompatibility (Ishihara et al. 1998; Ishihara 2000; Iwasaki et al. 2001) (Fig. 1).

Because the MPC unit is highly hydrophilic, poly(MPC) and amphiphilic MPC polymers can be dissolved in water. Every MPC polymer with hydrophobic monomer units could be dissolved in water, if the polymer contains more MCP units than hydrophobic monomer units. A MPC polymer composed of about 70 mol% of *n*-butyl methacrylate and with a molecular weight above 10^5 did not dissolve in water. However, the polymer became water-soluble as the molecular weight decreased. In our previous publication, the surface tension of an aqueous solution containing amphiphilic MPC polymers decreased with increasing polymer concentration (Ishihara et al. 1999). This phenomenon is due to the formation of an MPC polymer aggregate in water (Fig. 2).

The amphiphilic phospholipid polymer PMB30W can solubilize hydrophobic and water-insoluble molecules such as perylene and pyrene. Konno et al. (2003) reported the solubilization ability of PMB30W for the poorly water-soluble drug paclitaxel (PTX). The water solubility of PTX is less than 0.3 μg/mL. Because of its poor solubility in water and many other acceptable pharmaceutical solvents, specific emulsifying agents such as Cremophor EL® (Cre) are, in general, used to formulate PTX in commercial injection solutions. However, serious hypersensitivity reactions have been reported in some individuals because the amount of Cre

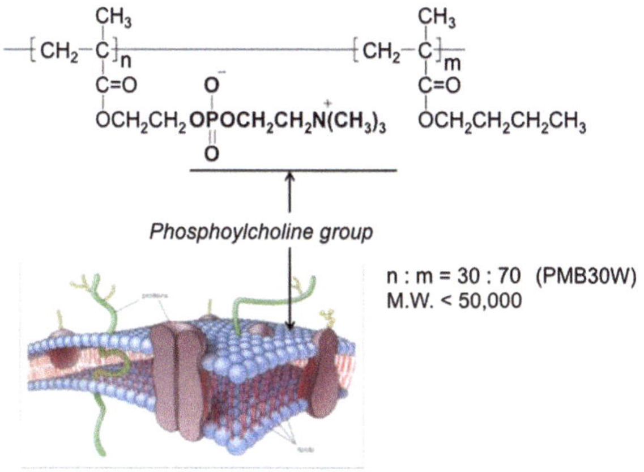

Fig. 1 Schematic representation of cell membrane and chemical structure of MPC polymer (PMB30W)

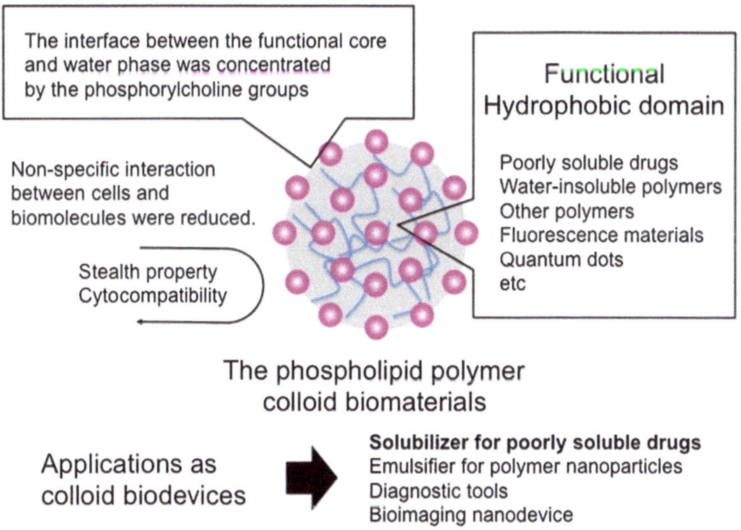

Fig. 2 Schematic illustration of PMB30W aggregate and characterization for colloid biomaterials

used in PTX formulations is significantly higher than that in any other marketed drug. Therefore, alternative dosage forms for the administration of PTX need to be developed to reduce undesirable side effects induced by Cre. The core polarity of PMB30W is the same as that of ethanol, which is a good solvent for PTX. The diameter of a PMB30W aggregate containing 1 mg/mL of PTX was 50 nm in an aqueous medium. The concentration of PTX in an aqueous solution of PMB30W

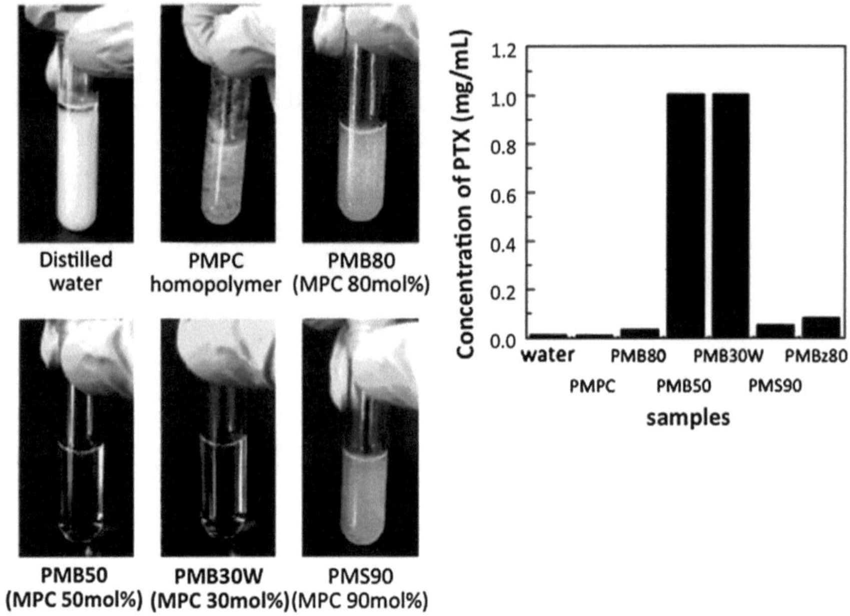

Fig. 3 Images of solubilized condition of PTX in the various amphiphilic MPC polymers

reached 5.0 mg/mL. The solution was transparent and stable for up to 1 month at room temperature, i.e., PTX did not precipitate during the storage period (Fig. 3).

Soma et al. (2009) reported the in vivo pharmacological activity of PTX solubilized in PMB30W. The authors used a peritoneal metastasis model in which nude mice were inoculated with the human gastric cancer-derived cell line MKN45P. The effect of intraperitoneal (i.p.) administration of PTX solubilized in PMB30W was then compared with that of conventional PTX dissolved in Cre. Drug accumulation in peritoneal nodules was evaluated by determining the intratumor PTX concentration and fluorescence microscopic observation. The results showed that the number of metastatic nodules and tumor volume were significantly decreased and the survival time was prolonged by treatment with PMB30 W-solubilized PTX compared to conventional PTX dissolved in Cre. Furthermore, the PTX concentration in disseminated tumors measured by high-performance liquid chromatography was higher in the PMB30W group than in the Cre group for up to 24 h after i.p. injection.

Kamei et al. (2011) reported the in vivo effectiveness of PMB30W as drug carrier. In their study, Oregon green-conjugated PTX was solubilized in PMB30W and the solubilized PTX showed perivascular accumulation in MKN45P tumors in the peritoneum at 24 h after intravenous (i.v.) injection. The amount of PTX detected in the tumor was markedly less than that in liver. In contrast, a larger amount of PTX accumulated in the peripheral area of disseminated nodules at 1 h after i.p. injection; the area increased gradually over time. The depth of PTX infiltration (measured from the tumor surface) reached 1 mm at 48 h after i.p. injection; the fluorescence intensity in the tumor was markedly greater than that in liver. Interestingly, PTX injected i.p. preferentially accumulated in relatively hypovascular areas, and many apoptotic tumor cells were observed near areas of PTX accumulation.

2 Bioconjugated MPC Polymer Nanoparticles

Immobilization of biomolecules is important in designing targeting drug carriers. The method used for immobilization under mild conditions should be simple. Furthermore, the biomolecules should maintain their specific bioactivity after immobilization. On the basis of the chemical structure of PMBs, which provide excellent cytocompatibility, new polymers are designed to conjugate biomolecules on the surface under mild conditions. The conjugation reaction of the biomolecules has to be carried out in an aqueous medium under the physiological pH range and temperature because of the low stability of biomolecules. Therefore, active ester groups are suitable because they can react with the amino groups in the biomolecules. A novel MPC polymer was synthesized on the basis of the chemical structure of PMB, poly(MPC-*co*-*n*-butyl methacrylate-*co*-*p*-nitrophenyloxycarbonyl poly(oxyethylene) methacrylate) (PMBN). The *p*-nitrophenyloxycarbonyl poly(oxyethylene) methacrylate unit has an active ester group in the side chain and can react with a specific biomolecule via condensation (Konno et al. 2004, 2006; Ito et al. 2006; Watanabe and Ishihara 2006) (Fig. 4).

Non-specific protein adsorption on nanoparticles was examined using PMBN modified poly(L-lactic acid) (PLA) nanoparticles (PMBN/PLA nanoparticles) and conventional polystyrene nanoparticles (Goto et al. 2008). Numerous albumin molecules were adsorbed on commercially available polystyrene nanoparticles. However, after reaction with glycine to avoid a reaction between albumin and *p*-nitrophenyloxycarbonyl poly(oxyethylene) methacrylate units on the nanoparticles, almost no adsorption on PMBN/PLA nanoparticles was observed. The amount of albumin adsorbed on PMBN/PLA nanoparticles was approximately 1/300 compared to that adsorbed on polystyrene nanoparticles. Thus, as a platform, PMBN/PLA nanoparticles showed excellent suppression of non-specific protein adsorption. Furthermore, Goto et al. (2008) reported PMBN/PLA nanoparticles that can penetrate cells noninvasively. The octaarginine peptide (R8) was immobilized on PMBN/PLA nanoparticles embedded with quantum dots for the evaluation of its cell penetrating ability. R8 is known as cell membrane-penetrating peptide. It has the ability to translocate through cell membranes in a manner that does not involve typical endocytic internalization pathways. R8-conjugated PMBN/PLA nanoparticles were clearly associated with HeLa cells and internalized to a perinuclear location. Preliminary experiments were carried out using other amino acid and octapeptide sequences to confirm the selectivity of the uptake by cells. However, only small amounts of octapeptides of glutamic acid (E8) and

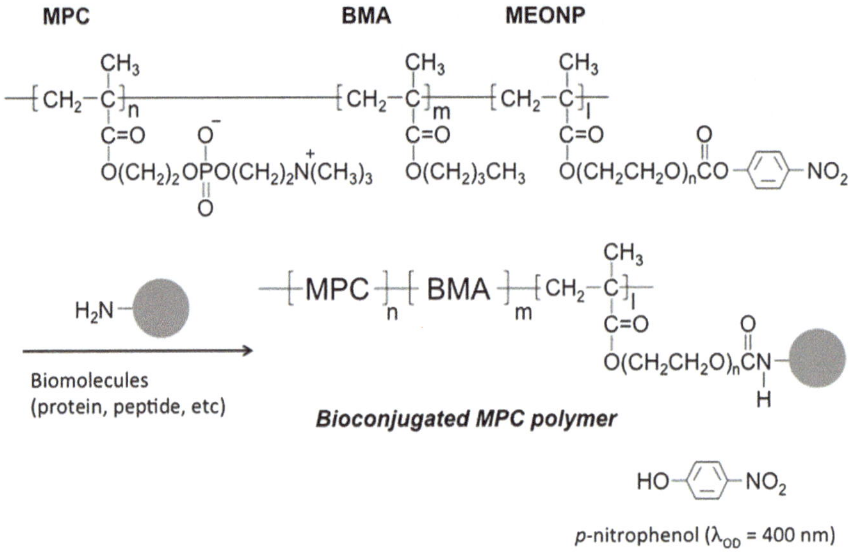

Fig. 4 Chemical structure of the MPC polymer for bioconjugation (PMBN)

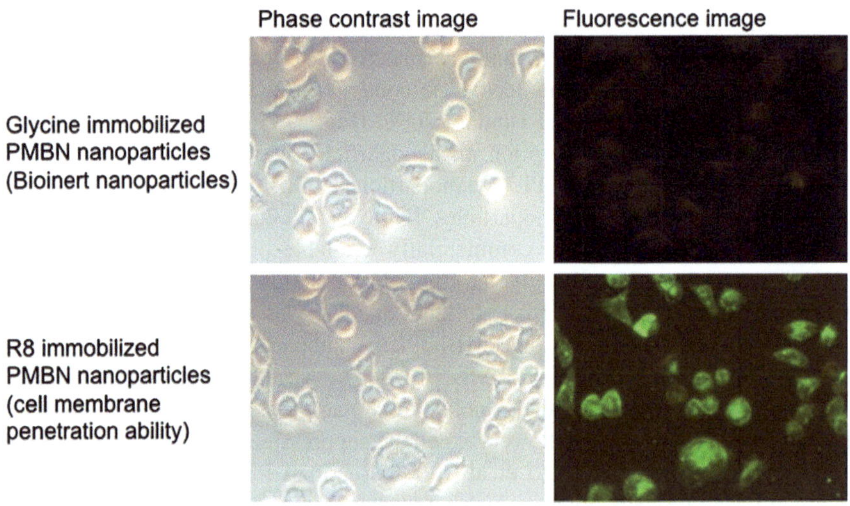

Fig. 5 Uptake of polymer nanoparticles covered with artificial cell membrane without and with R8 immobilization

asparagine (N8) were taken up by the cells. Octapeptides of tyrosine (Y8) and histidine (H8) were not effectively taken up by the cells. These results indicated that R8 could function as a cell-penetrating peptide even in close proximity to phosphorylcholine groups, which reduce translocation into the cytosol (Figs. 5, 6).

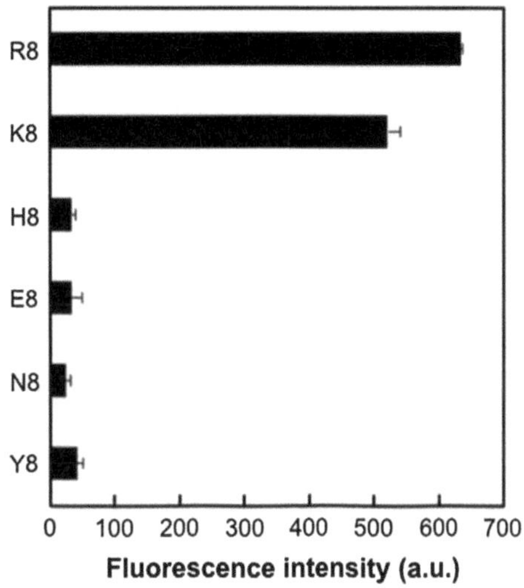

Fig. 6 Effect of amino acid residue of immobilized oligopeptide on cellular uptake

PMBN nanoparticles were applied in an in vivo experiment to evaluate their potential as targeting drug carrier. Miyata et al. (2009) reported PMBN conjugated with the preS1 domain of the hepatitis B surface antigen for targeting interleukin 6 and/or an immunoglobulin A binding protein. Conjugation of preS1 to PMBN resulted in a strongly enhanced synergistic inhibitory effect of PTX on HepG2 cells.

Shimada et al. (2009) reported the use of epidermal growth factor conjugated to PTX-containing PMBN particles and the growth inhibitory and antitumor effects of the obtained particles on cancer cells that overexpress epidermal growth factor receptors. Cytotoxicity and antitumor effects were significantly increased. Thus, PMBN nanoparticles are a strong candidate for delivering biomolecules, including drugs and differentiation reagents, and developing new applications in the field of cell engineering.

3 Cell Shuttle for Non-invasive Nanodevices

Amphiphilic MPC polymers were shown to have unique properties in biological milieus. In particular, the phospholipid polymer PMB30W is non-cytotoxic even though it has an amphiphilic nature. The amphiphilic nature of PMB30W is useful to solubilize poorly soluble drugs, including PTX, amphotericin B, all-*trans* retinoic acid, and so on. In addition, aggregates of PTX solubilized in PMB30W were shown to have good pharmacological activity in vivo. Recently, it was confirmed that water-soluble PMB permeates the cell membrane through a simple diffusion process (Goda et al. 2010; Kojima et al. 2011). When PMB was labeled with a

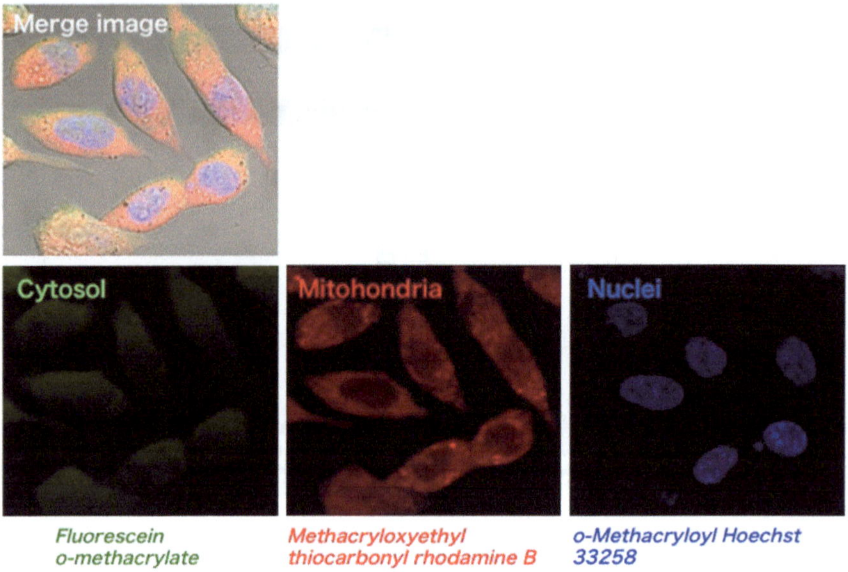

Fig. 7 Fluorescence PMB30W distributes in subcellular compartments of HepG2 cells based on the organelle specificity of the fluorescent dyes after direct penetration. Confocal images of multiple color staining of live HepG2 cells incubated with fluorescence PMB30W (1 mg/mL each) in serum-free medium for 30 min at 37 °C

specific fluorescence monomer via copolymerization and added to cell culture medium, uptake by the cells due to PMB diffusion could be observed within 1 min. Goda et al. (2010) reported the interaction between PMB and the cell membrane. When PMB30W labeled a with fluorescence monomer unit was added to hepatocytes, it was very rapidly taken up by the cells and the fluorescence intensity increased continuously. Noteworthy, localized fluorescence polymers were able to escape from the cytoplasm when the cell culture medium was replaced with fresh medium. Although the trends in polymer uptake and intracellular distribution were qualitatively similar in serum-free and serum-containing media, the comparable inhibition of the cellular uptake of PMB30W in serum-containing medium presumably reflects binding to serum proteins. Whether the fluorescent dyes located in the cytosol indicate the positions of PMBs or simply fluorescent dye that has been enzymatically cleaved from the polymer was confirmed by relaxation modes of the fluorescent dyes in the cytosol using fluorescence correlation spectroscopy. The results clearly indicated cytosolic entry of PMB30W without cleavage of the fluorescent dye from the polymer. Furthermore, the authors observed direct penetration of water-soluble fluorescent PMBs with higher weight-average molecular weight of up to 4.0×10^6 (Fig. 7).

4 Conclusion

On the basis of the presented fundamental findings, water-soluble MPC polymer aggregates and nanoparticles are widely applied to various colloid biomedical devices, e.g., hydrogels containing functional cells, multilayered hydrogels for controlled release of bioactive molecules, gene carriers, nanoparticles for bioimaging, and so on. It can be concluded that water-soluble and amphiphilic MPC polymers and their potential as colloid biomaterials would be useful for the development of powerful nanodevices to explore new applications in the fields of cell engineering and nanomedicine molecular science.

References

Goda T, Goto Y, Ishihara K (2010) Cell-penetrating macromolecules: direct penetration of amphipathic phospholipid polymers across plasma membrane of living cells. Biomaterials 31(8):2380–2387

Goto Y, Matsuno R, Konno T et al (2008a) Polymer nanoparticles covered with phosphorylcholine groups and immobilized with antibody for high-affinity separation of proteins. Biomacromolecules 9(3):828–833

Goto Y, Matsuno R, Konno T et al (2008b) Artificial cell membrane-covered nanoparticles embedding quantum dots as stable and highly sensitive fluorescence bioimaging probe. Biomacromolecules 9(11):3252–3257

Ishihara K (2000) New polymeric biomaterials-phospholipid polymers with a biocompatible surface. Front Med Biol Eng 10:83–95

Ishihara K, Ueda T, Nakabayashi N (1990) Preparation of phospholipid polymers and their properties as polymer hydrogel membrane. Polym J 22:355–360

Ishihara K, Nomura H, Mihara T et al (1998) Why do phospholipid polymers reduce protein adsorption? J Biomed Mater Res 39:323–330

Ishihara K, Iwasaki Y, Nakabayashi N (1999) Polymeric lipid nanosphere consisting of water-soluble poly (2-methacryloyloxyethyl phosphoryl choline-co-n-butyl methacrylate). Polym J 31:1231–1236

Ito T, Watanabe J, Takai M et al (2006) Dual mode bioreactions on polymer nanoparticles covered with phosphorylcholine group. Colloids Surf B Biointerfaces 50(1):55–60

Iwasaki Y, Nakabayashi N, Ishihara K (2001) Preservation of platelet function on 2-methacryloyloxyethyl phosphorylcholine graft polymer compared to various water-soluble graft polymers. J Biomed Mater Res 57:72–78

Kamei T, Kitayama J, Yamaguchi H et al (2011) Spatial distribution of intraperitoneally administrated paclitaxel nanoparticles solubilized with poly (2-methacryloxyethyl phosphorylcholine-co-n-butyl methacrylate) in peritoneal metastatic nodules. Cancer Sci 102(1):200–205

Kojima R, Kasuya M, Ishihara K et al (2011) Physicochemical delivery of amphiphilic copolymers to specific organelles. Polym J 43:718–722

Konno T, Watanabe J, Ishihara K (2003) Enhanced solubility of paclitaxel using water-soluble and biocompatible 2-methacryloyloxyethyl phosphorylcholine polymers. J Biomed Mater Res 65A:209–214

Konno T, Watanabe J, Ishihara K (2004) Conjugation of enzymes on polymer nanoparticles covered with phosphorylcholine groups. Biomacromolecules 5:342–347

Konno T, Ito T, Takai M et al (2006) Enzymatic photochemical sensing of hydrogen peroxide by polymer nanoparticles covered with artificial cell membrane. J Biomater Sci Polym Edn 17(12):1347–1357

Miyata R, Ueda M, Jinno H et al (2009) Selective targeting by preS1 domain of hepatitis B surface antigen conjugated with phosphorylcholine-based amphiphilic block copolymer micelles as a biocompatible drug delivery carrier for treatment of human hepatocellular carcinoma with paclitaxel. Int J Cancer 124(10):2460–2467

Moro T, Takatori Y, Ishihara K et al (2004) Surface grafting of artificial joints with a biocompatible polymer for preventing periprosthetic osteolysis. Nat Mater 3:829–836

Shimada T, Ueda M, Jinno H et al (2009) Development of targeted therapy with paclitaxel incorporated into EGF-conjugated nanoparticles. Anticancer Res 29(4):1009–1014

Singer SJ, Nicholson GL (1972) The fluid mosaic model of the structure of cell membranes. Science 175:720–731

Snyder TA, Tsukui H, Kihara S, Akimoto T, Litwak KN, Kameneva MV, Yamazaki K, Wagner WR (2007) Preclinical biocompatibility assessment of the EVAHEART ventricular assist device: coating comparison and platelet activation. J Biomed Mater Res A 81:85–92

Soma D, Kitayama J, Konno T et al (2009) Intraperitoneal administration of paclitaxel solubilized with poly(2-methacryloxyethyl phosphorylcholine-co-n-butyl methacrylate) for peritoneal dissemination of gastric cancer. Cancer Sci 100(10):1979–1985

Watanabe J, Ishihara K (2006) Sequential enzymatic reactions and stability of biomolecules immobilized onto phospholipid polymer nanoparticles. Biomacromolecules 7:171–175

Intracellular Protein Delivery Using Self-Assembled Amphiphilic Polysaccharide Nanogels

Asako Shimoda, Shin-ichi Sawada and Kazunari Akiyoshi

Abstract Intracellular delivery of exogenous proteins is a field of bioscience that has grown rapidly in recent years, driven by the potential clinical applications. We developed a protein nanocarrier composed of amphiphilic polysaccharide nanogels formed by self-assembly of ethylenediamine- and cholesteryl group-bearing pullulan ($CHP-NH_2$). The nanogel strongly interacts with cells allowing proteins to be internalized more effectively than with other carriers, such as cationic liposomes and a protein transduction domain-based amphiphilic peptide carrier. An interesting property of nanogels is that they can form a stable complex with proteins that are suitably sized suitable for intracellular uptake ($\sim$50 nm). Nanogels also act as artificial chaperones by preventing the aggregation of denatured protein and aid correct protein refolding. We also developed a cell-specific peptide (Arg-Gly-Asp; RGD)-modified nanogel (CHP-RGD) with greater potential for cell-specific, receptor-mediated delivery. This nanogel was effectively internalized into cells via integrin-mediated endocytosis, specifically clathrin-mediated endocytosis and macropinocytosis. Cell-specific peptide-modified polysaccharide nanogels are expected to have broad applications in drug delivery. In this chapter, we describe the results of recent studies and discuss future challenges for protein delivery using polysaccharide nanogels.

Keywords Cholesteryl group-bearing pullulan · Polysaccharide nanogels · Protein delivery

Abbreviations

RGD	Arg-Gly-Asp
BoHc	Botulinus toxin

A. Shimoda · S. Sawada · K. Akiyoshi (✉)
Department of Polymer Chemistry, Graduate School of Engineering, Kyoto University, Katsura Nishikyo-ku, Kyoto 615-8510, Japan
e-mail: akiyoshi@bio.polym.kyoto-u.ac.jp

A. Shimoda · S. Sawada · K. Akiyoshi
JST-ERATO, Tokyo, Japan

A. Prokop et al. (eds.), *Intracellular Delivery II*,
Fundamental Biomedical Technologies 7, DOI: 10.1007/978-94-017-8896-0_14,

CHP	Cholesteryl group-bearing pullulan
CHP-NH_2	Ethylenediamine-bearing CHP
PEG	Polyethylene glycol
PTD	Protein transduction domain
QD	Quantum dot
PEGSH	Thiol group-bearing PEG
W9 peptide	WP9QY

1 Introduction

Therapeutic proteins and peptides have received significant attention because of their potential to treat and prevent diseases. However, they are often unstable as a colloid, are rapidly cleared, and may induce serious side effects. To overcome these issues, nanocarrier-based drug delivery systems have been developed to improve their bioavailability, reduce their toxicity, and allow alternative routes of administration. Several types of nanocarriers have been reported, including hydrogels, microspheres and lipid-based materials. In this chapter, we focus on the potential for using hydrogel nanoparticles (nanogels) as protein delivery systems.

Hydrogels are polymeric three-dimensional networks with a substantial volume of aqueous solution. Nanogels are nanometer-sized hydrogel particles (<100 nm) that have attracted growing interest as possible drug delivery systems because of the unique properties of nano-sized hydrogel systems (Kabanov and Vinogradov 2009). Although many polymers have been used to develop hydrogels, polysaccharides have particular advantages for this purpose. They are available from various natural sources, including plants, microbes, and animals. Moreover, they are non-toxic, biocompatible, and biodegradable because they are natural biomaterials (Lee and Mooney 2001). Hence, polysaccharide hydrogels are ideal biomaterials for use in medical fields, including in tissue engineering and as drug delivery systems.

2 Design of Self-Assembled Nanogels Using Associating Polysaccharides

2.1 Development of Self-Assembled Nanogels Consisting of Cholesteryl Group-Bearing Pullulans as a Protein Drug Delivery System

Nanogels are generally prepared as chemically cross-linked nanogels or physically cross-linked nanogels (Hennink and Nostrum 2002). Various methods of preparing chemically cross-linked nanogels have been developed, including radical polymerization of polymers and reacting a low-molecular weight cross-linking agent with a polymer. Although these methods can be applied to most polymers, it is difficult to control the resulting network. Furthermore, many cross-linking agents are toxic. By contrast, physically cross-linked nanogels are formed by non-covalent interactions such as hydrogen bonds, van der Waals forces, electrostatic interactions, and hydrophobic interactions. The network is simple to prepare under mild conditions, and allows the hydrogels to stably trap the target compound. We reported that polysaccharides partially modified with hydrophobic molecules show a unique associative behavior. In particular, cholesteryl group-bearing pullulan (CHP) self-assembles into stable nanogels ($\sim$30 nm in diameter) in water. Because CHP nanogels contain multiple hydrophobic domains, they can trap various kinds of proteins by hydrophobic interaction and release them by exchange reactions with other proteins or following the addition of methyl-β-cyclodextrin, which is an acceptor for cholesterol (Akiyoshi et al. 1993). Interestingly, CHP nanogels display a chaperone-like activity and release encapsulated proteins in their native form. The CHP nanogels probably prevent the aggregation of isolated proteins and facilitate their refolding. These properties are important when designing protein carriers or effective protein reservoirs in drug delivery systems and in protein engineering (Sasaki and Akiyoshi 2010) (Fig. 1).

Because of their chaperone-like activity, CHP nanogels are useful carriers of therapeutic proteins, such as insulin, cytokines, antigen proteins in cancer vaccines, and nasal vaccines. Notably, the cancer antigen NY-ESO-1 complexed with CHP nanogels (CHP-NY-ESO-1) was effectively transported to lymph nodes and was internalized into antigen-presenting cells, including dendritic cells and macrophages (Ikuta et al. 2002; Kageyama et al. 2008). Several clinical trials of CHP-NY-ESO-1 are ongoing, which implies that nanogel-based vaccines have potent activity by inducing immune responses against the NY-ESO-1 antigen in cancer patients.

We have also used CHP nanogels as a peptide delivery system. WP9QY/W9 (W9 peptide), a tumor necrosis factor antagonist peptide, effectively inhibits bone resorption in many murine models, but it is rapidly degraded in serum and readily aggregates following administration. To generate a more efficient delivery system that controlled the release system of W9 peptide, we used CHP nanogel as a carrier. W9 peptide formed a stable complex with CHP nanogels without aggregation, and twice daily injection of W9/CHP inhibited bone resorption induced by low dietary Ca. These results suggest that CHP nanogels could be an effective carrier for peptide drugs (Alles et al. 2009).

2.2 Self-Assembled Nanogel Engineering

Rational design of hybrid nanogels offers the possibility of achieving specific functions by adjusting the molecular composition. In particular, it is possible to modify the hybrid nanogels with inorganic molecules, ionic or polymerizable groups, and targeting ligands such as peptides and antibodies. So far, many

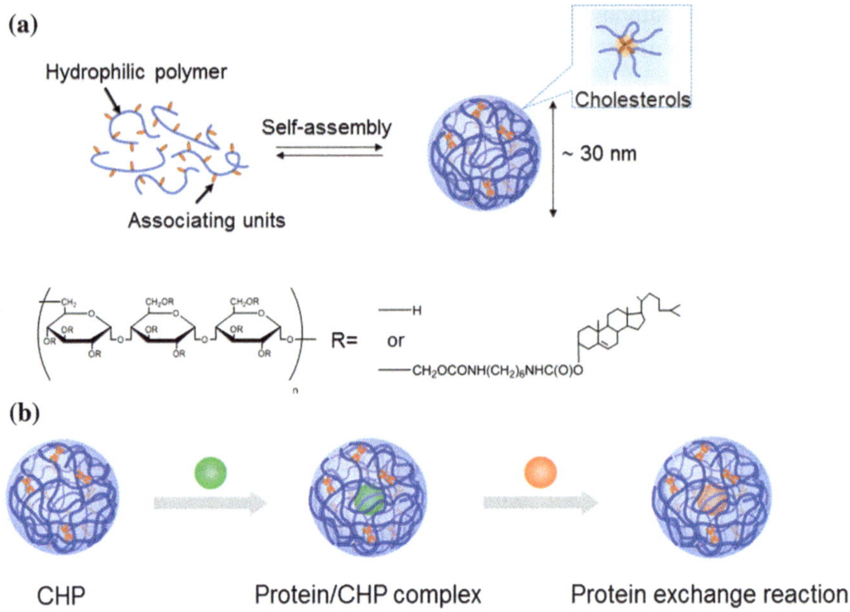

Fig. 1 Self-assembled nanogels composed of cholesteryl group-bearing pullulans (*CHP*). **a** Formation and chemical structure of the CHP nanogel. **b** Interaction between the CHP nanogel and protein

different functional nanogels have been developed, including stimulus-responsive nanogels, thermoresponsive nanogels, photoresponsive nanogels, and organic-inorganic hybrid nanogels. The bottom-up method of preparing nanogel-integrated hydrogels is an attractive way to control the hydrogel's nanostructure. This structure is formed by chemically cross-linking the nanogels with polyethylene glycol (PEG) derivatives and coating the surface of the nanogels with PEG. Acryloyl group-bearing CHP nanogels cross-linked with thiol group-bearing PEG (PEGSH) also form a nanogel assembly (Hasegawa et al. 2009; Shimoda et al. 2012a, b). In this system, the nanogels are gradually released from the hydrogel because the nanogels and the cross-linker dissociate under physiological conditions. Raspberry-like nanoparticles with a diameter of 50–150 nm and consisting of 70–160 nanogels formed under dilute concentrations of nanogels. These nanoparticles can be used as an injectable nanocarrier capable of sustained release of various proteins, such as cytokines, over a relatively long time. In addition, the elimination half-life ($t_{1/2}$) of nanoparticles was much longer than that of CHP nanogels following intravenous injection. These results indicate that nanogel-integrated hydrogels represent a novel sustained-release protein delivery system (Miyahara et al. 2012; Fujioka-Kobayashi et al. 2012).

3 Protein Delivery by Self-Assembled Polysaccharide Nanogels

3.1 Cationic Nanogels

The surface charge of nanogels is one of the important determinants of the efficiency of internalization into cells. Because the cellular membrane is negatively charged, cationic nanogels bind tightly to the cell surface. To improve the internalization of nanogels, a cationic group (ethylenediamine) was introduced to the CHP nanogels (CHP-NH_2 nanogels) (Ayame et al. 2008). CHP-NH_2 nanogels also formed monodisperse nanogels ($\sim$30 nm) capable of interacting with proteins. Moreover, CHP-NH_2 nanogels were less toxic and were more efficiently delivered into various cell types compared with other protein carriers, including cationic liposomes and protein transduction domain (PTD)-based carrier. We reported several applications of cationic nanogels as protein delivery systems, such as quantum dots (QDs). QDs are semiconductor nanocrystals that are often used as fluorescent probes for cellular tracking (Hasegawa et al. 2005). However, the uptake efficiency of QDs is very low and they readily form aggregates after internalization. To overcome these problems, we tested the potential for CHP-NH_2 nanogels for a QD delivery system. Internalization of the CHP-NH_2-protein-coated QD complex was more efficient than that of cationic liposomes. Hence, the CHP-NH_2 nanogel has great potential for in long-term cell imaging studies (Fig. 2).

In general, PTD-mediated protein carriers interact with heparan sulfate expressed on the surface of target cells. However, the cell surface expression of heparan sulfate is quite low on some cells, especially metastatic cells and lymphocytes. Therefore, alternative protein carriers are needed for cancer immunotherapy. We demonstrated that CHP-NH_2 nanogels efficiently delivered proteins into myeloma cells and primary $CD4^+$ T lymphocytes, which express low levels of heparan sulfate. Compared with other protein carriers, such as cationic liposomes and PTD-based peptides, the protein-CHP-NH_2 nanogel complex was effectively internalized into the cells with low heparan sulfate expression. This result was probably due to the mechanism of cell internalization, macropinocytosis. Interestingly, the CHP-NH_2 nanogel complexed with the anti-apoptotic protein Bcl-xL blocked apoptosis of these cells, indicating that the proteins delivered by CHP-NH_2 nanogels were capable of functionally regulating target cells (Watanabe et al. 2011) (Fig. 3).

We recently developed a novel nasal vaccine using cationic nanogels (Nochi et al. 2010; Yuki et al. 2012). We investigated the immune responses following nasal administration of complex of CHP-NH_2 nanogel and a C-terminal avirulent region of the heavy chain of botulinus toxin (BoHc) (Nochi et al. 2010). The serum total IgG and IgA antibody titers to botulinus toxin were significantly higher in the mice immunized with the CHP-NH_2–BoHc complex than in those immunized with

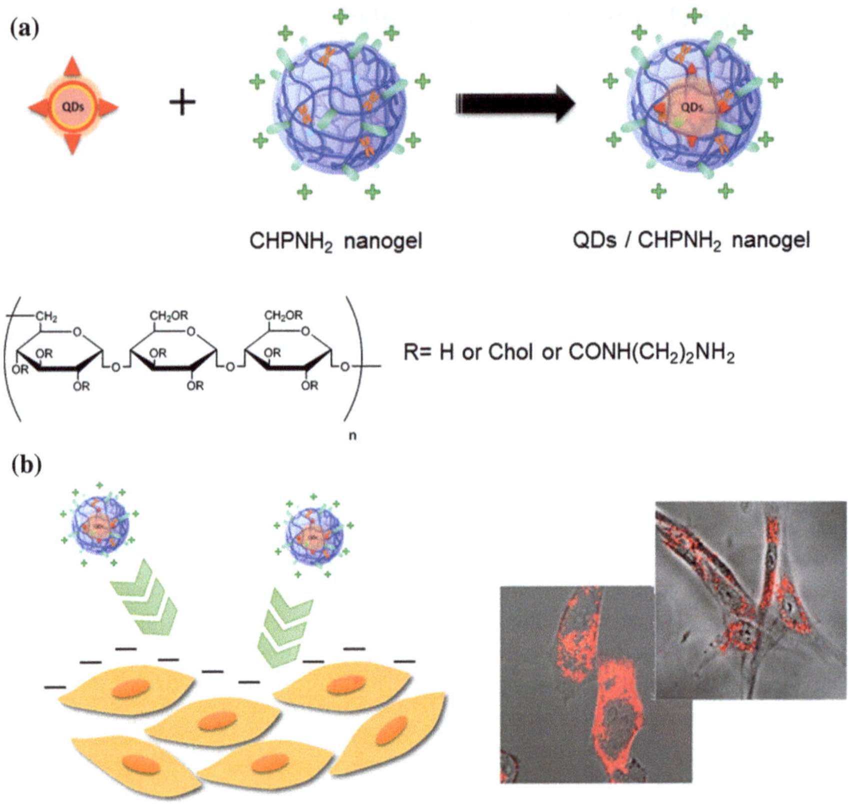

Fig. 2 Complexation of a cationic nanogel (CHP-NH$_2$) with protein-coated quantum dots (*QDs*). **a** Scheme and chemical structure of the CHP-NH$_2$-QD complex. **b** The CHP-NH$_2$-QD was effectively internalized into target cells (reprinted from Hasegawa et al. 2005)

BoHc alone. These findings indicate that transnasal administration of the CHP-NH$_2$–BoHc complex induces immune responses in the nasal mucous membrane. Additionally, the viability of mice immunized with the CHP-NH$_2$–BoHc complex was significantly greater than that of mice immunized with BoHc alone. Histological images revealed that the CHP-NH$_2$–BoHc complex was effectively delivered to the nasal mucosa and was retained for a long time after intranasal administration. Following internalization of the complex into the nasal epithelium, BoHc was gradually released from the CHP-NH$_2$ nanogel in the epithelial cells. BoHc released from the CHP-NH$_2$ nanogel was also effectively phagocytosed by CD11c$^+$ dendritic cells located in the epithelial layer and in the lamina propria of the nasal cavity. These results suggest that the CHP-NH$_2$ nanogel enhanced the effect of a vaccine antigen, and can be used as a safe delivery carrier for intranasal vaccines.

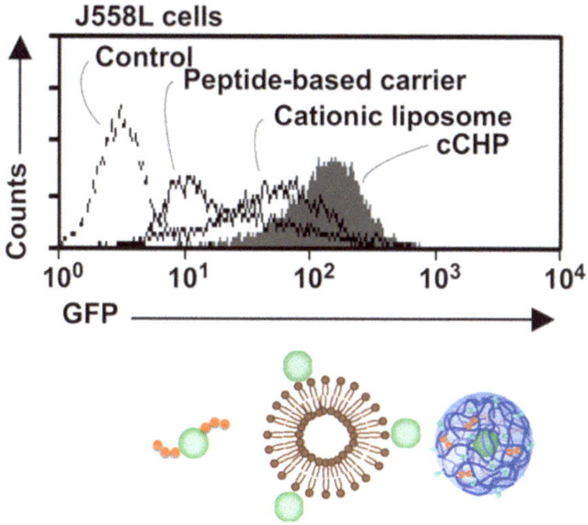

Fig. 3 Results of flow cytometry of the cellular uptake of proteins mixed with peptide-based carriers, cationic liposomes, and CHP-NH_2 nanogels (reprinted from Watanabe et al. 2011)

3.2 *RGD-Modified Nanogels*

CHP-NH_2 nanogels strongly interacted with cells and delivered various proteins much efficiently than other carriers. However, there are some limitations, including high cytotoxicity, which is correlated with the number of cationic groups, and low target site specificity. We developed a new nanogel with a targeting function using a ligand-modified protein delivery carrier. The cell recognition motif (RGD; Arg-Gly-Asp) was selected as the targeting ligand and evaluated the efficacy of cellular uptake (Shimoda et al. 2011). The nanogel exhibited much less toxicity compared with the CHP-NH_2 nanogel and was effectively internalized into cells via integrin receptor-mediated endocytosis. The fluorescence of proteins overlapped with the RGD-modified nanogel, indicated that the proteins were internalized as a complex with the nanogels. After 24 h, the proteins were released from the nanogels, probably via exchange reactions with other proteins. Interestingly, the cellular uptake of nanogels involved multiple pathways, including clathrin-mediated endocytosis, macropinocytosis, and caveolae-mediated endocytosis. The RGD-modified nanogels subsequently escaped from the endosomes after incubation for 24 h. These results indicate that polysaccharide nanogels with targeting peptides show great potential for intracellular protein delivery (Fig. 4, 5).

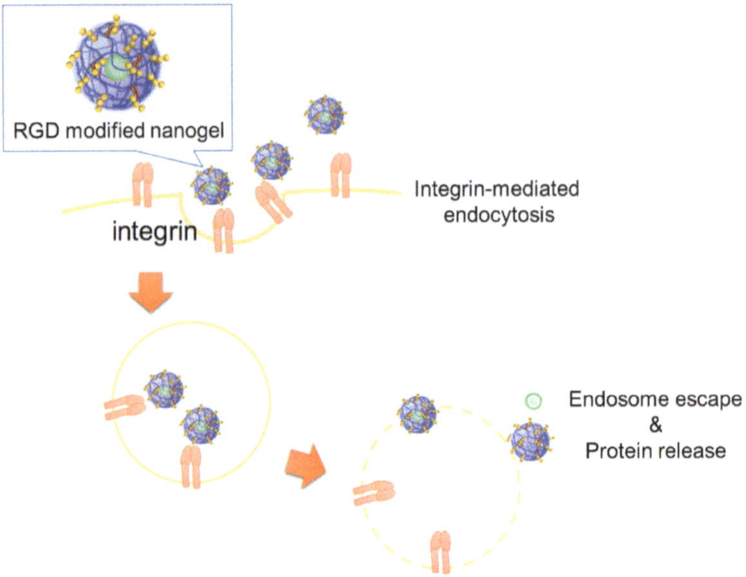

Fig. 4 RGD-modified nanogels were internalized into cells via integrin-mediated endocytosis

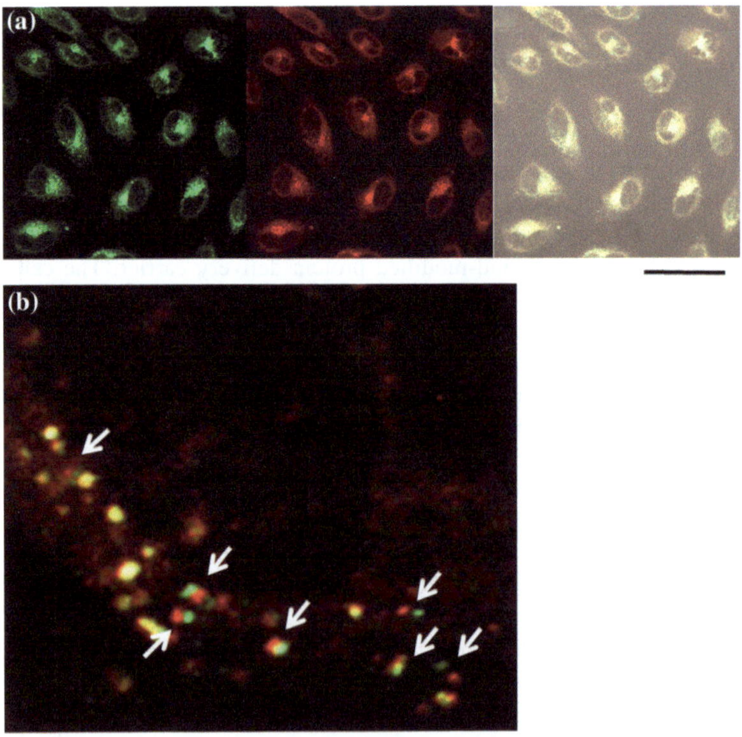

Fig. 5 Cellular uptake of protein/RGD-modified nanogel complexes. **a** Proteins internalize into the cells in the form of a complex with the nanogels. **b** Proteins were released from the nanogels after incubation for 24 h (reprinted from Shimoda et al. 2011)

4 Concluding Remarks

We developed polysaccharide nanogels with chaperone-like activity, a crucial property for protein delivery systems. In recent years, phase I clinical trials were completed using CHP nanogels trapping HER2 or NY-ESO-1 antigen proteins for cancer vaccination, and further clinical trials, including adoptive T cell therapy, are now underway.

CHP nanogels have been improved to provide a suitable template to fulfill various demands as protein delivery systems. CHP nanogels cross-linked with polyethylene glycol were developed as a sustained protein release system. The junction between the nanogel and the cross-linker degrades under physiological conditions, gradually releasing the individual nanogel molecule. Such nanogel cross-linked hydrogels are useful as a novel scaffold, including bone regeneration.

CHP with functional cationic groups and nanogels modified with cell recognition motifs have been proposed for intracellular protein delivery. Both nanogel systems were effectively internalized into cells and released their encapsulated proteins via exchange reaction. These nanogels can be used as imaging probes, for cancer therapy, and in immune regulation. Our integrated protein delivery systems will provide a new field for tailor-made functional hydrogel materials.

References

Akiyoshi K, Deguchi S, Moriguchi N et al (1993) Self-aggregates of hydrophobized polysaccharides in water. Formation and characteristics of nanoparticles. Macromolecules 26(12):3062–3068

Alles N, Soysa NS, Mian AH et al (2009) Polysaccharide nanogel delivery of a TNF-α and RANKL antagonist peptide allows systemic prevention of bone loss. Eur J Pharm Sci 37(2):83–88

Ayame H, Morimoto N, Akiyoshi K (2008) Self-assembled cationic nanogels for intracellular protein delivery. Bioconjug Chem 19(4):882–890

Fujioka-Kobayashi M, Ota MS, Shimoda A et al (2012) Cholesteryl group- and acryloyl group-bearing pullulan nanogel to deliver BMP2 and FGF18 for bone tissue engineering. Biomaterials 33(30):7613–7620

Hasegawa U, Nomura SM, Kaul SC et al (2005) Nanogel-quantum dot hybrid nanoparticles for live cell imaging. Biochem Bioph Res Co 331(4):917–921

Hasegawa U, Sawada S, Shimizu T et al (2009) Raspberry-like assembly of cross-linked nanogels for protein delivery. J Control Release 140(3):312–317

Hennink WE, Nostrum CF (2002) Novel crosslinking methods to design hydrogels. Adv Drug Deliv Rev 54(1):13–36

Ikuta Y, Katayama N, Wang L et al (2002) Presentation of a major histocompatibility complex class 1-binding peptide by monocyte-derived dendritic cells incorporating hydrophobized polysaccharide-truncated HER2 protein complex: implications for a polyvalent immuno-cell therapy. Blood 99(10):3717–3724

Kabanov AV, Vinogradov SV (2009) Nanogels as pharmaceutical carriers: finite networks of infinite capabilities. Angew Chem Int Edit 48(30):5418–5429

Kageyama S, Kitano S, Hirayama M et al (2008) Humoral immune responses in patients vaccinated with 1–146 HER2 protein complexed with cholesteryl pullulan nanogel. Cancer Sci 99(3):601–607

Lee KY, Mooney DJ (2001) Hydrogels for tissue engineering. Chem Rev 101(7):1869–1879

Miyahara T, Nyan M, Shimoda A et al (2012) Exploitation of a novel polysaccharide nanogel cross-linking membrane for guided bone regeneration (GBR). J Tissue Eng Regen M 6(8):666–672

Nochi T, Yuki Y, Takahashi H et al (2010) Nanogel antigenic protein-delivery system for adjuvant-free intranasal vaccines. Nat Mater 9:572–578

Sasaki Y, Akiyoshi K (2010) Nanogel engineering for new nanobiomaterials: from chaperoning engineering to biomedical applications. Chem Rec 10(6):366–376

Shimoda A, Sawada S, Akiyoshi K (2011) Cell specific peptide-conjugated polysaccharide nanogels for protein delivery. Macromol Biosci 11(7):882–888

Shimoda A, Sawada S, Kano A et al (2012a) Dual crosslinked hydrogel nanoparticles by nanogel bottom-up method for sustained-release delivery. Colloid Surf B 99:38–44

Shimoda A, Yamamoto Y, Sawada S, Akiyoshi K (2012b) Biodegradable nanogel-integrated hydrogels for sustained protein delivery. Macromol Res 20(3):266–270

Watanabe K, Tsuchiya Y, Kawaguchi Y et al (2011) The use of cationic nanogels to deliver proteins to myeloma cells and primary T lymphocytes that poorly express heparan sulfate. Biomaterials 32(25):5900–5905

Yuki Y, Kong I, Sato A et al (2012) Nanogel-based PspA intranasal vaccine prevents invasive disease and nasal colonization by *Streptococcus pneumoniae*. Infect Immun 81(5):1625–1634

Part III
Simulation for Delivery and Function

Molecular Dynamics Simulations of Polyplexes and Lipoplexes Employed in Gene Delivery

Deniz Meneksedag-Erol, Chongbo Sun, Tian Tang and Hasan Uludag

Abstract Gene therapy is an important therapeutic strategy in the treatment of a wide range of genetic disorders. Delivery of genetic materials into patient cells is limited since nucleic acids are vulnerable to degradation in extra- and intra-cellular environments. Design of delivery vehicles can overcome these limitations. Polymers and lipids are effective non-viral nucleic acid carriers; they can form stable complexes with nucleic acids known as polyplexes and lipoplexes. Despite the great amount of experimental work pursued on polymer or lipid based gene delivery systems, detailed atomic level information is needed for a better understanding of the roles the polymers and lipids play during delivery. This chapter will review molecular dynamics simulations performed on polyplexes and lipoplexes at critical stages of gene delivery. Interactions between various carriers and nucleic acids during the formation of polyplexes/lipoplexes, condensation and aggregation of nucleic acids facilitated by the carriers, binding of the polyplexes/lipoplexes to cell membrane, as well as their intracellular pathway are reviewed; and the gaps in the theoretical field are highlighted.

Keywords Molecular dynamics · Gene delivery · Non-viral carriers · Polyplex · Lipoplex · Plasmid DNA · siRNA

D. Meneksedag-Erol · H. Uludag (✉)
Department of Biomedical Engineering, Faculty of Medicine and Dentistry and Engineering, University of Alberta, Edmonton, AB, Canada
e-mail: huludag@ualberta.ca

C. Sun · T. Tang
Department of Mechanical Engineering, Faculty of Engineering, University of Alberta, Edmonton, AB, Canada

H. Uludag
Department of Chemical and Materials Engineering, Faculty of Engineering, University of Alberta, Edmonton, AB, Canada

H. Uludag
Faculty of Pharmacy and Pharmaceutical Sciences, University of Alberta, Edmonton, AB, Canada

A. Prokop et al. (eds.), *Intracellular Delivery II*,
Fundamental Biomedical Technologies 7, DOI: 10.1007/978-94-017-8896-0_15,

Abbreviations

RNAi	RNA interference
dsRNA	Double stranded RNA
RISC	RNA-induced silencing complex
siRNA	Short interfering RNA
DOPC	1,2-Dioleoyl-sn-glycero-3-phosphocholine
DOPE	Dioleoylphosphatidylethanolamine
DOTAP	1,2-Dioleoyl-3-trimethylammonium-propane
DMTAP	Dimyristoyltrimethylammonium propane
DMPC	Dimyristoylphosphatidylcholine
DPPC	Dipalmitoylphosphatidylcholine
PAMAM	Polyamidoamine
PBAE	Poly(beta-amino ester)
PEI	Polyethylenimine
PLL	Poly-L-lysine
CDP	Cyclodextrin-polycation
CME	Clathrin-mediated endocytosis
CvME	Caveolae/raft-mediated endocytosis
EGF	Epidermal growth factor
MD	Molecular dynamics
MM	Molecular mechanics
QM	Quantum mechanics
PME	Particle mesh Ewald
PBC	Periodic boundary conditions
DPD	Dissipative particle dynamics
US	Umbrella sampling
WHAM	Weighted histogram analysis method
DFT	Density functional theory
CG	Coarse-graining
PTI	Pancreatic trypsin inhibitor
ENM	Elastic network model
LJ	Lennard-Jones
MM-PBSA	Molecular mechanic/Poisson–Boltzmann surface area
DAP	1,3-Diaminopropane
DAPMA	*N*,*N*-Di-(3-aminopropyl)-*N*-(methyl)amine
TAP	Trimethylammonium
MC	Monte Carlo
PMF	Potential of mean force
CA	Caprylic acid
LA	Linoleic acid

1 Background on DNA and siRNA Delivery Systems

Gene therapy aims to modify expression of genes or correct abnormal genes by delivering genetic material into specific patient cells for treatment of wide range of genetic disorders, including cancer, inflammatory, metabolic, infectious cardiovascular and neurological diseases (Pack et al. 2005; Zhang et al. 2012). In gene-based therapy of cancers, targeting tumour suppressor pathways has been an attractive approach since the first clinical trial of p53 gene replacement for treatment of non-small lung carcinoma in 1996 (McNeish et al. 2004; Roth et al. 1996). Since then, 64 % of gene therapy trials have been focused on treatment of cancer diseases (J Gene Med 2013). The initial impetus behind gene therapy has been the desire to use functional DNA-based expression vectors (be in viral or non-viral form) to synthesize therapeutic proteins in situ. This approach relies on the delivery of exogenous DNA cassettes so as to tap into the local cellular protein synthesis machinery. The DNA in this case need to be cellularly internalized, trafficked to the nucleus and recruit the appropriate transcription factors for production of mRNAs for desired proteins. Such proteins have been intended for local activity, where the proteins are functional at the vicinity of the gene delivery site, as well as systemically, where the locally produced proteins are distributed throughout the organism to function systemically. With the discovery of RNA interference (RNAi) mechanism involving double stranded RNAs (dsRNAs) (Fire et al. 1998), the scope of gene therapy was expanded by relying on RNAi-based therapies (Zhang et al. 2012). RNAi is a post-transcriptional gene silencing process triggered by dsRNA molecules in animals and plants. Relatively long dsRNA molecules are cleaved by the enzyme Dicer, which belongs to Ribonuclease III family, into short RNA molecules, which are 21–22 nucleotides in length (Dominska and Dykxhoorn 2010; Elbashir et al. 2001). These dsRNA molecules consist of a passenger and a guide strand, which need to be dissociated into single strands to incorporate into RNA-induced silencing complex (RISC) via guide strand (Matranga et al. 2005). Passenger strand is released after entry into RISC and guide strand direct RISC to complementary sequence in target mRNAs, resulting in mRNA cleavage and gene silencing (Dominska and Dykxhoorn 2010). The practical (therapeutic) use of RNAi relies on short ($\sim$22 nucleotide pairs) synthetic dsRNAs intended to undertake silencing for a pharmacological effect and are named short interfering RNAs (siRNAs). siRNAs could be further chemically modified (e.g., conjugated with cholesterol) for improved delivery and/or pharmacological effect (Lorenz et al. 2004).

The great interest in deploying DNA and siRNA based gene therapeutics has been dampened to some degree due to many challenges encountered in the delivery of the agents and design of effective and safe gene therapy systems.

1.1 Complexation with Carrier Vectors and Condensation

While the first step in therapeutic delivery of DNA and RNA based nucleic acid reagents is systemic administration and extracellular penetration to target sites, we will consider cellular entry as the first step in delivery for the purposes of this manuscript. Significant limitations exist that prevent naked nucleic acids to be efficiently delivered into cells. The negatively charged and hydrophilic nucleic acids are not efficient to cross hydrophobic and negatively charged lipid bilayers of cell membranes (Zhang et al. 2012). Physical strategies that allow delivery of naked nucleic acids into cells, such as electroporation (Potter 1988), gene gun (Fynan et al. 1993) and ultrasound (Yoon and Park 2010), have been developed; but naked nucleic acids can lose their functionality in cells due to their vulnerability to cellular enzymes, ultimately leading to degradation by nucleases. This limitation stimulated the design of delivery systems in which nucleic acids are complexed and protected by delivery vehicles or carriers. Early attempts to design efficient delivery systems concentrated on viral vectors, since viruses are able to effectively transfer their genetic code to host cells. However, with the risk of high immunogenic responses, use of synthetic, non-viral carriers has garnered interest, which can provide a safer and effective alternative. Nucleic acids complexed with cationic polymers and lipids are referred to as 'polyplexes' and 'lipoplexes', respectively (Guo et al. 2010). Common examples of effective synthetic carriers include (1) cationic lipids such as 1,2-dioleoyl-sn-glycero-3-phosphocholine (DOPC), dioleoylphosphatidylethanolamine (DOPE), 1,2-dioleoyl-3-trimethylammonium-propane (DOTAP), dimyristoyltrimethylammonium propane (DMTAP), dimyristoylphosphatidylcholine (DMPC), dipalmitoylphosphatidylcholine (DPPC), and (2) cationic polymers such as polyamidoamine (PAMAM), poly(beta-amino ester) (PBAE), polyethylenimine (PEI), poly-L-lysine (PLL) and cyclodextrin-polycation (CDP) [reviewed in (Aliabadi et al. 2012)]. These carriers are able to bind to nucleic acids via electrostatic interactions between cationic moieties on carriers and anionic phosphate groups of nucleic acids, and condense (or aggregate) the nucleic acids into nanoparticles capable of being protected from degradation and their cellular uptake facilitated (Kircheis et al. 2001).

1.2 Interactions with Cell Membrane and Cellular Uptake

Cell surface binding of nucleic acid polyplexes and lipoplexes is triggered by ionic interactions between cationic carriers and anionic membrane proteins and/or anionic cell-surface 'receptors' such as heparan sulfates and integrins (Aliabadi et al. 2012; Elouahabi and Ruysschaert 2005; Zhang et al. 2012). After an effective cell surface binding, polyplexes and lipoplexes are internalized by endocytosis via a variety of mechanisms such as clathrin- and caveolin-1 independent, clathrin-mediated (CME), caveolae/raft-mediated (CvME) and macropinocytosis (Hillaireau and

Couvreur 2009). The individual mechanisms follow specific intercellular trafficking pathways that may affect the functionality of polyplexes and lipoplexes (Aliabadi et al. 2012). Carriers can also be designed in order to target specific cell types and ligands, such as asialoglycoprotein, epidermal growth factor (EGF), folate, integrin, lactose, mannose, and transferrin (Zhang et al. 2012). Many factors influence the uptake mechanism, such as the cargo (siRNA or DNA) and nature of carrier, the ligand targeted, and its intracellular trafficking pathway. Some studies suggested the CvME to be most conducive to transfection of complexes, while other studies suggested the CME to be the most efficient mechanism. Macropinocytosis was also proposed to be the most efficient mechanism in the transfection of complexes, in particular for larger size complexes that are not ideal for CvME and CME based mechanisms. These contradictory observations complicate identifying the optimal mechanism that polyplexes and lipoplexes follow in their cellular uptake and show the dependency of a functional effect on complex size and nature, as well as the cell types targeted for delivery (Hsu and Uludag 2012).

1.3 Intracellular Trafficking and Cytoplasmic Release

Upon endocytosis, intracellular trafficking of polyplexes/lipoplexes begins in early endosomes. Early endosomes generally fuse into late endosomes (pH $\sim$ 5–6) and transfer their content to lysosomes that have an acidic environment with pH as low as $\sim$4.5. Lipoplexes and polyplexes trapped in these acidic environments must efficiently escape into cytosol in order for the cargo not to be degraded and to reach its target (i.e., nucleus in the case of DNA and mRNA/RISC locations in cytoplasm in the case of siRNA) (Aliabadi et al. 2012; Dominska and Dykxhoorn 2010; Elouahabi and Ruysschaert 2005; Pack et al. 2005; Zhang et al. 2012). It is possible to enhance the endosomal escape of polyplexes/lipoplexes based on the use of fusogenic lipids and proteins, pH sensitive carriers, and photosensitive agents (Dominska and Dykxhoorn 2010). In the case of complexes with certain cationic polymers, such as PEI, endosomal escape is possible through the 'proton-sponge effect' (Boussif et al. 1995); protonation of secondary and tertiary amines of carriers in the low pH environment prevents further acidification, leading to swelling of PEI/nucleic acid particles due to increase in their radius of gyration. The influx of counterions into endosomes also follows to balance the H^+ concentration between the lysosomal compartment and cytoplasm. The resulting osmotic swelling of lysosomal compartment releases the complexes and other contents into the cytoplasm (Boussif et al. 1995; Dominska and Dykxhoorn 2010).

After the nucleic acids are released into cytoplasm, polyplexes/lipoplexes must be dissociated in order for the nucleic acids to be functional. Anionic endogenous molecules such as cytoplasmic RNA (mRNA, tRNA, miRNA, etc.) and glycosaminoglaycans such as heparin sulfate are thought to be involved in dissociation mechanism of lipoplexes and polyplexes (Moret et al. 2001). Electrostatic interactions between nucleic acids and polymeric carriers are thought to be weakened during the endosomal escape stage due to swelling and pH changes, hence facilitating the release of the nucleic acids from complexes (Aliabadi et al. 2012). After the release, the plasmid DNA (pDNA) has to reach nucleus for transcription and siRNA needs to incorporate into RISC and initiate mRNA cleavage for gene silencing. However naked nucleic acids also face cytoplasmic challenges on their way to pursue their functions since susceptibility to nucleases and cleaving agents may interfere with the pathway on their route and make them non-functional.

1.4 Application of Molecular Dynamics Simulations to Nucleic Acid Delivery

Despite the intensive experimental work on polymer and lipid based gene carriers, many issues related to their mechanisms of action cannot be directly addressed from experiments due to the lack of experimental tools at atomic resolution. Molecular dynamics (MD) is a simulation technique where intra- and intermolecular interactions are described by potential energy functions defined in empirical force fields, motion of atoms are obtained through the integration of equations of motion, and physical properties can be extracted via time average of the equilibrated systems. Since the introduction of the first simulation on a biological macromolecule in 1977 (McCammon et al. 1977), MD simulation has become a very useful tool in analyzing complex biological systems, and there has been an increasing interest in applying these simulations with the recent advances in computational methods and capability.

Recently, there has been a great amount of work pursued on MD simulations of gene delivery systems, and they shed light on many important interactions involved in the delivery pathway. On the complexation of nucleic acids with carrier vectors, MD simulations have provided insight into the mechanisms, dynamics and energetics of binding and complex formation, as well as how they are affected by the molecular composition and structure of the carrier vectors. On the uptake of complexes by cells, complementing the vast experimental work using imaging techniques, MD simulations have been run to study the translocation of delivery systems through lipid bilayers. These simulations have revealed residues that are responsible for interaction with cell membrane, thereby pointing some directions in how to enhance cellular attachment and penetration. On the intracellular behavior of the delivered nucleic acid complexes, with the help of MD simulations, the effects of pH and electrostatic interactions on endosomal escape are now better understood. Despite these existing works, there is still an urgent need for additional simulations to study many other aspects of nucleic acid delivery such as the dissociation of nucleic acids from the polyplexes and lipoplexes.

In the following sections of this manuscript, an introduction to the MD method and different simulation techniques will be first given, followed by a detailed review on the simulations performed to date on polyplexes and lipoplexes. Future perspectives will be provided on how MD simulations can further assist in our understanding and designing of effective polymer and lipid based gene carriers.

2 Molecular Dynamics Simulations

2.1 Fundamental Principles of Molecular Simulations

MD is a useful computational methodology for generating detailed atomistic information such as atomic coordinates and velocities in a system. The microscopic information of a system is connected to its macroscopic properties through the theoretical framework of 'statistical mechanics'. Statistical mechanics began with the studies by Maxwell and Boltzmann in late 1800s on the kinetic theory of gases. At the end of 19th century, Gibbs introduced a generalized formulation, statistical ensemble theory, which enabled the derivation of macroscopic thermodynamic properties from microscopic mechanical properties (Patria and Beale 2011). Classical mechanics constituted the basis of Gibbs' statistical ensemble theory since it was developed before the evolution of quantum mechanics.

To demonstrate the basic idea of statistical mechanics, let us consider a system constituted by N particles. The microscopic state of the system at time t can be described by a set of 6N quantities, namely the positions and momenta of the N particles $(\mathbf{q}_1(t),\ldots,\mathbf{q}_N(t),\mathbf{p}_1(t),\ldots,\mathbf{p}_N(t))$, where the vectors $\mathbf{q}_i(t)$and $\mathbf{p}_i(t)$ represent the position and momentum of particle i, respectively. These 6N quantities generate a multi-dimensional space called '*phase space*' with 6N axes (Tuckerman 2010; Wereszczynski and McCammon 2012). Any point in the phase space is called a phase point and it represents a microstate of the system. The total energy of the system written in terms of the particles' positions and momenta is the Hamiltonian:

$$\mathcal{H}(\mathbf{q}_1,\ldots,\mathbf{q}_N,\mathbf{p}_1,\ldots,\mathbf{p}_N)=\sum_{l=1}^{N}\frac{\mathbf{p}_i^2}{2m_i}+U(\mathbf{q}_1,\ldots,\mathbf{q}_N) \tag{1}$$

where the first term is the summation of the particles' kinetic energy, m_i being the mass of particle i, and the second term is the system's potential energy, usually as a function of the particles' positions. By knowing the Hamiltonian of the system, the equations of motion can be obtained in terms of the derivatives of the Hamiltonian, and they form a set of 6N first order differential equations:

$$\frac{d\mathbf{q}_i}{dt}=\frac{\partial\mathcal{H}}{\partial\mathbf{p}_i}=\frac{\mathbf{p}_i}{m_i};\quad\frac{d\mathbf{p}_i}{dt}=-\frac{\partial\mathcal{H}}{\partial\mathbf{q}_i}=-\frac{\partial U}{\partial\mathbf{q}_i}=\mathbf{F}_i(\mathbf{q}_1,\ldots,\mathbf{q}_N) \tag{2}$$

where $\mathbf{F}_i(\mathbf{q}_1, \ldots, \mathbf{q}_N)$ is the force acting on particle i. Time trajectory of a phase point can be obtained by integrating Eq. (2) with given initial conditions so that time evolution of the system can be monitored (Tuckerman 2010).

To relate the microstates of the system to its macroscopic state, it is necessary to introduce the concept of *ensemble*. An ensemble is a collection of points in the phase space that correspond to different microstates but share the same macroscopic state defined by measurable thermodynamic parameters such as pressure (P), temperature (T), volume (V), total mass (M) or total number of particles (N). There exist different ensembles in which different thermodynamic parameters are held fixed. In a *microcanonical ensemble*, which represents an isolated system, the macroscopic state of the system is given by a fixed number of particles N, a fixed volume V and a fixed amount of internal energy E. This ensemble is therefore also referred to as an NVE ensemble. The microcanonical ensemble is highly hypothetical, since it is difficult to find a truly isolated system in reality. As a result, other ensembles are defined to account for situations where the internal energy can vary. Commonly used ensembles include the *canonical ensemble*, which corresponds to a constant number of particles N, a constant volume V and a constant temperature T (NVT ensemble); the *isothermal-isobaric ensemble*, which has a fixed number of particles N, a fixed pressure P and a fixed temperature T (NPT ensemble, close to typical experimental conditions); and *grand canonical ensemble*, where the fixed macroscopic parameters are chemical potential μ, volume V and temperature T (μVT ensemble). Grand canonical ensemble is particularly useful to describe systems involving addition or removal of particles, such as in the cases of liquid-vapor equilibrium, capillary condensation etc. (Tuckerman 2010).

At thermodynamic equilibrium, each ensemble has a specific probability distribution $\rho(\mathbf{q}_1, \ldots, \mathbf{q}_N, \mathbf{p}_1, \ldots, \mathbf{p}_N)$ for the microstates. The *ensemble average* of a quantity $A(\mathbf{q}_1, \ldots, \mathbf{q}_N, \mathbf{p}_1, \ldots, \mathbf{p}_N)$ can be calculated from the probability distribution (Leach 2001; Salinas 2001):

$$\langle A \rangle = \frac{\int A(\mathbf{q}_1, \ldots, \mathbf{q}_N, \mathbf{p}_1, \ldots, \mathbf{p}_N)\rho(\mathbf{q}_1, \ldots, \mathbf{q}_N, \mathbf{p}_1, \ldots, \mathbf{p}_N)d\mathbf{q}_1, \ldots, d\mathbf{q}_N d\mathbf{p}_1, \ldots, d\mathbf{p}_N}{\int \rho(\mathbf{q}_1, \ldots, \mathbf{q}_N, \mathbf{p}_1, \ldots, \mathbf{p}_N)d\mathbf{q}_1, \ldots, d\mathbf{q}_N d\mathbf{p}_1, \ldots, d\mathbf{p}_N}. \quad (3)$$

The denominator in Eq. (3) is related to the so-called *partition function*, Z, by a non-dimensionalization factor. Partition function is one of the most important quantities in statistical mechanics. Given the type of ensemble, it essentially captures the number of accessible microscopic states in the phase space. Theoretically, once the partition function is known, all macroscopic thermodynamic quantities such as entropy, enthalpy and free energy can be determined from it (Tuckerman 2010). Practically, however, analytical calculation of the partition function is only possible for a few very simple systems. In addition, direct numerical computation of the partition function is extremely difficult due to the 6N-dimensional integration involved in its definition. In MD simulations, properties of the simulated system are determined from *time averages* (Eq. 4) over a period τ:

$$A_{avg} = \lim_{\tau \to \infty} \frac{1}{\tau} \int_{t=0}^{\tau} A(\mathbf{q}_1(t), \ldots, \mathbf{q}_N(t), \mathbf{p}_1(t), \ldots, \mathbf{p}_N(t))dt \tag{4}$$

The *Ergodic hypothesis* assumes that the time average (Eq. 4) is equivalent to the ensemble average (Eq. 3), which allows for the use of MD to evaluate macroscopic properties of the system from the motion of the atoms (Szasz 1996).

2.2 Classical Approach: All-Atom Molecular Dynamics

Classical approach to atomistic simulations is based on molecular mechanics (MM), where atoms as treated as soft balls and bonds as elastic sticks (Meller 2001). The intra- and intermolecular interactions in the simulated system is described by a *force field*, in which a mathematical expression of the potential energy is given in terms of geometrical variables such as atom distances and bond angles. The force acting on each atom is calculated from the negative gradient of potential energy function specified in the force field. Numerical integration of equations of motion from this force leads to time trajectory of the system (González 2011). MM and the development of force fields are associated with several approximations, such as the pair-wise additivity of the potential energy. These approximations were introduced in order to simplify the modeling of molecules in MD simulations; since predicting detailed atomistic structures from quantum mechanical (QM) approaches is computationally very expensive. The first force fields, systematic force fields, were developed by Shneior Lifson, Harold Scheraga and Norman Allinger in 1960s, while the most popular ones that are being used today for biomolecules, CHARMM (Brooks et al. 1983) and AMBER (Pearlman et al. 1995) were developed in 1980s (Monticelli and Tieleman 2013; Schlick 2010).

Time trajectory of a system in a classical MD simulation is generated by integrating the Newtonian equations of motion for every atom i:

$$\mathbf{a}_i = \frac{d^2\mathbf{q}_i}{dt^2} = \frac{\mathbf{F}_i}{m_i}, \tag{5}$$

where $\mathbf{a}_i$ is the acceleration of atom i. Given the initial coordinates $(\mathbf{q}_1(0), \ldots, \mathbf{q}_N(0))$ and velocities $\left(\frac{d\mathbf{q}_i}{dt}(0), \ldots, \frac{d\mathbf{q}_N}{dt}(0)\right)$ of all atoms in the system, Eq. (5) can be integrated. However, because the interparticle forces are nonlinear functions of positions, the integration needs to be done numerically. There are several integration algorithms for solving such equations, such as Verlet (Verlet 1967), velocity Verlet

(Swope et al. 1982), Leap-frog (Buneman 1967; Hockney 1970) and Beeman (Schofield 1973) algorithms. The choice of integration method depends on its applicability to the given system (Tuckerman 2010) and the required accuracy. In a typical MD simulation, at each time step, the forces acting on the atoms are calculated from the force field and the current positions of the atoms. Coordinates and velocities are solved and updated using the specified integration algorithm. Energy is calculated and averaged, and written as an output along with the new coordinates. Such a step is repeated until the desired simulation time is reached (Lindahl 2008).

Computationally the most costly part of an MD simulation is the calculation of forces. The computation of electrostatic interactions is particularly time-consuming due to the long-ranged nature of the Coulombic interactions. Several techniques have been developed in order to expedite the calculation of electrostatic forces, such as Ewald summation, lattice summation and fast multipole methods (Sagui and Darden 1999). Particle Mesh Ewald (PME) (Darden et al. 1993) is one of the Ewald summation methods, and is one of the most widely used methods for treating electrostatics. The PME method is based on creating a three-dimensional grid over the system, mapping the atomic charges onto the grid, and separating the summation of electrostatic interactions into a short-ranged part that converges quickly in real space and a long-ranged part that converges quickly in Fourier space. The use of the PME method requires the application of periodic boundary conditions (PBC), through which the system is made infinite-like by having exact replicas of the original system throughout a lattice. Along with PBC, PME has been shown to result in physically stable systems and accurate prediction of long range electrostatics (Phillips et al. 2005).

By solving Eq. (5), the MD simulation is implicitly run in a microcanonical ensemble. This is because the forces are calculated from the negative gradient of the potential energy function, which means that these are conservative forces, and hence the total energy of the system is fixed. To realize MD simulations for other ensembles, special techniques need to be introduced to hold other quantities, such as pressure and temperature, constant. To this end, many thermostats (to maintain constant temperature) and barostats (to maintain constant pressure) have been developed. Examples of thermostats include velocity re-scaling, Berendsen (Berendsen et al. 1984), Andersen (Andersen 1980), Nosé-Hoover (Hoover 1985; Nosé 1984), Langevin dynamics (Schneider and Stoll 1978) and dissipative particle dynamics (DPD) (Hoogerbrugge and Koelman 1992). For barostats, Berendsen (Berendsen et al. 1984) and Nosé-Hoover (Hoover 1985; Nosé 1984) are the most commonly used techniques (Tuckerman 2010).

In an MD simulation, some atoms may have high frequency motions that require very small time steps in the integration, hence slowing down the simulation. One way to increase the time efficiency is to allow larger time steps by constraining some intramolecular interactions, such as bond lengths and angles, thus preventing motions with very high frequency. There are several algorithms developed for this purpose; mostly used ones are SHAKE (Ryckaert et al. 1977), which is based on Verlet integration algorithm and used for constraining positions, and RATTLE (Andersen 1983), which is based on velocity Verlet algorithm and used for constraining both positions and velocities (Schlick 2010). Constraints/restraints in MD simulations can also be applied when the system has the high likelihood of being trapped in free energy wells, resulting in poor sampling. The umbrella sampling (US) method (Torrie and Valleau 1977) is one such example, in which a biasing potential is applied to restrain certain degrees of freedom of the molecules. To obtain sufficient sampling, a series of simulations are generally needed in US with different biasing potential applied in each simulation. The results are then unbiased and combined for analysis via statistical approaches such as the weighted histogram analysis method (WHAM) (Kumar et al. 1992).

Classical all-atom MD simulations provide all the detailed atomistic information regarding the system of interest: structural information such as molecular size (radius of gyration, etc.), molecular spacing, conformational changes (folding, bending, etc.), hydrogen bond forming and breaking; energetic information such as cohesion energy and free energy of binding. Certainly, the use of all-atom MD is limited by the time and length scales that can be handled within a practical period of time. The current state-of-the-art all-atom MD simulations can simulate less than a million atoms for less than one microsecond, which is typically much smaller compared with practical systems studied in experiments. Another challenge associated with all-atom MD is the accuracy of the force fields employed in these simulations (Lyubartsev et al. 2009; Schlick 2010). Nevertheless, with the fast growing computing power, great efforts on improving numerical efficiency, and extensive studies being conducted on developing advanced, more reliable force fields, these simulations demonstrate immense potential towards modeling systems of realistic time and length scales.

2.3 Hybrid Approach: QM/MM Simulations

Combination of QM approach with MM is originated from the need of compromise between maintaining a reasonable size of the simulated system and accurately describing the electronic structure for certain chemically active regions in the system. While MM is incapable of capturing the changes in electronic structures in these regions such as charge transfer or excitation of electrons, it is possible to take them into account via calculations at the QM level. On the other hand, QM computation for the entire domain of a large molecular system is not feasible, and the QM transformations often happen in a much smaller subset of the system. This motivated the creation of hybrid system models (QM/MM) in which the chemically active region is modeled with QM approach, while the surrounding environment is modeled with classical MM approach. Since the introduction of this approach by Warshel and Levitt in (Warshel and Levitt 1976), QM/MM method has been widely applied to simulation of biomolecular systems such as enzymatic systems (Senn and Thiel 2007). As well, a lot of research has been and is being performed at a fundamental level to improve the QM/MM methodology.

In QM/MM calculations, the system of interest is divided into two subsystems: a reactive part that has a small number of atoms to be described by QM and the rest of the system to be described by a force field (MM). For the QM part, *ab initio*, density functional theory (DFT), and semi-empirical approaches are typically used; while the MM part can be modeled with classical force fields such as CHARMM (Brooks et al. 1983), AMBER (Pearlman et al. 1995) and GROMOS (Scott et al. 1999; Monard and Merz 1999). To enable a QM/MM calculation, it is necessary to define an effective Hamiltonian that contains a potential energy function for the system. This potential energy function needs to properly describe the interactions within the QM region, within the MM region, as well as on the QM and MM interface. The last is the most challenging part in QM/MM modeling, as the interactions between QM and MM regions cannot be described at either QM or MM level alone. There have been several approaches for approximating the interactions between these two levels, and they are mainly categorized into two groups, namely subtractive coupling and additive coupling schemes (Groenhof 2013).

In subtractive coupling, the potential energy of the system, including both QM and MM regions, is first calculated at MM level ($U_{MM}(QM + MM)$). Then the potential energy of the QM region is calculated at QM level ($U_{QM}(QM)$) and added to $U_{MM}(QM + MM)$. Finally, the potential energy of the QM region is calculated at MM level ($U_{MM}(QM)$) and subtracted, i.e.,

$$U_{QM/MM} = U_{MM}(QM + MM) + U_{QM}(QM) - U_{MM}(QM) \tag{6}$$

The implementation of subtractive scheme is straightforward without the need to explicitly connect the QM and MM parts. However, to determine $U_{MM}(QM)$, a force field is required for the QM region and it should be adaptive to chemical changes (e.g., bond breaking or formation) occurring in that region, which is unlikely to be available in most cases (Groenhof 2013).

In additive coupling, the potential energy of the system is calculated by the summation of three terms: QM level energy of the QM region ($U_{QM}(QM)$), MM level energy of the MM region ($U_{MM}(MM)$) and QM-MM energy for the interactions between the two regions ($U_{QM-MM}(QM + MM)$):

$$U_{QM/MM} = U_{QM}(QM) + U_{MM}(MM) + U_{QM-MM}(QM + MM) \tag{7}$$

For this approach, most efforts have been spent on developing methods to evaluate $U_{QM-MM}(QM + MM)$, which explicitly defines the interactions between the QM and MM regions. Commonly used approaches include; for example mechanical embedding where the interactions between the QM and MM regions are treated with force field models, electrostatic embedding which improves mechanical embedding by adding polarization effects, and polarization embedding which provides further improvement on the treatment of polarization effects using several methods such as charge-on-a-spring (Lamoureux and Roux 2003), induced dipole (Warshel et al. 2006) and fluctuating charge models (Rappe and Goddard 1991), (Groenhof 2013).

Once the potential energy and hence the effective Hamiltonian of the QM/MM system is defined, it will allow the determination of wave function, energy and forces in the system, and an MD simulation can be run. As in classical MD, structural and energetic information can be obtained by unrestrained or restrained simulations. In addition, QM/MM simulations allow the investigation of electronic changes and chemical reactions in the system, which is not possible with classical MD. For example, the reaction pathway can be determined by generating the potential energy surface; the chemical reactivity of a system can be evaluated by calculating the free energy difference between reactants and product. Despite these advantages, the time scale for QM/MM simulations has been limited by the high computational cost of the QM calculations. While one can perform MD for hundreds of nanoseconds, QM/MM simulations can only be performed for hundreds of picoseconds at ab initio or DFT levels, although this can be extended 100 times with the application of semi-empirical approaches. Considering such limitations, most of the QM/MM simulations performed so far have focused on structural optimization (energy minimization) rather than unrestrained dynamics (Groenhof 2013).

2.4 Mesoscopic Approach: Coarse-Grained Simulations

To overcome the computational limits associated with classical MD simulations, there has been a great effort to develop novel approaches that has the ability to bridge atomistic and mesoscopic scales. Coarse-grained (CG) method represents one of such modeling approaches. The first CG model was proposed by Levitt and Warshel in (Levitt and Warshel 1975) for bovine pancreatic trypsin inhibitor (PTI). Since then many different CG models have been developed and applied to study protein folding, protein-protein interactions, membrane proteins, lipid bilayers, complex nucleic acids, nanocomposites, to name a few (Ingólfsson et al. 2013; Takada 2012).

In CG, instead of modeling every atom separately as in all-atom MD, a certain number of atoms are clustered into beads, namely CG sites (Voth 2009). The process of building a CG model consists of several important steps. The first step is to decide the resolution of the model, i.e., how many atoms will be clustered into one bead. In the second step, the location of the CG sites and their arrangement should be determined, with each CG site representing a relatively rigid region in the molecule. Afterwards, the interactions among the CG sites need to be described, and this is often done via parameterized effective potential energy functions, as in the all-atom MD simulations. For example, bonded interactions are generally described by elastic network models (ENM) and Lennard-Jones (LJ) like potentials can be used to model short-ranged non-bonded interactions. Parameterization of these potential energy functions can be achieved using a number of different approaches (e.g. force-matching, inverse Boltzmann fitting, etc.) in order to produce good agreement between the CG model and experimental data or higher level (QM, MM) calculations (Saunders and Voth 2013).

When the CG model for a system of interest is developed, CG simulations can be performed in a similar manner to that of classical MD simulations. Besides pure CG simulations, CG approach is often coupled with all-atom simulations to connect the information at atomistic and mesoscopic scales. Several methods facilitate building of such connection, including free energy method which determines key interactions from CG simulations and performs more accurate free energy calculation at atomistic scale, inverse mapping method which uses CG simulations to identify regions of the phase space not easily sampled by all-atom MD and allows additional MD simulations in these regions, and multiple-scale approach where the entire system is separated into MM and CG domains in a way similar to QM/MM simulations (Saunders and Voth 2013; Schlick 2010).

The greatest advantage of CG method is the reduction in the degrees of freedom of the systems, leading to larger length and time scales that can be simulated. While time scales are limited to ps in QM/MM and ns in all-atom MD, one can perform CG simulations in the ms range. As a result, direct comparison may be made with experimental observables (Voth 2009). It should be recognized, however, that treating several atoms as a group in CG models causes the loss of atomistic scale information, and can lead to inaccuracy for the data obtained from the simulation trajectories (Takada 2012). Careful evaluation of the quality of the CG model and the method of the simulation should be carried out when employing the CG approach.

3 Molecular Dynamics Simulations on the Formation of Polyplexes and Lipoplexes

In this section, we will provide a detailed review of simulations reported on the formation of polyplexes and lipoplexes. A list of the studies we included in the review is given in Table 1. The structures of the carriers reviewed in the scope of this manuscript are given in Fig. 1.

3.1 Complexation of Polymers or Lipids to Individual Nucleic Acids

All-atom and CG simulations have been used as efficient tools, respectively at atomistic and mesoscopic levels, in analyzing nucleic acid interactions with carriers, such as lipids, polyamines and polymers. A significant amount of efforts have been spent on understanding how the carriers bind and complex with individual

Table 1 Summary of studies on formation of polyplexes and lipoplexes

Interacting molecule	Investigated phenomena	Biomolecule simulated	Type of simulation	References
(a) Binding of carriers to nucleic acids				
Polyamines	DNA-polyamine interactions	dsDNA	All-atom MD	Korolev et al. (2001, 2002, 2003, 2004a, b)
Triazine dendrimers	Effect of dendrimer flexibility and DNA/siRNA comparison	dsDNA, siRNA	All-atom MD	Pavan et al. (2010c)
Spermine, DAP, DAPMA dendrons	Effect of dendrimer generation	dsDNA	All-atom MD	Jones et al. (2010)
Spermine dendrons	Effects of dendrimer flexibility and salt addition	dsDNA	All-atom MD	Pavan et al. (2009)
	Effect of dendrimer structural modification and DNA/siRNA comparison	dsDNA	All-atom MD	Pavan et al. (2010a, b)
	Effect of dendrimer structural modification	dsDNA	CG DPD	Posocco et al. (2010)
PAMAM	Effect of dendrimer flexibility	siRNA	All-atom MD	Jensen et al. (2010, 2011), Pavan et al. (2010a)
	Effects of dendrimer generation and DNA sequence on binding	ssDNA	All-atom MD	Maiti and Bagchi (2006)
	Effects of dendrimer generation, salt and H_2O on binding	dsDNA	All-atom MD	Nandy and Maiti (2011)
	Effect of dendrimer generation and binding of multiple dendrimers	siRNA	All-atom MD	Ouyang et al. (2010a, b), Vasumathi and Maiti (2010)
	Effect of dendrimer protonation ratio	siRNA	All-atom MD	Ouyang et al. (2011), Pavan et al. (2010d)
	Effects of dendrimer generation and protonation ratio	siRNA	All-atom MD	Karatasos et al. (2012)
	Effect of H_2O on binding	dsDNA	US MD	Mills et al. (2013)
PLL	Effect of dendrimer generation and binding of multiple dendrimers	siRNA	All-atom MD	Ouyang et al. (2010b)
	Interactions between DNA and PLL	dsDNA	All-atom MD	Ziebarth and Wang (2009)
	Effect of PLL architecture	dsDNA	All-atom MD	Elder et al. (2011)

(continued)

Table 1 (continued)

Interacting molecule	Investigated phenomena	Biomolecule simulated	Type of simulation	References
PEI	Interactions between DNA and PEI	dsDNA	All-atom MD	Ziebarth and Wang (2009)
	Effect of molecular weight, structure and protonation ratio	dsDNA	All-atom MD	Sun et al. (2011b, 2012a)
	Interactions with PEI and DNA/siRNA comparison	dsDNA, siRNA	All-atom MD	Zheng et al. (2012)
DMTAP, DMPC	Bilayer interactions	dsDNA	All-atom MD	Bandyopadhyay et al. (1999)
DMPC	Monolayer interactions	dsDNA	All-atom MD	Braun et al. (2003)
(b) Nucleic Acid Condensation				
Counterions	Valence effects	Polyanion	CG MD	Stevens (2001)
Polycation	Effects of number of polycations and charges	Polyanion	CG MC	Dias et al. (2003)
Copolymer chains	Effect of copolymer length and charge ratio	Polyanion	CG MD	Ziebarth and Wang (2010)
PAMAM	Effect of applied force	dsDNA	All-atom MD	Mills et al. (2010)
Polyanion + Fe(III)	Effect of Fe(III)	Polycation	CG MC	Jorge et al. (2012)
(c) Nucleic Acid Aggregation				
NaCl, KCl	Effect of Na^+ and K^+	2 dsDNAs	All-atom MD	Savelyev and Papoian (2007)
Polyamines	Effect of valence	2 dsDNAs	US MD	Dai et al. (2008)
PEI	Effect of PEI protonation ratio and charge ratio	2 dsDNAs	US MD	Bagai et al. (2013)
	Aggregation mechanism	Multiple dsDNAs	All-atom MD	Sun et al. (2011a)
	Effect of PEI lipid substitution	Multiple dsDNAs	All-atom MD	Sun et al. (2012b)
		Multiple siRNAs	All-atom MD	Sun et al. (2013)
Cationic lipids	Self-assembly into bilayer	dsDNA	CG MC	Farago et al. (2006)
Cationic and neutral lipids	Phase transition of packed DNA	dsDNA	CG MD	Farago and Gronbech-Jensen (2009)
DOPE, DOTAP	Phase transition of packed DNA	dsDNA	CG MD	Corsi et al. (2010)
Polycation	Interactions between oppositely charged polyions	Polyanion	CG MC	Hayashi et al. (2002, 2003, 2004)

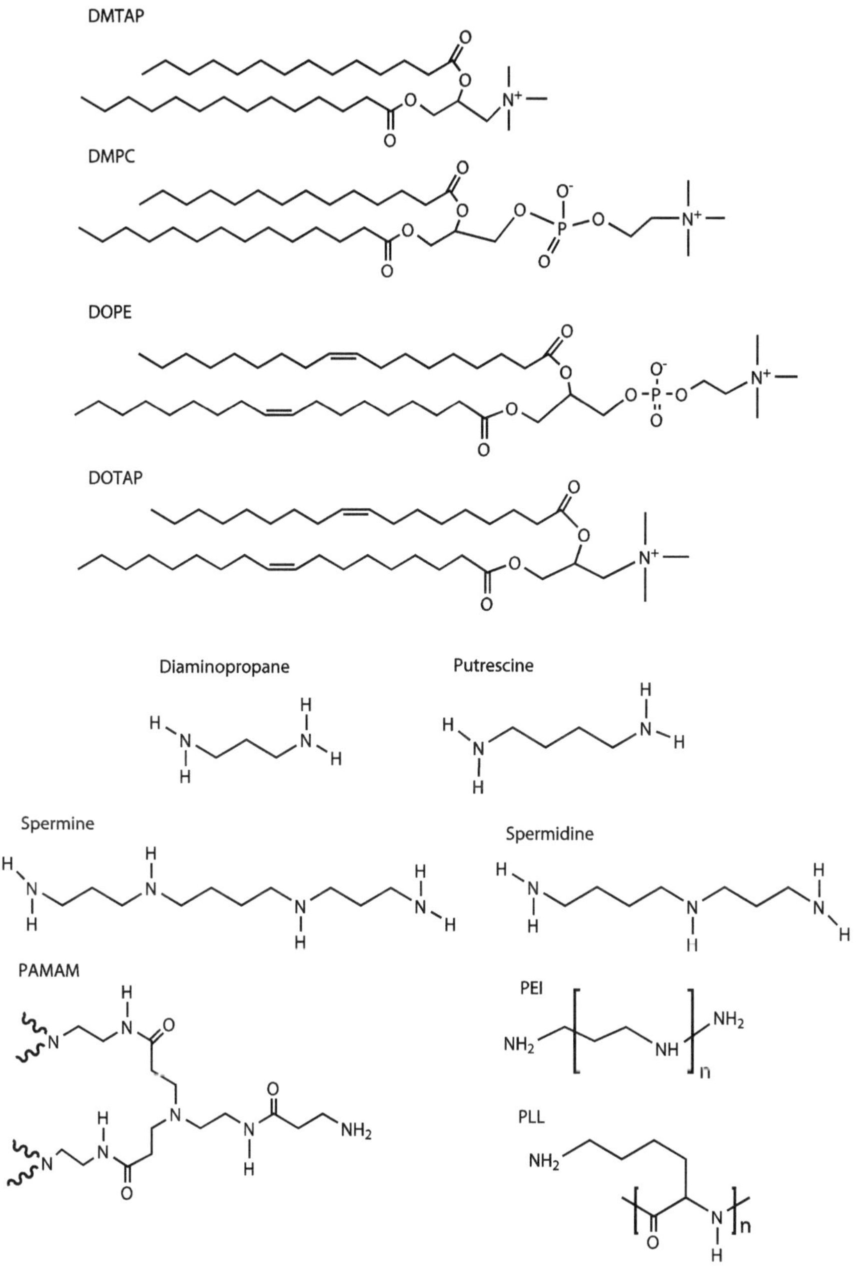

Fig. 1 Chemical structures of selected carriers

nucleic acids, and how the complexation is affected by carrier properties and the environment of complexation. Different types of carriers have been studied, as reviewed below.

3.1.1 Oligocations

Putrescine, spermidine and spermine are small organic oligocations which are naturally present in prokaryotic and eukaryotic cells (Pegg and McCann 1982). They were proposed to bind and stabilize DNA (Gosule and Schellman 1978); spermine and spermidine were reported to cause DNA or RNA precipitation in solution while putrescine lacks this ability (Razin and Rozansky 1959). DNA-oligocation interactions were investigated by Korolev et al. (2001, 2002, 2003, 2004a, b) in a series of MD studies. Interactions between DNA and putrescine (2^+), spermidine (3^+), spermine (4^+) and synthetic diaminopropane (2^+) were studied in the presence of H_2O and monovalent ions, Na^+ and Cl^-. These oligoamines were found to bind to DNA through their interaction with backbone O1P and O2P atoms. While the tri- and tetravalent spermidine and spermine were excluded from the major groove, the divalent oligoamines putrescine and diaminopropane were found in DNA major groove. Oligocations and Na^+ showed different roles in affecting the hydration of DNA. While Na^+ was reported to organize and attract H_2O in hydration of DNA, oligoamines were observed to repel H_2O from the hydration shell. This observation could be important in the interactions of nucleic acids with carriers in the presence of physiological ion composition.

3.1.2 Dendrimers and Dendrons

Many researchers have recently focused on dendrimers and dendrons, branched structures with repetitive units, as nucleic acid carriers due to their well-controlled chemical architecture. The influences of such parameters as generation number, backbone flexibility, protonation state and functionalization, on the complexation of dendrimers and dendrons with nucleic acids, have been widely investigated.

The effect of protonation was addressed by studying different dendrimer protonations corresponding to different pH environments (Karatasos et al. 2012; Ouyang et al. 2011; Pavan et al. 2010d). Molecular Mechanic/Poisson–Boltzmann Surface Area (MM-PBSA) analysis showed a more favorable dendrimer-siRNA binding at low pH (~ 5) compared to neutral pH (~ 7) conditions regardless of the dendrimer generation (Karatasos et al. 2012; Ouyang et al. 2011; Pavan et al. 2010d) and this was attributed to higher protonation ratio of dendrimers at lower pH. These studies also demonstrated that the electrostatic attraction between the cationic dendrimers and anionic nucleic acids played a crucial role in their complexation.

The effect of dendrimer flexibility on complexation was addressed in several studies (Jensen et al. 2010, 2011; Karatasos et al. 2012; Pavan et al. 2010a, c). For

PAMAM dendrimer, its rigidity was reported to increase with the increase in size/generation (Jensen et al. 2010, 2011; Pavan et al. 2010a). Pavan and coworkers studied complexation of G4 to G6 PAMAM dendrimers with siRNA. G4 dendrimer showed high binding affinity due to its flexible structure, while affinity towards siRNA was lost with G6 dendrimer's rigid spherical structure. G7 PAMAM simulations with siRNA also showed a rigid sphere behaviour (Jensen et al. 2010) where a lack of structural rearrangement in G7 dendrimers led to less than an optimal binding mode to siRNA. In a separate study (Jensen et al. 2011), comparison between G1, G4 and G7 PAMAM dendrimers showed that the increased rigidity associated with size increase caused higher entropic cost in binding to siRNA. Using MM-PBSA analysis, the free energy of binding between PAMAM dendrimers and siRNA was evaluated. G4 PAMAM was reported to be the most suitable carrier based on its strongest binding affinity as compared to G1 and G7. Rigidity increased with the increase in size from G4 to G7 and binding was weakened (Jensen et al. 2011). Although the flexibility of dendrimers was reported to contribute positively to complexation in most of the studies, a contradictory report was published by Pavan et al. (2010c). Their intent was to prove 'flexibility favors binding', however their results showed a beneficial effect of dendrimer rigidity. G2 triazine dendrimers designed to be flexible with an ethylene glycol chain resulted in a more compact and rigid structure due to the collapse of the flexible linkers in a salt environment, so that the number of interactions with DNA and siRNA were reduced. However, the rigid dendrimer designed with piperazine rings was found to establish more contacts with nucleic acids (Pavan et al. 2010c).

In many situations, the effects of protonation and flexibility are coupled. For example, G5 dendrimers showed a transition from flexible to rigid structures depending on the pH value (Pavan et al. 2010a), which was caused by different protonation ratios at different pH. The flexibility observed at neutral pH (pH $\sim$ 7.4) was lost with a decrease in pH to 5 or lower. At low pH, increase in dendrimer's cationic charge caused an increase in rigidity thus leading to lower affinity towards siRNA (Pavan et al. 2010a). As greater amount of charges facilitate binding through stronger electrostatic interactions but at the same time can make the dendrimer more rigid, there needs to be a delicate balance between charge and flexibility in order to achieve optimal binding. High generation dendrimers are large in size and weight due to the substitution of more building blocks to the low generation ones. They are more rigid but have more charges compared with low generation ones. While some work reported reduced binding affinity as the generation number increases, others reported that high generation PAMAM dendrimers bound more effectively to single stranded DNA (Maiti and Bagchi 2006), double stranded DNA (Nandy and Maiti 2011), and double stranded siRNA (Karatasos et al. 2012; Ouyang et al. 2010a; Vasumathi and Maiti 2010). Similarly, for 1,3-diaminopropane (DAP), *N*,*N*-di-(3-aminopropyl)-*N*-(methyl)amine (DAPMA) and spermine dendrons, Jones et al. (2010) reported an increase in binding affinity towards double stranded DNA with the increase in generation number. Since the dendrimer rigidity increases with the increase in generation number, this increased binding affinity should be considered in the context of increased dendrimer rigidity. Lower generations might show higher flexibility, but lack of necessary numbers of positive charges might cause ending up with lower binding affinity. The optimal threshold for the number of positive charges and flexibility should be determined in designing optimal dendrimers as nucleic acid carriers.

Functionalizing dendrimers/dendrons with end-groups can also lead to different binding capability to nucleic acids. Jones et al. (2010) studied the interactions between DNA and G1 and G2 dendrons with the end-groups spermine, DAP and DAPMA. MM-PBSA calculations indicated that enthalpic contribution to binding energy was related to surface charges of the dendron. When the surface is more charged, enthalpic contribution to binding energy became more favorable. As the protonable groups on each ligand in the scaffold is 3, 2 and 1 for spermine, DAPMA and DAP, respectively; spermine dendrons were selected to be the best candidates for DNA binding among both G1 and G2 dendrons based on their highest binding energy due to their largest surface charge. On the other hand, binding affinity of G2 DAPMA to DNA was found to be close to that of G1 and G2 spermine dendrons due to its favorable enthalpic interactions associated with ligand backfolding and DNA bending. In the work of Pavan et al. (2010b), a comparison was made between dendrons functionalized with non-degradable spermine and those functionalized with UV degradable spermine. UV degradable dendrons had the same enthalpic affinity towards DNA as non-degradable ones, but had a smaller entropic cost upon binding. UV modification on the structure with a photolabile linker introduced branches. Due to this branched structure, UV modified dendron could establish more stable interactions with DNA via more uniform vibrations of dendron amines and DNA phosphates, resulting in smaller cost in entropy (Pavan et al. 2010b). Posocco et al. modified spermine functionalized dendrons with hydrophobic cholesterol units in order to trigger self-assembly. They performed DPD simulations, a CG approach in which several atoms are lumped into a bead and the interactions between the beads (named DPD particles) are treated with soft potentials. The force acting on each DPD particle is a sum of conservative, dissipative and stochastic forces. Time trajectory can be obtained by solving Newton's equation of motions as in MD. According to their DPD simulations, self-assembly of first generation cholesterol-modified dendrons resulted in assembled dendrimer structures (due to cholesterol aggregation) that have higher charge density compared to second generation dendrons, thus more effective binding with DNA (Posocco et al. 2010).

Simulations with multiple PAMAMs revealed formation of more compact and stable DNA (Vasumathi and Maiti 2010) and siRNA (Ouyang et al. 2010b) complexes, in comparison to systems containing single dendrimers. When polyplexes become positively charged upon the saturation of nucleic acid charges, excess dendrimers would float around in free form (Ouyang et al. 2010b).

Physiological ions and aqueous environment affect the interactions between nucleic acids and carriers. Pavan et al. studied the effect of NaCl addition on complexation of DNA with spermine-functionalized G1 and G2 dendrimers. Increased salt concentration did not show a remarkable effect on binding affinity of

G2 towards DNA, while interactions with DNA were lost in G1 dendrimer-DNA system (Pavan et al. 2009). When salt was introduced to G5 PAMAM-DNA system, MM-PBSA analysis indicated a decrease in binding affinity due to the screening effect. However, the formed complexes were reported to be still very compact (Nandy and Maiti 2011). The screening effect from counterions therefore seems to become important in the case of small carriers; when the carriers are large enough to maintain strong interactions with nucleic acids; screening becomes negligible on the overall binding behavior. H_2O molecules were known to have an impact on nucleic acid-carrier interactions since they are released from the binding region while complexation occurs. Effect of H_2O molecules on DNA-G3 PAMAM interactions was studied by Mills and coworkers by US MD simulations. The results revealed the importance of H_2O ordering; it was found that H_2O molecules were capable of bridging DNA and dendrimer in highly charged dendrimer system (dendrimer carrying +32 charge in comparison to +16) (Mills et al. 2013). The PAMAM (G3 to G5) itself was reported to undergo swelling in the presence of H_2O molecules as a result of H_2O penetration into their structures, and the degree of penetration was proportional to dendrimer generation (Nandy and Maiti 2011). It is therefore important to simulate properly hydrated carriers for more realistic predictions.

The role of nucleic acid sequence in binding to PAMAM dendrimers was studied by Maiti and Bagchi with atomistic MD simulations. Vibrational density of states analysis showed the binding affinity difference between bases; the following order was found in binding to G4 dendrimer: poly(G) > poly(C) > poly(A) > poly(T) (Maiti and Bagchi 2006). Seeing such a difference after employing pure base sequences should not be surprising; however, whether the same result holds true with statistically mixed sequences (especially for longer nucleic acid chains) remains to be seen. In order to compare carrier binding to DNA and siRNA, Pavan et al. performed MD simulations on 21 bp long DNA (sequence: 5′-TCG AAG TAC TCA GCG TAA GTT-3′; 3′-AGC TTC ATG AGT CGC ATT CAA-5′) and siRNA (5′-UCG AAG UAC UCA GCG UAA G dTdT-3′; 3′-dTdT AGC UUC AUG AGU CGC AUU C-5′). Stronger dendron interactions were formed with the siRNA compared to DNA, due to siRNA's higher flexibility than DNA (Pavan et al. 2010b, c). This observation was concurrent to previous results seen with the PAMAM structures, where the rigidity of the carrier reduced the strength of binding.

3.1.3 Cationic Polymers

Despite the great interest in dendrimeric structures as gene carriers, their buffering capacity could be compromised at times (due to covalent modification of protonable residues in the dendritic core) and this led researchers to seek other types of carriers with superior buffering capacity. PEI is one of the most efficient nucleic acid carriers with high buffering capacity (Boussif et al. 1995). PEI-nucleic acid delivery systems were studied with atomistic simulations by Ziebarth and Wang (2009), our group (Sun et al. 2011b, 2012a) and Zheng et al. (2012). The initial study of PEI-DNA binding by Ziebarth and Wang (2009) reported the main interactions to be formed between the cationic PEI amines and DNA backbone phosphates. Our group focused on the effect of several PEI properties such as molecular weight, structure, and protonation ratio on PEI complexation with DNA. Sizes of PEIs simulated were 570 Da (Sun et al. 2011b) and 2 kDa (Sun et al. 2012a) with different PEI branching structures. The simulations with 570 Da PEI showed that the degree of branching did not have a significant influence on the binding pattern, while the protonation ratio played a key role. Decrease in protonation ratio resulted in less stable complexes due to the loss of direct H-bonding between PEI and DNA (Sun et al. 2011b). In contrast to 570 Da PEI, simulations with 2 kDa PEI (Sun et al. 2012a) showed very different binding pattern associated with linear and branched PEIs. Linear PEI was found to bind to DNA like cords due to its flexible structure, thus providing good coverage of the DNA surface. Branched PEIs, on the other hand, bound to DNA like beads, utilizing part of the PEI molecule to bind to a DNA molecule and potentially allowing interaction of individual PEIs with multiple DNAs. A better neutralization of DNA charges with 2 kDa PEI as compared to 570 Da PEI was observed. Zheng and coworkers studied a high molecular weight branched PEI (25 kDa) binding with DNA and siRNA. Structural and energetic analysis showed that strong electrostatic interactions were formed between DNA and cationic amines of PEI. While charge neutralization between oppositely charged PEI and DNA groups was apparent at the binding interface, several cationic PEI groups that were not contributing to binding were observed away from the interface. Simulations with one dsDNA (21 bp) and one 25 kDa PEI indicated that a stable DNA polyplex required more than one 25 kDa PEI molecule. Energetic comparison between DNA and siRNA complexation with 25 kDa PEI revealed stronger attraction of polymer to siRNA due to its more flexible structure (Zheng et al. 2012). This observation is in agreement with the previously reported comparison between DNA and siRNA in binding to dendrons (Pavan et al. 2010b, c).

Another polymeric carrier, PLL, was reported to be relatively inefficient in DNA charge neutralization (Ziebarth and Wang 2009) due to its linear structure compared to more densely charged PEI. All-atom MD simulations with different PLL structures probed the effect of PLL architecture on DNA binding mechanism. Comparison between linear PLL (uniform charges along the backbone) and grafted oligolysines (charged side-groups) on a neutral backbone showed less favorable binding of DNA with graft oligolysines, which was attributed to the steric hindrance caused by the hydrophobic backbone in graft architecture (Elder et al. 2011). PLL binding to siRNA was studied by Ouyang et al. (2010a, b) in a series of studies along with PAMAM dendrimers. They studied PLL^{4+} and PLL^{8+} systems that have the same charge as G0 and G1 PAMAM dendrimers, respectively. For the PLL-siRNA systems, the complexation behavior was found to be similar to that of the PAMAM-siRNA systems, suggesting that the electrostatic interactions dominated the binding affinities, rather than the molecular structural details.

3.1.4 Lipids

Cationic and neutral lipids are known for their ability to form complexes with nucleic acids to serve as effective carriers in nucleic acid delivery. A natural lipid structure contains a hydrophilic region, the lipid head group, and a hydrophobic region known as the lipid tail. When exposed to an aqueous environment, lipids self-assemble into a bilayer structure in which the hydrophilic head groups are exposed to water while the hydrophobic tails stay in the center region. DNA-lipid interactions were first studied by Bandyopadhyay and coworkers with computational approaches, where atomistic simulations were performed on a mixture of DMTAP and DMPC lipid bilayers interacting with DNA. Zwitterionic PC head groups were found to approach DNA and compete with cationic trimethylammonium (TAP) moieties in neutralizing the DNA phosphate charges (Bandyopadhyay et al. 1999). Braun et al. performed atomistic simulations on the interaction of a protonated DMPC monolayer with DNA. Although the DNA conserved its double helical structure through the simulation period, base pairing was affected by the interactions between the lipid head groups and DNA backbone phosphates as well as bases (Braun et al. 2003). These atomic level studies revealed the screening of the anionic DNA charges by lipid moieties.

3.2 Condensation and Aggregation

DNA condensation is defined as the dramatic decrease in DNA's volume triggered by packaging inside cellular environment. As an example, T4 bacteriophage's DNA occupies a volume of 4×10^9 nm^3 in solution; when it is packed in T4's head, its volume decreases to 5×10^5 nm^3 (Bloomfield 1996). This packaging in cellular environment can be reproduced in solution by the addition of cationic molecules to DNA. Attraction of multiple nucleic acid molecules to each other with the help of condensing agents such as cationic carriers results in formation of self-assembled (ordered) structures or aggregates (disordered structures). These aggregates could turn into spherical nanoparticles that are approximately 100 nm in diameter under practical conditions (Goula et al. 1998). While DNA condensation by cationic carriers is crucial for cellular uptake of DNA, the delivery of siRNA relies more on formation of aggregates. A number of simulations have been performed to study both condensation and aggregation of nucleic acids.

3.2.1 Condensation

Stevens carried out CG MD simulations on DNA-like polyelectrolytes to probe their condensation by di-, tri- and tetravalent counterions. While divalent counterions did not show any condensation, condensation was achieved with trivalent and tetravalent counterions. These results revealed the important interplay of enthalpic and entropic contributions to condensation. A valence of at least 3 was found to be required for electrostatics to overpower the entropic interactions and cause condensation (Stevens 2001). Dias et al. studied polyanion/polycation systems by CG Monte-Carlo (MC) simulations. The stochastic MC simulations follow a different algorithm than the deterministic MD methodology. Rather than obtaining a time trajectory of the system by integrating Newton's equations of motion in MD, MC method uses a random number generator to obtain new coordinates and performs trial moves (Schlick 2010). Despite the different underlying algorithms, MC simulations also try to sample the phase space as in MD, and for this reason there have been few MC simulations (mostly at the CG scale) on polyplexes and lipoplexes, which will be presented in this review. With the CG MC approach, condensation and collapse of polyanion chains were observed by increasing numbers of polycations or number of charges carried by each polycation. The systems with polycation/polyanion charge ratio below unity resulted in small polyanion chain deformation, but not in collapse or condensation (Dias et al. 2003). Ziebarth and Wang, using CG MD simulations, investigated polyanion condensation mediated by copolymer chains containing neutral hydrophilic and polycationic parts. Variation in copolymer length affected the condensation; successful condensation of polyanions was achieved with chains longer than 4 cationic blocks (Ziebarth and Wang 2010). In agreement with Dias et al. (2003), polyanion condensation was observed when the charge ratio $\geq$1 and only local bending/folding was seen for charge ratios below the unity (Ziebarth and Wang 2010). A CG MC study on compaction of DNA by PEI and Fe(III) (as a ternary system) was reported by Jorge and coworkers. Fe(III) ions were found to be located in the polyanion (corresponding to DNA) regions that were less occupied by polycation (corresponding to PEI) and they were proposed to induce folding of the polyanion chain (Jorge et al. 2012).

DNA condensation by G3 PAMAM dendrimers has been also studied with US MD simulations by Mills et al. (2010). Simulations were performed with 24 bp dsDNA and G3 PAMAM dendrimer with 32 protonated amine groups. Potential of mean force (PMF) calculations were carried out along a reaction coordinate defined as the center of mass distance between PAMAM and DNA. By taking the derivative of PMF with respect to the reaction coordinate, the interaction forces between DNA and dendrimer were calculated to retrieve the transition between extended (uncondensed) and condensed states. When a large force was applied, uncondensed state was favored; with the decrease in the applied force, DNA collapse (condensed state) was reported to be favorable (Mills et al. 2010).

3.2.2 Aggregation

Several studies focused on DNA aggregation by counterions (Savelyev and Papoian 2007), polyamines (Dai et al. 2008) and polymers (Bagai et al. 2013; Sun et al. 2011a, 2012b). Savelyev and Papoian performed all-atom MD simulations with two DNA molecules in the presence of NaCl and KCl. PMF analysis from US simulations was performed along a reaction coordinate, which was the center of mass distance between two DNA molecules. DNA repulsion when they approach each other was found to follow a steeper profile in the presence of KCl, which indicated that K^+ screening was less efficient than Na^+ screening for DNA charges (Savelyev and Papoian 2007). Dai and coworkers performed US MD simulations to evaluate the PMF of attraction between two DNAs in the presence of linear oligoamines putrescine (2^+), spermidine (3^+) and spermine (4^+). The results revealed more attraction between two DNA molecules with increasing valence of the oligoamine and the formation of transient ion bridges was observed (Dai et al. 2008). Our group (Bagai et al. 2013) reported attraction of two DNA molecules in the presence of 568 Da PEI with US MD simulations. According to PMF analysis, PEI had better capability of aggregating the DNAs compared with oligoamines, leading to larger depth in the PMF curve. In addition, more compact and stable aggregates were obtained when N/P ratio (ratio of the number of PEI Ns to the number of DNA Ps) of the system or protonation ratio of PEI was increased.

In a series of recent publications (Sun et al. 2011a, 2012b, 2013), we reported PEI mediated aggregation of multiple (>2) DNA and siRNA molecules. In the case of 568 Da PEI, DNA aggregation was facilitated by the formation of polyion bridges between DNA molecules and screening of DNA anionic charges (Sun et al. 2011a). Simulations performed on oleic acid modified PEI (831 Da) showed that the lipid moieties on the PEI can bring an additional mechanism of aggregation, namely, the association of lipid tails (Sun et al. 2012b). We further investigated the effect of lipid substitution on PEI-mediated siRNA aggregation (Sun et al. 2013). Simulations were performed for multiple siRNA molecules in the presence of caprylic acid (CA) and linoleic acid (LA) modified 1,874 Da PEI, as well as native (unmodified) PEI. Comparison among them showed more compact and stable siRNA aggregates in the case of lipid substituted systems (Sun et al. 2013). Also, LA association from different PEIs was found to be more stable than the association among shorter CA. Furthermore, the level of substitution was shown to affect the aggregation. At the level of one lipid substitution per PEI, several PEIs were found to stay in the solution instead of binding to the polyplex. However, when the substitution level was increased to three lipids per PEI, all PEIs could bind to the polyplex (see Fig. 2).

Lipid mediated DNA self-assembly was reported by Farago et al. in a series of publications (Farago and Gronbech-Jensen 2009; Farago et al. 2006; Corsi et al. 2010). CG MC simulations on randomly distributed lipids around DNAs revealed the self-assembly of lipids into a bilayer structure along with DNAs sandwiched between two lipid bilayers (Fig. 3a; Farago et al. 2006). In a further CG study (Farago and Gronbech-Jensen 2009), the self-assembly mechanism was studied with a much larger system by using a 10-fold higher number of lipid molecules (e.g., 800 cationic lipids, 1,600 neutral lipids along with 32 DNA molecules) than in (Farago et al. 2006). Charge density, φ_c, of the cationic and neutral lipid mixture, defined as the percentage of the cationic lipid in the mixture, was varied by changing the number of neutral lipids. An inverted hexagonal phase, in which DNA molecules are packed in cylindrical lipid micelles, was observed at

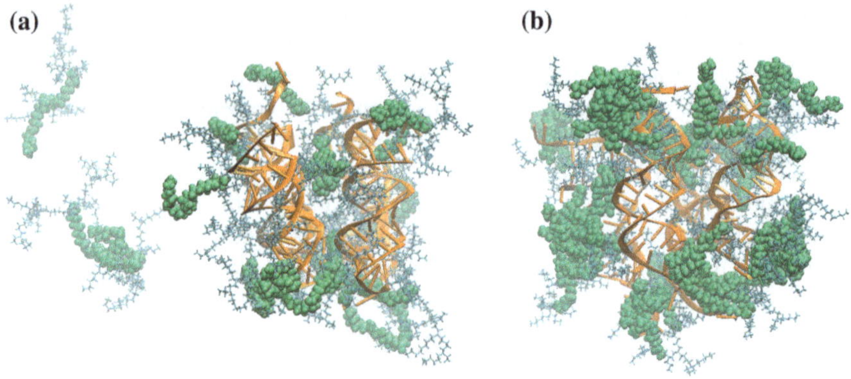

Fig. 2 Formation of polyplex from 4 siRNA and 18 LA-modifed PEI molecules: siRNAs are given in orange, while PEI and LA are *cyan* and *green*, respectively. **a** 1 LA substitution per PEI chain. **b** 3 LA substitution per PEI chain (Sun et al. 2013)

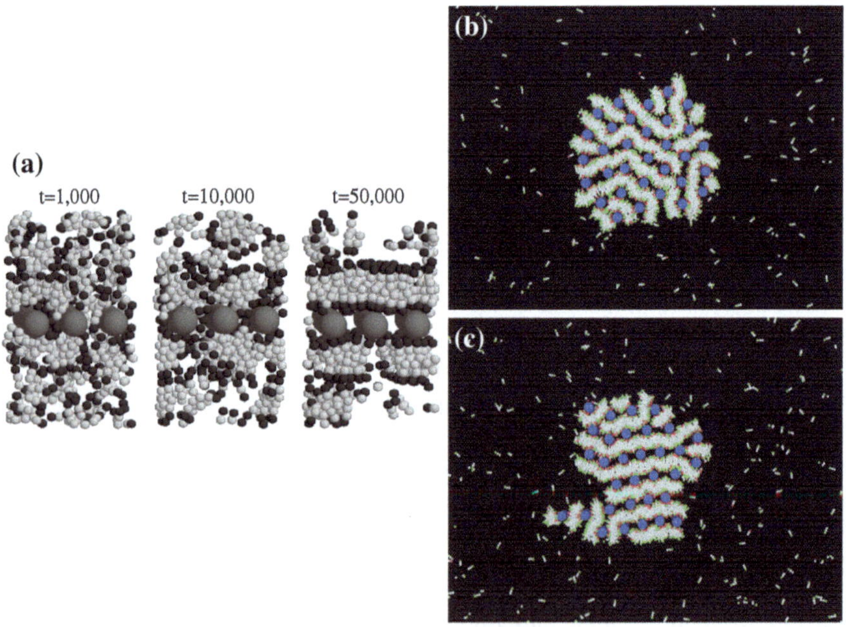

Fig. 3 Snapshots from CG MC simulations on self-assembly of cationic lipids and DNA. The evolution of the system is given in terms of the number of MC steps, t. Hydrophilic and hydrophobic regions of lipids are in *black* and *light gray* respectively, and DNAs are in *dark gray* (Farago et al. 2006). **b** Inverted hexagonal phase, $\kappa_s = 2.5$; and **c** Lamellar phase, $\kappa_s = 5$, of DNA-cationic lipid complexes Adapted with permission from Farago and Gronbech-Jensen (2009). Copyright 2013 American Chemical Society

intermediate charge densities ($4/15 \leq \varphi_c \leq 2/5$). At high charge densities (i.e. in a fully charged system, $\phi_c = 1$), a highly ordered square lattice form was visualized. Further, they modified the stiffness of the membrane by assigning an energy penalty for adjacent lipids to have different orientations. When the stiffness of the membrane increased (e.g. as the stiffness parameter κ_s changed from 2.5 to 5), transition from hexagonally packed DNA structure to a lamellar phase, in which DNA monolayers are sandwiched in between lipid bilayers, was observed (Fig. 3b, c). Increasing κ_s to >5 resulted in formation of lamellar phase only (Farago and Gronbech-Jensen 2009). Corsi et al. (2010) performed CG MD simulations on the assembly of DNA-DOPE and DNA-DOPE-DOTAP systems, where the experimentally observed transition of the DNA-DOPE-DOTAP system from fluid lamellar to inverse hexagonal phase was successfully reproduced. They further studied the phase transition of lipoplexes as a function of hydration level (number of H_2O molecules per lipid). Two H_2O molecules per lipid resulted in transition from lamellar to hexagonally packed DNAs in the presence of DOPE lipids, while addition of DOTAP to the system did not cause remarkable changes at the mesoscopic scale (Corsi et al. 2010).

In addition to the above works that studied particular carrier molecules, investigation of interactions between oppositely charged polyions has been attempted as a model of nucleic acid-carrier systems. Instead of modeling nucleic acid or carrier structures at atomistic level, simplifying nucleic acids as polyanions and cationic carriers as polycations allow researchers to investigate the aggregation mechanism with reduced computational costs. With that motivation, Hayashi et al. (2002, 2003, 2004) employed CG MC simulations on oppositely charged polyions. The effects of polyion's absolute charge, charge density and surrounding salt concentration on the aggregation mechanism were explored. All systems they studied had either 10 cationic polyions (Hayashi et al. 2002, 2003), or 20 cationic polyions (Hayashi et al. 2004) while the number of anionic polyions was varied so as to generate a fully cationic system (no anionic polyions) and systems with charge equivalencies (ratio between number of polyanions and number of polycations) of 20, 50, 90 and 100 % (Hayashi et al. 2002, 2003). Under salt-free conditions, aggregate formation was found to be variable with the increased charge equivalency. While formation of positively charged aggregates was favorable at 50 % charge equivalency, increasing charge equivalency up to 100 % led to the formation of only neutral aggregates. In addition, with 100 % charge equivalency, the formed aggregate tended to be small, i.e., the probability of having an aggregate formed with 1 polycation and 1 polyanion chain (1:1 aggregate) was higher than that of larger aggregates such as 2:2 or 3:3. Addition of salt shortened the Debye length of the solution, thereby reducing the electrostatic attraction between oppositely charged polyions. This led to formation of smaller aggregates (Hayashi et al. 2003). The same tendency was observed by decreasing the absolute charge of the polyions (Hayashi et al. 2004). These observations revealed the important interplay of electrostatic and entropic contributions to aggregate formation. Formation of large aggregates was electrostatically favorable due to the attractive forces, while smaller clusters were promoted to gain translational entropy (Hayashi et al. 2002, 2003, 2004). Although these CG MC simulations do not involve nucleic acids or carriers, the observed aggregation behavior should be applicable to nucleic acid delivery systems. They provided useful information on how one may better control the sizes of practical nucleic acid/carrier aggregates for delivery in cellular systems.

4 Molecular Dynamics Simulations on the Cell Membrane Interactions and Intracellular Pathway of Polyplexes and Lipoplexes

For internalization, polyplexes/lipoplexes have to go through the first delivery stage, namely the interaction with cell membranes prior to endocytosis. The first molecular level modeling regarding the attachment of dendrimer-DNA polyplexes to cell membranes was introduced by Voulgarakis et al. (2009). CG MC simulations with a G5 dendrimer-DNA complex showed that increasing the negative membrane charge density (reduced charge density in the range of 0 to 10) and decreasing Debye length of solution had a negative effect in cell membrane attachment. Specifically, when the charge density of membrane was increased, dendrimers were shown to dissociate from DNA (Voulgarakis et al. 2009), indicating the destabilizing effect of increasing membrane charge. Decreasing the Debye length had a similar effect on causing the instability of the complexes. These are important parameters that should be taken into consideration when evaluating the efficiency of cell attachment and delivery.

After attachment to cell membrane, the next step is internalization of polyplexes/lipoplexes, mainly by endocytosis. This is triggered by the formation of endosomes via cell membrane folding. The first CG model on endocytosis of DNA-nanovector complexes was introduced by Ding and Ma (2013). The nanovector used in their study consisted of polymer chains, each with protonable polyelectrolytes and representative surface ligands, while the modeled membrane was composed of oppositely charged lipids and receptors. Pre-complexed structures of dsDNA (modeled as rigid rods) and the nanovector were placed on top of the membrane and endocytosis (driven by ligand-receptor interaction) was investigated by varying length and protonation degree of polyelectrolyte, length and concentration of DNA, and membrane charge. Polymers with 1 polyelectrolyte molecule per chain prevented ligand-receptor interactions resulting in partial endocytosis, while those with 3 polyelectrolytes per chain caused full endocytosis. Variation in DNA length and concentration was also found to affect endocytosis. Furthermore, increasing membrane's negative charge resulted in better endocytosis compared to lower charged membranes (Ding and Ma 2013). This study indicated that for a target membrane, optimization of the delivery system (i.e., types of carriers and cargos) should be determined in order to achieve efficient endocytosis.

Following the internalization of polyplexes/lipoplexes is their intracellular pathway after endosomal escape. Although to our knowledge there has not been any atomic level computational work on this topic, Dinh et al. (2007) studied this phenomenon by developing a stochastic simulation model for 25 kDa PEI-DNA polyplexes, where the transition between distinct states, i.e. polyplexes bound to cell membrane, polyplexes inside endosomes, lysosomes, cytoplasm, and unpacked nucleic acids, were investigated. The intracellular pathways of the polyplexes were divided into two categories, namely transport and reaction events. Transport events were the diffusion related events while the reaction events were related to association or dissociation of polyplexes. Parameterization and validation for their computational model was achieved by experiments with the 25 kDa PEI-DNA polyplexes. In the stochastic simulations, they provided cell configurations, constructed based on different cell geometries obtained from experimental imaging techniques, as computational domains. After assigning the computational domains, the change in either cellular environment or polyplexes was updated with a given time step ($\Delta t = 0.2$ s). Transport events were treated by solving equations of motion, while the reaction events were modeled as first-order Markov processes. Ten thousand trajectories were sampled from 200 cells. The effects of several key properties, such as location and time of the endosomal escape, and cell shape on delivery were addressed. On the effect of endosomal escape location, the authors calculated the probability of a successful DNA delivery to nucleus when a polyplex escaped from endosomes at different distances from the nuclear boundary. The results showed the importance of escape location, indicated by 5 % probability of having a successful delivery when the escape occurred in the supranuclear region compared to 1 % probability in the case of an escape into cytoplasm. Further, the impact of escape time (defined as the length of period from transfection to the time when a polyplex escaped from endosomes and entered cytoplasm) was studied by calculating the probability that a polyplex with different escape time could reach the nucleus within 24 h after the escape. With an increase in escape time in the supranuclear region, the probability of reaching nucleus decreased, most likely due to lysosomal degradation. In the cytoplasmic region, the optimum escape time was found to be 12–13 h after transfection. Cell geometry (circularity and size) was also shown to be critical for delivery; greater delivery efficiency was observed when the cells were elongated and smaller. This observation was attributed to the fact that elongated and smaller cells has relatively larger perinuclear space; therefore the escape location is closer to the nucleus in these cells, leading to more efficient delivery (Dinh et al. 2007). Using a computational approach, this study provided clues for optimal conditions for efficient gene delivery to nucleus, which shed light onto the design of optimal delivery systems.

Finally, Jorge et al. investigated the decompaction of polyanion chains (representing DNA), which were pre-compacted by polycations (representing polycationic carriers), in the presence of heparin-like molecules. Heparin is a negatively charged sulfated anticoagulant which is known for its capability to release nucleic acids from their carrier vectors. Through CG MC simulations, the decompaction of polyanions was shown to be possible with heparin-like molecules, and the extent of decompaction was found to be higher in the presence of both polycations and Fe(III), which might indicate the importance of Fe(III) in facilitating nucleic acid release (Jorge et al. 2012).

5 Future Perspectives

In this chapter, we provided a detailed review on the simulation work performed to-date on the formation of polyplexes and lipoplexes, as well as their intracellular delivery stages. As can be seen from Sect. 3, there has been a great interest towards determining optimal carriers for nucleic acid delivery. Detailed atomic and mesoscopic level data regarding optimal conditions for complexation between various carriers and nucleic acids are beginning to be available in the literature. However, given the low amount of work pursued on the intracellular events in the delivery process, there is an urgent need to fill the gap in theoretical modeling on these crucial stages. All the reviewed works in Sect. 4 involved either mesoscopic or stochastic approaches, and to our knowledge, there has not been any atomic level work due to the significant computational time required for all atom simulations. The lack of atomic level information in mesoscopic approaches may lead inaccuracy in the observed results. With the growing computational power, mesoscopic time scales may be reached by atomistic simulations; thus accurate understanding could be achieved by direct comparison between atomistic trajectories and experimental data.

MD simulations are also providing unique insights into the atomistic structure of polyplexes and lipoplexes, especially with the multifunctional carriers composed of different chemical domains. Our own work has been initially driven by experimental considerations (Hsu and Uludag 2012), but recent theoretical modelling (Sun et al. 2011b, 2012a, b, 2013) is better explaining the outcomes observed in cellular systems. This is critical since it may facilitate the design of next-generation delivery systems based on known experimental and theoretical aspects of current delivery systems. It must be stated that most MD simulations have relied on existing experimental studies for validation and it might be misleading to employ results derived from systems where there is a discrepancy between experimental conditions and theoretical predictions (such as the hydration, salt concentrations, differences in molecular weight of employed species). Conducting simultaneous studies where the experimental and theoretical conditions are harmonized will be beneficial not only for better predictions, but also to improve the theoretical calculations if significant deviations from experimental data are observed.

References

Aliabadi HM, Landry B, Sun C et al (2012) Supramolecular assemblies in functional siRNA delivery: where do we stand? Biomaterials 33:2546–2569

Andersen H (1980) Molecular dynamics simulations at constant pressure and/or temperature. J Chem Phys 72:2384

Andersen HC (1983) Rattle: a "velocity" version of the shake algorithm for molecular dynamics calculations. J Comput Phys 52:24–34

Bagai S, Sun C, Tang T (2013) Potential of mean force of polyethylenimine-mediated DNA attraction. J Phys Chem B 117:49–56

Bandyopadhyay S, Tarek M, Klein ML (1999) Molecular dynamics study of a lipid-DNA complex. J Phys Chem B 103:10075–10080

Berendsen HJC, Postma JPM, van Gunsteren WF et al (1984) Molecular dynamics with coupling to an external bath. J Chem Phys 81:3684

Bloomfield VA (1996) DNA condensation. Curr Opin Struct Biol 6:334–341

Boussif O, Lezoualc'h F, Zanta MA et al (1995) A versatile vector for gene and oligonucleotide transfer into cells in culture and in vivo: polyethylenimine. Proc Natl Acad Sci USA 92:7297–7301

Braun CS, Jas GS, Choosakoonkriang S et al (2003) The structure of DNA within cationic lipid/DNA complexes. Biophys J 84:1114–1123

Brooks BR, Bruccoleri RE, Olafson BD et al (1983) Charmm—a program for macromolecular energy, minimization, and dynamics calculations. J Comput Chem 4:187–217

Buneman O (1967) Time-reversible difference procedures. J Comput Phys 1:517–535

Corsi J, Hawtin RW, Ces O et al (2010) DNA lipoplexes: formation of the inverse hexagonal phase observed by coarse-grained molecular dynamics simulation. Langmuir 26:12119–12125

Dai L, Mu Y, Nordenskiold L et al (2008) Molecular dynamics simulation of multivalent-ion mediated attraction between DNA molecules. Phys Rev Lett 100:118301

Darden T, York D, Pedersen L (1993) Particle mesh Ewald: an N·log(N) method for Ewald sums in large systems. J Chem Phys 98:10089–10092

Dias RS, Pais AACC, Miguel MG et al (2003) Modeling of DNA compaction by polycations. J Chem Phys 119:8150–8157

Ding HM, Ma YQ (2013) Design maps for cellular uptake of gene nanovectors by computer simulation. Biomaterials 34:8401–8407

Dinh AT, Pangarkar C, Theofanous T et al (2007) Understanding intracellular transport processes pertinent to synthetic gene delivery via stochastic simulations and sensitivity analyses. Biophys J 92:831–846

Dominska M, Dykxhoorn DM (2010) Breaking down the barriers: siRNA delivery and endosome escape. J Cell Sci 123(Pt 8):1183–1189

Elbashir SM, Harborth J, Lendeckel W et al (2001) Duplexes of 21-nucleotide RNAs mediate RNA interference in cultured mammalian cells. Nature 411:494–498

Elder RM, Emrick T, Jayaraman A (2011) Understanding the effect of polylysine architecture on DNA binding using molecular dynamics simulations. Biomacromolecules 12:3870–3879

Elouahabi A, Ruysschaert JM (2005) Formation and intracellular trafficking of lipoplexes and polyplexes. Mol Ther 11:336–347

Farago O, Gronbech-Jensen N (2009) Simulation of self-assembly of cationic lipids and DNA into structured complexes. J Am Chem Soc 131:2875–2881

Farago O, Gronbech-Jensen N, Pincus P (2006) Mesoscale computer modeling of lipid-DNA complexes for gene therapy. Phys Rev Lett 96:018102

Fire A, Xu S, Montgomery MK et al (1998) Potent and specific genetic interference by double-stranded RNA in Caenorhabditis elegans. Nature 391:806–811

Fynan EF, Webster RG, Fuller DH et al (1993) DNA vaccines: protective immunizations by parenteral, mucosal, and gene-gun inoculations. Proc Natl Acad Sci USA 90:11478–11482

González MA (2011) Force fields and molecular dynamics simulations. Collect SFN 12:169–200
Gosule LC, Schellman JA (1978) DNA condensation with polyamines I. Spectroscopic studies. J Mol Biol 121:311–326
Goula D, Remy JS, Erbacher P et al (1998) Size, diffusibility and transfection performance of linear PEI/DNA complexes in the mouse central nervous system. Gene Ther 5:712–717
Groenhof G (2013) Introduction to QM/MM simulations. In: Monticelli L, Salonen E (eds) Methods in molecular biology biomolecular simulations: methods and protocols. Springer Science Business Media, New York, pp 43–66
Guo P, Coban O, Snead NM et al (2010) Engineering RNA for targeted siRNA delivery and medical application. Adv Drug Deliv Rev 62:650–666
Hayashi Y, Ullner M, Linse P (2002) A Monte Carlo study of solutions of oppositely charged polyelectrolytes. J Chem Phys 116:6836–6845
Hayashi Y, Ullner M, Linse P (2003) Complex formation in solutions of oppositely charged polyelectrolytes at different polyion compositions and salt content. J Phys Chem B 107:8198–8207
Hayashi Y, Ullner M, Linse P (2004) Oppositely charged polyelectrolytes. Complex formation and effects of chain asymmetry. J Phys Chem B 108:15266–15277
Hillaireau H, Couvreur P (2009) Nanocarriers' entry into the cell: relevance to drug delivery. Cell Mol Life Sci 66:2873–2896
Hockney RW (1970) The potential calculation and some applications. Meth Comput Phys 9:135–211
Hoogerbrugge PJ, Koelman JMVA (1992) Simulating microscopic hydrodynamic phenomena with dissipative particle dynamics. Europhys Lett 19:155
Hoover WG (1985) Canonical dynamics: equilibrium phase-space distributions. Phys Rev A 31(3):1695–1697
Hsu CY, Uludag H (2012) Cellular uptake pathways of lipid-modified cationic polymers in gene delivery to primary cells. Biomaterials 33:7834–7848
Ingólfsson HI, Lopez CA, Uusitalo JJ et al (2013) The power of coarse graining in biomolecular simulations. WIREs Comput Mol Sci
Jensen LB, Mortensen K, Pavan GM et al (2010) Molecular characterization of the interaction between siRNA and PAMAM G7 dendrimers by SAXS, ITC, and molecular dynamics simulations. Biomacromolecules 11:3571–3577
Jensen LB, Pavan GM, Kasimova MR et al (2011) Elucidating the molecular mechanism of PAMAM-siRNA dendriplex self-assembly: effect of dendrimer charge density. Int J Pharm 416:410–418
Jones SP, Pavan GM, Danani A et al (2010) Quantifying the effect of surface ligands on dendron-DNA interactions: insights into multivalency through a combined experimental and theoretical approach. Chemistry 16:4519–4532
Jorge AF, Dias RS, Pais AA (2012) Enhanced condensation and facilitated release of DNA using mixed cationic agents: a combined experimental and Monte Carlo study. Biomacromolecules 13:3151–3161
Journal of Gene Medicine (2013) [cited 2013 May 06]. Available from: www.wiley.co.uk/genmed/clinical
Karatasos K, Posocco P, Laurini E et al (2012) Poly(amidoamine)-based dendrimer/siRNA complexation studied by computer simulations: effects of pH and generation on dendrimer structure and siRNA binding. Macromol Biosci 12:225–240
Kircheis R, Wightman L, Wagner E (2001) Design and gene delivery activity of modified polyethylenimines. Adv Drug Deliv Rev 53:341–358
Korolev N, Lyubartsev AP, Nordenskiold L et al (2001) Spermine: an "invisible" component in the crystals of B-DNA. A grand canonical Monte Carlo and molecular dynamics simulation study. J Mol Biol 308:907–917
Korolev N, Lyubartsev AP, Laaksonen A et al (2002) On the competition between water, sodium ions, and spermine in binding to DNA: a molecular dynamics computer simulation study. Biophys J 82:2860–2875

Korolev N, Lyubartsev AP, Laaksonen A et al (2003) A molecular dynamics simulation study of oriented DNA with polyamine and sodium counterions: diffusion and averaged binding of water and cations. Nucleic Acids Res 31:5971–5981

Korolev N, Lyubartsev AP, Laaksonen A et al (2004a) Molecular dynamics simulation study of oriented polyamine- and Na-DNA: sequence specific interactions and effects on DNA structure. Biopolymers 73:542–555

Korolev N, Lyubartsev AP, Laaksonen A et al (2004b) A molecular dynamics simulation study of polyamine- and sodium-DNA. Interplay between polyamine binding and DNA structure. Eur Biophys J 33:671–682

Kumar S, Rosenberg JM, Bouzida D et al (1992) The weighted histogram analysis method for free-energy calculations on biomolecules. I. The method. J Comput Chem 13:1011–1021

Lamoureux G, Roux B (2003) Modeling induced polarization with classical Drude oscillators: theory and molecular dynamics simulation algorithm. J Chem Phys 119:3025–3039

Leach AR (2001) Molecular modeling principles and applications. Pearson Education Limited, Great Britain

Levitt M, Warshel A (1975) Computer-simulation of protein folding. Nature 253:694–698

Lindahl ER (2008) Molecular dynamic simulations. In: Kukol A (ed) Methods in molecular biology molecular modeling of proteins. Humana Press, pp 3–23

Lorenz C, Hadwiger P, John M et al (2004) Steroid and lipid conjugates of siRNAs to enhance cellular uptake and gene silencing in liver cells. Bioorg Med Chem Lett 14:4975–4977

Lyubartsev A, Tu YQ, Laaksonen A (2009) Hierarchical multiscale modelling scheme from first principles to mesoscale. J Comput Theor Nanosci 6:951–959

Maiti PK, Bagchi B (2006) Structure and dynamics of DNA-dendrimer complexation: role of counterions, water, and base pair sequence. Nano Lett 6:2478–2485

Matranga C, Tomari Y, Shin C et al (2005) Passenger-strand cleavage facilitates assembly of siRNA into Ago2-containing RNAi enzyme complexes. Cell 123:607–620

McCammon JA, Gelin BR, Karplus M (1977) Dynamics of folded proteins. Nature 267:585–590

McNeish IA, Bell SJ, Lemoine NR (2004) Gene therapy progress and prospects: cancer gene therapy using tumour suppressor genes. Gene Ther 11:497–503

Meller J (2001) Molecular dynamics. In: Encyclopedia of life sciences. Nature Publishing Group

Mills M, Orr B, Banaszak Holl MM et al (2010) Microscopic basis for the mesoscopic extensibility of dendrimer-compacted DNA. Biophys J 98:834–842

Mills M, Orr BG, Banaszak Holl MM et al (2013) Attractive hydration forces in DNA-dendrimer interactions on the nanometer scale. J Phys Chem B 117:973–981

Monard G, Merz KM (1999) Combined quantum mechanical/molecular mechanical methodologies applied to biomolecular systems. Acc Chem Res 32:904–911

Monticelli L, Tieleman DP (2013) Force fields for classical molecular dynamics. In: Monticelli L, Salonen E (eds) Methods in molecular biology biomolecular simulations: methods and protocols. Springer Science Business Media, New York, pp 197–213

Moret I, Esteban Peris J, Guillem VM et al (2001) Stability of PEI-DNA and DOTAP-DNA complexes: effect of alkaline pH, heparin and serum. J Control Release 76:169–181

Nandy B, Maiti PK (2011) DNA compaction by a dendrimer. J Phys Chem B 115:217–230

Nosé S (1984) A molecular dynamics method for simulations in the canonical ensemble. Mol Phys 52:255–268

Ouyang D, Zhang H, Herten DP et al (2010a) Structure, dynamics, and energetics of siRNA-cationic vector complexation: a molecular dynamics study. J Phys Chem B 114:9220–9230

Ouyang D, Zhang H, Parekh HS et al (2010b) Structure and dynamics of multiple cationic vectors-siRNA complexation by all-atomic molecular dynamics simulations. J Phys Chem B 114:9231–9237

Ouyang D, Zhang H, Parekh HS et al (2011) The effect of pH on PAMAM dendrimer-siRNA complexation: endosomal considerations as determined by molecular dynamics simulation. Biophys Chem 158:126–133

Pack DW, Hoffman AS, Pun S et al (2005) Design and development of polymers for gene delivery. Nat Rev Drug Discov 4:581–593

Patria RK, Beale PD (2011) Statistical mechanics. Elsevier, United States

Pavan GM, Danani A, Pricl S et al (2009) Modeling the multivalent recognition between dendritic molecules and DNA: understanding how ligand "sacrifice" and screening can enhance binding. J Am Chem Soc 131:9686–9694

Pavan GM, Albertazzi L, Danani A (2010a) Ability to adapt: different generations of PAMAM dendrimers show different behaviors in binding siRNA. J Phys Chem B 114:2667–2675

Pavan GM, Kostiainen MA, Danani A (2010b) Computational approach for understanding the interactions of UV-degradable dendrons with DNA and siRNA. J Phys Chem B 114:5686–5693

Pavan GM, Mintzer MA, Simanek EE et al (2010c) Computational insights into the interactions between DNA and siRNA with "rigid" and "flexible" triazine dendrimers. Biomacromolecules 11:721–730

Pavan GM, Posocco P, Tagliabue A et al (2010d) PAMAM dendrimers for siRNA delivery: computational and experimental insights. Chemistry 16:7781–7795

Pearlman DA, Case DA, Caldwell JW et al (1995) Amber, a package of computer-programs for applying molecular mechanics, normal-mode analysis, molecular-dynamics and free energy calculations to simulate the structural and energetic properties of molecules. Comput Phys Commun 91:1–41

Pegg AE, McCann PP (1982) Polyamine metabolism and function. Am J Physiol 243:C212–C221

Phillips JC, Braun R, Wang W et al (2005) Scalable molecular dynamics with NAMD. J Comput Chem 26:1781–1802

Posocco P, Pricl S, Jones S et al (2010) Less is more—multiscale modelling of self-assembling multivalency and its impact on DNA binding and gene delivery. Chem Sci 1:393–404

Potter H (1988) Electroporation in biology: methods, applications, and instrumentation. Anal Biochem 174:361–373

Rappe AK, Goddard IWA (1991) Charge equilibration for molecular dynamics simulations. J Phys Chem 95:3358–3363

Razin S, Rozansky R (1959) Mechanism of the antibacterial action of spermine. Arch Biochem Biophys 81:36–54

Roth JA, Nguyen D, Lawrence DD et al (1996) Retrovirus-mediated wild-type p53 gene transfer to tumors of patients with lung cancer. Nat Med 2:985–991

Ryckaert JP, Ciccotti G, Berendsen HJC (1977) Numerical integration of the cartesian equations of motion of a system with constraints: molecular dynamics of n-alkanes. J Comput Phys 23:327–341

Sagui C, Darden TA (1999) Molecular dynamics simulations of biomolecules: long-range electrostatic effects. Annu Rev Biophys Biomol Struct 28:155–179

Salinas SRA (2001) Introduction to statistical physics. Springer, New York

Saunders MG, Voth GA (2013) Coarse-graining methods for computational biology. Annu Rev Biophys 42:73–93

Savelyev A, Papoian GA (2007) Inter-DNA electrostatics from explicit solvent molecular dynamics simulations. J Am Chem Soc 129:6060–6061

Schlick T (2010) Molecular modeling and simulation: an interdisciplinary guide. Springer, New York

Schneider T, Stoll E (1978) Molecular-dynamics study of a three-dimensional one-component model for distortive phase transitions. Phys Rev B 17:1302–1322

Schofield P (1973) Computer simulation studies of the liquid state. Comput Phys Commun 5:17–23

Scott WRP, Hunenberger PH, Tironi IG et al (1999) The GROMOS biomolecular simulation program package. J Phys Chem A 103:3596–3607

Senn HM, Thiel W (2007) QM/MM methods for biological systems. Atomistic approaches in modern biology: from quantum chemistry to molecular simulations 268: 173–290

Stevens MJ (2001) Simple simulations of DNA condensation. Biophys J 80:130–139

Sun C, Tang T, Uludag H (2011a) Molecular dynamics simulations of PEI mediated DNA aggregation. Biomacromolecules 12:3698–3707

Sun C, Tang T, Uludag H et al (2011b) Molecular dynamics simulations of DNA/PEI complexes: effect of PEI branching and protonation state. Biophys J 100:2754–2763

Sun C, Tang T, Uludag H (2012a) Molecular dynamics simulations for complexation of DNA with 2 kDa PEI reveal profound effect of PEI architecture on complexation. J Phys Chem B 116:2405–2413

Sun C, Tang T, Uludag H (2012b) Probing the effects of lipid substitution on polycation mediated DNA aggregation: a molecular dynamics simulations study. Biomacromolecules 13:2982–2988

Sun C, Tang T, Uludag H (2013) A molecular dynamics simulation study on the effect of lipid substitution on polyethylenimine mediated siRNA complexation. Biomaterials 34:2822–2833

Swope WC, Andersen HC, Berens PH et al (1982) A computer simulation method for the calculation of equilibrium constants for the formation of physical clusters of molecules: application to small water clusters. J Chem Phys 76:637

Szasz D (1996) Boltzmann's ergodic hypothesis, a conjecture for centuries? Studia Scientiarum Mathematicarum Hungaria 31:299–322

Takada S (2012) Coarse-grained molecular simulations of large biomolecules. Curr Opin Struct Biol 22:130–137

Torrie GM, Valleau JP (1977) Nonphysical sampling distributions in Monte Carlo free-energy estimation: umbrella sampling. J Comput Phys 23:187–199

Tuckerman ME (2010) Statistical mechanics: theory and molecular simulation. Oxford University Press, New York

Vasumathi V, Maiti PK (2010) Complexation of siRNA with dendrimer: a molecular modeling approach. Macromolecules 43:8264–8274

Verlet L (1967) Computer "Experiments" on classical fluids. I. Thermodynamical properties of Lennard-Jones molecules. Phys Rev 159:98–103

Voth GA (2009) Coarse-graining of condensed phase and biomolecular systems. CRC Press/ Taylor and Francis Group, Boca Raton

Voulgarakis NK, Rasmussen KO, Welch PM (2009) Dendrimers as synthetic gene vectors: cell membrane attachment. J Chem Phys 130:155101

Warshel A, Levitt M (1976) Theoretical studies of enzymic reactions: dielectric, electrostatic and steric stabilization of the carbonium ion in the reaction of lysozyme. J Mol Biol 103:227–249

Warshel A, Sharma P, Kato M et al (2006) Electrostatic basis for enzyme catalysis. Chem Rev 106:3210–3235

Wereszczynski J, McCammon JA (2012) Statistical mechanics and molecular dynamics in evaluating thermodynamic properties of biomolecular recognition. Q Rev Biophys 45:1–25

Yoon CS, Park JH (2010) Ultrasound-mediated gene delivery. Expert Opin Drug Deliv 7:321–330

Zhang XX, McIntosh TJ, Grinstaff MW (2012) Functional lipids and lipoplexes for improved gene delivery. Biochimie 94:42–58

Zheng M, Pavan GM, Neeb M et al (2012) Targeting the blind spot of polycationic nanocarrier-based siRNA delivery. ACS Nano 6:9447–9454

Ziebarth J, Wang Y (2009) Molecular dynamics simulations of DNA-polycation complex formation. Biophys J 97:1971–1983

Ziebarth J, Wang Y (2010) Coarse-grained molecular dynamics simulations of DNA condensation by block copolymer and formation of core-corona structures. J Phys Chem B 114:6225–6232

Computational Studies of Highly PEG-ylated Sterically Stabilized Micelles: Self-Assembly and Drug Solubilization

Petr Král and Lela Vuković

Abstract Self-assembled micelles of block copolymers, containing controllable physical, chemical, and biological properties, are strong candidates for new drug delivery platforms. Here, we summarize our studies of structure, dynamics and drug solubilization in micelles self-assembled from highly PEG-ylated block copolymers. First, we examined sterically stabilized micelles (SSM) formed by self-assembled phospholipids (DSPE–PEG_{2000}) in pure water and isotonic HEPES-buffered saline solution. The observed micelle sizes of 2–15 nm were shown to largely depend on the solvent and the lipid concentration used. Computational modeling showed that micelle sizes are determined by the interactions of their charged $-PO_4^-$ groups with the present counterions. Second, we studied solubilization of prototypical therapeutic molecules, a drug bexarotene and a vasoactive intestinal peptide (VIP), in SSM, as observed in experiments. Free energy calculations revealed that molecules of bexarotene can reside at the micellar ionic interface of the PEG corona or in the alkane core center, where several bexarotene molecules can cluster and self-stabilize. Charged molecules, such as VIP, can be stabilized at the SSM ionic interface by Coulombic coupling between their positively charged residues and the $-PO_4^-$ groups of the lipids. The performed studies illustrate that atomistic simulations can reveal drug solubilization character in nanocarriers and be used in efficient optimization of novel nanomedicines.

Keywords PEG · PEG-ylated phospholipids · Micelles · Micellar nanocarriers · Drug solubilization · Bexarotene · Therapeutic peptides

---

P. Král (✉) · L. Vuković
Department of Chemistry, University of Illinois at Chicago, 845 W, Taylor Street, Chicago, IL 60607, USA
e-mail: pkral@uic.edu

L. Vuković
e-mail: lvukov1@illinois.edu

A. Prokop et al. (eds.), *Intracellular Delivery II*,
Fundamental Biomedical Technologies 7, DOI: 10.1007/978-94-017-8896-0_16,

Abbreviations

PEG	Poly(ethylene) glycol
SSM	Sterically stabilized micelles
DSPE–PEG2000	1,2-distearoyl-sn-glycero-3-phosphatidylethanolamine-N-[methoxy(polyethylene glycol) 2000]
HEPES	Hydroxyethyl piperazineethanesulfonic acid
VIP	Vasoactive intestinal peptide
MD	Molecular dynamics

1 Introduction

Lipids, surfactants, and amphiphilic block copolymers in aqueous media can assemble into numerous stable structures, such as micelles (Belsito et al. 2000; Hristova and Needham 1995; Patra and Král 2011; Zhang et al. 1996), disks (Johnsson and Edwards 2003), vesicles (Jones 1995; Kaler et al. 1989; Luo and Eisenberg 2001), and bilayers (Nagle and Tristram-Nagle 2000; van Meer et al. 2008; Titov et al. 2009), many of which can have important biomedical applications. Poly(ethylene glycol) (PEG) lipid conjugates, assembled alone or with other copolymers and cargo molecules, have broad applicability in biomedical formulations.

In this chapter, we review our recent computational and experimental characterization of the sterically stabilized micelles (SSM) self-assembled from copolymers containing long poly(ethylene glycol) (PEG) alkane conjugates. The SSM are assembled from a widely used PEG-ylated phospholipid, DSPE–PEG_{2000}, 1,2-distearoyl-sn-glycero-3-phosphatidylethanolamine-N-[methoxy(polyethylene glycol)$_{2000}$]. The use of DSPE–PEG_{2000} in drug delivery formulations has been wide-spread due to its many favorable properties, such as low toxicity, biocompatibility, and ease of excretion (Zalipsky 1995).

2 Structure and Dynamics of Highly PEG-ylated Micelles

Experimentally Observed Micelle Sizes. DSPE–PEG_{2000} solvated in water or ionic solution can aggregate into micelles, which can be used as drug delivery vehicles. The experimental results shown in Fig. 1 [from (Vuković et al. 2011)] show the observed intensity weighted size distributions and average hydrodynamic diameters of the experimentally prepared DSPE–PEG_{2000} micellar assemblies equilibrated in water (a–e) and ionic solution (f–j). As the monomer concentration increases from c = 5 to 40 mM, the hydrodynamic diameters of SSM in water

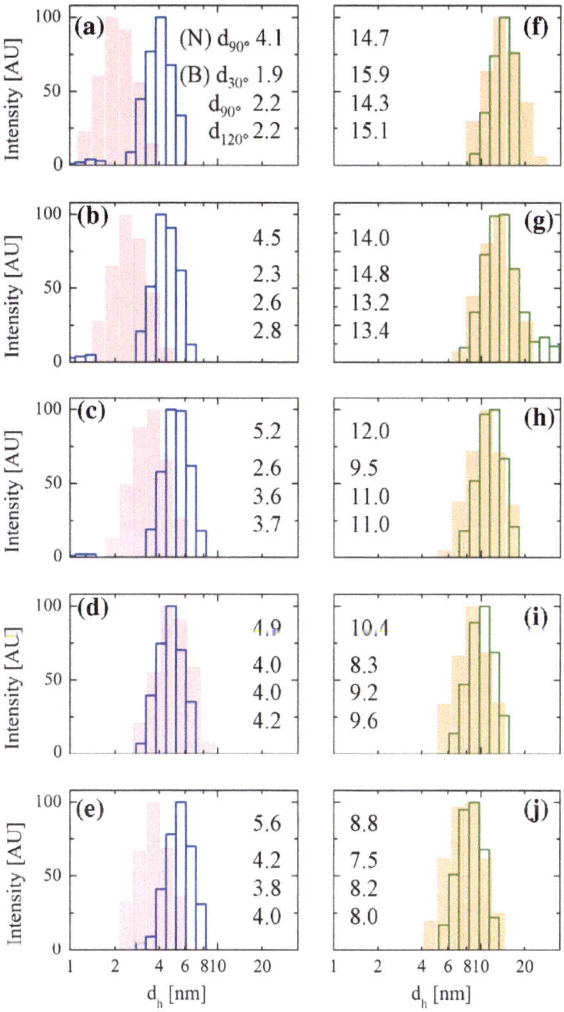

Fig. 1 Experimental intensity weighted size distributions of SSM (*left*) in pure water at concentrations: **a** 5 mM, **b** 10 mM, **c** 20 mM, **d** 30 mM, **e** 40 mM and (*right*) in HEPES buffered saline at concentrations: **f** 5 mM, **g** 10 mM, **h** 20 mM, **i** 30 mM, **j** 40 mM. *Line histograms* show data obtained by the NICOMP instrument at the 90° angle, and *shadow histograms* show data obtained by the Brookhaven instrument at the 90° angle. Average hydrodynamic diameters, d_h, obtained by the NICOMP instrument (*N*) at the 90° angle, and by the Brookhaven instrument (*B*) at 30°, 90° and 120° angles, are shown in each plot. Reprinted (adapted) with permission from Vuković L, et al. (2011), Copyright (2011) American Chemical Society

slowly increase from d_h ~4.1 to 5.6 nm (NICOMP instrument), or from d_h ~2.2 to 4.0 nm (Brookhaven instrument). This size dependence may be caused by the fact that the effective ionic strength of the solutions (micelle screening) increases with the monomer concentration, as the monomer counterions become crowded in the limited space between the micelles).

In ionic solutions, the observed micelles always have narrow size distributions, whose peaks shift from the diameter of d_h ~15 nm at c = 5 mM to d_h = 8 nm at c = 40 mM (Fig. 1f–j). The results in Fig. 1 indicate that the presence of ions is of great importance for the micelle sizes.

MD Simulations of Equilibrated Micelles. In order to better understand the experimentally observed results, we simulated individual equilibrated molecular aggregates formed by a different number of DSPE–PEG_{2000} monomers in water and ionic solution. Figure 2a–e shows micelles formed by 8, 10, 15, 20, and 50 DSPE–PEG_{2000} monomers equilibrated in water for 5−16 ns. Note that the size of the 10-monomer SSM in water is in rough agreement with the SSM sizes observed in water by both DLS instruments at the monomer concentrations of $c > 30$ mM (Vuković et al. 2011). On the other hand, a 90-monomer SSM equilibrated in 0.166 M NaCl solution, shown in Fig. 2f, is a model of the SSM formed in HEPES buffered saline. The 90-monomer estimate used in the preparation of the micelle shown in Fig. 2f is based on the small angle neutron scattering measurements of SSM in HEPES buffered saline at monomer concentration of $c = 5$ mM (Arleth et al. 2005; Ashok et al. 2004).

Structure and Dynamics of the Micelle Core. In all the equilibrated micelles, three unique regions can be recognized: core, ionic interface, and PEG corona. The micelle cores in Fig. 2a–f are formed by aggregated alkane blocks, shown in Fig. 2g (same order). At smaller aggregation numbers, the micelles reorganize in sub-nanosecond timescales, and their cores gain a spherical shape. Larger micelles ($N_{agg} > 20$) reorganize in several nanoseconds and their cores become more ellipsoidal/oblate. In the case of the 90-monomer micelle formed in ionic solution, the observed oblate shape is in agreement with the previous measurements of SSM with the aggregation number of $N_{agg} \sim 93$, obtained by small-angle neutron scattering (Arleth et al. 2005).

Structure and Dynamics of the Ionic Interface. The cores of all the SSM are covered by the charged phosphate groups (–PO_4^-) from the DSPE–PEG_{2000} monomers, which are screened by the Na^+ counterions freely present in our (neutral) systems. Screening of the ionic interface is of great importance for the micelle stabilization. In Fig. 3, we show the molarity of the ionic species present in the 90-monomer SSM in 0.166 M NaCl solution and the 10-monomer SSM in water (inset) as a function of the radial distance from the micelle center. In the 10-monomer SSM, the distribution of the –PO_4^- groups has a single peak localized at $r \sim 1.7$–1.8 nm (spherical shape of the core). In the 90-monomer SSM, the core has an oblate shape, which results in two peaks of the –PO_4^- groups, localized at $r \sim 2.2$ and $r \sim 3.1$ nm. In both cases, the –PO_4^- groups are largely screened by the Na^+ counterions in the Stern layer, located in the region of the PO_4^- groups. The following diffuse ionic atmosphere (Israelachvili 1992) maintains the neutrality of the whole molecular complex.

Structure and Dynamics of the Micelle Corona. The local density and conformations of PEG chains forming the SSM corona also depend on the aggregation number, N_{agg}. As seen in Fig. 2a–f, the PEG chains in all the studied SSM can transiently form clumps or remain isolated. We analyze the chain dynamics in more detail for the 10-monomer in water and for the 90-monomer SSM in 0.166 M NaCl solution, shown in Fig. 2b and f, respectively.

In Fig. 4, we show equilibrium fluctuations of the local PEG thickness in the micellar PEG corona, d_{PEG}, as a function of the inclination angle θ and the

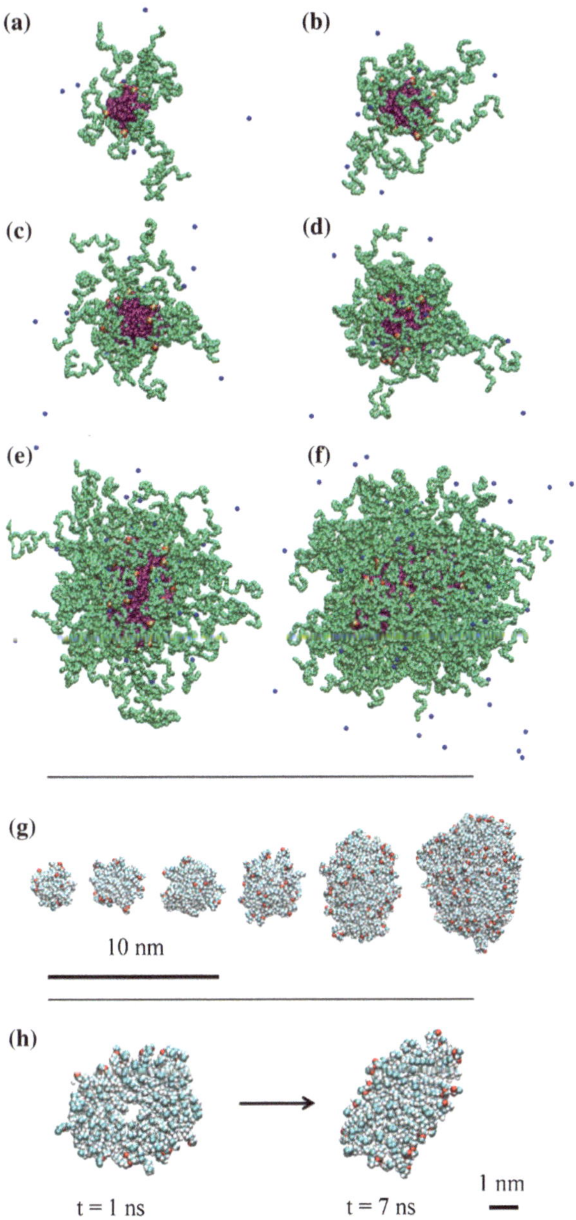

Fig. 2 Snapshots of equilibrated micelles (T = 300 K) containing the following number of monomers N_{agg}: **a** 8, **b** 10, **c** 15, **d** 20, **e** 50—water, and **f** 90—0.166 M NaCl solution. **g** Equilibrated alkyl cores of the micelles in (**a–f**) are shown in the same order. Water molecules are not shown for clarity. Images (**a–g**) are shown at the same scale [scale bar shown in (**g**)]. **h** Relaxation of the 90-monomer SSM core in 0.166 M NaCl solution, shown on a 1 nm thick cross-section. Reprinted (adapted) with permission from (Vuković L, et al. (2011), Copyright (2011) American Chemical Society

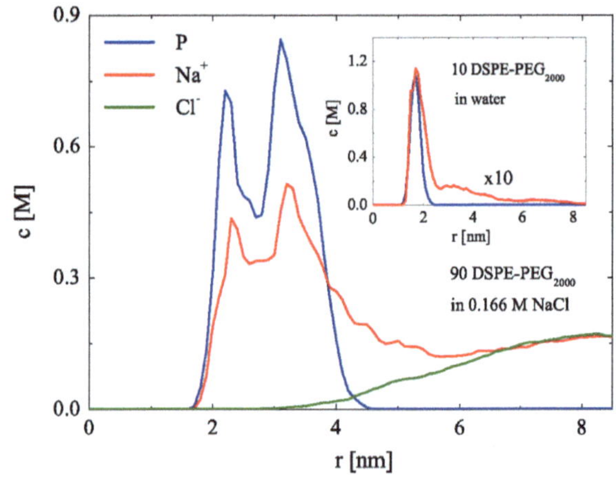

Fig. 3 Molarity of P atom ($-PO_4^-$ group on DSPE–PEG_{2000} monomers), Na^+ and Cl^- as a function of the radial coordinate, r, with respect to the SSM center of mass for 90-monomer SSM in 0.166 M NaCl solution. (inset) Molarity of P and Na^+ for 10-monomer micelle in water. Reprinted (adapted) with permission from (Vuković L, et al. (2011), Copyright (2011) American Chemical Society

azimuthal angle Φ (spherical coordinates with the origin at the SSM center of mass). In Fig. 4a–h, we observe that in the 10-monomer SSM a large fraction ($\sim$30 %) of the core is always fully exposed to water, in contrast to the 90-monomer SSM that only occasionally exposes the core to the ionic solution. In the four snapshots of 10-monomer SSM (Fig. 4), most of the PEG chains fluctuate but remain folded at the core, while one or two chains can occasionally protrude away from the core ($d_{PEG} > 5$ nm). On the other hand, the 90-monomer SSM in 0.166 M NaCl has a more homogeneous PEG corona.

In Fig. 5, we show the average density distributions, ρ(r), of the hydrophobic core groups, the PEG corona groups, and water for (top) the 10-monomer SSM in water and (bottom) the 90-monomer SSM in 0.166 M NaCl solution. In the 10-monomer SSM, shown in Fig. 5 (top), the hydrophobic core has a relatively sharp boundary at $r = 1.7-1.8$ nm. It is followed by the $-PO_4^-$ groups and a narrow PEG layer of the thickness of PEG $\sim$ 0.5 nm. The individual PEG chains are highly coiled, which leads to "mushroom-like" PEG conformations, observed on flat surfaces (membranes) at low PEG densities (Lee et al. 2008). The 90-monomer SSM, shown in Fig. 5 (bottom), has a rather different structure of its layers, due to the oblate core terminated at $r = 2.5-4$ nm. The onset of the PEG layer is broader than in the 10-monomer SSM, due to the core ellipticity. The congested PEG chains have more "brush-like" conformations, as seen on flat surfaces in the limit of high surface coverage (Lee et al. 2008). The average thickness of the PEG corona, $d_{PEG} \sim 3.8$ nm, is in good agreement with a value of $d_{PEG} = 3.5$ nm (Johnsson and Edwards 2003), found in DSPE–PEG_{2000} micelles in the 0.15 M NaCl solution. Water fills the space between the PEG chains for both 10-monomer and 90-monomer SSM.

Comparison of the Experimental and Theoretical Micelle Sizes. The sizes of the micelles modeled in MD simulations can be compared to the experimentally obtained sizes (Fig. 1). Comparison of theoretical effective sizes, d_{tot}, with the experimental hydrodynamic sizes, d_h, shown in Fig. 1, indicates that micelles

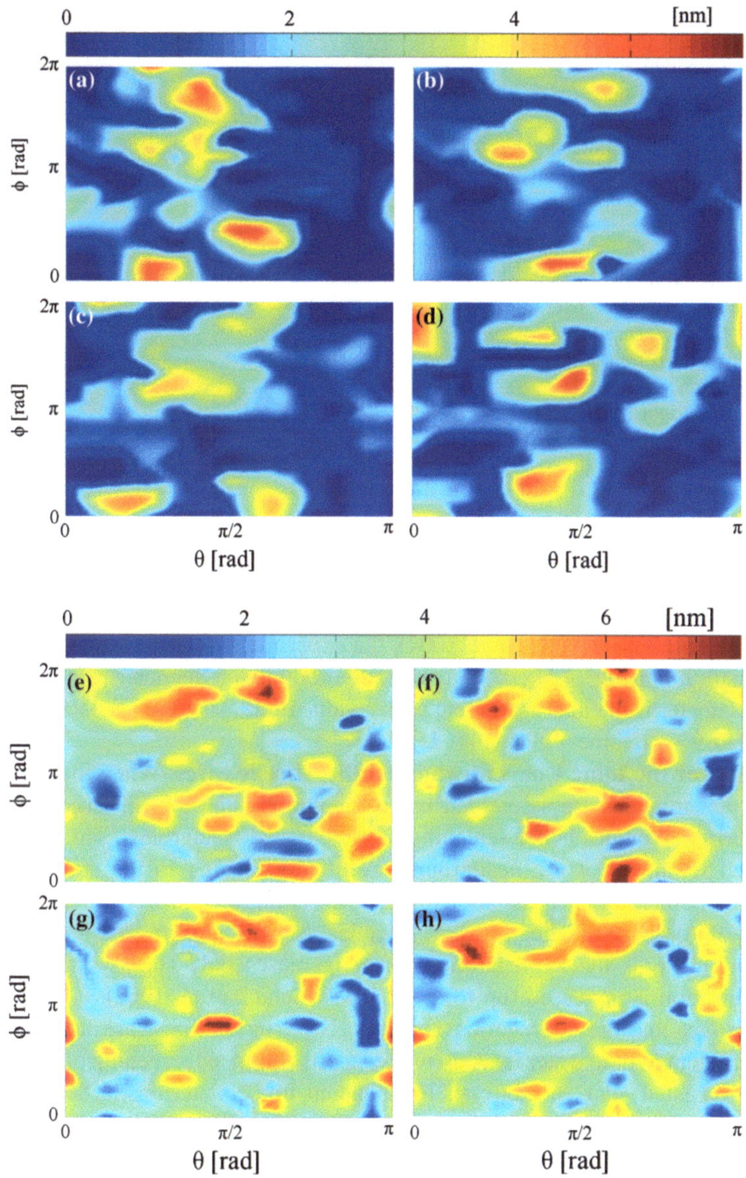

Fig. 4 Local thickness of the PEG corona, d_{PEG}, for SSM with $N_{agg} = 10$ in water, at **a** 3 ns, **b** 6 ns, **c** 10 ns, **d** 16 ns, and for SSM with $N_{agg} = 90$ in 0.166 M NaCl solution, at **e** 7 ns, **f** 8 ns, **g** 9 ns, and **h** 10 ns, as a function of the inclination angle, θ, and the azimuthal angle, Φ. Reprinted (adapted) with permission from (Vuković L, et al. (2011), Copyright (2011) American Chemical Society

observed in pure water with diameters of $d_h = 4-5$ nm should contain $N_{agg} < 8$ monomers. On the other hand, in the buffered solutions the experimental SSM sizes are always having $d_h > 8$ nm, as seen in Fig. 1 (bottom), indicating

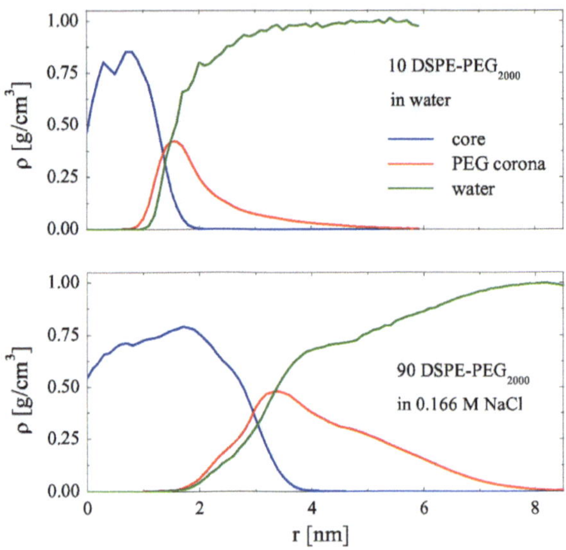

Fig. 5 Density distribution as a function of the radial coordinate, r, with respect to the SSM center of mass for 10-monomer SSM in pure water (*top*) and 90-monomer SSM in 0.166 M NaCl solution (*bottom*). Core and PEGPEG distributions are averaged over last 4 ns of the simulations, while the water distribution is averaged over last 0.1–0.2 ns. Reprinted (adapted) with permission from (Vuković L, et al. (2011), Copyright (2011) American Chemical Society

$N_{agg} > 20$. At the DSPE–PEG$_{2000}$ monomer concentration of $c = 5$ mM in buffer (Arleth et al. 2005; Ashok et al. 2004), the experimentally observed SSM have an aggregation number of $N_{agg} \sim 90$ and a hydrodynamic diameter of $d_h \sim 15$ nm. This diameter is in close agreement with a value of $d_{tot} = 13.9$ nm obtained for the modeled 90-monomer SSM (Fig. 2f).

Effect of the Ionic Concentration on the Micelle Sizes. When ionic lipid and surfactant micelles are assembled in ionic solutions, the overall large number of counterions provides better screening of the charged headgroups. This results in a reduced repulsion of the headgroups, eventually leading to the stabilization of more monomers in each assembled structure (Almgren and Lofroth 1981; Mazer et al. 1976). On the other hand, in micelles assembled from non-ionic (neutral) monoalkyl-PEGs the aggregation number increases only slightly when the salt concentration is increased (up to $c = 1.3$ M) (Schick et al. 1962).

The fact that the aggregation numbers, N_{agg}, in our SSM are larger in ionic solution than in pure water is likely related to the enhanced screening of the charged and aggregated –PO_4^- headgroups by abundant free counterions in the ionic solutions (Almgren and Lofroth 1981; Mazer et al. 1976). This possibility was clarified by theoretical examination on how 10-monomer PEG-ylated micelles in water and 0.166 M NaCl solution are screened by the ion distributions formed around the ionic –PO_4^- groups (Vuković et al. 2011). The analyses of P–Na^+, P–Cl^-, and P–P radial distribution functions (Vuković et al. 2011), obtained from MD simulations, confirm the hypothesis that the presence of electrolytes provides better stabilization of the PEG-ylated SSM by decreasing the Coulombic repulsion of the –PO_4^- groups and allowing more monomers to be accommodated in the SSM. Our MD simulations also showed that PEG can further stabilize the SSM by forming stable counterion configurations close to the –PO_4^- groups (Vuković et al. 2011).

3 Drug and Peptide Stabilization in Highly PEG-ylated Micelles

While a lot is understood about pharmacological properties and physiological effects of lipid-based drug delivery systems, much less is understood about the microscopic mechanisms of stabilization of therapeutic agents within them (Rane and Anderson 2008). Most predictive strategies to determine drug solubility in lipid environments do not take into account the microscopic nature of the carriers (Anderson et al. 1980; Avdeef and Testa 2002; Fernandes De Oliveira et al. 2005; Hawker 1995; Huynh et al. 2008; Lipinski et al. 2001; Patel et al. 2008; Patton et al. 1984; Rane and Anderson 2008; Totrov 2004). For example, water/octanol partition coefficients of drugs are widely used to estimate drug solubility. Here, we use molecular dynamics simulations, an approach that can provide atomistic detail into drug solubility inside the nanocarriers. SSM nanocarrier can solubilize hydrophobic drugs, amphiphilic drugs, and peptides (Krishnadas et al. 2003; Koo et al. 2005; Onyuksel et al. 1999, 2009). Next, we summarize molecular dynamics studies of stabilization modes for bexarotene drug and the human vasoactive intestinal peptide (VIP) and the discovery of their preferred locations inside SSM.

Gibbs Energy Profiles of Bexarotene in SSM. Experiments have determined that 11 bexarotene molecules are encapsulated per SSM (Vuković et al. 2013). In order to examine where 11 bexarotene drugs could reside within the SSM, we have calculated solvation free energy profiles, ΔG(d), of bexarotene inside SSM.

In Fig. 6 we plot ΔG(d) profile for bexarotene in 10-monomer SSM in water. Here, the bexarotene has a minimum in ΔG(d) at the ionic interface, and it has a moderate barrier for transfer into the aqueous phase, $\Delta\Delta G = 5$ kcal/mol with respect to the interface. In Fig. 6, we also show ΔG(d), of a single bexarotene drug along the minor axis of 90 monomer SSM in 0.17 M NaCl solution. This ΔG(d) profile is relatively flat, with several shallow local minima. The lowest minimum is again observed at the ionic interface, at $d = 2.1$ nm. The free energy profiles of bexarotene are in rough agreement in all the studied systems; the global minima in ΔG(d) are at interfaces of alkane and aqueous regions. In initial stages of equilibration, the empty 90-monomer nanocarrier had a vacancy in the core, ~ 1 nm in diameter. During the course of the simulation, the core eventually acquired an oblate shape, as the vacancy became splashed (Vuković et al. 2011). In order to examine if this vacancy could potentially host several drugs, we have prepared systems with 3 and 5 bexarotene molecules relaxed in the vacancy within the SSM core. During the relaxation of these systems, the drug molecules oriented their –COOH groups inwards and formed a cluster that is stabilized by a hydrogen bond network.

In Fig. 7a, we show the arrangement of 5 drug molecules within the alkyl core, and the hydrogen bond network which they form. In order to check if bexarotene prefers the core over the ionic interface, when there is a cluster of drugs within the core, we obtain ΔG(d) for one of the bexarotene molecules as a function of its distance from the SSM center (drug cluster). In Fig. 6, we show that when there are multiple bexarotene molecules present in the core, the drug develops a minimum in

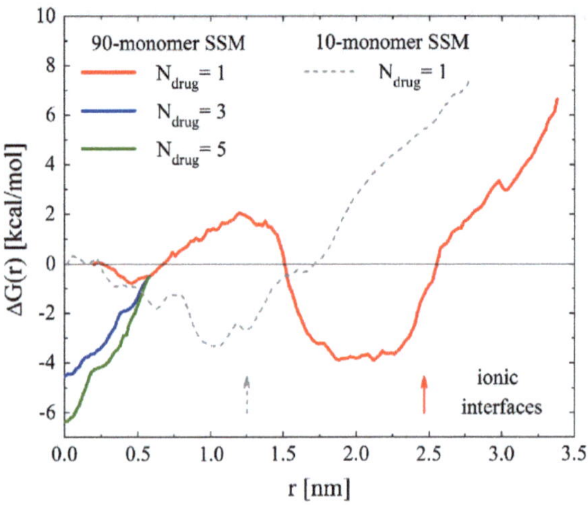

Fig. 6 Free energy profiles of bexarotene in SSM-10 (water) and SSM-90 (0.16 M NaCl). Whole solvated micelles loaded with drugs were present in the free energy calculations; different numbers of drugs are considered in the SSM-90 core. The *vertical arrows* show the positions of ionic interfaces in the two SSMs. Reprinted (adapted) with permission from (Vuković L, et al. (2013), Copyright (2013) American Chemical Society

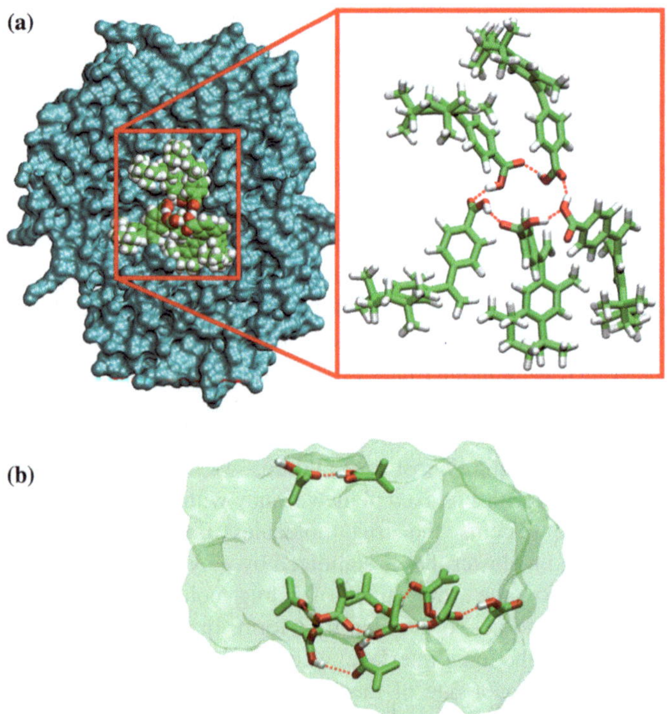

Fig. 7 **a** A snapshot of the cluster of 5 bexarotene molecules formed inside the SSM core. A hydrogen bond network between –COOH groups is highlighted. **b** The cluster of 11 bexarotene drugs (transparent background) forms 2 hydrogen bond networks (highlighted in licorice representation for clarity) in the SSM core after 15 ns of equilibration. Reprinted (adapted) with permission from (Vuković L, et al. (2013), Copyright (2013) American Chemical Society

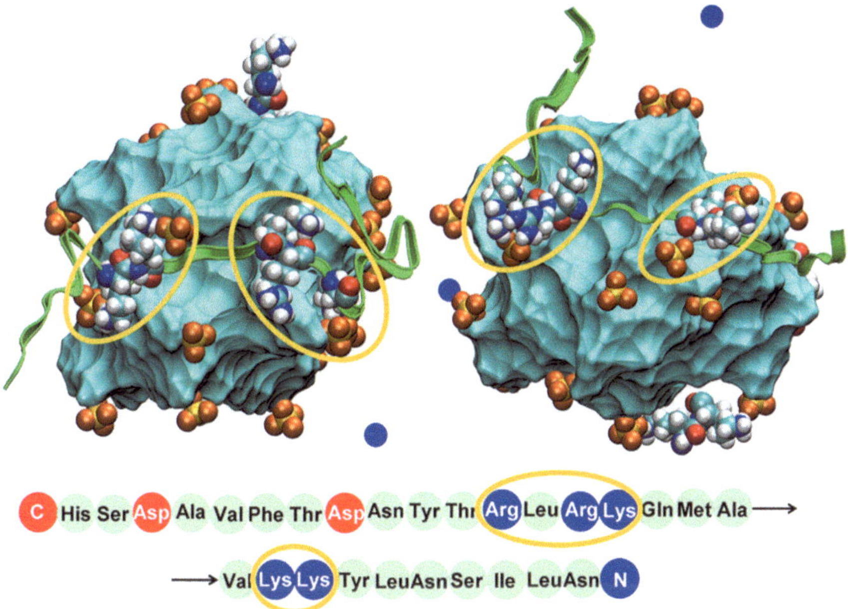

Fig. 8 Snapshots from MD simulation of a 20-monomer SSM loaded with 2 VIPs. The alkyl core is shown as a cyan surface, on which $-PO_4^-$ groups coordinate two positive regions on each of the VIP peptides. VIP sequence is shown at the bottom. Reprinted (adapted) with permission from (Vuković L, et al. (2013), Copyright (2013) American Chemical Society

the core center, with either the same ΔG(d) or lower by ΔΔG(d) = −1.5 kcal/mol with respect to the ionic interface, when there are 3 and 5 drugs present in the drug cluster. The development of this free energy minimum in the core with increasing number of drugs most explains the large loading capacity of SSM, which was experimentally determined as $N_{bex} = 11$.

To examine the hypothesis that all 11 drug molecules are located in the core of the 90-monomer SSM nanocarrier, we performed further MD calculations and equilibrated a cluster of 11 drugs in the alkyl core. In Fig. 7b, we show the shape of the equilibrated drug cluster. We observe that 9 drugs form one network, while the remaining 2 drugs associate with each other. This indicates potential crowding of the drugs in the SSM core, since a cluster of 11 drug molecules seems to be too large to form a single hydrogen bond network. These observations show that drugs can potentially be located both in the core and at the ionic interface, the regions where the drug has minima in free energy profiles.

Stabilization of VIP in SSM. Next, we examined the stabilization of a therapeutic VIP peptide in SSM. VIP amino acid sequence, listed in Fig. 8 (bottom), shows two clusters of positively charged residues (3 Lys and 2 Arg) in the central region of VIP, two negative residues closer to the carboxy terminus, and neutral residues elsewhere. To study the conformation of multiple VIP molecules loaded on a single SSM, we equilibrate 2 VIP molecules complexed with a 20-monomer

SSM. The composition of VIP-SSM system in water was chosen according to the experimental findings (phosphate buffer) (Vuković et al. 2013). The peptides are initially placed on opposite sides of SSM within 0.7 nm of the SSM core, and the system is equilibrated for over 30 ns. Over the time, we observe that VIP molecules become closely coordinated by the $-PO_4^-$ groups, as shown for both VIPs in Fig. 8. During the simulation course, $-PO_4^-$ groups first migrate towards 2 VIPs, and then distribute more homogeneously on the alkane core surface, with several of $-PO_4^-$ groups staying close to the positive VIP residues. The quantitative analysis of VIP interactions with $-PO_4^-$ groups show that among the 5 positive residues on each VIP, there are at least 3 of them which are almost fully coordinated by counterions (~ 1 $-PO_4^-$ group within $r = 0.7$ nm of the residue center of mass) (Vuković et al. 2013). The obtained results indicate a large role of Coulombic interactions in stabilization of cationic peptides in SSM.

4 Conclusion

In this chapter, we presented the results of the experimental and theoretical characterization of DSPE–PEG_{2000} assemblies in pure water and HEPES buffered saline. We also examined the factors which can affect drug solubility in PEG-ylated phospholipid nanocarriers. Our simulations of the SSM revealed that the inflatable SSM core, complex ionic interface and highly fluctuating corona form suitable nesting sites for the drugs and other carried molecules. The overall picture that emerges from quantitative free energy calculations is that amphiphilic drugs can be solvated in alkane core and at the ionic interface. Our analysis of therapeutic peptide VIP also clearly shows that the PEG corona can also be effective in solubilizing therapeutic agents. The observed behavior of highly PEG-ylated micelles is of crucial importance for the design of new nanomedicines and other nanoconstructs with versatile applications.

References

Almgren M, Lofroth J-E (1981) Determination of micelle aggregation numbers and micelle fluidities from time-resolved fluorescence quenching studies. J Colloid Interf Sci 81:486–499

Anderson BD, Rytting JH, Higuchi T (1980) Solubility of polar organic solutes in nonaqueous systems: role of specific interactions. J Pharm Sci 69:676–680

Arleth L, Ashok B, Onyuksel H, Thiyagarajan P, Jacob J, Hjelm RP (2005) Detailed structure of hairy mixed micelles formed by phosphatidylcholine and PEGylated phospholipids in aqueous media. Langmuir 21:3279–3290

Ashok B, Arleth L, Hjelm RP, Rubinstein I, Onyuksel H (2004) In vitro characterization of PEGylated phospholipid micelles for improved drug solubilization: effects of PEG chain length and PC incorporation. J Pharm Sci 93:2476–2487

Avdeef A, Testa B (2002) Physicochemical profiling in drug research: a brief survey of the state-of-the-art of experimental techniques. Cell Mol Life Sci 59:1681–1689

Belsito S, Bartucci R, Montesano G, Marsh D, Sportelli L (2000) Molecular and mesoscopic properties of hydrophilic polymer-grafted phospholipids mixed with phosphatidylcholine in aqueous dispersion: interaction of dipalmitoyl n-poly(ethylene glycol) phosphatidylethanolamine with dipalmitoylphosphatidylcholine studied by spectrophotometry and spin-label electron spin resonance. Biophys J 78:1420–1430
Fernandes De Oliveira CA, Werneck Guimaraes CR, De Mello H, Echevarria A, De Alencastro RB (2005) A molecular dynamics study on liquid 1-octanol. Part 3. Evaluating octanol/water partition coefficients of novel thrombin inhibitors via free-energy perturbations. Int J Quantum Chem 102:542–553
Hawker DW (1995) Application of regular solution theory to solubility in lipids and partitioning involving lipids. Toxicol Environ Chem 50:39–49
Hristova K, Needham D (1995) Phase behavior of a lipid/polymer-lipid mixture in aqueous medium. Macromolecules 28:991–1002
Huynh L, Grant J, Leroux J-C, Delmas P, Allen C (2008) Predicting the solubility of the anticancer agent docetaxel in small molecule excipients using computational methods. Pharm Res 25:147–157
Israelachvili J (1992) Intermolecular and surface forces. Academic Press, London
Johnsson M, Edwards K (2003) Liposomes, disks, and spherical micelles: aggregate structure in mixtures of gel phase phosphatidylcholines and poly(ethylene glycol)-phospholipids. Biophys J 85:3839–3847
Jones MN (1995) The surface properties of phospholipid liposome systems and their characterisation. Adv Colloid Interfac 54:93–128
Kaler E, Murthy A, Rodriguez B, Zasadzinski J (1989) Spontaneous vesicle formation in aqueous mixtures of single-tailed surfactants. Science 245:1371–1374
Koo OM, Rubinstein I, Onyuksel H (2005) Camptothecin in sterically stabilized phospholipid micelles: a novel nanomedicine. Nanomedicine 1:77–84
Krishnadas A, Rubinstein I, Onyuksel H (2003) Sterically stabilized phospholipid mixed micelles: in vitro evaluation as a novel carrier for water-insoluble drugs. Pharm Res 20:297–302
Lee H, Venable RM, MacKerell AD Jr, Pastor RW (2008) Molecular dynamics studies of polyethylene oxide and polyethylene glycol: hydrodynamic radius and shape anisotropy. Biophys J 95:1590–1599
Lipinski CA, Lombardo F, Dominy BW, Feeney PJ (2001) Experimental and computational approaches to estimate solubility and permeability in drug discovery and development settings. Adv Drug Deliver Rev 46:3–26
Luo L, Eisenberg A (2001) Thermodynamic stabilization mechanism of block copolymer vesicles. J Am Chem Soc 123:1012–1013
Mazer NA, Benedek GB, Carey MC (1976) An investigation of the micellar phase of sodium dodecyl sulfate in aqueous sodium chloride solutions using quasielastic light scattering spectroscopy. J Phys Chem 80:1075–1085
Nagle JF, Tristram-Nagle S (2000) Structure of lipid bilayers. BBA—Rev Biomembranes 1469:159–195
Onyuksel H, Ikezaki H, Patel MP, Gao X, Rubinstein I (1999) A novel formulation of VIP in sterically stabilized micelles amplifies vasodilation in vivo. Pharm Res 16:155–160
Onyuksel H, Mohanty PS, Rubinstein I (2009) Vip-grafted sterically stabilized phospholipid nanomicellar 17-allylamino-17-demethoxy geldanamycin: a novel targeted nanomedicine for breast cancer. Int J Pharm 365:157–161
Patel S, Lavasanifar A, Choi P (2008) Application of molecular dynamics simulation to predict the compatability between water-insoluble drugs and self-associating poly(ethylene oxide)-b-poly(-caprolactone) block copolymers. Biomacromolecules 9:3014–3023
Patra N, Král P (2011) Controlled self-assembly of filled micelles on nanotubes. J Am Chem Soc 133:6146–6149
Patton JS, Stone B, Papa C, Abramowitz R, Yalkowsky SH (1984) Solubility of fatty acids and other hydrophobic molecules in liquid trioleoylglycerol. J Lipid Res 25:189–197

Rane SS, Anderson BD (2008) What determines drug solubility in lipid vehicles: is it predictable? Adv Drug Deliv Rev 60:638–656
Schick MJ, Atlas SM, Eirich FR (1962) Micellar structure of nonionic detergents. J Phys Chem 66:1326–1333
Titov AV, Král P, Pearson R (2009) Sandwiched graphene-membrane superstructures. ACS Nano 4:229–234
Totrov M (2004) Accurate and efficient generalized born model based on solvent accessibility: derivation and application for logP octanol/water prediction and flexible peptide docking. J Comput Chem 25:609–619
van Meer G, Voelker DR, Feigenson GW (2008) Membrane lipids: where they are and how they behave. Nat Rev Mol Cell Biol 9:112–124
Vuković L, Khatib FA, Drake SP, Madriaga A, Brandenburg KS, Král P, Onyuksel H (2011) Structure and dynamics of highly PEG-ylated sterically stabilized micelles in aqueous media. J Am Chem Soc 133:13481–13488
Vuković L, Madriaga A, Kuzmis A, Banerjee A, Tang A, Tao K, Shah N, Král P, Onyuksel H (2013) Solubilization of therapeutic agents in micellar nanomedicines. Langmuir 29:15747–15754
Zalipsky S (1995) Functionalized poly(ethylene glycols) for preparation of biologically relevant conjugates. Bioconjugate Chem 6:150–165
Zhang L, Yu K, Eisenberg A (1996) Ion-induced morphological changes in crew-cut aggregates of amphiphilic block copolymers. Science 272(5269):1777–1779

Toward Intracellular Delivery and Drug Discovery: Stochastic Logic Networks as Efficient Computational Models for Gene Regulatory Networks

Peican Zhu, Jinghang Liang and Jie Han

Abstract Biological functions are regulated through the interactions among genes, proteins and other molecules in a cell. Among various approaches to modeling gene regulatory networks (GRNs), Boolean networks (BNs) and its probabilistic extension, probabilistic Boolean networks (PBNs), have been effective means; in particular, PBNs consider molecular and genetic noise, so they provide significant insights into the understanding of the dynamics of GRNs. The applications of PBNs, however, are hindered by the complexities involved in the computation of the state transition matrix and steady-state distribution of a PBN. This chapter discusses stochastic logic networks as computationally efficient gene network models. Initially, stochastic Boolean networks (SBNs) are presented as a novel implementation of PBNs. SBNs are based on the notions of stochastic logic and stochastic computation. To further exploit the simplicity of logical models, a multiple-valued network employs gene states that are not limited to binary values, thus providing a finer granularity in the modeling of GRNs. Subsequently, stochastic multiple-valued networks (SMNs) are presented for modeling the effects of noise and gene perturbation in a GRN. These novel logical models provide accurate and efficient simulations of probabilistic Boolean and multiple-valued networks (PBNs and PMNs). The analysis of a p53–Mdm2 network and a WNT5A network shows that the stochastic logic networks are efficient in evaluating the network dynamics and steady state distribution of gene networks under random gene perturbation. These techniques are potentially useful in the investigation of intracellular delivery and drug discovery.

Keywords Stochastic Boolean network · Stochastic multiple-valued network · Gene perturbation · Gene activity profile · State transition matrix · Time-frame expansion · Steady state distribution analysis · Stochastic computation · Random permuted sequences · Non-Bernoulli sequence · Regulatory gene network

---

P. Zhu · J. Liang · J. Han (✉)
Department of Electrical and Computer Engineering, University of Alberta,
Edmonton, AB T6G 2V4, Canada
e-mail: jhan8@ualberta.ca

A. Prokop et al. (eds.), *Intracellular Delivery II*,
Fundamental Biomedical Technologies 7, DOI: 10.1007/978-94-017-8896-0_17,

Abbreviations

GRN	Gene regulatory network
BN	Boolean network
GAP	Gene activity profile
PBN	Probabilistic Boolean network
SBN	Stochastic Boolean network
PMN	Probabilistic multiple-valued networks
SMN	Stochastic multiple-valued network
STM	State transition matrix
SSD	Steady state distribution
MCMC	Markov chain Monte Carlo
MC	Monte Carlo
FSM	Finite state machine
EL	Equal or larger
ES	Equal or smaller
TB	Ternary buffer
TI	Ternary inverter
DSBs	Double strand breaks

1 Introduction

In a cell, biological functions are implemented through the interactions among genes, proteins and other molecules. However, gene networks are noisy due to the effect of stochastic fluctuations in genetic interactions (Elowitz et al. 2002). Various methodologies have been proposed to model the interactions among genes (de Jong 2002). These models can be classified into three broad categories: logical models, continuous models and stochastic models at the single-molecule level (Karlebach and Shamir 2008). As a classic logical model, Boolean networks (BNs) provide a qualitative analysis of the network dynamics (Kauffman 1969; Glass and Kauffman 1973; Huang 1999). Albeit simplistic, BNs have been shown to be efficient in the modelling of gene regulatory networks (GRNs) by taking advantages of low complexity and a minimum requirement on the quality (and quantity) of experimental data (Pandey et al. 2010). To account for the intrinsic noise in genetic and molecular interactions, probabilistic Boolean networks (PBNs) have been developed as a generalization of BNs (Shmulevich et al. 2002a, b; Shmulevich and Dougherty 2010). In a PBN, the inherent stochastic nature of molecular and genetic interactions dictates that the next state of target genes is predicted by several BNs with various probabilities. The evolution of such a system is thus a Markov chain and the state transitions can be described by a transition probability matrix. A steady-state analysis further tells whether a PBN will evolve into a stable target state in the presence of random gene perturbations, thereby providing valuable information for developing intervention-based therapeutic approaches (Shmulevich et al. 2002c, 2003; Dougherty et al. 2010; Faryabi et al. 2009; Karlebach and Shamir 2010).

The computation of the steady-state distribution of a PBN, however, presents a challenge. In a PBN with n genes and N Boolean networks, the complexity to compute the state transition matrix (STM) is $O(nN2^{2n})$ (Shmulevich et al. 2002b) and it is more difficult to compute the steady-state distribution (SSD). This complexity is reduced to $O(nN2^n)$ for a sparse STM (Zhang et al. 2007) and can further be reduced (to the same order, but with a smaller N) by ignoring the Boolean networks with probabilities below certain threshold (Ching et al. 2007). Methodologies have also been developed by eliminating genes (Ivanov et al. 2007) and using optimal control policies (Qian et al. 2009) to reduce computational complexity. State reduction techniques have been used for network intervention (Qian et al. 2010) and to reduce the model complexity of context-sensitive PBNs (Pal 2010). Nevertheless, it remains a difficult problem to reduce the computational complexity of a PBN without a compromise on the accuracy of an evaluation.

The Boolean simplification, furthermore, may incur an accuracy loss in the modeling of complex biological networks such as a random Boolean network (Harvey and Bossomaier 1997; Kitano 2001). To address this, an approach using multiple-valued variables introduces an increased level of granularity and can thus be more accurate in the modeling of a GRN (Thomas and D'Ari 1990; Morris et al. 2010; Dubrova 2006; Garg et al. 2007). For examples, three states of the protein p53 are considered in Abou-Jaoude et al. (2009), Murrugarra et al. (2012) and multiple-valued gene nodes are analyzed in a T-helper network (Garg et al. 2007). Moreover, deterministic multiple-valued networks are analyzed in Li and Cheng (2010). A multiple-valued analysis provides a tradeoff between the simplicity of Boolean networks and the complexity of differential equation based approaches (Morris et al. 2010). Multiple-valued networks have also been studied in chemical reactions (Adamatzky 2003) and cognitive sciences (Volker and Conrad 1998).

When gene expressions are discretized into multiple values, they are considered to be not only affected by the presence of activating or repressing proteins, but also by the absence of a protein (Aldana et al. 2003). Random and probabilistic multiple-valued networks (PMNs) have respectively been studied in Dubrova (2006) and Kim et al. (2002), for providing insights into the long run behavior of a network with noise. For a k-valued network of n genes with N network functions, however, a $k^n \times k^n$ matrix is required for an accurate analysis of the steady state distribution (SSD), resulting in a complexity of $O(nNk^{2n})$ by a PMN analysis in the computation of the STM. This also requires a memory usage in the order of at least $O(k^{2n})$. Since the size of an STM (and the required memory) increases exponentially with the number of genes, the analysis of a network with a higher quantization level presents even a greater challenge. This prevents the use of an accurate analysis in the evaluation of large networks. For a network with an increased number of genes, a Markov chain Monte Carlo (MCMC) method is often used to estimate the SSD of a PBN (Shmulevich et al. 2003) and its multiple-valued extension, PMNs (Kim et al. 2002). An MCMC simulation is considered to produce an accurate result when a sufficient number of simulations are performed to produce a stable output; however, this number is usually required to be very large, due to the slow convergence of the MCMC method (Rosenthal 1995), thus incurring a very long simulation time.

As an application of BNs, logic circuits have been used to simulate genetic networks (McAdams and Shapiro 1995). Circuit diagnosis techniques have been utilized to identify the most vulnerable molecules in cellular networks (Abdi et al. 2008). Synchronous simulation of Boolean networks has been proposed for the analysis of biological regulatory networks (Kervizic and Corcos 2008). An unreliable logic circuit usually behaves probabilistically and thus becomes an instance of PBNs. Initially proposed for reliable circuit design (von Neumann 1956; Gaines 1969), stochastic computation has been demonstrated in several physical and biological applications (Adar et al. 2004; Benenson et al. 2004).

In this chapter, stochastic logic networks are presented for efficient implementations of PBNs and PMNs (where the quantization level of a gene's state is not limited to binary). These implementations of a PBN and PMN are referred to as a stochastic Boolean network (SBN) and a stochastic multiple-valued network (SMN), respectively. As in stochastic computation, SBNs/SMNs employ random streams of binary/multiple values to represent probabilities and computation is performed by stochastic logic. Due to stochastic fluctuations, however, the computational results obtained by an SBN/SMN are not deterministic, but probabilistic. In Han et al. (2013), Liang and Han (2012), it has been shown that the use of non-Bernoulli sequences of random permutations of fixed numbers of 1's and 0's as initial inputs reduces the stochastic fluctuation and produces more accurate results than using Bernoulli sequences. In a k-valued SMN, similarly, randomly permuted sequences of fixed numbers of the k values are used to reduce the required computational complexity. It is shown in simulation results that the use of the non-Bernoulli and randomly permuted sequences increases the computational efficiency and allows for a tunable tradeoff between accuracy and efficiency. The required complexity for computing the STM of a k-valued network is reduced from $O(nNk^{2n})$ to $O(nLk^n)$, where L, determined by the minimum length of stochastic sequences for achieving a desired accuracy, increases slower than N.

Nevertheless, the analysis of the SSD is challenging due to the size of the STM required. However, the SSD analysis of a PBN/PMN resembles that of a finite state machine (FSM), due to their common underlying Markovian nature. An FSM is often implemented by a sequential circuit, which can be unrolled into a series of identical combinational modules by a so-called time-frame expansion in the spatial domain. A time-frame expansion of an SBN/SMN, hence, is used for an SSD analysis, which makes the SBN/SMN approach very efficient in the analysis of complex GRNs. Simulation results show that the proposed SBN/SMN approach produces very accurate results for small networks compared to a theoretical analysis. For large networks, the SBN/SMN approach using the time-frame expansion technique is more efficient than a simulation-based MCMC method. It is shown that the stochastic approach reveals the oscillatory dynamics of a p53–Mdm2 network (Murrugarra et al. 2012) with random gene perturbation, and that it accurately and efficiently predicts the SSD of a ternary WNT5A network (Kim et al. 2002) with gene perturbation. Hence, the stochastic logic networks can be used to predict the long run behavior of a cell under different intervention strategies. This feature can be useful in the study of intracellular delivery and discovery of drugs.

2 Probabilistic Boolean/Multiple-Valued Networks

A probabilistic network of n genes is defined by $G(V,F)$, with a node set $V = \{x_1, x_2, \ldots, x_n\}$ and a list of sets of predictor functions $F = \{F_1, F_2, \ldots, F_n\}$ (Shmulevich et al. 2002a). If the state of gene i is quantized into k levels, then $x_i \in \{0, \ldots, k-1\}$ for $i \in \{1, 2, \ldots, n\}$. For $k = 2$, a network is referred to as a probabilistic Boolean network (PBN), where V is a set of binary-valued nodes and $x_i = 1$ (or 0) indicates that gene i is (or not) expressed; for $k = 3$, it is considered as a ternary network (Kim et al. 2002). At time t, the state of a network can be described by a vector, $\boldsymbol{x}(t) = (x_1(t), x_2(t), \ldots, x_n(t))$, where the state of a gene is given by $x_i(t) \in \{0, 1, \ldots, k-1\}$ for $i \in \{1, 2, \ldots, n\}$. A network state is also referred to as a gene activity profile (GAP). For a k-valued network of n genes, hence, there are a total of k^n network states or GAPs. A GAP is also given as a decimal index. For a ternary network of n genes, a GAP is indexed by:

$$d = \sum_{i=1}^{n} x_i(t) \cdot 3^{i-1} + 1, \qquad (1)$$

where x_i is the state of the ith gene, $i \in \{1, 2, \ldots, n\}$.

For gene i ($i \in \{1, 2, \cdots, n\}$), the set of predictor functions is given by $F_i = \left\{f_1^{(i)}, f_2^{(i)}, \ldots, f_{l(i)}^{(i)}\right\}$, with each predictor function $f_{j(i)}^{(i)}\colon \{0, 1, \ldots, k-1\}^n \rightarrow \{0, 1, \ldots, k-1\}$, where $l(i)$ is the number of possible predictor functions for gene i and $l(i)$ is usually a small number (Ching et al. 2007; Guelzim et al. 2002). Due to the stochastic behavior, the next state of gene i is determined by all of its predictor functions in F_i, i.e., $f_1^{(i)}, f_2^{(i)}, \ldots, f_{l(i)}^{(i)}$ with probabilities $c_1^{(i)}, c_2^{(i)}, \ldots, c_{l(i)}^{(i)}$.

If the predictor functions are independent, there are $N = \prod_{i=1}^{n} l(i)$ possible realizations of the network, each of which is referred to as a context. Assume that the jth context is represented as $f^j = \left(f_{j(1)}^{(1)}, f_{j(2)}^{(2)}, \ldots, f_{j(n)}^{(n)}\right)$, where each $f_{j(i)}^{(i)}\colon \{0, 1, \ldots, k-1\}^n \rightarrow \{0, 1, \ldots, k-1\}$, for $1 \leq j(i) \leq l(i)$, is a predictor function of gene i; the next state of a gene is determined by both the present state and the selected context.

In a PBN, the probability that the jth BN is selected is:

$$P_j = \prod_{i=1}^{n} c_{j(i)}^{(i)}, \tag{2}$$

where $c_{j(i)}^{(i)}$ is the probability that the Boolean function $j(i)$ is selected for gene i. By a different selection of the BNs during a state transition, the genes can reach a different state from their present state.

A multiple-valued network can be modeled by a Markov chain (Kim et al. 2002), so the next state of gene i, x_i ($x_i \in \{0, 1, \ldots, k-1\}$ in a k-valued network) is given by:

$$x_i^{(t+1)} = \begin{cases} 0 & \text{with } C_i^0\left(\boldsymbol{S}^{(t)}\right) = \Pr\left(x_i^{(t+1)} = 0|\boldsymbol{S}^{(t)}\right) \\ 1 & \text{with } C_i^1\left(\boldsymbol{S}^{(t)}\right) = \Pr\left(x_i^{(t+1)} = 1|\boldsymbol{S}^{(t)}\right) \\ & \vdots \\ k-1 & \text{with } C_i^{k-1}\left(\boldsymbol{S}^{(t)}\right) = \Pr\left(x_i^{(t+1)} = k-1|\boldsymbol{S}^{(t)}\right) \end{cases} \tag{3}$$

where $C_i^0\left(\boldsymbol{S}^{(t)}\right) + C_i^1\left(\boldsymbol{S}^{(t)}\right) + \cdots + C_i^{k-1}\left(\boldsymbol{S}^{(t)}\right) = 1$. Thus, the transition probability from the network state (or GAP) $\boldsymbol{S}^{(t)}$ at time t to $\boldsymbol{S}^{(t+1)}$ at $t+1$ is given by:

$$Pr\left(\boldsymbol{S}^{(t)} \rightarrow \boldsymbol{S}^{(t+1)}\right) = \prod_{i=1}^{n} C_i^{x_i^{(t+1)}}. \tag{4}$$

Using the decimal indices of GAPs by (1), the state transition of a ternary network is described by the state transition matrix (STM) as follows:

$$\boldsymbol{A} = \begin{bmatrix} Pr(1|1) & Pr(2|1) & \cdots & \cdots & Pr(3^n|1) \\ Pr(1|2) & Pr(2|2) & \cdots & \cdots & Pr(3^n|2) \\ \cdots & \cdots & \cdots & \cdots & \cdots \\ \cdots & \cdots & \cdots & \cdots & \cdots \\ Pr(1|3^n) & Pr(2|3^n) & \cdots & \cdots & Pr(3^n|3^n) \end{bmatrix} \tag{5}$$

In $\boldsymbol{A}$, each entry indicates the conditional probability that the network transitions from a present state into a next state. For N realizations of the network, $\boldsymbol{A}$ can be obtained as $\boldsymbol{A} = \sum_{j=1}^{N} P_j \boldsymbol{A}_j$, where P_j ($P_j = \prod_{i=1}^{n} c_{j(i)}^{(i)}$) is the probability that the jth realization of the network emerges and $\boldsymbol{A}_j$ is the STM resulting from the jth realization (Shmulevich et al. 2002a). Hence, the STM can be derived for a multiple-valued network with a complexity of $O(nNk^{2n})$, where N is the number of possible realizations of the network and k is the quantization level of the gene states.

External stimuli cause random gene perturbations that make the dynamics of a network an ergodic Markov chain (Shmulevich et al. 2002c). In an ergodic Markov chain, all states are communicated and thus a steady state distribution (SSD) exists in a network. Since a perturbed gene has $k - 1$ possible states, there are $(k - 1)^{n_0}$ states for n_0 perturbed genes ($n_0 \in \{1, 2, \ldots, n\}$); hence, each of the perturbed states in $\boldsymbol{S}^{(t+1)}$ is selected with a probability of $[1/(k - 1)]^{n_0}$. The event that no gene is perturbed, occurs with a probability of $(1 - p)^n$. Hence, $\boldsymbol{S}^{(t+1)}$ is determined by the selected context if no perturbation exists, i.e. $\Pr\{\boldsymbol{S}^{(t)} \to \boldsymbol{S}^{(t+1)}\} = \left(\prod_{i=1}^{n} C_i^{x_i^{t+1}}\right)$. If n_0 genes are perturbed, $\boldsymbol{S}^{(t)} \to \boldsymbol{S}^{(t+1)}$ occurs with probability $p^{n_0} \cdot (1 - p)^{n - n_0} \cdot [1/(k - 1)]^{n_0}$. Following (Kim et al. 2002), therefore, the state transition probability from $\boldsymbol{S}^{(t)}$ to $\boldsymbol{S}^{(t+1)}$ in a perturbed k-valued network is given by:

$$Pr\left\{\boldsymbol{S}^{(t)} \to \boldsymbol{S}^{(t+1)}\right\} = \left(\prod_{i=1}^{n} C_i^{x_i^{(t+1)}}\right) \cdot (1 - p)^n + p^{n_0} \cdot (1 - p)^{n - n_0} \cdot p_0^{n_0} \cdot 1\left[\boldsymbol{S}^{(t)} \neq \boldsymbol{S}^{(t+1)}\right], \tag{6}$$

with

$$n_0 = \sum_{i=1}^{n} 1\left(x_i^{(t)} \neq x_i^{(t+1)}\right), \tag{7}$$

$$p_0 = 1/(k - 1), \tag{8}$$

where p is the perturbation rate, n_0 is the number of perturbed genes, p_0 is the probability that a gene will change to a new state if perturbed, and $1(\cdot)$ is an indicator function; $1(\boldsymbol{S}^{(t)} \neq \boldsymbol{S}^{(t+1)}) = 1$ if $\boldsymbol{S}^{(t)} \neq \boldsymbol{S}^{(t+1)}$ and $1(\boldsymbol{S}^{(t)} \neq \boldsymbol{S}^{(t+1)}) = 0$ otherwise. Using (6), a perturbed STM or perturbation matrix (Ching et al. 2007; Liang and Han 2012) can be obtained for further analysis of the SSD. Similarly, the STM can be calculated for a PBN when $k = 2$.

3 Stochastic Logic Networks

3.1 Stochastic Computation for Boolean and Multiple-Valued Logic

In stochastic computation, probabilities are encoded into random binary bit streams. Information is carried in the statistics of the binary streams and processed by stochastic logic (Gaines 1969). Usually, a probability is represented by a proportional number of bits, e.g. the mean number of 1's in a bit sequence. In Boolean logic, for example, an inverter computes the complement of a probability while the multiplication of probabilities is implemented by an AND gate with independent inputs. Thus, stochastic computation performs a probabilistic analysis in the real

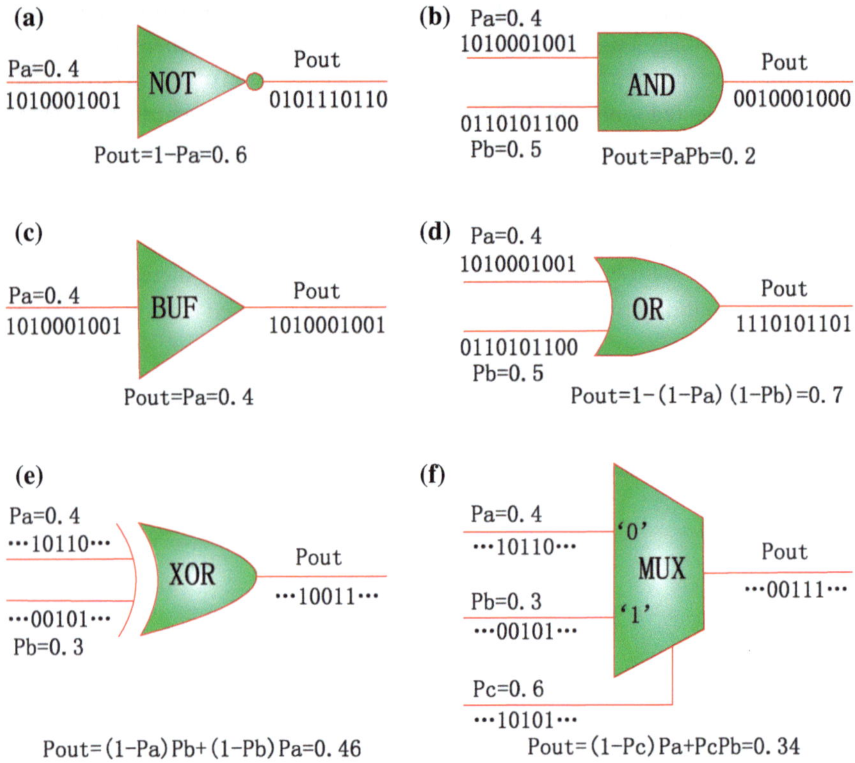

Fig. 1 Stochastic logic. **a** a NOT gate, **b** an AND gate, **c** a buffer, **d** an OR gate, **e** an XOR gate and, **f** a multiplexer

domain. Due to inevitable stochastic fluctuations, the computational result by stochastic logic is not deterministic but probabilistic. However, stochastic fluctuations can be reduced through the use of non-Bernoulli sequences of random permutations of fixed numbers of 1's and 0's as initial inputs. This produces more accurate results than using Bernoulli sequences (Han et al. 2013). Signal correlations are efficiently handled in a stochastic network by the bit-wise dependencies encoded in the random binary streams, thus making it an efficient approach to computing probabilities (Han et al. 2013).

Figure 1 shows an inverter (NOT), an AND, a buffer, an OR, an XOR gate and a multiplexer. While an XOR gate performs a controlled inversion, a multiplexer takes one of its inputs as output according to the values of the control bits. For the 2-to-1 multiplexer of Fig. 1f, for example, its output takes the value of its input 'a' or 'b' when the control bit 'c' is 0 or 1. Similarly, a stochastic multiplexer chooses one of its inputs as output according to the distributions of 0's and 1's and thus the probability of 0 and 1 encoded in the random sequences of the control bits. For a sequence length of 1,000 bits, for example, an input probability of 0.4 indicates

Fig. 2 The stochastic encoding of a ternary signal using a sequence of 10 values

"0121212112" for $\begin{cases} P(0) = 0.1 \\ P(1) = 0.5 \\ P(2) = 0.4 \end{cases}$

that approximately 400 1's are in the random sequence of the input 'a', as shown in Fig. 1f. If the random input sequences are independent, the output of the multiplexer is expected to be $P_a(1 - P_c) + P_b P_c = 0.34$, which means that approximately 340 1's are expected in the output sequence.

Stochastic computation is also applicable to the probabilistic analysis of multiple-valued signals. For a k-valued signal, the probability of each value is given in a vector $P = [p_{k-1}, p_{k-2}, \ldots, p_1, p_0]$, with $\sum_{i=0}^{k-1} p_i = 1$. This probability vector can be encoded into a multiple-valued stochastic sequence. An example is shown in Fig. 2 for a ternary signal.

Multiple-valued logic includes the buffer, inverter, MIN (minimum), MAX (maximum) and rotator; some are defined as follows (Li and Cheng 2010):

(1) A multiple-valued buffer:

$$BUF(A) = A,$$

(2) A multiple-valued inverter:

$$INV(A) = (k - 1) - A,$$

(3) A multiple-valued rotator $\emptyset$:

$$\emptyset(A) = \begin{cases} A + 1 & A \neq k - 1 \\ 0 & A = k - 1 \end{cases}.$$

The following new logic operators are further defined:

(4) A multiple-valued equal or larger (EL) than a operator, where a is a constant number between 0 and k-1.

$$EL(A \geq a) = MAX(A, a),$$

(5) A multiple-valued equal or smaller (ES) than a operator, where a is a constant number between 0 and k-1.

$$ES(A \leq a) = MIN(A, a).$$

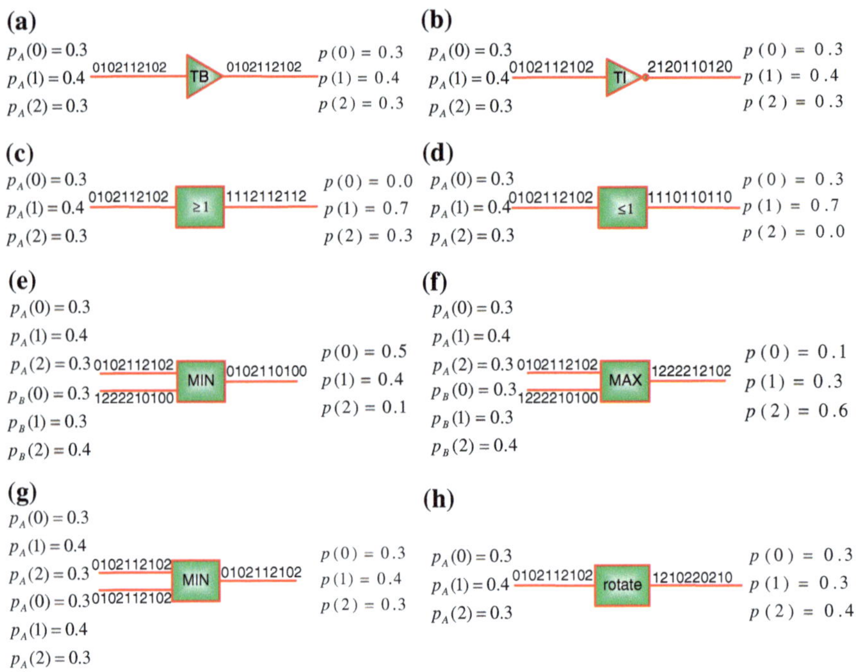

Fig. 3 Stochastic logic: **a** a ternary buffer (*TB*); **b** a ternary inverter (*TI*); **c** an EL operator; **d** an ES operator; **e** a ternary MIN with independent inputs; **f** a ternary MAX with independent inputs; **g** a ternary MIN with totally dependent inputs; **h** a ternary rotate gate. A probabilistic computation is performed through stochastic logic operations by encoding signal probabilities into random sequences

Several ternary stochastic processing elements are shown in Fig. 3, including a buffer, an inverter, an EL operator, an ES operator, a MIN, a MAX, a rotate gate.

For the ternary MIN logic, if the two inputs are independent with probabilities $A = [0.3\ \ 0.4\ \ 0.3]$ and $B = [0.5\ \ 0.4\ \ 0.1]$, the output probabilities are expected to be $p(2) = p_A(2) \cdot p_B(2) = 0.3 \times 0.1 = 0.03$, $p(0) = p_A(0) + p_B(0) - p_A(0) \cdot p_B(0) = 0.65$, and $p(1) = 1 - p(0) - p(2) = 0.32$. This function can be implemented by the ternary MIN gate, as shown in Fig. 3e, using stochastic sequences. For a sequence length of 10,000 values, the output sequence is expected to have approximately 6,500 0's, 3,200 1's and 300 2's. For the ternary rotate logic, if the input's signal probability is given by $A = [0.3\ \ 0.4\ \ 0.3]$, the output's signal probability is expected to be $p(0) = p_A(2) = 0.3$, $p(1) = p_A(0) = 0.3$ and $p(2) = p_A(1) = 0.4$. This function can be implemented by the ternary rotate gate with the use of stochastic sequences (Fig. 3h).

However, these numbers are not deterministic but probabilistic, due to inherent stochastic fluctuations. For stochastic Boolean networks, compared to the use of Bernoulli sequences of independently generated binary bits (such as in a coin-flipping experiment), the effect of the fluctuation can be significantly reduced through the use of non-Bernoulli sequences of random permutations of fixed numbers of 1's and 0's for initial input probabilities (Liang and Han 2012). For stochastic multiple-valued networks, stochastic sequences of random permutations of fixed numbers of the multiple values, hereafter referred to as randomly permuted sequences, are used for encoding initial input probabilities. The use of randomly permuted sequences reduces the amount of stochastic fluctuations in a network. It will be shown in the Results and Discussion Section that the effect of fluctuation is negligible when a reasonable sequence length is used in the simulation.

3.2 Stochastic Logic Networks Without Perturbation

A general structure of the stochastic Boolean network (SBN) is defined as follows. As discussed previously, the next state of a gene is determined by the present state of its input genes and a set of predictor functions according to their occurring probabilities. In an SBN, these probabilities are represented by random binary bit streams and the selection of the Boolean functions is implemented by a multiplexer with properly generated control sequences.

In the general case that multiple quantization levels are considered, a stochastic multiple-valued network (SMN) can be constructed to model a multiple-valued gene network. In an SMN, probabilities are encoded into randomly permuted multiple-valued sequences and the selection of the predictor functions is also implemented by a multiplexer. A general structure of an SBN or SMN is shown in Fig. 4 for a single gene.

If the next state of gene i is determined by $l(i)$ predictor functions, the number of control bits of the multiplexer is given by $\log_2(l(i))$. Usually, a function has only a few input variables and the number of possible predictor functions is generally small (Ching et al. 2007; Guelzim et al. 2002). By a multiplexer with control bits $S_1 \sim S_m$, a function is selected in the jth BN for gene i with probability $c_{j(i)}^{(i)}$. Assume that a network transfers from state $\boldsymbol{S}^{(t)}$ to $\boldsymbol{S}^{(t+1)}$ in a context (or network function), then the transition probability for $\boldsymbol{S}^{(t)} \to \boldsymbol{S}^{(t+1)}$ is given by the probability of selecting this context. This indicates that when all the genes are considered, the SBN and SMN models in Fig. 4 accurately implements the functions of (2) and (4).

3.3 Stochastic Logic Networks with Perturbation

Under external stimuli, a gene's state can be perturbed by a small chance during a transition (Shmulevich et al. 2002c). Assume $\mathbf{x} = (x_1, x_2, \ldots x_n)$ represents the current state of an n-gene network at time t and γ is the vector that indicates the

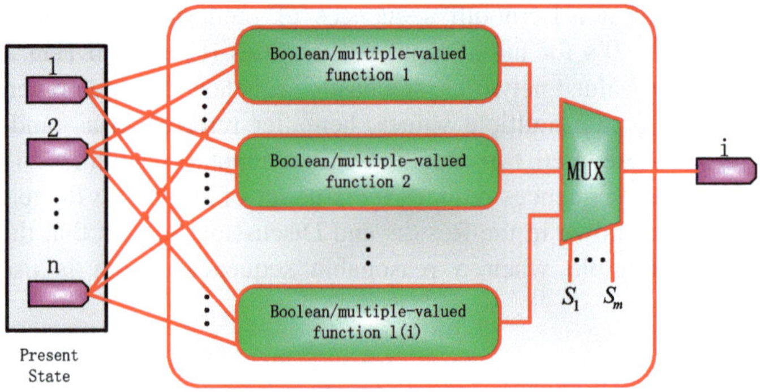

Fig. 4 A stochastic Boolean or multiple-valued network (SBN or SMN) without perturbation (for a single gene *i*). The control sequences S_1~S_m of the multiplexer (MUX) probabilistically determine the selection of the predictor functions

effect of random perturbation, the next state $\mathbf{x}'$ is given by Shmulevich et al. (2002c):

$$x' = \begin{cases} x \oplus \gamma & \textit{with a probability of } 1 - (1-p)^n \\ f_k(x) & \textit{with a probability of } (1-p)^n \end{cases} \tag{9}$$

where $\oplus$ is the modulo 2 of additions and $f_k()$ is the function of the kth Boolean network at time t. The effect of perturbation to the state transition matrix can then be described by a matrix called the perturbation matrix (Ching et al. 2007). The perturbation matrix is determined by the number of genes and the gene perturbation probability p. It is usually computed by a (complex) analytical approach.

However, the effect of perturbation can be readily accounted for in an SBN. Figure 5 illustrates a general model of SBNs with perturbation. As perturbation introduces a probabilistic inversion to the state of a gene, XOR gates are used to implement the addition modulo 2 of the perturbation vector and the present state. The probability that either a Boolean function works or a perturbation works (given in (9)) is computed by a stochastic n-input OR gate. This probability is then encoded into the output sequence of the OR gate and used as the control sequence of a bus multiplexer. If the perturbation vectors ('Pert 1' … 'Pert n' in Fig. 5) are all 0's, which means there is no perturbation, then the output sequence of the OR gate contains all 0's, which subsequently determines that the next state is given by the original SBN without perturbation; otherwise, the next state is determined by the perturbation probability encoded in the output sequence of the stochastic OR gate. Per the stochastic functions of XOR, OR and the multiplexer, the next state is given as the output of the SBN with perturbation, by:

$$x' = (x \oplus \gamma) \cdot (1 - (1-p)^n) + f_k(x) \cdot (1-p)^n \tag{10}$$

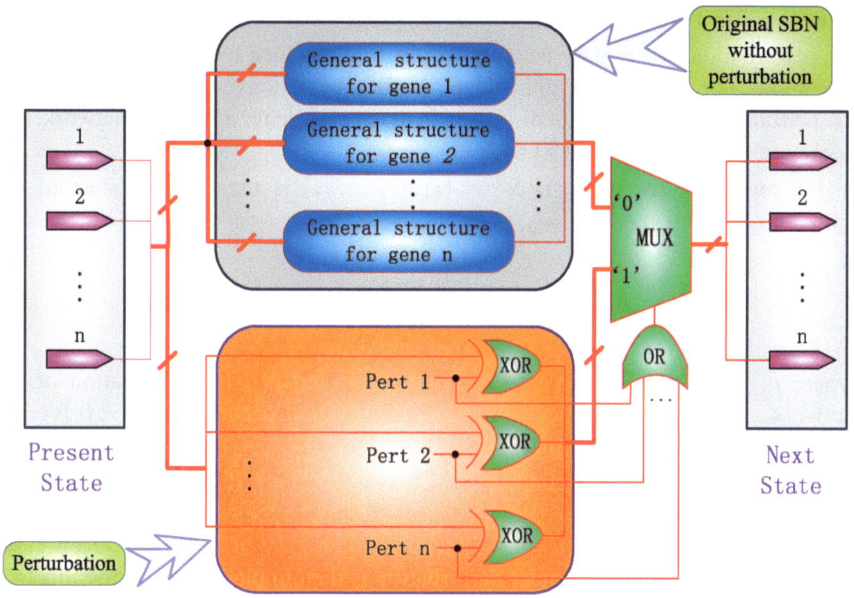

Fig. 5 An SBN with perturbation

Table 1 State transition rules for a gene in a ternary network under perturbation. Adapted from Zhu and Han (2013) and © IEEE

Current state (x)	Perturbation (γ)	Next state ($x' = modulo(x+\gamma, 3)$)
0	1	1
	2	2
1	1	2
	2	0
2	1	0
	2	1

which is equivalent to (9). This indicates that a PBN with perturbation can be accurately implemented by an SBN with perturbation.

In a k-valued network of n genes, a similar perturbation flag vector γ is used to indicate whether a gene is to be perturbed. Assume that the network goes from state $\boldsymbol{S}^{(t)}$ to $\boldsymbol{S}^{(t+1)}$ under perturbation. If each gene is to be perturbed with a probability p, the probability that the next state is totally determined by a network function (i.e., no perturbation occurs) is $(1-p)^n$. When a perturbation occurs, the state of the perturbed gene transitions to a different state: this new state is determined by the present state and the value in the perturbation flag vector γ. Without the loss of generality, a set of transition rules can be determined, as shown in Table 1 for a ternary network. The set of rules in Table 1 can be implemented by sum and modulo operations; for $\boldsymbol{S}^{(t)} = (0,0,0,1,1,1,2,2,2)$ and $\gamma = (0,1,2,0,1,2,0,1,2)$, as an example, the next state is given by $S^{(t+1)} = modulo\big((S^{(t)} + \gamma), 3\big) = (0, 1, 2, 1, 2, 0, 2, 0, 1)$. Hence, the perturbation in a ternary network can be implemented by the sum and modulo operations. For a network of higher levels, similar operations can be implemented for the perturbation (although not discussed in detail), while for a Boolean network, this operation is simplified to an XOR gate.

For an SMN, therefore, if $S^{(t)} = (x_1, x_2, \ldots, x_n)$ is the GAP or state of the network at time t; the next state $S^{(t+1)}$ is given by:

$$S^{(t+1)} = \begin{cases} modulo(sum(S^{(t)}, \gamma), k) & with\ 1 - (1 - p)^n, \\ f(x_1, x_2, \ldots, x_n) & with\ (1 - p)^n, \end{cases} \tag{11}$$

where p is the perturbation rate for each gene and $f(\cdot)$ is a realization of the network at time t. (11) indicates that no perturbation occurs, i.e., $\gamma_i = 0$ for any $i \in \{1, \ldots, n\}$, with a probability of $(1 - p)^n$. In this case, the next state $S^{(t+1)}$ is determined by the selected context (or network function). If gene i is perturbed, γ_i in γ is assigned to be m ($m \neq 0$) with a probability of $1/(k - 1)$; the gene's state x_i is then changed from j to m ($m \neq j$) with a probability of $1/(k - 1)$ (Kim et al. 2002). This state transition under perturbation is then implemented by the function of $modulo(sum(S^{(t)}, \gamma), k)$. In a network of n genes, if n_0 genes are to be perturbed, this indicates that the perturbation flag vector γ contains n_0 non-zero values and $n - n_0$ zeros. For each zero, the current state of the corresponding gene remains, as shown in the aforementioned example. For the n_0 non-zero values, a different set of values leads to a different next state of the perturbed genes. For random gene perturbation, each set occurs with a probability of $[1/(k - 1)]^{n_0}$, so the network transition from the present state to a particular next state, i.e. $S^{(t)} \rightarrow S^{(t+1)}$, occurs with a probability of $p^{n_0} \cdot (1 - p)^{n - n_0} \cdot [1/(k - 1)]^{n_0}$. Since a perturbed state is considered to be different from the present state, i.e. $S^{(t+1)} \neq S^{(t)}$, under perturbation, the probability of the state transition of (11) is given by (6).

To account for the perturbation effect, a modified SMN is shown in Fig. 6. The probability that the multiple-valued network is left without perturbation or that a perturbation takes effect, is determined by the output of an n-input MAX gate.

In the SMN in Fig. 6, gene perturbation is considered as follows. Since a random gene perturbation probabilistically changes the state of a gene, the modules of sum and modulo k operations are used to implement the perturbation function (of the perturbation vector and the genes' current states). The jth perturbation vector, Per_j, consists of a number of i's, $i = 0, 1, \ldots, (k - 1)$; for instance, if an L-bit sequence Per_i is used to indicate the perturbation rate p in a ternary network and let $M = L \cdot p$, then there are $L - M$ 0's, $M/2$ 1's and $M/2$ 2's in the sequence.

This indicates that if a gene at state i is perturbed, the new state can be any $j (j \neq i)$ with an equal probability of $1/(\mathrm{k} - 1)$. Hence, if n_0 genes are perturbed, a perturbed state is chosen with a probability of $[1/(\mathrm{k} - 1)]^{n_0}$. The probability that either an original multiple-valued function works or a perturbation occurs (by (11))

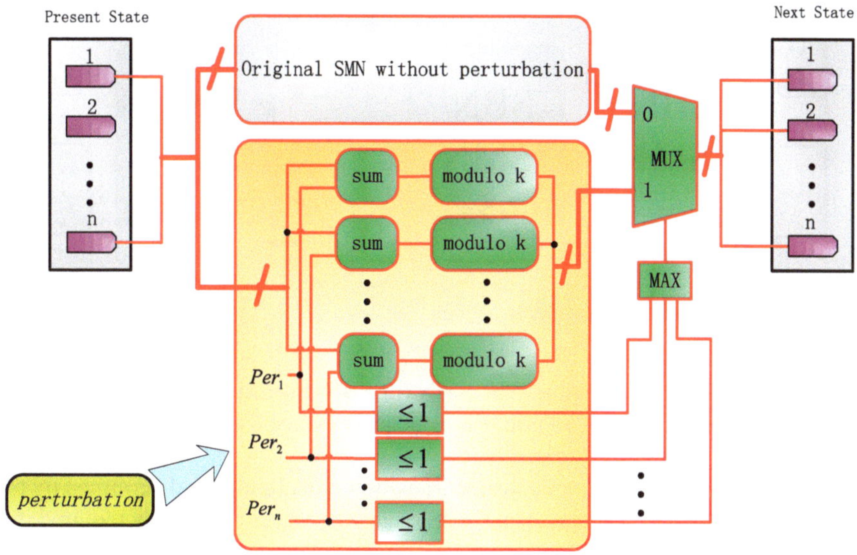

Fig. 6 An SMN with perturbation. Gene perturbation is implemented by the sum and modulo k functions of the perturbation vector and the present state. Adapted from Zhu and Han (2013) and © IEEE

is implemented by the output sequence of an n-input MAX gate. This sequence is then used as the control sequence of a bus (or multiple-bit) multiplexer. If no perturbation occurs, the perturbation vectors $('Per'_1, 'Per'_2, \ldots, 'Per'_n$ in Fig. 6) consist of all 0's, and thus the output sequence of the MAX gate will contain all 0's. The next state is subsequently given by the original SMN without perturbation; otherwise, the ES operators will produce 1s for the input of the MAX gate, thus the next state is determined by the perturbation probability encoded in the output sequence of the MAX gate. From this analysis, it can be seen that the SMN model implements the function of (11) and thus computes the transition probability of (6). This indicates that it accurately implements a probabilistic multiple-valued network with perturbation.

3.4 State Transition Matrix and Steady State Analysis

In the simulation of an SBN/SMN, each input combination results in output sequences that contain information about the transition probability from this input to every output (or next state). For a deterministic input (i.e. the present state), the proportions of the numbers of the next states encoded in the output sequences return the statistics as the transition probabilities in a row in the state transition matrix (STM). Hence, all the transition probabilities for this input can be generated in a single run. For a probabilistic multiple-valued network (PMN) with k levels

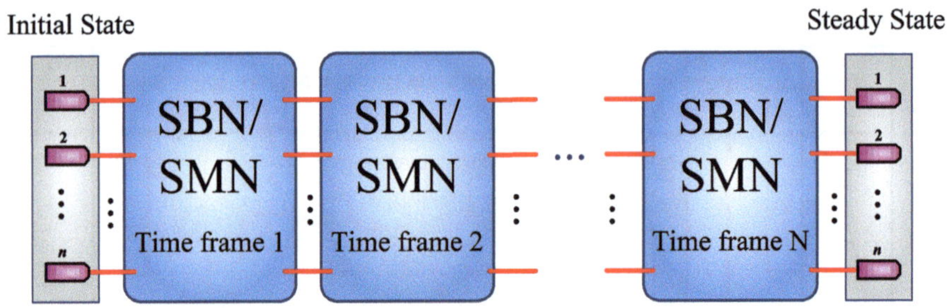

Fig. 7 A time-frame expanded stochastic logic network

(or a PBN if $k = 2$) and n genes, the SMN (or SBN) needs to be run for each of the k^n input states and an $O(n)$ number of sequences need to be generated for the control signals of the multiplexers.

The accuracy in the computed state transition probabilities is determined by the length of the stochastic sequences. Since longer sequences are usually required in a larger network for achieving an evaluation accuracy, a factor, L, is used here to account for the computational overhead required by using a longer stochastic sequence. For a k-valued network of n genes, a complexity of $O(nLk^n)$ results for computing the STM at a desired accuracy. As shown in the simulation results in (Liang and Han 2012; Zhu and Han 2013), the required minimum sequence length increases slower with the numbers of genes than the number of possible networks, N, which generally increases exponentially with the number of genes in a network. Therefore, the complexity of using an SMN (or SBN when $k = 2$) to compute the STM, i.e., $O(nLk^n)$, is smaller than $O(nNk^{2n})$ of an accurate analysis (Kim et al. 2002). This difference becomes significant for a large gene network.

Given the size of the STM of a PBN, the analysis of the steady-state distribution is challenging for using both analytical and simulative approaches. The Markovian nature of a PBN/PMN makes its analysis similar to that of a finite state machine (FSM). An FSM is equivalent to a sequential circuit implementation. By a time-frame expansion, a sequential circuit can be unrolled into a series of identical combinational modules connected in the spatial domain. Using a similar technique, the temporal operation of a stochastic logic network can be transformed into a spatial operation of identical SBNs or SMNs connected in series. This is shown in Fig. 7. This spatial extension of an SBN/SMN can be used for the steady-state analysis and the required iterations of the network are determined by the number of state transitions before reaching a steady state.

A steady-state analysis using a time-frame expanded network starts with an initial input state, generates the random bit sequences for the inputs and control bits of multiplexers, and then propagates the stochastic signals through the expanded SBN or SMN structure. This process is equivalent to an analytical procedure of multiplying the input probabilities with the powers of the state transition matrix (STM). Finally, a small variance threshold is used to determine whether the system has reached a steady state. The steady-state distribution is then obtained from the output sequences at the end of the operation.

In the above process, the speed of convergence to a steady state is dependent on a number of factors, including the length of random bit sequences, the variance threshold value and the perturbation rate. In practice, a sequence length that is long enough to have a resolution of at least two magnitudes smaller than the threshold value, is used to guarantee that the convergence is not dominated by stochastic fluctuations. It is shown later that the analysis using an extended network structure provides an alternative and efficient way of estimating the steady-state distribution of a PBN or PMN without resorting to the STM.

A general multiple-valued network (with any k) can be analyzed by the time-frame expanded SMN approach. The simulation results in Zhu and Han (2013) reveal that, while the SMN approach takes longer time than a Markov chain analysis (Kim et al. 2002) for small networks, it becomes faster in the analysis of large networks. Although the evaluation accuracy slightly decreases with the increase of the discretization level, k, a better accuracy is obtained when longer stochastic sequences are used.

Further investigations on the memory usage of the stochastic approach show that the Markov chain analysis (Kim et al. 2002) requires less memory than SMN for small networks with a low quantization level, k, whereas the required memory outgrows that of the SMN approach in the analysis of a larger network with a larger k (Zhu and Han 2013). In fact, the required memory by the Markov chain analysis increases exponentially with the number of genes and depends heavily on k, because of the increased size of transition matrices in an analysis. On the other hand, the memory required by the time frame expanded SMN approach is mainly determined by the sequence length (L) and number of genes (n), while the quantization level (k) has little impact. Therefore, the Markov chain analysis incurs a significantly longer run time than the stochastic approach in the analysis of networks with larger n and k.

4 Results and Discussion

4.1 The p53–Mdm2 Network

p53 is a tumour suppressor gene that plays an important role in preventing the development and progression of tumour cells (Weinberg 2006; Vogelstein et al. 2000). In a p53 network, signaling pathways are triggered by DNA damages and external factors such as chemotherapeutic drugs and ultraviolet light. For instance, DNA double strand breaks (DSBs) activate pathways that involve the p53 and Mdm2 genes (Fig. 8) (Lahav et al. 2004; Batchelor et al. 2009). In response to DSBs, the ATM kinase is first stimulated and the Chk2 is then stimulated by ATM. These activated kinases subsequently induce an increase in the concentration level

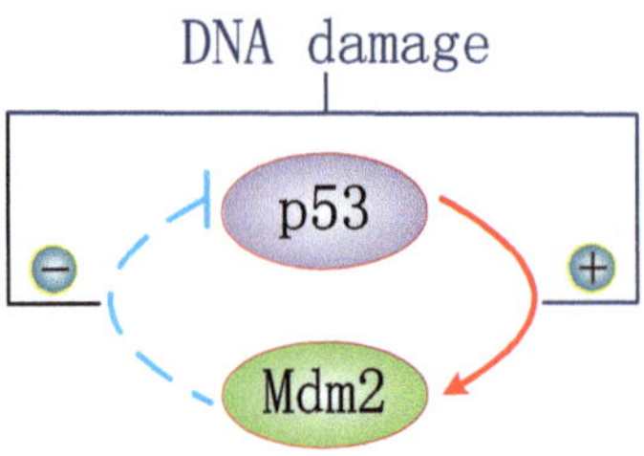

Fig. 8 The P53 –Mdm2 network. Adapted from Lahav et al. (2004)

Table 2 State transition probabilities of the p53–Mdm2 network

Present state	Next state probability			
p53, Mdm2(or, x_1x_2)	p53		Mdm2	
	0	1	0	1
00	0.01	0.99	0.99	0.01
01	0.1	0.9	0.9	0.1
10	0.9	0.1	0.1	0.9
11	0.5	0.5	0.5	0.5

Table 3 Truth table of the PBN for the p53–Mdm2 network

x_1x_2	$f_1^{(1)}$	$f_2^{(1)}$	$f_3^{(1)}$	$f_4^{(1)}$	$f_1^{(2)}$	$f_2^{(2)}$	$f_3^{(2)}$	$f_4^{(2)}$
00	1	1	1	0	0	0	0	1
01	1	1	0	0	0	0	1	1
10	0	0	1	1	1	1	0	0
11	0	1	1	1	1	0	0	0
$c_j^{(i)}$	0.5	0.4	0.09	0.01	0.5	0.4	0.09	0.01

of p53 and a decrease in the interactions between p53 and Mdm2. The increase in the p53 protein level and its transcription activity promote the expression of the Mdm2 gene, which in turn proceeds to trigger the degradation and destruction of p53. This prior knowledge enables us to come up with the transition rules for the p53–Mdm2 interactions, as shown in Table 2. Based on these rules, an independent PBN of the two genes p53 and Mdm2 can be established: V = (x_1, x_2) with the function classes $F_1 = \left\{f_1^{(1)}, f_2^{(1)}, f_3^{(1)}, f_4^{(1)}\right\}$ and $F_2 = \left\{f_1^{(2)}, f_2^{(2)}, f_3^{(2)}, f_4^{(2)}\right\}$. The state transitions of this PBN are given in the truth table of Table 3

In Table 3, the leftmost column indicates the present state of the genes p53 and Mdm2. The internal entries in the table indicate whether a function will result in a logical 1 or 0 at the next state of each gene. The row on the bottom shows the probability of each transition by a function. Given an initial state of '01,' for example, the next state of the genes can be '00' with a probability of $(0.09 + 0.01) \times (0.5 + 0.4) = 0.09$, '01' with a probability of $(0.09 + 0.01) \times$

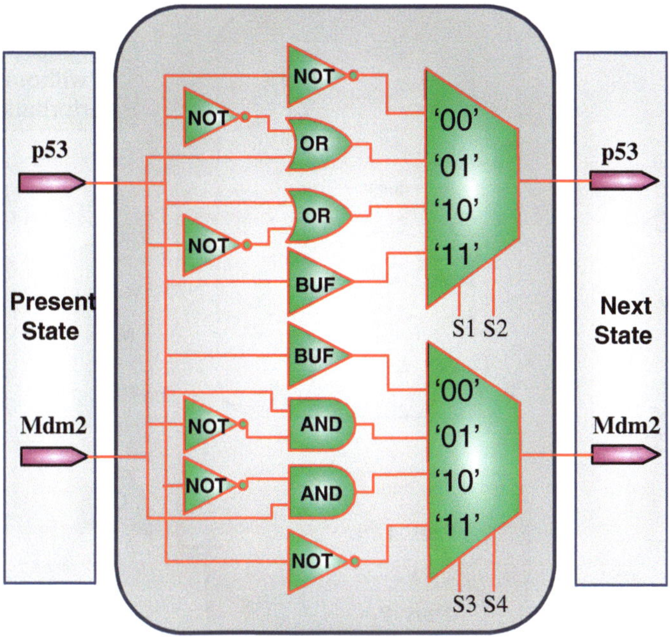

Fig. 9 An SBN for the p53–Mdm2 network (without perturbation)

$(0.09 + 0.01) = 0.01$, '10' with a probability of $(0.5 + 0.4) \times (0.5 + 0.4) = 0.81$ or '11' with a probability of $(0.5 + 0.4) \times (0.09 + 0.01) = 0.09$. A PBN is determined by the truth table of Table 3 and its state transition matrix (STM) can be computed as:

$$A_{PBN} = \begin{bmatrix} 0.0099 & 0.0001 & 0.9801 & 0.0099 \\ 0.0900 & 0.0100 & 0.8100 & 0.0900 \\ 0.0900 & 0.8100 & 0.0100 & 0.0900 \\ 0.2500 & 0.2500 & 0.2500 & 0.2500 \end{bmatrix}. \tag{12}$$

For this PBN, an SBN can be constructed using stochastic multiplexers and random binary bit streams as information carriers, as shown in Fig. 9. As discussed previously, the control binary sequences determine the probability that each Boolean network is selected. For example, as the Boolean functions for the p53 gene occur with probabilities 0.5, 0.4, 0.09 and 0.01, the binary bit sequences for the control vectors 'S1S2' to the multiplexer are generated with a probability of 0.5 to be '00,' a probability of 0.4 to be '01', a probability of 0.09 to be '10' and a probability of 0.01 to be '11.' Then the output bit sequences are read out and decoded into (transition) probabilities. With a sequence length of 10000 bits, the STM is obtained as follows:

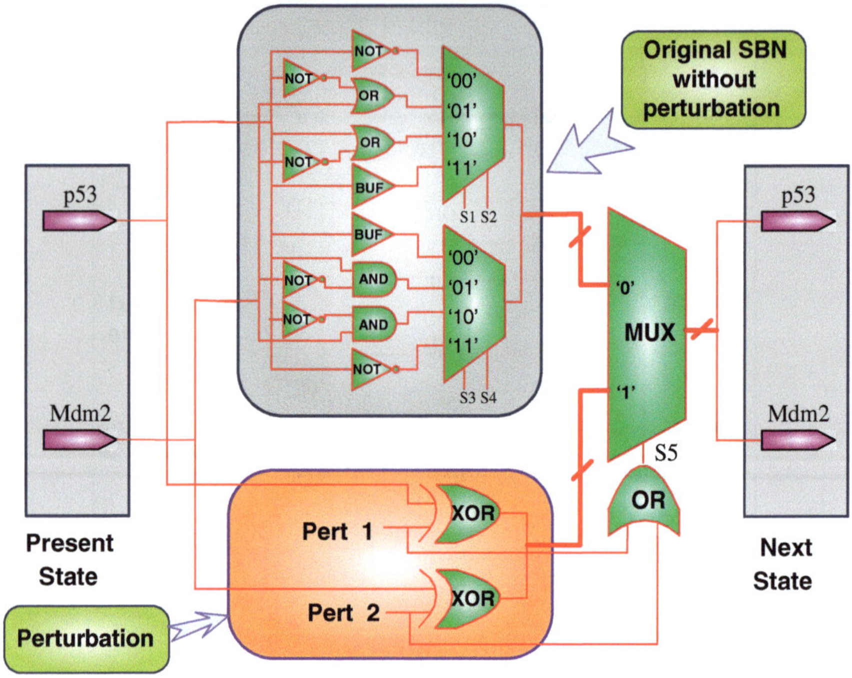

Fig. 10 An SBN for the p53–Mdm2 network (with perturbation)

$$A_{SBN} = \begin{bmatrix} 0.0097 & 0.0003 & 0.9803 & 0.0097 \\ 0.0899 & 0.0101 & 0.8101 & 0.0899 \\ 0.0904 & 0.8096 & 0.0096 & 0.0904 \\ 0.2511 & 0.2489 & 0.2489 & 0.2511 \end{bmatrix}. \quad (13)$$

The difference between (12) and (13) is evaluated using the following norms: $\|\cdot\|_1$ and $\|\cdot\|_\infty$, which specify the maximum absolute value of the summed differences of columns and rows of the two matrices respectively, and $\|\cdot\|_2$, which is a measure on the average difference of all the entries in these matrices. For (12) and (13), we obtain $\|A_{SBN} - A_{PBN}\|_1 = 0.0018$, $\|A_{SBN} - A_{PBN}\|_2 = 0.0024$ and $\|A_{SBN} - A_{PBN}\|_\infty = 0.0044$, which indicate that the SBN structure accurately computes the STM of the PBN.

With random gene perturbation, an SBN with perturbation can be constructed, as shown in Fig. 10. If the stochastic OR outputs a '1' (indicated by S5 in Fig. 10), which means that at least one of the p53 and Mdm2 are perturbed, the multiplexer is then switched to the perturbation network. If the output of the OR is 0, the multiplexer is switched to the original SBN and the network works as the one in Fig. 9 without perturbation.

A similar procedure can be used to compute the STM of the SBN with perturbation—the result is shown in (14) for a perturbation probability of 0.01:

$$\tilde{A}_{SBN} = \begin{bmatrix} 0.0097 & 0.0100 & 0.9705 & 0.0098 \\ 0.0975 & 0.0106 & 0.7946 & 0.0973 \\ 0.0998 & 0.7921 & 0.0082 & 0.0999 \\ 0.2444 & 0.2565 & 0.2551 & 0.2440 \end{bmatrix}. \tag{14}$$

Compared to the analytical result by a method based on (15):

$$\tilde{A}_{PBN} = \begin{bmatrix} 0.0097 & 0.0100 & 0.9705 & 0.0098 \\ 0.0981 & 0.0098 & 0.7940 & 0.0981 \\ 0.0981 & 0.7940 & 0.0098 & 0.0981 \\ 0.2451 & 0.2549 & 0.2549 & 0.2450 \end{bmatrix}, \tag{15}$$

the differences between (14) and (15) are revealed in the measures of $\|\tilde{A}_{SBN} - \tilde{A}_{PBN}\|_1 = 0.0032$, $\|\tilde{A}_{SBN} - \tilde{A}_{PBN}\|_2 = 0.0030$ and $\|\tilde{A}_{SBN} - \tilde{A}_{PBN}\|_\infty = 0.0042$. These show that the proposed approach using an SBN can accurately and efficiently compute the STM. The differences in these results come from the stochastic fluctuation, which is an intrinsic property of stochastic computation. Additional simulation results show that the fluctuations are generally small. A steady state analysis using (14) further confirms the p53–Mdm2 oscillatory dynamics observed in experiments.

4.2 A Multiple-Valued p53–Mdm2 Network

The dynamic behavior of a p53 network has been studied by using various Boolean models (Abou-Jaoude et al. 2009; Ciliberto et al. 2005) and an oscillatory behavior of the p53 and Mdm2 has been observed (Lahav et al. 2004; Batchelor et al. 2009). A four-node network has been analyzed in Abou-Jaoude et al. (2009), Murrugarra et al. (2012) with "DNA damage" as one of the nodes. As DNA damage (such as double strand breaks) is one of the major factors that activate the p53 network (Vogelstein et al. 2000; Lahav et al. 2004; Batchelor et al. 2009), a three-node network that excludes the DNA damage as an external factor, as shown in Fig. 11, is considered in this section for an application of the SMN model. Let X_1 denote the gene p53, cytoplasmic p53 and nucleic p53 (i.e. protein p53), and X_2 and X_3 denote the cytoplasmic Mdm2 and nucleic Mdm2, respectively. Protein p53 activates the cytoplasmic Mdm2 that has a positive effect on the nuclear Mdm2. Thus, protein p53 promotes nucleic Mdm2 indirectly through the cytoplasmic Mdm2. At the same time, p53 down-regulates nucleic Mdm2 by directly inhibiting the nuclear translocation of p53 (Abou-Jaoude et al. 2009; Murrugarra et al. 2012).

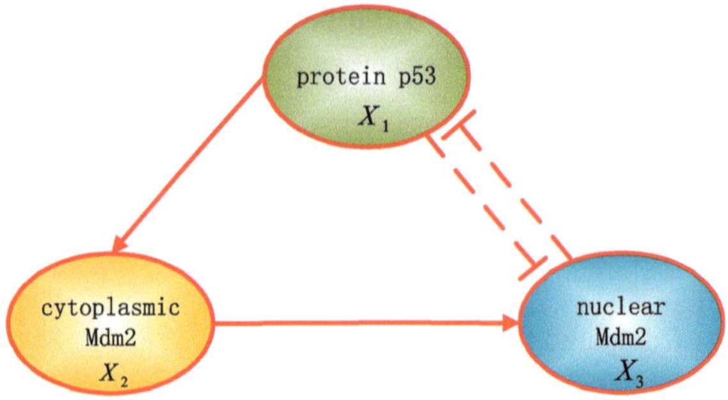

Fig. 11 The multiple-valued p53–Mdm2 network under DNA damage. Adapted from Abou-Jaoude et al. (2009), Murrugarra et al. (2012), Zhu and Han (2013) and © IEEE

Table 4 State transitions of X_2. Adapted from Zhu and Han (2013) and © IEEE

X_1	X^1 ($\geq$1)	X^2 (rotate)	X_2
0	1	2	0
1	1	2	0
2	2	0	1

Based on these interactions, an SMN for the p53 network is established as follows: $V = \{X_1, X_2, X_3\}$, where X_1 has ternary values, each of which indicates a different concentration level of the p53 protein (low, medium and high) (Abou-Jaoude et al. 2009), while X_2 and X_3 are binary nodes, with the ternary functional sets $F_1 = \left\{f_1^{(1)}, f_2^{(1)}\right\}$, $F_2 = \left\{f_1^{(2)}, f_2^{(2)}\right\}$, and $F_3 = \left\{f_1^{(3)}, f_2^{(3)}\right\}$. Given their truth tables (Murrugarra et al. 2012), these functions can be implemented by multiple-valued logic gates. For the gene node X_2 (i.e. cytoplasmic Mdm2), for example, the state transitions are shown in the first and last columns in Table 4. These transitions can be implemented by an ES operator and two rotate gates, as shown in Fig. 12. The intermediate states during the transitions are shown in Table 4.

Similarly, the implementation functions for the other genes X_1 and X_3 can be determined from their truth tables as well (in Tables 5 and 6 respectively).

While the state transition in (Murrugarra et al. 2012) is dependent on the current state and the state after transition, random state transitions are considered in this work, as in Shmulevich et al. (2002a, c), Shmulevich and Dougherty (2010), Kim et al. (2002). Under this assumption, the present state is transitioned into a next state with a transition probability when perturbation occurs. The selection probabilities are shown in Table 7 for the predictor functions.

For the p53–Mdm2 network in Fig. 11, an SMN can be constructed for implementing its functions, as shown in Fig. 13. For this three-gene network, a two-input multiplexer is used for each gene to probabilistically select a function with the selection probability encoded in the control sequence. For the update functions, $f_1^{(i)}$ ($i \in \{1, 2, 3\}$) is for the state transition due to interactions with other genes or the change of the current state, while $f_2^{(j)}$ ($j \in \{1, 2, 3\}$) indicates the preservation of the current state. In this model, the effect of asynchronicity (Luo and Wang 2013) is implicitly considered at each step of the state updating process.

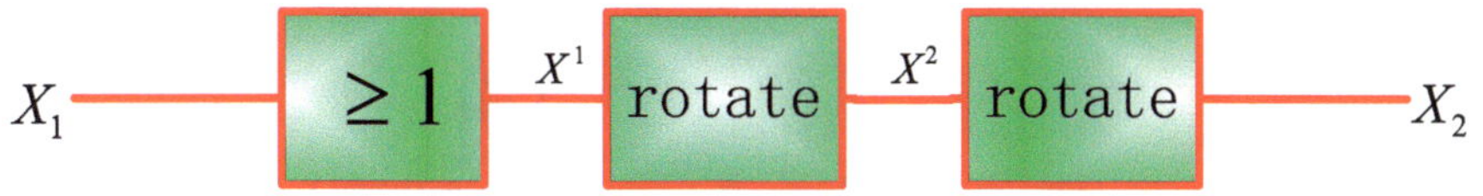

Fig. 12 A stochastic multiple-valued network (SMN) for the p53-Mdm2 network under DNA damage. Adapted from Zhu and Han (2013) and © IEEE

Table 5 Truth table for X_1. Adapted from Murrugarra et al. (2012), Zhu and Han (2013) and © IEEE

X_3	X_1	X_1
0	0	1
0	1	2
0	2	2
1	0	0
1	1	0
1	2	1

Table 6 Truth table for X_3. Adapted from Murrugarra et al. (2012), Zhu and Han (2013) and © IEEE

X_1	X_2	X_3
0	0	0
0	1	1
1	0	0
1	1	1
2	0	0
2	1	1

Table 7 The selection probabilities of predictor functions for the multiple-valued p53–Mdm2 network. Adapted from Zhu and Han (2013) and © IEEE

f^1	f^2	f^3
0.95	0.95	0.95
0.05	0.05	0.05

For each input state, the output sequences are read out and decoded into (transition) probabilities.

The p53 SMN model is used to compute the state transition matrix (STM) for this network, which is compared to the STM obtained by a Markov chain analysis. The norms $\|\cdot\|_1$, $\|\cdot\|_2$, and $\|\cdot\|_\infty$ are then used to measure the differences of the STMs obtained by the different methods. Let $\boldsymbol{A}_{SMN}$ and $\boldsymbol{A}_{MCA}$ be the STMs obtained by the SMN and a Markov chain analysis; the difference between these two matrices is then given by $\Delta\boldsymbol{A} = \boldsymbol{A}_{SMN} - \boldsymbol{A}_{MCA}$. For the multiple-valued p53–Mdm2 network with no perturbation, we obtain $\|\Delta\boldsymbol{A}\|_1 = 0.0049$, $\|\Delta\boldsymbol{A}\|_2 = 0.0023$ and $\|\Delta\boldsymbol{A}\|_\infty = 0.0021$ by using a sequence length of 10,000 values for the SMN.

The STM of the p53–Mdm2 network under perturbation can similarly be computed using an SMN with perturbation (by implementing the SMN in Fig. 13 into that of Fig. 6). The STMs obtained by different approaches are illustrated in

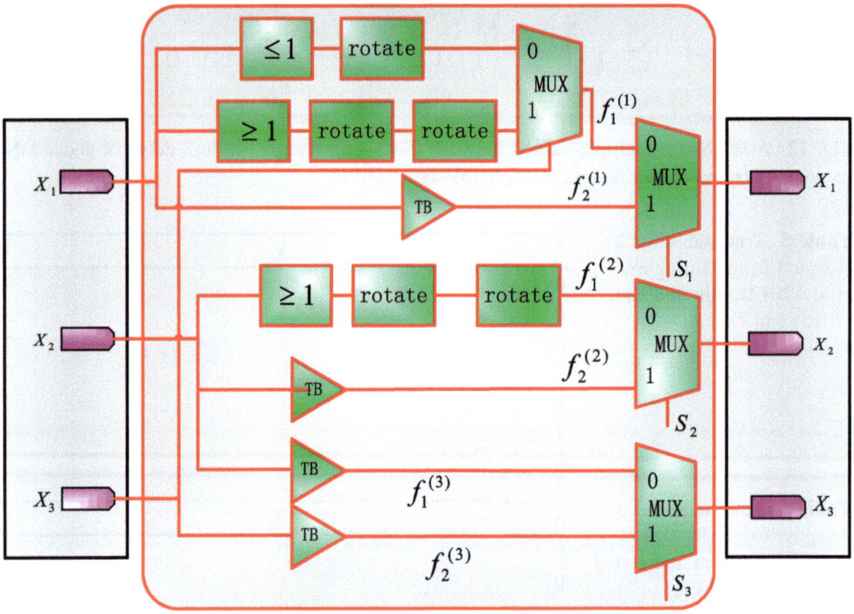

Fig. 13 A stochastic multiple-valued network (*SMN*) for the p53–Mdm2 network under DNA damage. Adapted from Zhu and Han (2013) and © IEEE

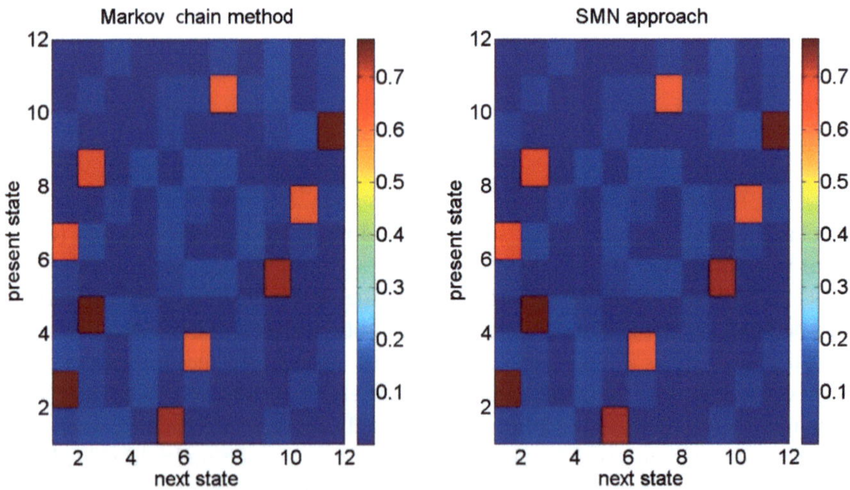

Fig. 14 State transition matrices (*STMs*) obtained by the Markov chain (Kim et al. 2002) and SMN approaches for the p53–Mdm2 network. Sequence length: L = 10,000 values; perturbation rate: $p = 0.1$. Adapted from Zhu and Han (2013) and © IEEE

Table 8 Norms of the difference between the STMs obtained by Markov chain analysis (MCA) and the SMN approach for the p53–Mdm2 network, $\Delta A_{MCA-SMN}$. Adapted from Zhu and Han (2013) and © IEEE

$p = 0$			
L (values)	1,000	10,000	100,000
$\|\|\Delta A_{MCA-SMN}\|\|_1$	0.0091	0.0049	7.6500×10^{-4}
$\|\|\Delta A_{MCA-SMN}\|\|_2$	0.0091	0.0023	8.1496×10^{-4}
$\|\|\Delta A_{MCA-SMN}\|\|_\infty$	0.0183	0.0021	0.0016
Average time for MCA (s)	0.00522		
Average time for SMN (s)	0.06804	0.57853	5.72595
$p = 0.1$			
L (values)	1,000	10,000	100,000
$\|\|\Delta A_{MCA-SMN}\|\|_1$	0.0368	0.0097	0.0030
$\|\|\Delta A_{MCA-SMN}\|\|_2$	0.0210	0.0061	0.0016
$\|\|\Delta A_{MCA-SMN}\|\|_\infty$	0.0401	0.0105	0.0032
Average time MCA (s)		0.01538	
Average time SMN (s)	0.05937	0.64545	5.96927

p perturbation rate; *L* sequence length for the stochastic approach

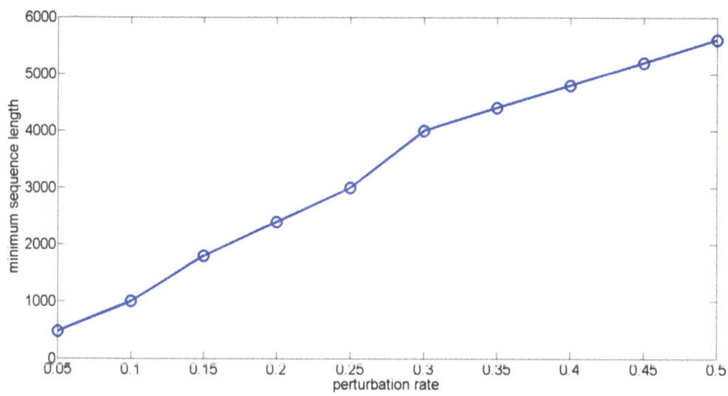

Fig. 15 The relationship between the minimum sequence length required for computing the STM (with an accuracy requirement of $\|\cdot\|_2 = 0.02$) and the perturbation rate for the multiple-valued p53–Mdm2 network. Adapted from Zhu and Han (2013) and © IEEE

Fig. 14, while the norms of the differences, $\|\Delta A\|_1$, $\|\Delta A\|_2$ and $\|\Delta A\|_\infty$, are shown in Table 8 for using different sequence lengths. The average run time is also provided for both approaches.

As revealed in Table 8, the difference between the STMs computed using the SMN and the Markov chain analysis decreases with the increase of sequence length *L*. For the same accuracy requirement, as can be seen, a larger sequence length is needed for a higher perturbation rate. This relationship between the sequence length and perturbation rate is further shown in Fig. 15. However, the computational inaccuracy due to the inherent stochastic fluctuation in stochastic computation is generally small and negligible. Hence, the proposed SMN approach can accurately

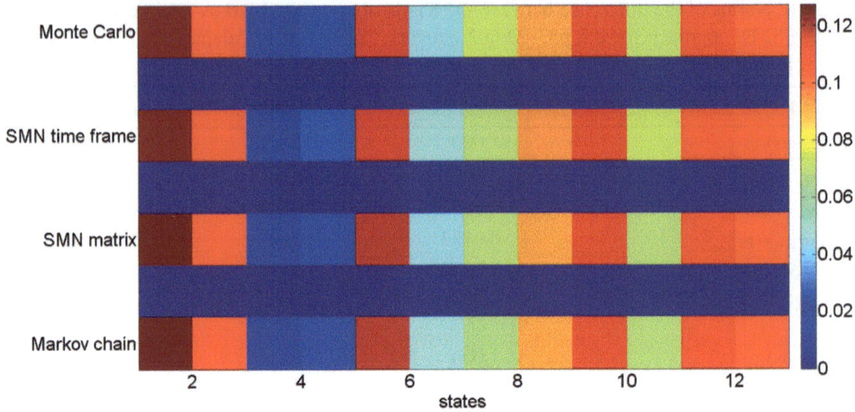

Fig. 16 Steady state distributions (*SSDs*) of the multiple-valued p53 network after 30 state transitions with an initial state of 000. The X-axis indicates the network state, and the Y-axis is for the different approaches. The color bar on the right shows the values of the SSD. Perturbation rate: $p = 0.1$; sequence length or simulation runs: 10,000. Adapted from Zhu and Han (2013) and © IEEE

and efficiently compute the STM of a probabilistic multiple-valued network (PMN) with or without perturbation.

A probabilistic network with random perturbation evolves as an ergodic Markov chain (Shmulevich et al. 2002c), because the non-zero perturbation rate makes all the states in the network connected. Hence, a steady state exists in a network with perturbation. The steady state distribution (SSD) for the p53 network under DNA damage is obtained by using different approaches, as shown in Fig. 16.

As shown in Fig. 16, all approaches produce similar SSDs. In fact, the difference between the results by the SMN and the accurate Markov chain analysis is negligible when reasonably long stochastic sequences are used (such as those of 10,000 values). Using the STM computed by an SMN approach or the time frame expanded SMN approach results in a very accurate approximation of the SSD compared to the rigorous Markov chain analysis. A further analysis shows that the relative error is less than approximately 0.2 % for the stochastic approach. Individual gene expressions are shown in Fig. 17 for a single simulation of 30 transitions. It can be seen that the likely expression levels of p53 and nuclear Mdm2 follow an oscillatory pattern as analytically (Murrugarra et al. 2012) and experimentally (Geva-Zatorsky et al. 2006) shown previously.

4.3 A WNT5A Network

Next, a WNT5A network (Kim et al. 2002) is used to illustrate the efficiency of the stochastic multiple-valued network (SMN) model and the time-frame expansion technique. A ten-gene network is derived from the predictive relationships in

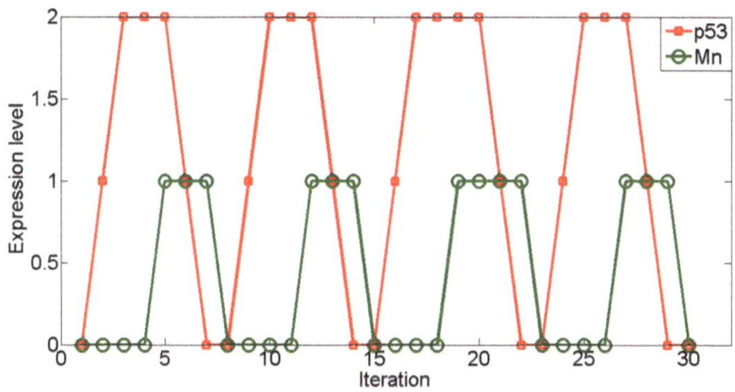

Fig. 17 Individual gene expressions for the p53 network generated from a single simulation of 30 iterations with an initial state of 011. X-axis indicates the iteration number and Y-axis shows the expression level of p53 or nuclear Mdm2. Adapted from Zhu and Han (2013) and © IEEE

Table 9 Selection probabilities of predictor functions for a WNT5A network. Estimated from Kim et al. (2002), Zhu and Han (2013) and © IEEE

Target	Predictor f_1	Select prob.	Predictor f_2	Select prob.	Predictor f_3	Select prob.
pirin	WNT5A	0.6	STC2	0.2	HADHB	0.2
WNT5A	pirin	0.6	S100P	0.2	RET-1	0.2
S100P	WNT5A	0.33	RET-1	0.33	Synuclein	0.34
RET-1	pirin	0.43	WNT5A	0.24	S100P	0.33
MMP-3	S100P	0.43	RET-1	0.25	HADHB	0.32
PHO-C	MART-1	0.33	Synuclein	0.33	STC-2	0.34
MART-1	pirin	0.44	WNT5A	0.28	MMP-3	0.28
HADHB	pirin	0.3	WNT5A	0.4	MMP-3	0.3
Synuclein	pirin	0.25	S100P	0.25	MART-1	0.5
STC2	pirin	0.35	WNT5A	0.3	PHO-C	0.35

Table 9. The selection probabilities of the predictor functions are also given in Table 9 [estimated from (Kim et al. 2002)]. Figure 18 shows a detailed structure of the network with double (or single)—headed arrows indicating the bi (or uni)—directional relationships of gene pairs. While the number of output arcs varies, every node (or gene) has three input arcs in Fig. 18.

For the 10-gene ternary WNT5A network, it requires a state transition matrix (STM) of $3^{10} = 59{,}049$ columns and rows for an accurate analysis. This makes it difficult, if not impossible, to estimate the steady state of an SMN using a matrix-based analysis. In general, it is difficult to analyze a large gene network, due to its excessive computational overhead. A Monte Carlo (MC) method has been used in (Kim et al. 2002) for evaluating the steady state distribution (SSD) of a network with perturbation. However, the MC method is very time consuming due to the slow convergence typically encountered in an MC simulation.

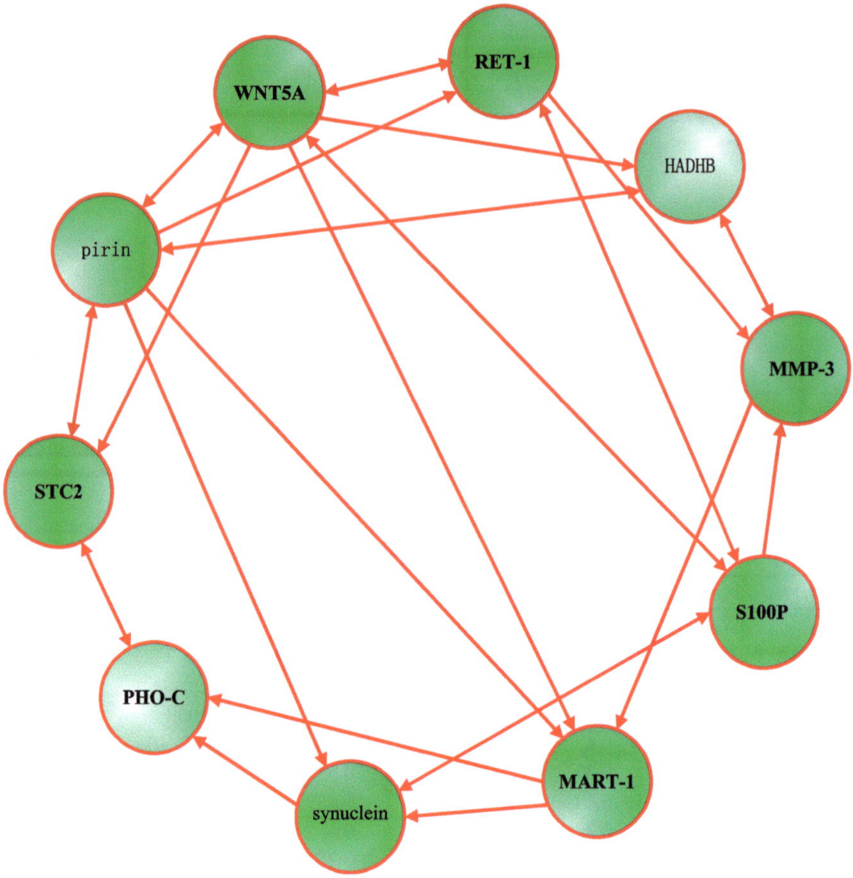

Fig. 18 A ternary WNT5A network with gene interactions. Adapted from Kim et al. (2002), Zhu and Han (2013) and © IEEE

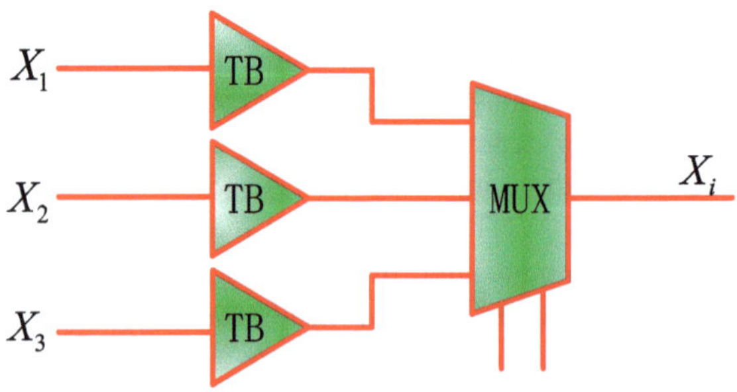

Fig. 19 An SMN module for gene i in the ternary WNT5A network, with the predictor function implemented by a ternary buffer. Let $G_i = (X_1, X_2, X_3)$ be the input vector for gene i; the input vector for each of the genes in the ternary WNT5A network is given by: $G_{WNT5A} = (pirin, S100P, RET - 1)$; $G_{pirin} = (WNT5A, STC2, HADHB)$; $G_{RET-1} = (pirin, WNT5A, S100P)$; $G_{HADHB} = (pirin, WNT5A, MMP - 3)$; $G_{MMP-3} = (S100P, RET - 1, HADHB)$; $G_{S100P} = (WNT5A, RET - 1, Synuclein)$; $G_{MART-1} = (pirin, WNT5A, MMP - 3)$; $G_{Synuclein} = (pirin, S100P, MART - 1)$; $G_{PHO-C} = (MART - 1, Synuclein, STC2)$;$G_{STC2} = (pirin, WNT5A, PHO - C)$. Adapted from Zhu and Han (2013) and © IEEE

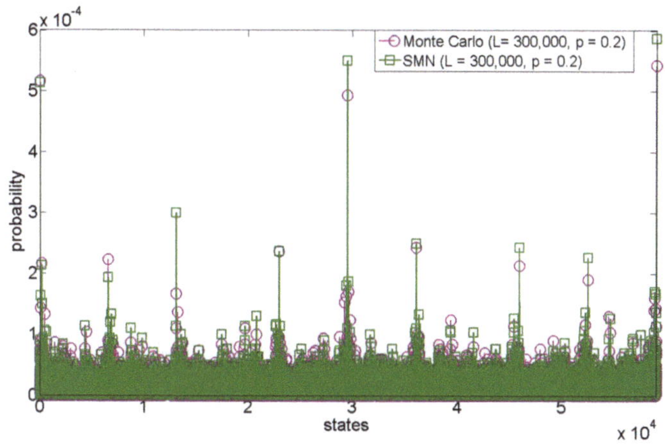

Fig. 20 SSDs of the ternary WNT5A network using the SMN model and Monte Carlo (MC) simulation with perturbation rate $p = 0.2$ and sequence length or simulation runs $L = 300{,}000$. Adapted from Zhu and Han (2013) and © IEEE

However, an SMN model can be constructed for the ternary WNT5A network, as shown in Fig. 19. For this SMN, the SSD can be estimated using the aforementioned time-frame expansion technique and compared with the MC simulation (Kim et al. 2002). By the time-frame expansion technique, the temporal operation of an SMN is laid out into a series of identical SMN modules in the spatial domain (as in Fig. 7). The required iterations of the SMN are determined by the number of state transitions before reaching a steady state. As in Zhang et al. (2007), a steady state is considered to have been reached if the discrepancy between two adjacent simulations is smaller than a threshold or the number of simulations has reached a maximum value. The state or GAP of the WNT5A network can be represented by a ternary vector as $(x_1, x_2, \ldots, x_{10})$, or its decimal index. The SSDs of the network with all of the 59,049 states, obtained using the SMN and the MC method (Kim et al. 2002), are shown in Fig. 20.

The norms of the differences of the SSDs obtained using the time frame expanded SMN approach with different sequence lengths and the MC method are shown in Table 10. As can be seen, the time-frame expanded SMN technique efficiently evaluates the SSD of the WNT5A network and produces very accurate results compared to the Monte Carlo simulation (Kim et al. 2002). The average run time reveals the efficiency of the SMN approach. This is because the use of randomly permuted sequences results in a faster convergence than in the MC simulation. The use of longer stochastic sequences further improves the accuracy of evaluation and remains more efficient by several orders of magnitude than the MC method. Albeit at a higher memory cost than the MC simulation (shown in Table 10), the SMN approach requires much less memory than an accurate approach such as a Markov chain analysis (Zhu and Han 2013). Since it is difficult to compute the STM or SSD of

Table 10 Norms of the difference between the SSDs obtained by the time frame expanded SMN technique and Monte Carlo (MC) simulation for the ternary WNT5A network with perturbation rate $p = 0.2$. Adapted from Zhu and Han (2013) and © IEEE

L (values)	3 k	30 k	300 k	3000 k
$\|\|\boldsymbol{SSD}_{MC} - \boldsymbol{SSD}_{SMN}\|\|_1$	1.8827	1.3291	0.4915	0.1605
$\|\|\boldsymbol{SSD}_{MC} - \boldsymbol{SSD}_{SMN}\|\|_2$	0.0258	0.0082	0.0026	8.5342×10^{-4}
$\|\|\boldsymbol{SSD}_{MC} - \boldsymbol{SSD}_{SMN}\|\|_\infty$	1.0000×10^{-3}	2.6667×10^{-4}	1.3333×10^{-4}	5.6333×10^{-5}
Average time for MC (s)	98.4768	981.159	9731.04	97336.5
Average time for SMN (s)	0.47811	4.23694	58.9336	673.928
Required memory for MC (M Byte)	2.7117	10.0076	51.2108	599.607
Required memory for SMN (M byte)	9.8083	40.9368	373.299	3696.5

The average run time is also shown. *L*: sequence length for the stochastic approach or the number of simulation runs for the MC method; $\boldsymbol{SSD}_{MC}$ and $\boldsymbol{SSD}_{SMN}$ respectively denote the steady state distributions obtained by the MC simulation and the time frame expanded SMN technique; a maximum number of 50 iterations is applied to the steady state evaluation

a large GRN by using an accurate analysis, a time-frame expanded SMN provides an alternative method to evaluate the SSD of a large network with a tunable tradeoff between accuracy and efficiency by using stochastic sequences of different lengths.

5 Conclusion

Stochastic logic networks are presented as an efficient approach to modeling the effects of noise in gene regulatory networks (GRNs). In stochastic Boolean and multiple-valued networks (SBNs and SMNs), the state transition matrix can be accurately and efficiently computed with a complexity of $O(nLk^n)$, where n is the number of genes in a network, k is the quantization level of a gene's state ($k = 2$ for SBNs) and L is a factor determined by the stochastic sequence length. Since L increases slower with n than the number of network functions, N, this result is an improvement compared to the previous result of $O(nNk^{2n})$ for an accurate analysis. The use of randomly permuted sequences further increases computational efficiency and allows for a tunable tradeoff between accuracy and efficiency. A steady state analysis using a time-frame expansion technique has shown a significant speedup compared to an accurate Markov chain analysis and produced very accurate results compared to Monte Carlo simulation.

Stochastic networks are constructed for the analysis of a p53–Mdm2 network; furthermore, SMNs are constructed for the analysis of a ternary WNT5A network under gene perturbation. Simulations have revealed the oscillatory dynamics of the p53–Mdm2 network with random gene perturbation. The stochastic approach can also efficiently predict the steady state distribution of the WNT5A network with gene perturbation. Hence, the stochastic logic networks are useful in evaluating the effects of gene perturbation and, potentially, helpful in drug discovery for an intervention-based gene therapy.

Acknowledgment This work was supported in part by the Natural Sciences and Engineering Research Council of Canada (NSERC) in a Discovery Grant.

References

Abdi A, Tahoori MB, Emamian ES (2008) Fault diagnosis engineering of digital circuits can identify vulnerable molecules in complex cellular pathways. Sci Signal 1(42):ra10

Abou-Jaoude W, Ouattara D, Kaufman M (2009) From structure to dynamics: frequency tuning in the p53–mdm2 network: I. logical approach. J Theor Biol 258(4):561–577. doi:10.1016/j.jtbi.2009.02.005

Adamatzky A (2003) On dynamically non-trivial three-valued logics: oscillatory and bifurcatory species. Chaos Solit Fract 18:917–936

Adar R, Benenson Y, Linshiz G, Rosner A, Tishby N, Shapiro E (2004) Stochastic computing with biomolecular automata. PNAS 101(27):9960–9965

Aldana M, Coopersmith S, Kadanoff LP (2003) Boolean dynamics with random couplings. http://arXiv.org/abs/adap-org/9305001

Batchelor E, Loewer A, Lahav G (2009) The ups and downs of p53: understanding protein dynamics in single cells. Nature Rev Cancer 9:371–377

Benenson Y, Gil B, Ben-Dor U, Adar R, Shapiro E (2004) An autonomous molecular computer for logical control of gene expression. Nature 429:423–429

Ching W, Zhang S, Ng M, Akutsu T (2007) An approximation method for solving the steady-state probability distribution of probabilistic Boolean networks. Bioinformatics 23:1511–1518

Ciliberto A, Novak B, Tyson JJ (2005) Steady states and oscillations in the p53–Mdm2 network. Cell Cycle 4:486–493

de Jong H (2002) Modeling and simulation of genetic regulatory systems: a literature review. J Comput Biol 9(1):67–103. doi:10.1089/10665270252833208

Dougherty ER, Pal R, Qian X, Bittner ML, Datta A (2010) Stationary and structural control in gene regulatory networks: basic concepts. Int J Syst Sci 41(1):5–16

Dubrova E (2006) Random multiple-valued networks: theory and applications. In: Proceedings of international symposium on multiple-valued logic (ISMVL '06), pp 27–33, May 2006

Elowitz MB, Levine AJ, Siggia ED, Swain PS (2002) Stochastic gene expression in a single cell. Science 297:1183–1186

Faryabi B, Vahedi G, Datta A, Chamberland JF, Dougherty ER (2009) Recent advances in intervention in Markovian regulatory networks. Curr Genomics 10(7):463–477

Gaines BR (1969) Stochastic computing systems. Adv Inf Syst Sci 2:37–172

Garg A, Mendoza L, Xenarios I, DeMicheli G (2007) Modeling of multiple valued gene regulatory networks. In: Proceedings of 29th IEEE International Conference on Engineering in Medicine and Biology Society (EMBC '07), pp. 1398–1404, Aug 2007

Geva-Zatorsky N, Rosenfeld N, Itzkovitz S, Milo R, Sigal A, Dekel E, Yarnitzky T, Liron Y, Polak P, Lahav G, Alon U (2006) Oscillations and variability in the p53 system. Mol Syst Biol 2:0033. doi: 10.1038/msb4100068

Glass L, Kauffman S (1973) The logical analysis of continuous non-linear biochemical control networks. J Theor Biol 39:103–129

Guelzim N, Bottani S, Bourgine P, Kepes F (2002) Topological and causal structure of the yeast transcriptional regulatory network. Nat Genet 31:60–63

Han J, Chen H, Liang J, Zhu P, Yang Z, Lombardi F (2013) A stochastic computational approach for accurate and efficient reliability evaluation. IEEE Trans Comput (in press)

Harvey I, Bossomaier T (1997) Time out of joint: attractors in asynchronous random Boolean networks. In: Husbands P, Harvey I. (eds) Proceedings of 4th European conference on artificial life (ECAL97). MIT Press, New York, pp 67–75

Huang S (1999) Gene expression profiling, genetic networks, and cellular states: an integrating concept for tumorigenesis and drug discovery. J Mol Med 77:469–480

Ivanov I, Pal R, Dougherty ER (2007) Dynamics preserving size reduction mappings for probabilistic Boolean networks. IEEE Trans Signal Process 55(5):2310–2322

Karlebach G, Shamir R (2008) Modelling and analysis of gene regulatory networks. Nat Rev Mol Cell Biol 9:770–780

Karlebach G, Shamir R (2010) Minimally perturbing a gene regulatory network to avoid a disease phenotype: the glioma network as a test case. BMC Syst Biol 4:15

Kauffman SA (1969) Metabolic stability and epigenesis in randomly constructed genetic nets. Theor Biol 22:437–467

Kervizic G, Corcos L (2008) Dynamical modeling of the cholesterol regulatory pathway with Boolean networks. BMC Syst Biol 2:99

Kim S, Li H, Dougherty ER et al (2002) Can Markov chain models mimic biological regulation? J Biol Syst 10(4):337–357

Kitano H (2001) Foundations of systems biology. MIT Press, Massachusetts

Lahav G, Rosenfeld N, Sigal A, Geva-Zatorsky N, Levine AJ, Elowitz MB, Alon U (2004) Dynamics of the p53–Mdm2 feedback loop in individual cells. Nat Genet 36:147–150

Li Z, Cheng D (2010) Algebraic approach to dynamics of multivalued networks. Int J Bifurcat Chaos 20(3):561–582

Liang J, Han J (2012) Stochastic Boolean networks: an efficient approach to modeling gene regulatory networks. BMC Syst Biol 6:113

Luo C, Wang X (2013) Dynamics of random Boolean networks under fully asynchronous stochastic update based on linear representation. PLoS ONE 8(6):e66491. doi:10.1371/journal.pone.0066491

McAdams HH, Shapiro L (1995) Circuit simulation of genetic networks. Science 269(5224):650

Morris MK, Saez-Rodriguez J, Sorger PK, Lauffenburger DA (2010) Logic-based models for the analysis of cell signaling networks. Biochemistry 49:3216–3224

Murrugarra D, Veliz-Cuba A, Aguilar B, Arat S, Laubenbacher R (2012) Modeling stochasticity and variability in gene regulatory networks. EURASIP J Bioinform Syst Biol 1:5

Pal R (2010) Context-sensitive probabilistic Boolean networks: steady-state properties, reduction, and steady-state approximation. IEEE T Signal Proces 58(2):879–890

Pandey S, Wang R, Wilson L, Li S, Zhao Z, Gookin T, Assmann S, Albert R (2010) Boolean modeling of transcriptome data reveals novel modes of heterotrimeric G-protein action. Mol Syst Biol 372. doi:10.1038/msb.2010.28

Qian X, Ivanov I, Ghaffari N, Dougherty ER (2009) Intervention in gene regulatory networks via greedy control policies based on long-run behavior. BMC Syst Biol 3:61

Qian X, Ghaffari N, Ivanov I, Dougherty ER (2010) State reduction for network intervention in probabilistic Boolean networks. Bioinformatics 26(24):3098–3104

Rosenthal JS (1995) Minorization conditions and convergence rates for Markov chain Monte Carlo. J Am Stat Assoc 90:558–566

Shmulevich I, Dougherty ER (2010) Probabilistic Boolean networks: the modeling and control of gene regulatory networks. ociety for Industrial & Applied Mathematics, U.S

Shmulevich I, Dougherty ER, Zhang W (2002a) From Boolean to probabilistic Boolean networks as models of genetic regulatory networks. In: Proceedings of IEEE, vol 90, pp 1778–1792

Shmulevich I, Dougherty ER, Kim S, Zhang W (2002b) Probabilistic Boolean networks: a rule-based uncertainty model for gene regulatory networks. Bioinformatics 18:261–274

Shmulevich I, Dougherty ER, Zhang W (2002c) Gene perturbation and intervention in probabilistic Boolean networks. Bioinformatics 18(10):1319–1331

Shmulevich I, Gluhovsky I, Hashimoto RF, Dougherty ER, Zhang W (2003) Steady-state analysis of genetic regulatory networks modelled by probabilistic Boolean networks. Comp Funct Genom 4:601–608. doi:10.1002/cfg.342

Thomas R, D'Ari R (1990) Biological feedback. CRC Press, Boca Raton

Vogelstein B, Lane D, Levine AJ (2000) Surfing the p53 network. Nature 408:307–310

Volker LG, Conrad M (1998) The role of weak interactions in biological systems: the dual dynamic model. J Theor Biol 193:287–306

von Neumann J (1956) Probabilistic logics and the synthesis of reliable organisms from unreliable components. In: Shannon CE, McCarthy J (eds) Automata studies. Princeton University Press, Princeton, pp 43–98

Weinberg RA (2006) The biology of cancer, 1st edn. Garland Science, New York

Zhang S et al (2007) Simulation study in probabilistic Boolean network models for genetic regulatory networks. Int J Data Min Bioinformatics 1:217–240

Zhu P, Han J (2013) Stochastic multiple-valued gene networks. IEEE Trans Biomed Circuits Syst 8(1):42–53

Zhu P, Han J (2014) Asynchronous stochastic boolean networks as gene network models. J Comput Biol (in press)

Zhu P, Liang J, Han J (2014) Gene perturbation and intervention in context-sensitive stochastic boolean networks. BMC Syst Biol (in press)

Part IV
Nanocarriers for Drug Discovery and Treatment

Nanodiamonds as Intracellular Probes for Imaging in Biology and Medicine

Jitka Slegerova, Ivan Rehor, Jan Havlik, Helena Raabova, Eva Muchova and Petr Cigler

Abstract In recent years, diamond nanoparticles have received a great deal of attention due to their unique photophysical and biological properties. Nanodiamonds (NDs) show low toxicity and are considered to be a highly biocompatible carbon nanomaterial useful in a wide range of applications. Thanks to their ability to accommodate nitrogen-vacancy (N-V) color centers, NDs are a prime example of non-photobleachable fluorescent labels and nanosensors. Here, we present a survey of ND applications in biology and medicine with an emphasis on bioimaging. We focus on distinguishing the properties of detonation NDs and high-pressure high-temperature (HPHT) NDs and describing their physicochemical properties, structure and possible modifications by small molecules and biomolecules. We summarize and critically evaluate in vitro and in vivo data on ND toxicity and biocompatibility, cellular internalization, localization and targeting by surface-attached ligands. We discuss current achievements in bioimaging using fluorescent NDs and the potential of NDs in diagnostics and drug delivery.

Keywords Nanodiamond · Fluorescence · Nitrogen-vacancy center · Intracellular probe · Bioimaging · Biocompatibility · Targeting · Cellular internalization · Drug delivery

Abbreviations

ATRP	Atom-transfer radical-polymerization
CVD	Chemical vapor deposition
DND	Detonation nanodiamond
DOX	Doxorubicin hydrochloride
ESR	Electron spin resonance
FND	Fluorescent nanodiamond

J. Slegerova · I. Rehor · J. Havlik · H. Raabova · E. Muchova · P. Cigler (✉)
Laboratory of Synthetic Nanochemistry, Institute of Organic Chemistry and Biochemistry AS CR v. v. i., Flemingovo nam. 2 160 00 Prague 6, Czech Republic
e-mail: cigler@uochb.cas.cz

A. Prokop et al. (eds.), *Intracellular Delivery II*,
Fundamental Biomedical Technologies 7, DOI: 10.1007/978-94-017-8896-0_18,

GH	Growth hormone
GHR	Growth hormone receptor
HCPT	10-Hydroxycamptothecin
HPHT	High-pressure high-temperature
MTT	3-[4,5-dimethylthiazol-2-yl]-2,5-diphenyltetrazolium bromide
MWNT	Multi-walled carbon nanotubes
ND	Nanodiamond
ND-FA	Nanodiamond conjugated with folic acid
ND-Tx	Nanodiamond conjugated with chlorotoxin
N-V center	Nitrogen-vacancy center
PAH	Polyallylamine
PEI-800	Polyethyleneimine-800
SWNT	Single-walled carbon nanotubes
TE-DNDs	Targeted epirubicin-loaded DNDs
ZPL	Zero phonon line

1 Introduction

Nanodiamond (ND) particles are known since the 1950s: they were first produced in the USSR by detonation of explosives (Danilenko 2004). The research and potential applications of diamond nanoparticles have been largely overlooked and applications of NDs in biology and medicine began only about one decade ago. However, since then, we have witnessed a boom in ND basic research and development of ND-based applications in various fields.

The properties and applications (Mochalin et al. 2011; Krueger 2008, 2010, 2011; Holt 2007; Schrand et al. 2009), preparation (Kharisov et al. 2010), chemistry (Krueger and Lang 2012; Krueger 2008b, 2008c) and photophysics of NDs (Wrachtrup et al. 2013; Aharonovich et al. 2011; Jelezko and Wrachtrup 2006) and their use in bioimaging (Vaijayanthimala and Chang 2009; Liu et al. 2012; Hui et al. 2010; Barnard 2009), drug delivery (Zhu 2012) and nanoscale medicine (Ho 2009, 2010; Man and Ho 2012) have been recently reviewed. Here, we present a comprehensive survey of current research related to use of NDs as bioimaging probes and to their biological properties. We also critically assess the biological and toxicological differences between individual types of NDs. In addition to this main focus, we introduce the basics of ND structure, physico-chemical properties and available synthetic pathways for the attachment of various molecules to ND surfaces. Finally, we discuss the potential of ND use in drug delivery.

2 Physical and Chemical Properties of Nanodiamonds

2.1 Classification, Structure and Preparation of Nanodiamonds

NDs belong to a broad family of nanocarbon materials that includes fullerenes, tubes, onions, rods, platelets and graphene (Shenderova et al. 2002; Geim and Novoselov 2007). ND particles are typically polyhedra comprising a sp^3-hybridized diamond core coated with a sp^2-hybridized graphite shell or by amorphous carbon with dangling bonds with terminal functional groups.

NDs can be classified by size as nanocrystalline (<100 nm) or ultrananocrystalline (<10 nm) (Williams 2011). They also can be sorted into three basic types based on the method of their synthesis: chemical vapor deposition (CVD) NDs, high-pressure high-temperature (HPHT) NDs and detonation NDs (DNDs). A transmission electron microscopy image of DNDs and HPHT NDs (Rehor and Cigler 2014) is shown in Fig. 1.

DNDs are synthesized by controlled detonation of explosives with a negative oxygen balance (Dolmatov 2007). The resulting detonation soot contains primary particles that are 4–5 nm in diameter, which form clusters ranging from 100 to 200 nm in diameter (Krüger et al. 2005). These cluster have been successfully de-aggregated by milling DND suspensions with ceramic microbeads and by micro-bead assisted ultrasonic disintegration (Krüger et al. 2005; Ozawa et al. 2007; Eidelman et al. 2005; Liang et al. 2009).

HPHT NDs are prepared from larger, micron-sized crystals grown from graphite at 5 GPa at temperatures higher than 1,400 °C. HPHT NDs contain up to 300 ppm of nitrogen impurities, which give rise, after proper treatment, to highly fluorescent NV centers. These fluorescent NDs are widely applicable as cellular labels or markers (Schrand et al. 2009; Boudou et al. 2009).

CVD NDs are grown as nanocrystalline and ultracrystalline films, which can be used to create biocompatible coatings with superior mechanical and wear resistant properties. CVD NDs also can be used for biosensor applications after proper surface modification with biomolecules. The grain size varies from 5 nm to several micrometers, and the particles can display different morphologies and crystallinities (Philip et al. 2003; Butler and Sumant 2008; Gracio et al. 2010).

2.2 Colloidal Properties of Nanodiamonds

To handle and process NDs, a stable colloidal dispersion is generally required. The colloidal behavior of NDs prepared by different synthetic methods varies considerably. While HTHP and CVD nanoparticles tend to agglomerate moderately, DNDs usually form strongly bound agglomerates if no countermeasures are taken

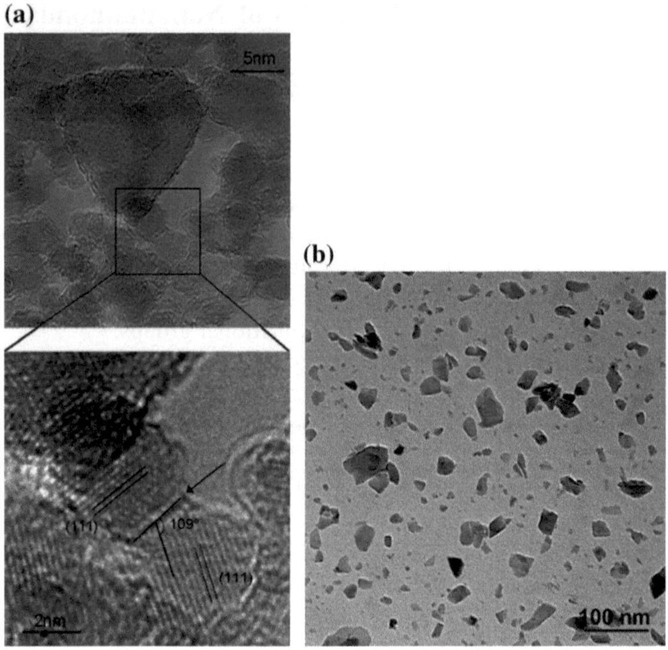

Fig. 1 Transmission electron microscopy (*TEM*) images of **a** DNDs adapted from Chang et al. (2011) and **b** HPHT NDs. Courtesy of Royal Society of Chemistry

(Krüger et al. 2005). The colloidal properties are further determined by ND size, origin and the chemical groups present on the nanoparticle surface.

A key parameter that describes the stability of a colloid and its tendency toward agglomeration is called zeta potential. A colloid is considered stable when zeta potential is lower than −30 mV or higher than 30 mV. NDs can form colloids with zeta potentials ranging from −50 mV to +50 mV, depending on their terminal groups.

For NDs to be useful in bioapplications, they must show colloidal stability in solutions of electrolytes. Electrolyte molecules disturb the equilibrium between attractive van der Waals forces and repulsive Coulomb forces (which act between the charged colloidal particles). Destabilization inevitably leads to unwanted particle aggregation in buffers and biological liquids (Dahoumane et al. 2009; Sreenivasan et al. 2011). Stable NDs can be prepared by modifying the particles with charged groups that strengthen electrostatic forces (Liang et al. 2011). However, in this case, the particles bear substantive charge, which can compromise their use in certain applications. A preferred approach involves modification of the ND surface with high molecular weight molecules [e.g., polymers (Zhao et al. 2011) or proteins (Dahoumane et al. 2009)] to introduce steric stabilization. The high molecular weight molecules can be either covalently bound (Sreenivasan et al. 2011) or adsorbed (Maitra et al. 2008; Nguyen et al. 2007) on the surface.

2.3 Chemical and Macromolecular Surface Modification

Chemical treatment and subsequent functionalization of the ND surface have a critical impact on the properties and behavior of NDs (Mochalin et al. 2011; Krueger and Lang 2012).

The properties and reactivity of DNDs, HPHT NDs and CVD NDs vary significantly, due in part to the different sizes, shapes and chemical compositions of the particles (Krueger and Lang 2012; Barnard 2009). The surface of ND particles and particle aggregates is typically oxidized during the production process (the only exception being CVD NDs, whose surfaces are initially hydrogenated; however, subsequent isolation treatment leads to an oxidized surface). The ND surface is typically covered with a variety of functional groups, including carboxyl, lactone, ketone, hydroxy and sometimes alkyl groups, as well as a considerable amount of graphitic material (Jiang and Xu 1995; Mochalin et al. 2009; Tu et al. 2006). Surfaces containing a variety of groups are not suitable for selective or precise functional tailoring. Indeed, initial chemical pretreatment helps achieve desired surface homogeneity (a surface containing only one functional group).

Various functional organic groups can be attached to ND surfaces. Treatment with a wide range of reactants can produce esters, amides, hydroxyls, silane groups, and other functionalities (see Fig. 2). A full description of all possible surface modifications is beyond the scope of this chapter; for further reading, we recommend the comprehensive review of Krueger and Lang (2012).

Polymer grafting to ND surfaces is another attractive way to improve and homogenize ND properties. Both "grafting to" and "grafting from" approaches have shown to be reliable methods for producing NDs with surface polymers, although the "grafting from" approach remains more popular. Atom-transfer radical-polymerization (ATRP) has been shown to be an efficient method for growing polymer brushes on ND surfaces terminated with an initiator (Dahoumane et al. 2009; Li et al. 2006; Zhang et al. 2012a). Methacrylamide copolymers were introduced as highly biocompatible, protein resistant coating enabling bioorthogonal attachment of various molecules by click chemistry (Rehor et al. 2014a). Polymer molecules covering the ND surface are hydrophilic and highly flexible, and they efficiently prevent aggregation.

Recently, novel NDs with silica coatings have been introduced (Rehor et al. 2014b, von Haartman et al. 2013; Prabhakar et al. 2013; Bumb et al. 2013). This coating provides a platform for subsequent chemical treatment based on silica chemistry. It enables selective attachment of various biomolecules and radically improves the colloidal stability of the particles.

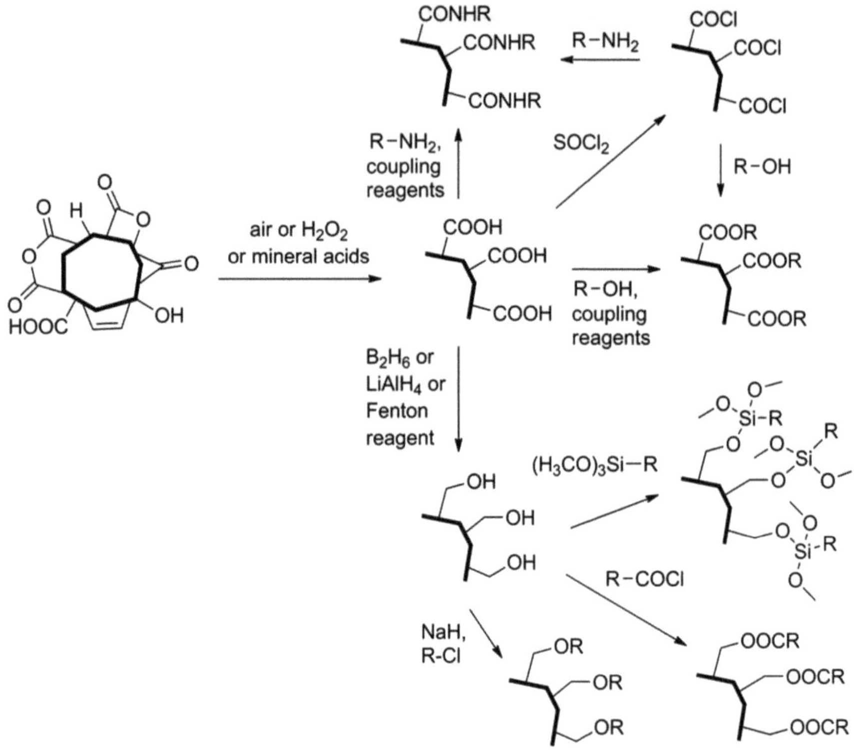

Fig. 2 Overview of frequently used chemical modifications of oxidized NDs. Notably, the ND always bears a mixture of various moieties because the reactions never reach 100 % completion

2.4 Biomolecules on the ND Surface

Connecting biomolecules to ND surfaces is desirable for numerous applications, including cellular targeting, internalization and drug and gene delivery.

Both non-covalent and covalent linkages are suitable for biomolecule attachment to ND surfaces. Non-covalent linkages can easily be created by incubating NDs in a solution containing the biomolecule. For example, lysozyme (Nguyen et al. 2007; Chung et al. 2006; Perevedentseva et al. 2007, 2011; Wu 2010), bovine serum albumin (Li et al. 2010; Tzeng et al. 2011; Wang et al. 2011), cytochrome (Huang and Chang 2004), toxins (Liu et al. 2008; Puzyr' et al. 2007) and DNA (Purtov et al. 2008) have been adsorbed on ND surfaces. DNA can also bind to pre-treated particles, such as poly-L-lysine (Fu et al. 2007) and thionine (Martín et al. 2010a) coated NDs. The major disadvantage of adsorption is the non-specificity of the interaction. In other words, NDs bind almost every protein in solution. [Adsorption has been successfully applied, however, for high-affinity non-specific capture of proteins from solution (Kong et al. 2005)]. Covalent grafting, on the other hand, creates a stronger, much more specific connection between a molecule and the ND surface (Huang and Chang 2004). These connections are usually realized via amide bonds; either an amino group of a biomolecule reacts with a carboxyl group on the ND surface (Dahoumane et al. 2009; Sreenivasan et al. 2011; Fu et al. 2007, 2012; Cheng et al. 2007; Wei et al. 2010; Zhang et al. 2009a; Mkandawire et al. 2009; Mohan et al. 2010; Li and Zhou 2010; Weng et al. 2009, 2012) or vice versa (Hens et al. 2008). Silane linkers can be also used to connect biomolecules to the ND surface by forming a covalent amide bond (Krüger et al. 2006; Krueger et al. 2008) or by the reaction of isocyanate with an amino group (Martín et al. 2010a).

When attaching a biomolecule to the ND surface, special care must be taken to maintain its bioactivity (Vaijayanthimala and Chang 2009). Various spacers can be inserted between a biomolecule and the ND to reduce steric constraints and retain the function of the biomolecule. This technique is often used to ensure that enzymes with active sites that may experience steric hindrance near the ND surface can bind and process substrate (Cao 2005). In addition, spacers can also suppress non-specific interactions (Yeap et al. 2008) and protein conformational changes caused by strong biomolecule-particle interactions (Nguyen et al. 2007; Vertegel et al. 2004). Because enzymatic activity decreases as surface coverage is lowered, the ND surface can be blocked with supplementary proteins to create a more "crowded" environment. For example, use of this method caused the activity of ND-bound lysozyme to increase from 60 to 70 % (Nguyen et al. 2007).

2.5 Optical Properties of Nanodiamonds

The key property of NDs that enables their use in bioimaging applications (Aharonovich et al. 2011; Hui et al. 2010; Chang et al. 2008; Liu et al. 2007) is their photoluminescence. NDs host many types of luminescent centers, and their nitrogen-vacancy (NV) centers have been particularly well-studied. NV centers emit red fluorescence in the visible range, which is favorable for optical bioimaging (Weissleder and Ntziachristos 2003). Moreover, they are completely resistant to photobleaching and photoblinking (Gruber et al. 1997; Tisler et al. 2011).

NV centers consist of a nearest-neighbor pair of a nitrogen atom, naturally present as a lattice defect, and a lattice vacancy (Fig. 3). NV centers are typically created by irradiation of NDs with high energy particles (electrons, protons, helium ions, etc.) followed by vacuum annealing (Chang et al. 2008; Neugart et al. 2007; Havlik et al. 2013). Irradiation of NDs creates vacancies (Slepetz et al. 2010), which migrate during annealing and get trapped by nitrogen atoms. Two charge states of the NV centre—negative NV^- and neutral NV^0—have been identified; each has different photoluminescence and spin properties. One NV can exist in both states depending on its surroundings (Wrachtrup et al. 2013; Kratochvílová et al. 2010).

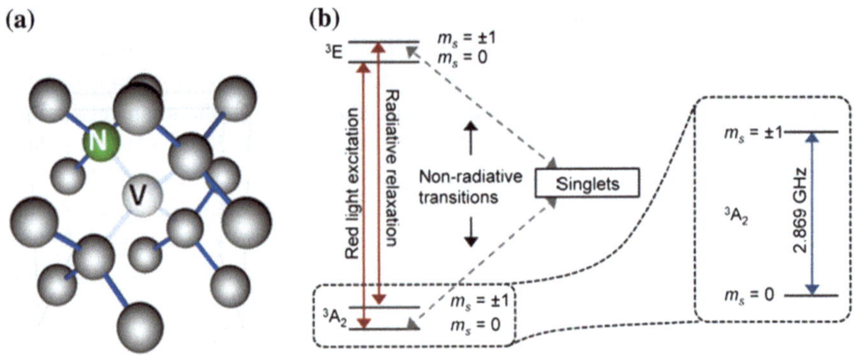

Fig. 3 **a** Schematic picture of a NV center. *N* is substitutional nitrogen, *V* is a carbon vacancy. Adapted from Balasubramanian et al. (2008), **b** Electronic level structure of NV depicting radiative and non-radiative transitions. Adapted from Hegyi and Yablonovitch (2013). Courtesy of the Nature Publishing Group and the American Society of Chemistry

NV center contains a three-level emission system (see Fig. 3). This means that, in addition to the desired optical transition, the centers have a parallel dark decay channel through a metastable state. Fluorescence intensity can thus be switched on and off by populating sublevels with magnetic spin numbers $m_s = 0$ and $m_s = \pm 1$ (see Fig. 3b) by optical pumping. This optically pumped spin polarization can be destroyed by a resonant microwave field (at 2.87 GHz), which inverts populations between spin sublevels.

In an external magnetic field, NV^- centers operate as atom-sized scanning probe vector magnetometers (Maze et al. 2008; Balasubramanian et al. 2008; Zhao et al. 2012) and can detect even weak magnetic or electric fields, for example, a field produced by a single electron located tens of nanometers away.

Low brightness is a major drawback of NV centers that limits their wider applications. Methods to boost fluorescence currently are being intensively investigated (Havlik et al. 2013; Chi et al. 2011; Schietinger et al. 2009).

In addition to the red fluorescence of NV color centers, NDs also emit green fluorescence (Davies et al. 1992; Opitz et al. 2010). This fluorescence arises from the excitation of N–V–N color centers. Similar to NV centers, N–V–N centers are thermally stable and do not photobleach or photoblink. Diamonds containing these color centers are, for example, used as a lasing medium for room temperature color center lasers.

2.5.1 Nanodiamonds as Optical Chemosensors

Recent research has opened the possibility of using FNDs as an optical sensing device. Sensing is based on switching between the neutral (NV^0, ZPL emission wavelength of 575 nm) and negatively charged (NV^-, ZPL of 638 nm) states (Iakoubovskii et al. 2000; Gaebel et al. 2006). The population conversion between the two charge states enables radiometric, double-color reading of electronic changes in the vicinity of NV centers.

The charge state of NV centers is influenced by distance from the diamond surface (particle size), surface termination and electron donor (i.e., nitrogen) content (Petrakova et al. 2012). In bulk, NV^- centers prevail. However, close to the diamond surface, NV^- centers become unstable and turn to the NV^0 state (Fu et al. 2010). The charge state of NV near the surface can be converted from NV^0 to NV^- by various types of oxidation reactions (Fu et al. 2010; Rondin et al. 2010). The most significant differences in fluorescence spectra are observed for H- and O-terminated HPHT NDs because of their high surface/volume ratio (Petrakova et al. 2012).

The influence of various surface terminations on NV luminescence has been explained and theoretically modeled by means of quantum chemistry. For details, see (Kratochvílová et al. 2010; Petráková et al. 2011, 2012; Hu et al. 2013; Pinto et al. 2011, 2012).

3 Behavior and Impact of Nanodiamonds on Biological Systems

3.1 Biocompatibility of Nanodiamonds

According to William's broad definition, biocompatibility refers to the "ability of a material to perform with an appropriate host response in a specific situation" (Williams 1987). Biocompatibility is, however, contextual; it is related to a specific interaction and use of a biomaterial. The results and conclusions of many studies concerning the biocompatibility of NDs are not directly comparable because they involve different evaluation methods and criteria to judge the material's biocompatibility. Furthermore, the results depend primarily on the type of particle studied (i.e., DNDs or HPHT NDs), as well as their concentrations and surface modifications. Biocompatibility also may vary by cell type. This chapter emphasizes the importance of these and other factors on the behavior of NDs in biological systems. We also focus on distinguishing the biological impact of individual ND types. However, we emphasize that, after purification and post-synthetic modifications, NDs have low toxicity and are considered to be among the most biocompatible types of carbon nanostructures (Huang et al. 2007; Schrand et al. 2007a, b, 2011; Liu et al.2007; Xing et al. 2011; Zhang et al. 2012b).

3.1.1 In Vitro Biocompatibility of DNDs

Concerns about toxicity of NDs stem from their small size and ability to enter cells and localize in critical organelles (Schrand et al. 2009). Schrand et al. (2007a) reported that DNDs (both raw DNDs and modified DND-COOH, DND-COONa

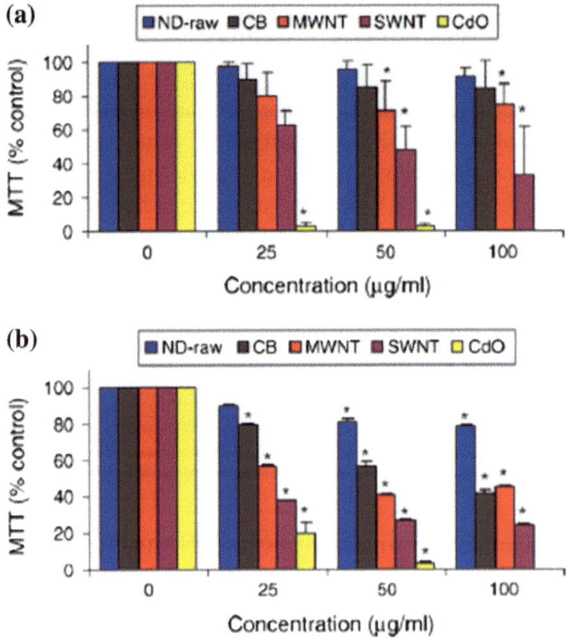

Fig. 4 Effect of various carbon nanomaterials on cytotoxicity in (**a**) neuroblastoma cells or (**b**) macrophages. (*ND-raw* = DND, *CB* = carbon black, *MWNT* = multi-walled carbon nanotubes, *SWNT* = single-walled carbon nanotubes, *CdO* = cadmium oxide). Biocompatibility decreases as follows: ND > CB > MWNT > SWNT > CdO. The cells were treated with nanoparticles in a concentration range of 0–100 μg/ml for 24 h. The cell viability was measured by MTT assay. Figure adapted from Schrand et al. (2007b)

and DND-SO_3Na) are generally biocompatible in vitro in a variety of cell types. Various techniques, such as 3-[4,5-dimethylthiazol-2-yl]-2,5-diphenyltetrazolium bromide (MTT) assay were used. In this method, eventual reduction of MTT corresponds to lower cellular viability based on mitochondrial function. The biocompatibility and low cytotoxicity of DNDs was shown. No disruption of mitochondrial membrane permeability, morphological alterations or viability changes (using luminescence measurement of ATP production) were observed with 5–100 μg/ml DNDs. Furthermore, DNDs did not induce generation of reactive oxygen species (ROS) and did not cause oxidative stress leading to membrane dysfunction, protein degradation and DNA damage (Schrand et al. 2007a). Similar studies have also been performed with other carbon nanomaterials, including carbon black and single- and multi-walled carbon nanotubes (SWNTs and MWNTs) (Schrand et al. 2007b). The DNDs showed higher biocompatibility than carbon black, MWNTs and SWNTs (Schrand et al. 2007b) (Fig. 4). Huang et al. (2007) reported no change in expression of genes serving as indicators of inflammation and protection against apoptosis in macrophages and colorectal cancer cells incubated with DNDs.

Several recent studies have indicated that DNDs can be toxic under certain conditions (Li et al. 2010; Martín et al. 2010a, b; Bakowicz-Mitura et al. 2007; Horie et al. 2012; Karpukhin et al. 2011; Marcon et al. 2010; Puzyr et al. 2004; Solarska et al. 2010, 2012a, b; Solarska-Ściuk et al. 2013; Xing et al. 2011; Zhang et al. 2012b). The observed toxicities were concentration-dependent, time-dependent or both; notably, however, DNDs have much lower toxicity than other

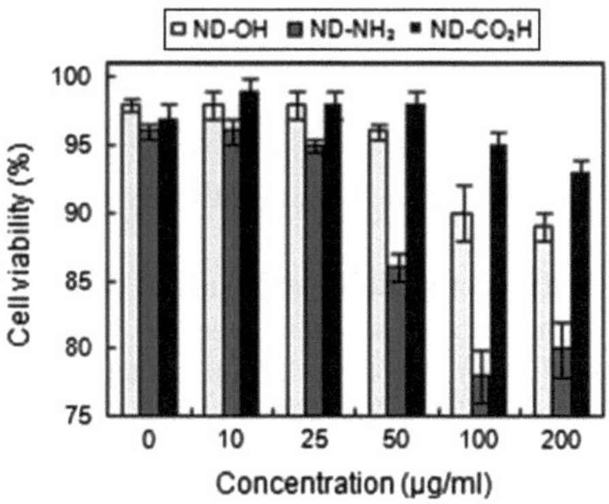

Fig. 5 Effect of DNDs with various surface functionalities on human embryonic kidney cell (HEK293) survival (based on trypan blue dye exclusion assays). Cells were treated with different concentrations (10–200 μg/ml) of functionalized NDs for 24 h. Adapted from Marcon et al. (2010)

carbon materials, such as MWNTs (Liu et al. 2007; Schrand et al. 2007b; Xing et al. 2011; Zhang et al. 2012b).

Harmful effects of DNDs were first reported in 2004. Destruction of human red and white blood cells following incubation with DNDs was observed (Puzyr et al. 2004). In addition, raw DNDs incubated with human neoplastic cells were shown to influence gene expression (Bakowicz-Mitura et al. 2007). Karpukhin et al. (2011) observed concentration-dependent toxicities in neutrophils, such as ROS formation and phagocytosis of oxidized DNDs. Data suggests that surface charge is a crucial determinant of toxicity, due to the affinity of cationic nanoparticles for the negatively charged cell membrane. The cytotoxicity of modified DNDs (at concentrations higher than 50 μg/ml) toward human embryonic kidney cells can be ranked as follows: $-NH_2 \gg -OH > -COOH$ (Fig. 5) (Marcon et al. 2010).

The influence of surface charge, or more precisely zeta potential, and DND concentration on cell viability and growth was also reported by Horie et al. (2012). Toxicity studies on modified DNDs were performed with polyaniline-modified DNDs in human embryonic kidney cells (where toxicity was observed in a concentration-dependent manner) (Villalba et al. 2012). In various studies, Fenton-treated DNDs showed different levels of toxicity, ranging from biocompatibility in HeLa cells (Martín et al. 2010b) to stimulation of ROS production and other effects in HUVEC-ST cells (Solarska et al. 2010; Solarska-Ściuk et al. 2013).

Serum proteins may also influence the biological response to raw DNDs. The toxicity of DNDs toward a variety of cell types depends on the presence of serum in the medium (Li et al. 2010). Other studies showed that DNDs are nontoxic in serum-free media (Schrand et al. 2007a, b).

A DND genotoxicity study was performed on mouse embryonic stem cells, which are highly sensitive to DNA damage (Xing et al. 2011). DNA damage caused by both raw and oxidized DNDs was assessed by observing the activation of DNA repair proteins such as p53, DNA double band breaks markers (MOGG-1), DNA repair markers (Rad51) and other markers. Oxidized DNDs (containing mainly carboxyl groups) caused more DNA damage and cell differentiation than raw DNDs (Fig. 5). This effect is probably caused by differences in colloidal properties. While oxidized DNDs form rather stable colloids, raw DNDs tend to aggregate, which inhibits the entry of DNDs into cells. However, both types of DNDs caused much less DNA damage than MWNTs (Xing et al. 2011).

3.1.2 In Vitro Biocompatibility of HPHT NDs

The biocompatibility of oxidized HPHT NDs was first examined in 2005 in human kidney cells by Chang's group (Yu et al. 2005) and confirmed in subsequent studies by the same lab and others (Fu et al. 2007; Vial et al. 2008; Vaijayanthimala et al. 2009; Faklaris et al. 2008; Lee et al. 2013; Blaber et al. 2013; Fang et al. 2011; Liu et al. 2009). Most studies assessed biocompatibility by MTT assay or by measuring levels of enzymatic activity. For instance, Liu et al. (2009) showed that NDs are non-cytotoxic during division and differentiation of lung cancer cells and embryonic fibroblasts. NDs do not interfere with expression of genes or proteins that regulate cell cycle progression, spindle formation and chromosome segregation. NDs also do not alter long-term (10 days) cell growth (Liu et al. 2009). Fang et al. (2011) reported no significant alteration in growth or proliferation of a variety of cell types incubated with NDs for 8 days. The biocompatibility is not influenced by surface modification with peptides (Vial et al. 2008), proteins or polymers (Lee et al. 2013).

On the other hand, Faklaris et al. (2008) noted that biocompatibility toward HeLa cells is dose-dependent; cell viability decreases at ND concentrations equal or higher than 50 μg/ml. In addition, Weng et al. (2012) observed an approximately two-fold reduction in the proliferation rate of HeLa cells treated with 10 μg/ml HPHT NDs modified with amino groups or transferrin. Marcon et al. (2010) obtained analogous results with DNDs. However, the toxicity of HPHT NDs, similar to that of DNDs, is much lower than that of other carbon nanomaterials, such as MWNTs (Liu et al. 2007).

3.1.3 Comparative In Vitro Biocompatibility Studies of Different ND Types

The comparative toxicities of different ND types have been intensively studied (Liu et al. 2007; Thomas et al. 2012; Lin et al. 2012; Chao et al. 2007; Burleson et al. 2009). ND toxicity depends on many factors, including production process, concentration and incubation time. Smaller DNDs show higher toxicity toward human cells than larger HPHT NDs (Fig. 6), evidenced by effects on cell viability, proliferation, apoptosis and metabolic activity (Liu et al. 2007; Thomas et al. 2012; Lin et al. 2012; Chao et al. 2007; Burleson et al. 2009). This phenomenon can be explained not only by the different sizes (and surface areas) of the particles but also by the presence of reactive disordered carbon structures on the surfaces of DNDs, which contrasts with the highly ordered (and oxidized) sp^3 carbons on HPHT NDs (Thomas et al. 2012; Lin et al. 2012). Similarly, DNDs have been

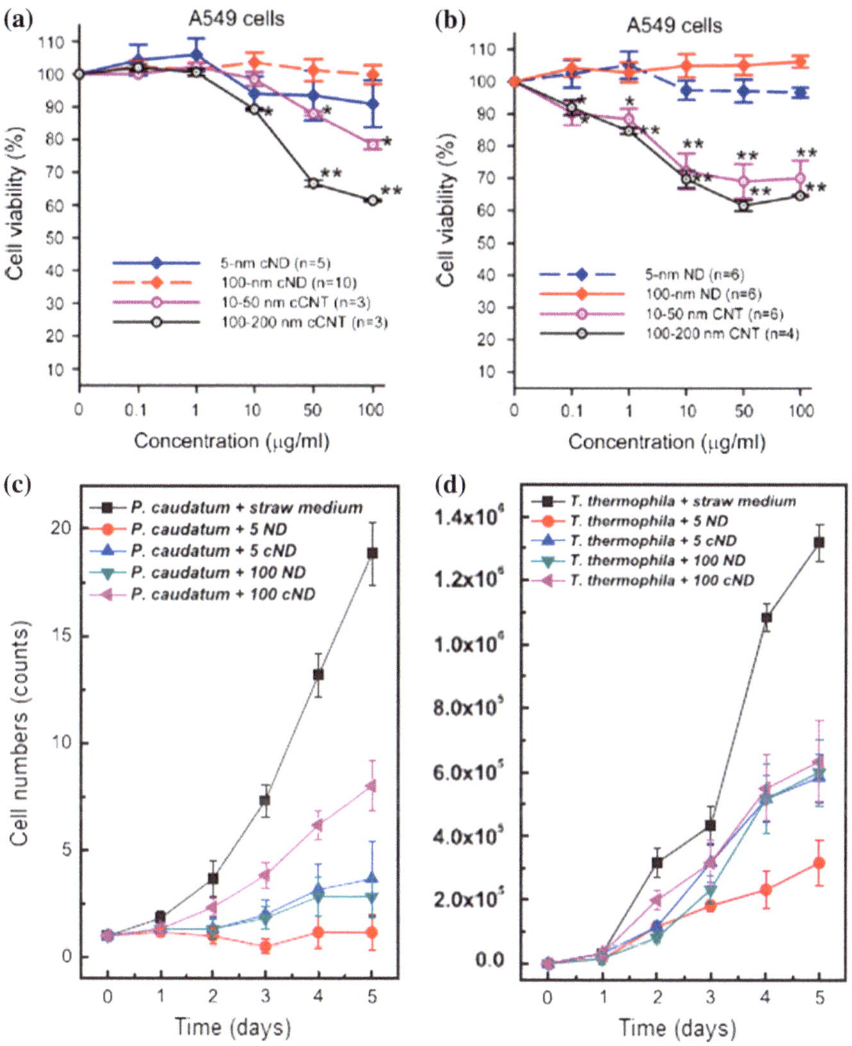

Fig. 6 Comparison of the biocompatibilities of DNDs and HPHT NDs incubated with (**a**, **b**) A549 cells and (**c**, **d**) microorganisms (*P. caudatum* and *T. thermophila*). **a**, **b** A549 cells were treated for 4 h in the presence or absence of different concentrations (0.1–100 μg/ml) of various types of NDs: 5 nm-cND = carboxylated DNDs, 100 nm-cND = carboxylated HPHT NDs, cCNT = carbon nanotubes (10–50 nm or 100–200 nm). Cell viability was evaluated with MTT assay. **c**, **d** Dependence of microorganism cell number on time of incubation with different types of NDs (at a concentration of 20 μg/ml): 5 ND = DND, 5 cND = carboxylated DND, 100 ND = HPHT ND, 100 cND = carboxylated HPHT NDs. The results presented are the average of 6 experiments. Non-carboxylated NDs are considered more toxic than carboxylated NDs. Figure adapted from Lin et al. (2012) and Liu et al. (2007)

found to be more toxic than HPHT NDs in microorganisms, presumably for the same reasons (Fig. 6) (Lin et al. 2012).

Other studies compared the toxicities of DNDs and CVD NDs (Solarska et al. 2012a, b). CVD ND grains are larger than DNDs and show lower cytotoxicity toward HUVEC-ST cells, show lower biological activity and induce apoptosis in fewer cells. These findings support the hypothesis that origin and size are determinants of nanodiamond cytotoxicity.

3.2 Cellular Internalization and Exocytosis of Nanodiamonds

Both DNDs (Li et al. 2010; Liu et al. 2007; Schrand et al. 2007a; Thomas et al. 2012; Chao et al. 2007) and HPHT NDs (Fu et al. 2007; Liu et al. 2007; Neugart et al. 2007; Faklaris et al. 2008; Thomas et al. 2012; Chao et al. 2007) can spontaneously enter cultured cells. Their uptake is affected by many factors; the following paragraphs emphasize the most important ones.

Particle size is a limiting factor for cellular internalization (Weng et al. 2012; Vaijayanthimala et al. 2009; Lee et al. 2013; Alhaddad et al. 2012). While large aggregates cannot be easily internalized, smaller DNDs enter cells efficiently. Surface charge and modifications of NDs also influence the uptake (Weng et al. 2012; Marcon et al. 2010). Positively charged nanoparticles can easily penetrate the negatively charged cell membrane and are internalized rapidly. DNDs modified with amino groups provide an example of such behavior (Marcon et al. 2010). On the other hand, negatively charged nanoparticles (for example DND–COOH) are also internalized, possibly by nonspecific binding or clustering on cationic sites of the plasma membrane followed by endocytosis.

In addition, molecules absorbed on ND surfaces can affect internalization (Marcon et al. 2010). NDs are typically incubated in cell media, and the ND surface is covered with serum proteins. The rate of cellular uptake in the presence or absence of serum is, however, determined by many factors. Weng et al. (2012) showed that the interaction of HPHT ND-COOH with cells is negligible. In serum-supplemented medium, the negatively charged ND surface is repelled by the negatively charged cells and internalization remains low. When no serum is available (and cells are starved), the particles are internalized at a higher rate. Chang's group conducted a similar study with PEG-modified HPHT NDs in HeLa cells (Zhang et al. 2009a). In the absence of serum, the particles were internalized at a higher rate than if serum was absorbed on the ND surface. Vaijayanthimala et al. (2009) observed agglomeration of HPHT NDs in the presence of serum and found a 90 % decrease in internalization of HPHT NDs, similar to the results of Weng et al. (2012).

Internalization of NDs with different polymers and proteins adsorbed on their surfaces was measured (Dahoumane et al. 2009; Sreenivasan et al. 2011; Lee et al.

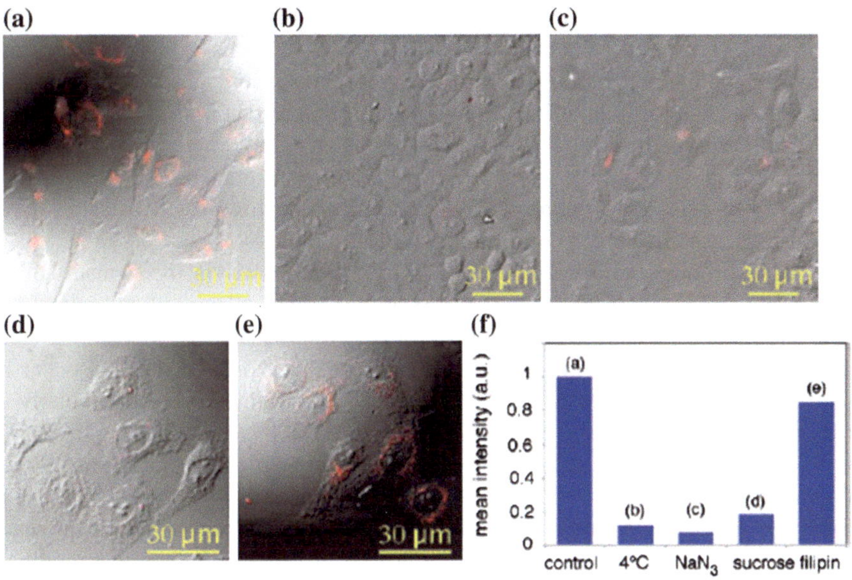

Fig. 7 Nanodiamond uptake by HeLa cells through endocytosis. Merged photoluminescence confocal raster scans (*red channel*) and differential interference contrast (*DIC*) images of HPHT NDs (20 μg/ml) incubated with HeLa cells for 2 h **a** at 37 °C (control) and **b** at 4 °C, or at 37 °C after pretreatment with either **c** NaN_3 (10 mM), **d** sucrose (0.45 M) or **e** filipin (5 μg/ml). **f** Mean photoluminescence intensity of NDs per cell for the different cell treatments is normalized to that of control cells. Figure adapted from Faklaris et al. (2009a)

2013). These coatings helped stabilize the particles at physiological salt concentration. Particle aggregation was suppressed and particles exhibited enhanced intracellular uptake.

The ND internalization pathway has been studied primarily with HPHT NDs. To reveal the exact mechanism, cells were incubated at low temperature in an ATP-depleted environment (in sodium azide or 2-deoxyglucose) (Vaijayanthimala et al. 2009; Faklaris et al. 2009a). These conditions reduced the uptake of NDs significantly, indicating that internalization depends on temperature and the energetic status of a cell. This clearly implies that NDs are internalized by endocytosis (see Fig. 7). The observation has been confirmed by several groups (Yu et al. 2005; Vial et al. 2008; Fang et al. 2011).

The major, and best-understood, route for endocytosis in most cells is a clathrin-mediated pathway. In this pathway, ligands first bind to a cell surface receptor and then are internalized. When NDs were modified with molecules that facilitate receptor-mediated endocytosis such as a cationic polymer (Alhaddad et al. 2012), poly-L-lysine (Vaijayanthimala et al. 2009) or transferrin (Weng et al. 2012), the uptake was enhanced. In contrast, in the presence of inhibitors such as dynasore (which inhibits dynamin, a protein required for clathrin-coated vesicle formation), sucrose or phenyl arsine oxide (which disrupt the formation of clathrin vesicles), the uptake was reduced to a level comparable to that of control cells (Schrand et al.

2011; Vaijayanthimala et al. 2009; Faklaris et al. 2009a). In addition, the blocking compound filipin did not reduce the uptake of NDs in cells, which excludes the possibility of internalization via caveolae endocytosis (Fig. 7) (Vaijayanthimala et al. 2009; Faklaris et al. 2009a). These results confirm that many NDs enter cells via a clathrin-mediated pathway. Actin filaments and microtubules are also involved in this process, as was shown by a decrease in uptake in the presence of inhibitors of actin- and microtubule-dependent endocytosis (cytochalasin D and nocodazole, respectively) (Vaijayanthimala et al. 2009).

NDs, especially surface-modified NDs, can enter cells via mechanisms other than a clathrin-mediated pathway. For example, HPHT NDs modified with folic acid internalize by caveolin-dependent endocytosis (Zhang et al. 2009a). Macropinocytosis is an additional uptake pathway observed for NDs that can occur in parallel with endocytosis in internalization of non-modified NDs, particularly if they are aggregated (Weng et al. 2012; Liu et al. 2009; Alhaddad et al. 2012). HPHT NDs modified with amino groups also internalize by macropinocytosis (Weng et al. 2012). Liu et al. (2009) observed uptake of NDs by both clathrin-mediated endocytosis and macropinocytosis followed by the formation of macropinosomes. Alhaddad et al. (2012) showed that polyethylenimine-modified HPHT NDs localize mostly in large macropinosome vesicles but also in endosomes. When an inhibitor of macropinocytosis, amiloride (inhibitor of Na^+/H^+ ATPase), was employed, the cellular uptake was reduced, which again indicates that NDs can internalize via macropinocytosis (Alhaddad et al. 2012).

The uptake mechanism can, in certain cases, even define ND function in a cell. For example, siRNA gene silencing occurred only if siRNA-modified polymer-coated HPHT NDs were internalized by macropinocytosis (Alhaddad et al. 2012).

The short-term exocytosis of HPHT NDs from cells is low compared, for example, to that of quantum dots (Fang et al. 2011). Long-term exocytosis (after 6 days) of 15–30 % was observed in three different cell types (cancer and stem cells). Relation between the shape of NDs and their intracellular fate and exocytosis (with respect of the presence of sharp corners and edges) was recently revealed (Chu et al. 2014).

3.3 Cellular Localization

After internalization, both DNDs and HPHT NDs localize inside the cell (Fu et al. 2007; Liu et al. 2007, 2009; Neugart et al. 2007; Huang et al. 2007; Vial et al. 2008; Chao et al. 2007). Some studies imply that a fraction of NDs is entrapped and localized in membrane-bound compartments called endosomes (Schrand et al. 2011; Horie et al. 2012; Alhaddad et al. 2012; Faklaris et al. 2009a). These vesicles are responsible for transport of extracellular materials. During endocytosis, ND-containing extracellular material is engulfed by the cell membrane and trapped in a newly created endosome. Inside the cell, endosomes containing NDs can either transform into lysosomes or fuse with existing lysosomes, as has been

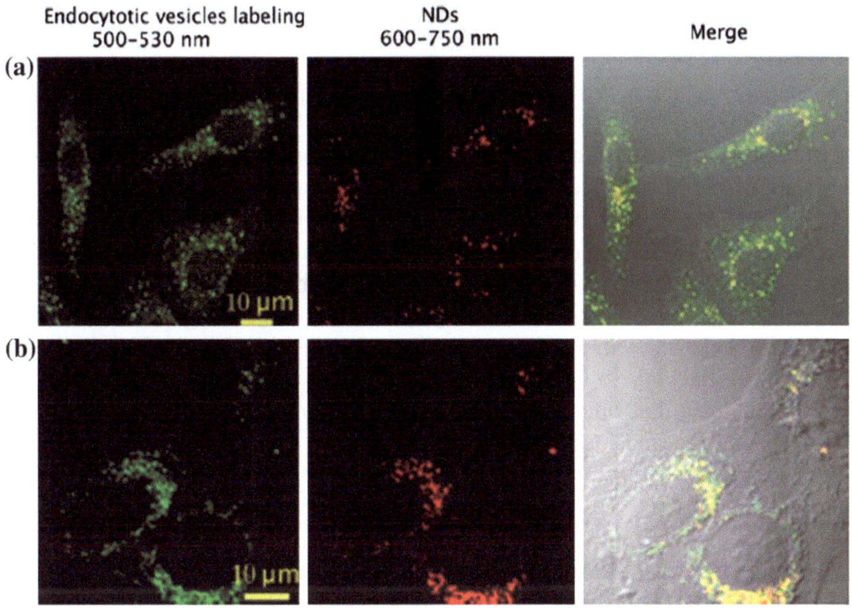

Fig. 8 Localization of HPHT NDs (10 μg/ml) in HeLa cells. **a** Colocalization study of HPHT NDs with (**a**) early endosomes and (**b**) lysosomes labeled with fluorescent dyes. HPHT NDs colocalized with endosomes or lysosomes appear in yellow in the merged fluorescence scans. Figure adapted from Faklaris et al. (2009a)

shown by colocalization experiments with dyed lysosomes and NDs (Fig. 8) (Schrand et al. 2011; Lin et al. 2012; Alhaddad et al. 2012; Faklaris et al. 2009a). The interaction between endosomes and lysosomes during the early stages of the ND uptake pathway was observed in an experiment with brefeldin A (a known lactone antibiotic that interferes with the merging of endosomes and lysosomes). Cells treated with brefeldin A exhibited increased accumulation of NDs, likely caused by the prevention of merging of early endosomes with lysosomes in the natural degradation process (Schrand et al. 2011).

NDs can be liberated from endosomes soon after internalization and localize in the cytosol (Faklaris et al. 2008; Chu et al. 2014). The fact that NDs do not stay inside the endosomes is promising for bioapplications such as drug delivery.

The smallest particles (approximately 5 nm) can localize in the cytoplasm (Schrand et al. 2011; Faklaris et al. 2008, 2009a). It remains unclear if these very small NDs are directly internalized via passive transport across the cellular membrane or if they are released from endosomes.

NDs have not been found in the cell nucleus (Schrand et al. 2011; Yu et al. 2005; Liu et al. 2009; Alhaddad et al. 2012; Faklaris et al. 2009a). DNDs modified with thionine, which have been directly observed in the nucleus, seem to be the only exception (Martín et al. 2010a).

3.4 Cellular Targeting by Surface-Modified Nanodiamonds

The major goal of nanomedicine for cancer therapy is to enhance therapeutic efficiency and selectively target cancerous cells. Unfortunately, the utility of certain chemotherapy drugs is compromised by poor intracellular uptake, limited circulation stability and collateral damage to normal cells. Cellular targeting by nanoparticle conjugates provides an intriguing opportunity to improve cancer therapeutics. NDs have circumvented many individual challenges in cancer therapy, including bypassing chemoresistance and enabling controlled delivery and intracellular tracking. A common strategy to achieve targeting at a cellular level is to functionalize the ND surface with biomolecules whose receptors are overexpressed in the cells to be targeted.

Biomolecules that bind to specific cell surface receptors have been used for targeting of both DNDs and HPHT NDs (Table 1). In addition, functionalization with biomolecules fully preserves the unique optical properties of NDs, which enables bioimaging. The following text introduces several case studies in which NDs were employed in targeting.

Human epidermal growth factor receptor (EGFR), a receptor tyrosine kinase, is a well-studied target for anticancer drug delivery systems. EGFR is overexpressed in more than 30 % of all solid tumors (including lung, colorectal, brain and breast cancers). DNDs modified with anti-EGFR monoclonal antibodies can selectively bind to these receptors (Zhang et al. 2011).

Different attempts to target cancer cells have been made with conjugates of both DNDs and HPHT NDs with transferrin (ND-Tf) (Li and Zhou 2010; Weng et al. 2009; Weng et al. 2012; Cheng et al. 2013). The assumption that ND-Tf conjugates specifically target transferrin receptors on cancer cells was confirmed by preincubation with free Tf. An increase in the free Tf concentration causes a decrease in ND-Tf cellular uptake, as measured by flow cytometry (Li and Zhou 2010). The receptor-mediated uptake of ND-Tf was further confirmed using confocal fluorescence spectroscopy (Weng et al. 2012). Fluorescence spectra and lifetimes of pristine and surface-modified NDs were very similar, confirming the negligible effects of chemical surface modifications (Weng et al. 2012).

Some studies indicated that the production of growth hormone (GH) from endocrine and autocrine systems can stimulate cancer development via the GH/GHR (growth hormone receptor) signal transduction pathway. GH/GHR is likely to participate in pathogenesis of human colorectal cancer (Cheng et al. 2007). Indeed, identification of the GHR level in tumor cells is crucial for monitoring the status of cancer development. HPHT NDs covalently modified with GH were successfully used as a specific probe for in vitro observation of GHRs on a single cell level under physiological conditions (Cheng et al. 2007). GH on the ND surface interacted with GHR present on the surface of A549 human lung epithelial cells. The interaction and penetration of particles into the cells were observed by recording ND's unique spectroscopic signal via confocal Raman mapping.

Table. 1 Modifications of NDs with biomolecules to target specific receptors on different cell types

ND type	Surface of ND	Connection	Targeting group	Receptor	Cell type	Mechanism of uptake	References
DND	Modified with thiol groups	Disulfide bond	Anti-EGFR monoclonal antibodies	Human epidermal growth factor receptor	Human breast adenocarcinoma (MDA-MB-231)	Receptor-mediated endocytosis	Zhang et al. (2011)
DND	Modified with gold/silver	Disulfide bond	Transferrin	Transferrin receptor	Human hepatoma cell line (J5)	Receptor-mediated endocytosis	Cheng et al. (2013)
DND	Carboxylated	Amide bond	Mitochondria/actin antibodies	Mitochondria and actin	HeLa cells	Transfection—intracellular targeting	Mkandawire et al. (2009), Opitz et al. (2010)
HPHT ND	Carboxylated	Amide bond	Transferrin	Transferrin receptor	HeLa cells	Receptor-mediated endocytosis (clathrin-dependent)	Li and Zhou (2010), Weng et al. (2009, 2012)
HPHT ND	Carboxylated	Amide bond	Fish growth hormone	Growth hormone receptor	Human adenocarcinomic lung epithelial cells (A549)	Endocytosis	Cheng et al. (2007)
HPHT ND	Carboxylated	Amide bond	Chlorotoxin-like peptide BmK-CT	Matrix metalloproteases MMP-2	Rat glioma cells (C6)	Receptor-mediated uptake	Fu et al. (2012)
HPHT ND	Carboxylated	Adsorption	α-bungarotoxin	α7-nicotinic acetylcholine receptor	*Xenopus laevis* oocytes human adenocarcinomic lung epithelial cells (A549)	Unspecified	Liu et al. (2008)
HPHT ND	Modified with $PEG\text{-}NH_2$	Amide bond	Folic acid	Folate receptor	HeLa cells	Receptor-mediated endocytosis (caveolin-dependent)	Zhang et al. (2009a)

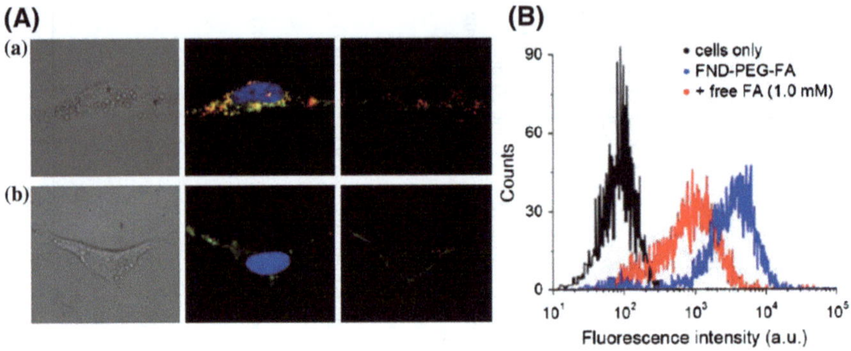

Fig. 9 *A* Images of ND-FA internalized by HeLa cells with or without free FA in the media. **a** ND–FA, **b** ND–FA pretreated with free FA. The images displayed are differential interference contrast (*left*), epifluorescence (*middle*), and confocal fluorescence (*right*) images of the same cells with their membranes and endoplasmic reticulums stained in *green* with wheat germ agglutinin Alexa Fluor 488 conjugates, and nuclei stained in blue with Hoechst 33258. *B* Typical raw data from flow cytometry analysis. All the cells were incubated with NDs for 3 h at a particle concentration of 10 μg/ml. Figure adapted from Zhang et al. (2009a)

Fu et al. (2012) showed targeting of glioma cells by conjugates of HPHT NDs and toxins. Particles composed of HPHT NDs and chlorotoxin (ND-Tx) (a specific chloride channel blocker that inhibits the enzymatic activity of metalloproteases) were prepared and directly visualized in tumors by confocal microscopy. Similar targeting was described in the work of Liu et al. (2008). Their particles contained non-covalently bound α-bungarotoxin (a neurotoxin derived from *Bungarus multicinctus* that specifically blocks the α7-nicotinic acetylcholine receptor). They revealed that the conjugate binds receptors on the cell membrane of oocytes and A549 human lung cancer cells. In addition, the conjugate was able to execute its biological function and block the activation of a7-nicotinic acetylcholine receptor.

HPHT NDs also can be conjugated with folic acid (ND-FA), receptors for which are over-expressed on cancer cells (Zhang et al. 2009a). These particles can be used for single-particle tracking in three dimensions over a real-time span of more than 5 min during endocytosis. The folic acid was bound covalently to the surface of HPHT NDs via a PEG-diamine linker. Activated carboxyl groups on the ND surface reacted with one end of the linker; the other end was connected to the activated carboxyl group of folic acid. The uptake of such ND-FA particles was enhanced up to 20-fold compared to unmodified particles. The amount of ND-FA internalized was reduced when the HeLa cells were pretreated with free FA, as can be seen in confocal microscopy images and from data obtained by flow cytometry (Fig. 9). This pre-incubation experiment provided evidence for internalization through receptor-mediated endocytosis and for the specificity of the interaction.

DNDs modified with antibodies have been shown to specifically target intracellular structures such as mitochondria or actin filaments after being transfected into HeLa cells using 4th-generation dendrimers, cationic liposomes and cell-penetrating protamine sulfate (Mkandawire et al. 2009; Opitz et al. 2010).

4 Bioimaging Using Fluorescent Nanodiamonds

Visualization and tracking of sub-cellular structures and biomolecules in cells are key issues for cell biologists. Most biomolecules are not fluorescent and must be modified with a fluorescent label that can be detected by a fluorescence microscope, flow cytometer or some other fluorescence-reading instrument. The ideal fluorescent label should meet a certain set of requirements, such as chemical stability and solubility, convenient size, minimal interference, low toxicity and high brightness. The most frequently used labels are fluorescent proteins, organic dyes and semiconductor nanocrystals (quantum dots) (Resch-Genger et al. 2008; Sahoo 2012).

Imaging of fluorescence probes in living organisms is an intriguingly demanding task, putting even more requirements on the properties of fluorescent probes. The surrounding tissue can shield a probe's emission and is also a source of autofluorescence, which originates from photoexcitation of endogenous fluorophores such as collagens, retinols, retinoic acids, porphyrins, flavins and NADHs. A useful fluorescent probe should therefore emit in a different spectral region. Light scattering and tissue absorption furthermore limit the optical window (spectral window I for imaging or tissue imaging window) in living organisms (Weissleder and Ntziachristos 2003).

Fluorescent NDs (FNDs) are a promising and exciting alternative to existing probes. NDs are far more biocompatible than other carbon materials, and they do not contain any toxic elements (unlike quantum dots). ND fluorescence, which arises from NV centers, is extremely stable and resistant towards photobleaching with an almost unlimited emission time. Particles bigger than 5 nm do not even photoblink (Bradac et al. 2010), which makes NDs an ideal candidate for imaging. The emission wavelength of FNDs is in the near infrared region, where both autofluorescence and tissue absorption (Fig. 10) are low. Moreover, FNDs have long fluorescence lifetimes (for NV color centers, it is in range approximately 10–20 ns) compared to organic dyes and cellular components with low nanosecond lifetimes (Weng et al. 2009; Faklaris et al. 2008). Such a long lifetime allows for filtering of short radiating signals that do not originate from NV centers (for example, by fluorescence lifetime imaging microscopy, or FLIM) (Kuo et al. 2013).

The major disadvantage of FNDs is their low fluorescence intensity compared to the same mass concentration of molecular dyes. In vivo fluorescence imaging requires very bright FNDs, and only HPHT NDs are considered sufficiently bright.

4.1 NDs as Fluorescent Probes for Intracellular Tracking

The suitability of NDs as in vitro fluorescent labels was introduced by Chang's group (Yu et al. 2005), who observed the cellular uptake of NDs by confocal

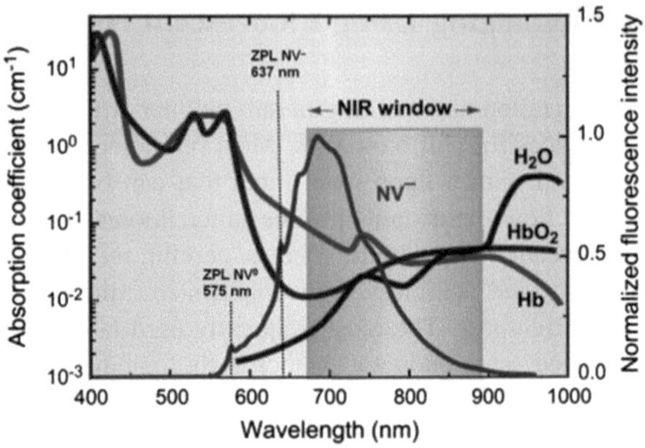

Fig. 10 Comparison of the fluorescence spectrum of NV centers in FNDs with the near-infrared (*NIR*) window of biological tissues. The *black*, *dark gray* and *light gray curves* are the absorption spectra of H_2O, oxygen-bound hemoglobin (*HbO_2*) and hemoglobin (*Hb*), respectively. The absorption spectra were adapted from Weissleder and Ntziachristos (2003)

fluorescence microscopy. ND fluorescence has been observed to be well-separated from the spectral region of cellular endogenous fluorescence (Fu et al. 2007; Zhang et al. 2009a; Li and Zhou 2010; Weng et al. 2009; Chang et al. 2008; Yu et al. 2005). Nowadays, FNDs are already established as labels for in vitro tracking detected by confocal microscopy or flow cytometry (Prabhakar et al. 2013; Liu et al. 2008, 2010; Fu et al. 2007, 2012; Zhang et al. 2009a; Mkandawire et al. 2009; 2010; Weng et al. 2009, 2012; Chang et al. 2008; Neugart et al. 2007; Opitz et al. 2010; Yu et al. 2005; Faklaris et al. 2008, 2009a, 2009b; Fang et al. 2011; Chao et al. 2007; Wee et al. 2009).

Because of their stable fluorescence with no photobleaching or photoblinking, NDs can be used for long-term tracking of both fast and slow events in cells (Fu et al. 2007; Zhang et al. 2009a; Mkandawire et al. 2009; Chang et al. 2008; Faklaris et al. 2008, 2009a, b; Fang et al. 2011). For example, Fang et al. (2011) used NDs for long-term labeling and tracking of division, proliferation and differentiation of stem cells detected by confocal microscopy. In another study, single 35 nm NDs were tracked inside a HeLa cell over a time span of more than 200 s (Chang et al. 2008). Analysis of the three-dimensional trajectory (Fig. 11) enabled determination of the diffusion coefficient for internalized ND within an endosome. The trajectory and diffusion motion of NDs in HeLa cells has been confirmed by other groups (Fu et al. 2007; Zhang et al. 2009a; Neugart et al. 2007). Long-term tracking experiments are highly promising because they can reveal details about intracellular therapeutic activities such as drug delivery or viral infection.

Although many groups have achieved outstanding observations using FNDs, which can be in some cases compared to QDs and dyes (Fu et al. 2007; Faklaris et al. 2009b), the widespread use of FNDs is still somewhat limited. The major issue is their low brightness (related to the weight of the material) (Havlik et al. 2013). Smaller NDs (<50 nm) can be detected only by their fluorescence, which has a sufficiently high signal-to-background ratio, and not by backscattered light of the excitation laser (Faklaris et al. 2009a). Only few groups, such as Neugart et al.,

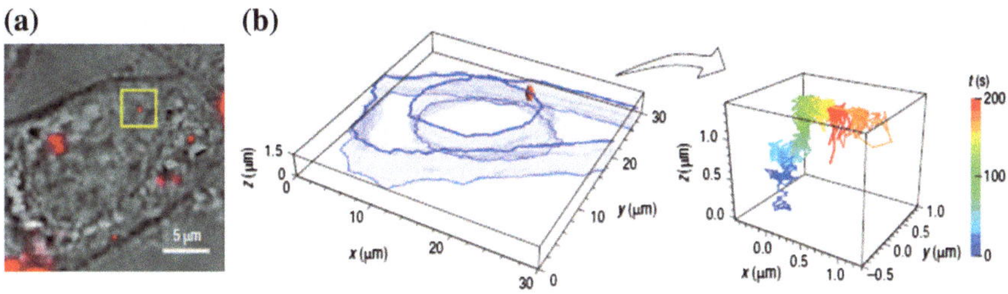

Fig. 11 Three-dimensional tracking of single 35-nm HPHT NDs in a live HeLa cell. **a** Bright-field and epifluorescence (*red pseudo-color*) images of the cell after ND uptake. **b** Three-dimensional reconstruction (*left panel*), showing the boundaries of the nucleus and the cytoplasm of the cell. Three-dimensional trajectory (shown in *pseudo-color*, *right panel*) and displacements of a single FND [labeled with a yellow box in (**a**)] inside the cell over a time span of 200 s. Figure adapted from Chang et al. (2008)

have claimed to be able to observe single diffusing NDs in living HeLa cells, even though most particles contained no more than two defects (Neugart et al. 2007).

Besides detection of NV centers, the intrinsic fluorescence of NDs has been used for imaging both HPHT NDs and DNDs by confocal microscopy or flow cytometry. In these studies, NDs were mostly excited at 488 nm, and their fluorescence was collected from 510 to 530 nm (Liu et al. 2007, 2008, 2010; Mkandawire et al. 2009; Opitz et al. 2010; Chao et al. 2007). The green light emission (Liu et al. 2007) most likely arose from natural defects and impurities and the large surface-to-volume ratio of nanoparticles. Similar to red fluorescent NDs, the uptake of 70-nm green fluorescent NDs in HeLa cells depends on their concentration and incubation time, as was shown by both confocal microcopy and flow cytometry (Wee et al. 2009).

4.2 Comparative In Vivo Biocompatibility, Biodistribution and Clearance of Different ND Types

Before NDs can be used for in vivo imaging, their in vivo biocompatibility, biodistribution and clearance have to be evaluated. The results from evaluations using animal models reflect the potential impact of NDs on the environment and human health.

DNDs are produced as a powder that can be easily spread in air; therefore, the effect of DNDs on the respiratory system is of primary interest. In the majority of studies, the material has been delivered into mice by intratracheal instillation. When DNDs were administered into mice, the animals' lungs, liver, kidneys and hematological system were adversely affected, but their body weight did not change and no other pathological changes were observed (Zhang et al. 2010). In addition, no pulmonary toxicity was observed for low doses of DNDs (0.1 and

1 mg/kg) (Yuan et al. 2010). Some toxicological effects, such as inflammation or tissue damage, were observed only in the lungs (Zhang et al. 2010) (intratracheal instillation most influences the lung). The studies mentioned have revealed that even though NDs are considered to have low pulmonary toxicity, they may still produce side effects in living organisms if the concentration exceeds a certain range.

Other studies have dealt with different routes of ND administration. Injecting DNDs into a mouse did not change the animal's blood parameters (interleukin-6, indicating lack of systemic inflammation, and alanine transferase, corresponding to a lack of influence on liver function) (Chow et al. 2011). Additionally, Puzyr et al. (2007) reported that no inflammatory symptoms were observed in mice when they were subcutaneously exposed to NDs for three months. The results indicated that the manner of administration does not significantly alter ND biocompatibility. Biocompatibility of chemically modified DNDs also has been evaluated. Carboxylated DNDs are considered the most toxic; they can cause embryotoxicity, teratogenicity and malformation in *Xenopus laevis* embryos (Marcon et al. 2010). Strikingly, carboxylated NDs were found more toxic than ND-NH_2 and ND-OH, in contrast to their in vitro toxicity.

Before NDs can be used in living organisms, their distribution, metabolism and excretion must be evaluated. DNDs accumulate mostly in the lungs, spleen, kidneys and liver of mice. The highest retention of DNDs has been observed in mouse lungs after injection (Rojas et al. 2011) or intratracheal instillation (Zhang et al. 2010) and in the liver after injection (Chow et al. 2011; Wei et al. 2012). High concentrations of DNDs were also reported in the animals' blood and heart, whereas no significant concentrations were measured in the brain (Zhang et al. 2010). However, DNDs can likely be redistributed (Zhang et al. 2010; Wei et al. 2012).

DNDs are eliminated slowly from the tissues (Zhang et al. 2010; Wei et al. 2012). Smaller DNDs that are not trapped in organs are excreted into the urinary tract of mice and rats (Rojas et al. 2011). In mice, whole-body clearance was observed within 10 days (Chow et al. 2011). If particles with a larger size were removed by filtration, DND uptake in the lung and spleen was completely inhibited; the particles were found in the bladder and were eventually excreted (Rojas et al. 2011).

HPHT NDs show similar properties to DNDs. Chang et al. reported that no in vivo toxicity or any apparent side effects, such as stress response, were observed after microinjection of HPHT NDs into *C. elegans*. In addition, the entire embryonic development of *C. elegans* containing FNDs appeared normal (Mohan et al. 2010). The long-term biocompatibility of HPHT NDs in rats has been shown; no significant difference between control and FND-treated organisms was observed over a 5-month period (Vaijayanthimala et al. 2012). Similar to DNDs, HPHT NDs modified with BSA accumulated mostly in mouse liver, lungs and spleen if administered by injection. The elimination of NDs from tissues is not time-dependent; HPHT NDs were entrapped at a constant level over 28 days (Yuan et al. 2009).

4.3 Fluorescence Imaging In Vivo

The first in vivo ND imaging was performed in 1-mm-long transparent worms by Mohan et al. (2010), who fed *Caenorhabditis elegans* fluorescent HPHT NDs. They found that stable dispersions of dextran- or serum albumin-coated NDs were absorbed inside the intestinal cells, with very few remaining in the gut lumen (the same was true for unmodified NDs) (Mohan et al. 2010). NDs were then successfully documented by fluorescence microscopy. Furthermore, NDs were also microinjected into the distal gonads of gravid hermaphrodites. NDs were incorporated into oocytes, and because the fluorescence signal persisted throughout the entire embryogenesis, it was possible to investigate the event in real-time. These experiments have unambiguously shown that long-term tracking of ND fluorescence is possible under in vivo conditions (Mohan et al. 2010). Notably, no harmful effects were observed throughout the study.

The results from *C. elegans* were encouraging, but observing FNDs fluorescence in mammals is more demanding. Complications stem from large adsorption and scattering of light in bulky bodies and strong tissue auto-fluorescence. Only recently were FNDs visualized in a mouse and rat with a fluorescence camera (Vaijayanthimala et al. 2012). NDs administrated to a rat both by subcutaneous (s.c.) and intraperitoneal (i.p.) injection were successfully observed. This is particularly noteworthy because rats' skin is relatively thick and strongly absorbs light (Vaijayanthimala et al. 2012). The researchers claimed that the visualization of NDs was successful more than 1 month after administration (reported for s.c. injection), confirming the promise of NDs as long-term fluorescent labels. Neither optical degradation of NDs nor toxic effects on animals were observed, even at huge doses (5 mg/kg repetitively administered i.p.).

Chang's group made a real breakthrough in FND in vivo imaging. They not only visualized particles in the place of administration, but also tracked them throughout the body (Vaijayanthimala et al. 2012). Vaijayanthimala et al. intradermally injected 100 nm serum albumin-coated FNDs into a mouse foot paw. FNDs were observed to migrate into the axillary lymph node over several days, and their concentration gradually increased in the node. In the experiment, a remarkably low dose of 40 μg NDs was administered, demonstrating the extremely high sensitivity of the experimental setup. This experiment is a model for sentinel lymph node mapping, which is widely used in cancer diagnosis and treatment (Jakub et al. 2003). These outstanding in vivo imaging results are based on mastering the preparation of extraordinarily bright FNDs (Chang et al. 2008) and designing elaborate imaging protocols that subtract background auto-fluorescence (Vaijayanthimala et al. 2012).

Background signals from surrounding tissue can be further suppressed by two approaches based on the unique optical properties of FNDs. The first approach exploits the long radiative lifetime of NV centers (>10 ns) compared to sources of autofluorescence (1–4 ns). This was established experimentally with *C. elegans* fed with 100 nm FNDs (Kuo et al. 2013). While standard fluorescence acquisition suffered from a strong background signal, FLIM allowed researchers to distinguish between these two types of signals.

The second approach utilizes the three-level emission system of NV centers (Fig. 3) (Maze et al. 2008; Igarashi et al. 2012). Irradiation of FNDs with resonant frequency decreases their fluorescence intensity, while all other fluorescence sources remain unchanged (Maze et al. 2008; Balasubramanian et al. 2008; I-garashi et al. 2012; Hall et al. 2012; McGuinness et al. 2011). Background-free imaging of FNDs was obtained in *C. elegans* (per oral administration) as well as in mouse (i.p. injection of 10 μg NDs) (Igarashi et al. 2012). Hegyi and Yablonovitch applied gradient magnetic fields (in a somewhat similar manner as in MRI) and achieved 3D images of FNDs in a chicken breast phantom. Although this is not "real" in vivo imaging, the elaborate technology and extremely promising results are noteworthy.

In summary, in vivo imaging of FNDs still represents a challenging task, even though huge progress, such as tracking FNDs in a mouse body, has been made recently.

5 Drug Delivery

A suitable drug carrier for intracellular delivery should have sufficient loading capacity in proportion to the weight of the carrier, strong binding of active molecules to the surface and a functional mechanism of targeted release. NDs comply with all these requirements, thanks to their enormous surface area, good biocompatibility and easy functionalization with biomolecules. Although the possible advantages of simultaneous imaging and drug delivery using NDs have not been yet utilized, we summarize here important results achieved in ND drug delivery research. NDs are mainly employed as carriers for delivery of small molecules with an emphasis on chemotherapeutic agents. Active substances are almost exclusively attached to the particle surface via non-covalent bond interactions, which allows easier carrier formation. Another intensively developing area of research explores NDs as potential intracellular nucleic acid carriers.

5.1 *Small Molecules*

Ho's group made a pioneering step in use of NDs as possible drug carriers (Huang et al. 2007). The authors established non-covalent bonding of doxorubicin hydrochloride (DOX), an apoptosis-inducing drug widely used for chemotherapy, to the oxidized surface of NDs. The design of these nanoparticles is based on ionic interactions between negatively charged groups on the DND surface and the positively charged amino group of doxorubicin. The resulting particles form clusters. The rate of ND-cluster motion into living cells was outlined by confocal microscopy of DNDs with FITC-linked poly-L-lysine physically adsorbed on their surface and by TEM of ND-DOX composites. Both techniques revealed that DNDs were instantly attached to the cell surface and were slowly internalized into the cytoplasm. The amount of bound DOX increases with ionic strength and basicity of the environment due to strengthening of the electrostatic interaction (Adnan et al. 2011; Yan et al. 2012). Stable DOX sequestering also increased blood circulation halftime 10-fold from 0.83 to 8.43 h and suppressed DOX efflux from cancer cells. Such features enable efficient treatment of liver and mammary tumors with significantly decreased toxicity compared to a standard DOX treatment (Fig. 12) (Chow et al. 2011) ND conjugates with cisplatin and 10-hydroxycamptothecin (HCPT) show a similar particle architecture (Li et al. 2010; Guan et al. 2010). Conjugation of DOX-ND with cell-penetrating TAT peptide (HIV trans-activator of transcription protein) enhances intracellular delivery and blocks premature release (Li et al. 2011a). DOX can be bound to FNDs which, unlike DNDs, allow long-term fluorescence localization of particles within the cell. Li et al. (2011b) showed that FND-DOX particles with diameters of $\sim$140 nm can be delivered inside HeLa cells efficiently via the clathrin-dependent endocytosis route with similar kinetics as non-adsorbed DOX.

Another possible application of NDs is complexation with poorly water-soluble drugs to enhance their dispersive properties in water. A study conducted on the drugs purvalanol A (a cyclin-dependent kinase inhibitor) and 4-hydroxytamoxifen (an antagonist of the estrogen receptor in breast tissue) is an inspiring example of how water-insoluble compounds can be applied to treatment-relevant scenarios (Chen et al. 2009).

Particles with covalently bound drugs represent an alternative group of carriers. The first functional covalent linkage of drugs with DNDs was demonstrated with paclitaxel, which successfully induced both mitotic arrest and apoptosis in A549 human lung carcinoma cells (Liu et al. 2010).

An interesting hybrid carrier consists of a conjugate of NDs with non-covalently bound epirubicin coated with a lipid double layer containing a small amount of biotinylated lipids. The lipids are connected to biotinylated EGFR-targeting (epidermal growth factor receptor) antibodies via a streptavidin anchor (Fig. 13) (Moore et al. 2013). The uptake of these particles in vitro was higher than for antibody-free particles (2.9-fold higher in MDA-MB-231 cell line). In in vivo experiments, mice bearing luciferase-expressing MDA-MB-231 tumor xenografts were treated with 150 μg of epirubicin or an equivalent amount of targeted epirubicin-loaded DNDs (TE-DNDs). All mice treated by unmodified epirubicin suffered from drug-related diseases and died within 2 weeks, whereas mice treated by unmodified epirubicin TE-DNDs survived 7 weeks with nearly complete tumor regression.

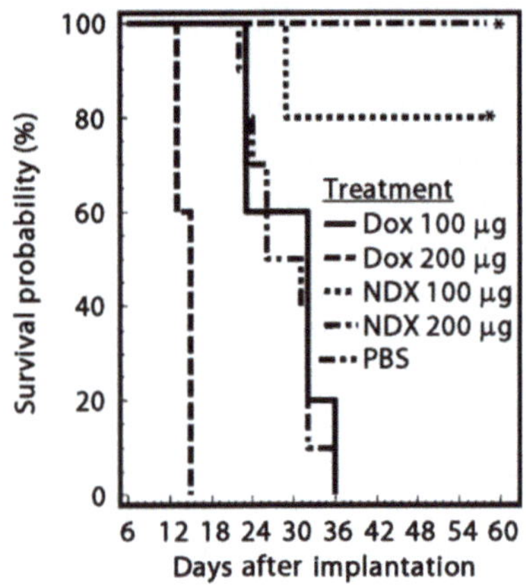

Fig. 12 ND delivery of doxorubicin (*DOX*) inhibits tumor growth in a murine mammary carcinoma model (4T1 cells). Survival plot for tumor-bearing mice treated with PBS (n = 7), DOX (100 mg) (n = 10), DND-doxorubicin conjugate (NDX; 100 mg of DOX equivalent) (n = 10), DOX (200 mg) (n = 5) or NDX (200 mg of DOX equivalent) (n = 5) by tail vein injection every 6 days. *P < 0.003). Adapted from Chow et al. (2011)

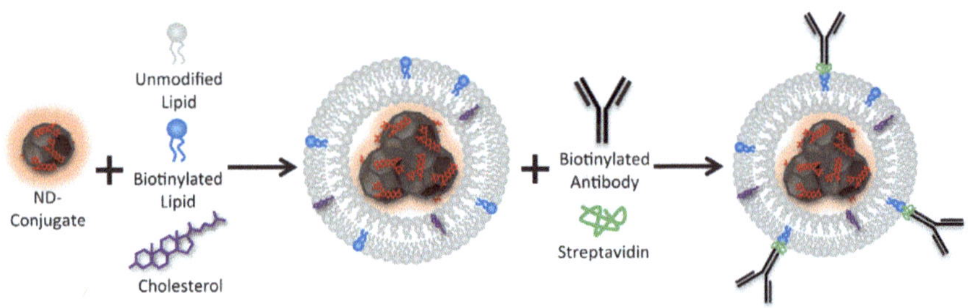

Fig. 13 Nanodiamond-lipid hybrid particles synthesized by rehydration of lipid thin films are targeted using biotinylated antibodies and streptavidin crossbridges. Adapted from Moore et al. (2013)

5.2 Biomolecules

The potential use of ND as carriers of biomolecules was illustrated by pH-controlled release of bovine insulin bound on ND surfaces by Shimkunas et al. (2009). The use of NDs for gene delivery is limited by the negative charge and high molecular weight of bounded nucleic acids. One possible way to overcome these obstacles is modifying ND particles with positively charged polyethyleneimine-800 (PEI-800). These particles can electrostatically bind a negatively charged plasmid DNA (Fig. 14) (Chen et al. 2010, Zhang et al. 2009b). In comparison with unmodified NDs, NDs containing PEI-800 internalize more efficiently. The surface

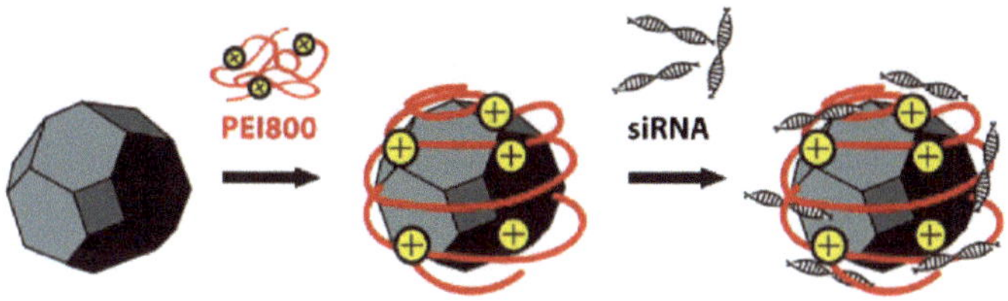

Fig. 14 Preparation of ND-PEI-siRNA particles. Adapted from Chen et al. (2010)

functionalization is critical because particles with more positive surface potential are attracted to the negatively charged plasma membrane.

In a similar manner, NDs coated with non-covalently bound cationic polymers can be used for intracellular transporting of siRNA with a view to in vivo anti-cancer nucleic-acid drug delivery (Alhaddad et al. 2012; Chen et al. 2010; Alhaddad et al. 2011). In comparison to the common transfection reagent lipofectamine, DNDs functionalized with PEI and PAH (polyallylamine) showed similar siRNA transfection efficiency with lower cytotoxicity. The uptake pathways of siRNA:ND-PAH and siRNA:ND-PEI particles differ. While clathrin-mediated endocytosis is a dominant internalization pathway for PAH particles, PEI particles enter cells via macropinocytosis. The affinity of siRNA to ND-PAH was reported to be larger than its affinity to ND-PEI, indicating that siRNA dissociates only slowly from ND-PAH and its biological activity is suppressed (Alhaddad et al. 2012).

Different surface architectures capable of intracellular plasmid transfer have been introduced recently for DNDs with covalently bound triethylamine (Martín et al. 2010a) and lysine (Badea et al. 2012) functional groups or for cationic 2-(dimethylamino)ethyl methacrylate NDs.

6 Conclusions

Nanodiamonds are a vast group of nanomaterials ranging in size from a few nanometers to hundreds of nanometers. Although their physico-chemical and biological properties are often generalized, many behaviors are size- and surface-specific. For example, the smallest DNDs are ideal candidates for drug delivery, but their photophysical and imaging properties are much less useful compared to those of the larger HPHT NDs. Thanks to their high surface-to-volume ratio, DNDs are chemically the most reactive and least stable NDs, while HPHT NDs are more difficult to chemically modify. Large differences can also be found in their interaction with cells, in vitro and in vivo toxicities, cellular uptake and fate.

In the past few years, ND research enabled development of several unique applications in bioimaging and drug delivery. Thanks to rapid progress in ND chemical modifications, novel ND coating procedures and improvement of ND optical properties, one can envision quick development in construction of extremely photostable nanosensors, highly selective targeted delivery of drugs, super-resolved subcellular in vivo bioimaging and ND plasmonics.

Acknowledgements The work was supported by GACR project P108/12/0640 and MSMT CR grant No. LH11027.

References

Adnan A, Lam R, Chen H, Lee J, Schaffer DJ, Barnard AS, Schatz GC, Ho D, Liu WK (2011) Atomistic simulation and measurement of pH dependent cancer therapeutic interactions with nanodiamond carrier. Mol Pharm 8:368–374. doi:10.1021/mp1002398

Aharonovich I, Greentree AD, Prawer S (2011) Diamond photonics. Nat Photonics 5:397–405. doi:10.1038/nphoton.2011.54

Alhaddad A, Adam M-P, Botsoa J, Dantelle G, Perruchas S, Gacoin T, Mansuy C, Lavielle S, Malvy C, Treussart F, Bertrand J-R (2011) Nanodiamond as a vector for siRNA delivery to Ewing sarcoma cells. Small 7:3087–3095. doi:10.1002/smll.201101193

Alhaddad A, Durieu C, Dantelle G, Le Cam E, Malvy C, Treussart F, Bertrand J-R (2012) Influence of the internalization pathway on the efficacy of siRNA delivery by cationic fluorescent nanodiamonds in the Ewing sarcoma cell model. PLoS ONE 7:e52207. doi:10.1371/journal.pone.0052207

Badea I, Kaur, Michel, Chitanda, Maley, Yang, Borondics, Verrall (2012) Lysine-functionalized nanodiamonds: synthesis, physiochemical characterization, and nucleic acid binding studies. Int J Nanomedicine 3851. doi:10.2147/IJN.S32877

Bakowicz-Mitura K, Bartosz G, Mitura S (2007) Influence of diamond powder particles on human gene expression. Surf Coatings Technol 201:6131–6135. doi:10.1016/j.surfcoat.2006.08.142

Balasubramanian G, Chan IY, Kolesov R, Al-Hmoud M, Tisler J, Shin C, Kim C, Wojcik A, Hemmer PR, Krueger A, Hanke T, Leitenstorfer A, Bratschitsch R, Jelezko F, Wrachtrup J (2008) Nanoscale imaging magnetometry with diamond spins under ambient conditions. Nature 455:648–651. doi:10.1038/nature07278

Barnard AS (2009) Diamond standard in diagnostics: nanodiamond biolabels make their mark. Analyst 134:1751. doi:10.1039/b908532g

Blaber SP, Hill CJ, Webster RA, Say JM, Brown LJ, Wang S-C, Vesey G, Herbert BR (2013) Effect of labeling with iron oxide particles or nanodiamonds on the functionality of adipose-derived mesenchymal stem cells. PLoS ONE 8:e52997. doi:10.1371/journal.pone.0052997

Boudou J-P, Curmi PA, Jelezko F, Wrachtrup J, Aubert P, Sennour M, Balasubramanian G, Reuter R, Thorel A, Gaffet E (2009) High yield fabrication of fluorescent nanodiamonds. Nanotechnology 20:235602. doi:10.1088/0957-4484/20/23/235602

Bradac C, Gaebel T, Naidoo N, Sellars MJ, Twamley J, Brown LJ, Barnard AS, Plakhotnik T, Zvyagin AV, Rabeau JR (2010) Observation and control of blinking nitrogen-vacancy centres in discrete nanodiamonds. Nat Nanotechnol 5:345–349. doi:10.1038/nnano.2010.56

Bumb A, Sarkar SK, Billington N, Brechbiel MW, Neuman KC (2013) Silica encapsulation of fluorescent nanodiamonds for colloidal stability and facile surface functionalization. J Am Chem Soc 135:7815–7818. doi:10.1021/ja4016815

Burleson T, Yusuf N, Stanishevsky A (2009) Surface modification of nanodiamonds for biomedical application and analysis by infrared spectroscopy. J Ach Mat Manufac Eng 37:258–263

Butler JE, Sumant AV (2008) The CVD of nanodiamond materials. Chem Vap Depos 14:145–160. doi:10.1002/cvde.200700037

Cao L (2005) Immobilised enzymes: science or art? Curr Opin Chem Biol 9:217–226. doi:10.1016/j.cbpa.2005.02.014

Chang Y-R, Lee H-Y, Chen K, Chang C-C, Tsai D-S, Fu C-C, Lim T-S, Tzeng Y-K, Fang C-Y, Han C-C, Chang H-C, Fann W (2008) Mass production and dynamic imaging of fluorescent nanodiamonds. Nat Nanotechnol 3:284–288. doi:10.1038/nnano.2008.99

Chang L-Y, Ōsawa E, Barnard AS (2011) Confirmation of the electrostatic self-assembly of nanodiamonds. Nanoscale 3:958. doi:10.1039/c0nr00883d

Chao JI, Perevedentseva E, Chung PH, Liu KK, Cheng CY, Chang CC, Cheng CL (2007) Nanometer-sized diamond particle as a probe for biolabeling. Biophys J 93:2199–2208

Chen M, Pierstorff ED, Lam R, Li S-Y, Huang H, Osawa E, Ho D (2009) Nanodiamond-mediated delivery of water-insoluble therapeutics. ACS Nano 3:2016–2022. doi:10.1021/nn900480m

Chen M, Zhang X-Q, Man HB, Lam R, Chow EK, Ho D (2010) Nanodiamond vectors functionalized with polyethylenimine for siRNA delivery. J Phys Chem Lett 1:3167–3171. doi:10.1021/jz1013278

Cheng C-Y, Perevedentseva E, Tu J-S, Chung P-H, Cheng C-L, Liu K-K, Chao J-I, Chen P-H, Chang C-C (2007) Direct and in vitro observation of growth hormone receptor molecules in A549 human lung epithelial cells by nanodiamond labeling. Appl Phys Lett 90:163903. doi:10.1063/1.2727557

Cheng L-C, Chen HM, Lai T-C, Chan Y-C, Liu R-S, Sung JC, Hsiao M, Chen C-H, Her L-J, Tsai DP (2013) Targeting polymeric fluorescent nanodiamond-gold/silver multi-functional nanoparticles as a light-transforming hyperthermia reagent for cancer cells. Nanoscale 5:3931–3940. doi:10.1039/C3NR34091K

Chi Y, Chen G, Jelezko F, Wu E, Zeng H (2011) Enhanced photoluminescence of single-photon emitters in nanodiamonds on a gold film. IEEE Photonics Technol Lett 23:374–376. doi:10.1109/LPT.2011.2106488

Chow EK, Zhang X-Q, Chen M, Lam R, Robinson E, Huang H, Schaffer D, Osawa E, Goga A, Ho D (2011) Nanodiamond therapeutic delivery agents mediate enhanced chemoresistant tumor treatment. Sci Transl Med 3:73ra21

Chu Z, Zhang S, Zhang B, Zhang C, Fang C-Y, Rehor I, Cigler P, Chang H-C, Lin G, Liu R, Li Q (2014) Unambiguous observation of shape effects on cellular fate of nanoparticles. Sci Reports 4:4495. doi:10.1038/srep04495

Chung P-H, Perevedentseva E, Tu J-S, Chang CC, Cheng C-L (2006) Spectroscopic study of bio-functionalized nanodiamonds. Diam Relat Mater 15:622–625. doi:10.1016/j.diamond.2005.11.019

Dahoumane SA, Nguyen MN, Thorel A, Boudou J-P, Chehimi MM, Mangeney C (2009) Protein-functionalized hairy diamond nanoparticles. Langmuir 25:9633–9638. doi:10.1021/la9009509

Danilenko VV (2004) On the history of the discovery of nanodiamond synthesis. Phys Solid State 46:595–599

Davies G, Lawson SC, Collins AT, Mainwood A, Sharp SJ (1992) Vacancy-related centers in diamond. Phys Rev B 46:13157. doi:10.1103/PhysRevB.46.13157

Dolmatov VY (2007) Detonation-synthesis nanodiamonds: synthesis, structure, properties and applications. Russ Chem Rev 76:339–360. doi:10.1070/RC2007v076n04ABEH003643

Eidelman ED, Siklitsky VI, Sharonova LV, Yagovkina MA, Vul' AY, Takahashi M, Inakuma M, Ozawa M, Ōsawa E (2005) A stable suspension of single ultrananocrystalline diamond particles. Diam Relat Mater 14:1765–1769. doi:10.1016/j.diamond.2005.08.057

Faklaris O, Garrot D, Joshi V, Druon F, Boudou JP, Sauvage T, Georges P, Curmi PA, Treussart F (2008) Detection of single photoluminescent diamond nanoparticles in cells and study of the internalization pathway. Small 4:2236–2239

Faklaris O, Joshi V, Irinopoulou T, Tauc P, Sennour M, Girard H, Gesset C, Arnault JC, Thorel A, Boudou JP (2009a) Photoluminescent diamond nanoparticles for cell labeling: study of the uptake mechanism in mammalian cells. ACS Nano 3:3955–3962

Faklaris O, Garrot D, Treussart F, Joshi V, Curmi P, Boudou J-P, Sauvage T (2009b) Comparison of the photoluminescence properties of semiconductor quantum dots and non-blinking diamond nanoparticles. Observation of the diffusion of diamond nanoparticles in living cells. ArXiv Prepr. ArXiv09042648

Fang C-Y, Vaijayanthimala V, Cheng C-A, Yeh S-H, Chang C-F, Li C-L, Chang H-C (2011) The exocytosis of fluorescent nanodiamond and its use as a long-term cell tracker. Small 7:3363–3370. doi:10.1002/smll.201101233

Fu CC, Lee HY, Chen K, Lim TS, Wu HY, Lin PK, Wei PK, Tsao PH, Chang HC, Fann W (2007) Characterization and application of single fluorescent nanodiamonds as cellular biomarkers. Proc Natl Acad Sci 104:727–732

Fu K-MC, Santori C, Barclay PE, Beausoleil RG (2010) Conversion of neutral nitrogen-vacancy centers to negatively charged nitrogen-vacancy centers through selective oxidation. Appl Phys Lett 96:121907. doi:10.1063/1.3364135

Fu Y, An N, Zheng S, Liang A, Li Y (2012) BmK CT-conjugated fluorescence nanodiamond as potential glioma-targeted imaging and drug. Diam Relat Mater 21:73–76. doi:10.1016/j.diamond.2011.10.010

Gaebel T, Domhan M, Wittmann C, Popa I, Jelezko F, Rabeau J, Greentree A, Prawer S, Trajkov E, Hemmer PR, Wrachtrup J (2006) Photochromism in single nitrogen-vacancy defect in diamond. Appl Phys B 82:243–246. doi:10.1007/s00340-005-2056-2

Geim AK, Novoselov KS (2007) The rise of graphene. Nat Mater 6:183–191. doi:10.1038/nmat1849

Gracio JJ, Fan QH, Madaleno JC (2010) Diamond growth by chemical vapour deposition. J Phys Appl Phys 43:374017. doi:10.1088/0022-3727/43/37/374017

Gruber A, Dräbenstedt A, Tietz C, Fleury L, Wrachtrup J, von Borczyskowski C (1997) Scanning confocal optical microscopy and magnetic resonance on single defect centers. Science 276:2012–2014. doi:10.1126/science.276.5321.2012

Guan B, Zou F, Zhi J (2010) Nanodiamond as the pH-responsive vehicle for an anticancer drug. Small 6:1514–1519. doi:10.1002/smll.200902305

Hall LT, Beart GCG, Thomas EA, Simpson DA, McGuinness LP, Cole JH, Manton JH, Scholten RE, Jelezko F, Wrachtrup J, Petrou S, Hollenberg LCL (2012) High spatial and temporal resolution wide-field imaging of neuron activity using quantum NV-diamond. Sci Reports. doi:10.1038/srep00401

Havlik J, Petrakova V, Rehor I, Petrak V, Gulka M, Stursa J, Kucka J, Ralis J, Rendler T, Lee S-Y, Reuter R, Wrachtrup J, Ledvina M, Nesladek M, Cigler P (2013) Boosting nanodiamond fluorescence: towards development of brighter probes. Nanoscale 5:3208–3211. doi:10.1039/c2nr32778c

Hegyi A, Yablonovitch E (2013) Molecular imaging by optically detected electron spin resonance of nitrogen-vacancies in nanodiamonds. Nano Lett 13:1173–1178. doi:10.1021/nl304570b

Hens SC, Cunningham G, Tyler T, Moseenkov S, Kuznetsov V, Shenderova O (2008) Nanodiamond bioconjugate probes and their collection by electrophoresis. Diam Relat Mater 17:1858–1866. doi:10.1016/j.diamond.2008.03.020

Ho D (2009) Beyond the sparkle: the impact of nanodiamonds as biolabeling and therapeutic agents. ACS Nano 3:3825–3829. doi:10.1021/nn9016247

Ho D (2010) Nanodiamonds: applications in biology and nanoscale medicine. Springer, Berlin

Holt KB (2007) Diamond at the nanoscale: applications of diamond nanoparticles from cellular biomarkers to quantum computing. Philos Trans R Soc Math Phys Eng Sci 365:2845–2861. doi:10.1098/rsta.2007.0005

Horie M, Komaba LK, Kato H, Nakamura A, Yamamoto K, Endoh S, Fujita K, Kinugasa S, Mizuno K, Hagihara Y, Yoshida Y, Iwahashi H (2012) Evaluation of cellular influences induced by stable nanodiamond dispersion; the cellular influences of nanodiamond are small. Diam Relat Mater 24:15–24. doi:10.1016/j.diamond.2012.01.037

Hu W, Li Z, Yang J, Hou J (2013) Nondecaying long range effect of surface decoration on the charge state of NV center in diamond. J Chem Phys 138:034702. doi:10.1063/1.4775364

Huang L-CL, Chang H-C (2004) Adsorption and immobilization of cytochrome *c* on nanodiamonds. Langmuir 20:5879–5884. doi:10.1021/la0495736

Huang H, Pierstorff E, Osawa E, Ho D (2007) Active nanodiamond hydrogels for chemotherapeutic delivery. Nano Lett 7:3305–3314. doi:10.1021/nl071521o

Hui YY, Cheng CL, Chang HC (2010) Nanodiamonds for optical bioimaging. J Phys Appl Phys 43:374021

Iakoubovskii K, Adriaenssens GJ, Nesladek M (2000) Photochromism of vacancy-related centres in diamond. J Phys Condens Matter 12:189–199. doi:10.1088/0953-8984/12/2/308

Igarashi R, Yoshinari Y, Yokota H, Sugi T, Sugihara F, Ikeda K, Sumiya H, Tsuji S, Mori I, Tochio H, Harada Y, Shirakawa M (2012) Real-time background-free selective imaging of fluorescent nanodiamonds in vivo. Nano Lett 12:5726–5732. doi:10.1021/nl302979d

Jakub JW, Pendas S, Reintgen DS (2003) Current status of sentinel lymph node mapping and biopsy: facts and controversies. Oncologist 8:59–68. doi:10.1634/theoncologist.8-1-59

Jelezko F, Wrachtrup J (2006) Single defect centres in diamond: a review. Phys Status Solidi 203:3207–3225. doi:10.1002/pssa.200671403

Jiang T, Xu K (1995) FTIR study of ultradispersed diamond powder synthesized by explosive detonation. Carbon 33:1663–1671. doi:10.1016/0008-6223(95)00115-1

Karpukhin AV, Avkhacheva NV, Yakovlev RY, Kulakova II, Yashin VA, Lisichkin GV, Safronova VG (2011) Effect of detonation nanodiamonds on phagocyte activity. Cell Biol Int 35:727–733. doi:10.1042/CBI20100548

Kharisov BI, Kharissova OV, Chávez-Guerrero L (2010) Synthesis techniques, properties, and applications of nanodiamonds. Synth React Inorg, Met Org, Nano Met Chem 40:84–101

Kong XL, Huang LCL, Hsu C-M, Chen W-H, Han C-C, Chang H-C (2005) High-affinity capture of proteins by diamond nanoparticles for mass spectrometric analysis. Anal Chem 77:259–265. doi:10.1021/ac048971a

Kratochvílová I, Kovalenko A, Taylor A, Fendrych F, Řezáčová V, Vlček J, Záliš S, Šebera J, Cígler P, Ledvina M, Nesládek M (2010) The fluorescence of variously terminated nanodiamond particles: quantum chemical calculations. Phys Status Solidi 207:2045–2048. doi:10.1002/pssa.201000012

Krueger A (2008a) New carbon materials: biological applications of functionalized nanodiamond materials. Chem Eur J 14:1382–1390. doi:10.1002/chem.200700987

Krueger A (2008b) The structure and reactivity of nanoscale diamond. J Mater Chem 18:1485. doi:10.1039/b716673g

Krueger A (2008c) Diamond nanoparticles: jewels for chemistry and physics. Adv Mater 20:2445–2449. doi:10.1002/adma.200701856

Krueger A (2010) Nanodiamond. Carbon Mater. Nanotechnol. Wiley-VCH Verlag GmbH & Co. KGaA, pp 329–388

Krueger A (2011) Beyond the shine: recent progress in applications of nanodiamond. J Mater Chem 21:12571. doi:10.1039/c1jm11674f

Krueger A, Lang D (2012) Functionality is key: recent progress in the surface modification of nanodiamond. Adv Funct Mater 22:890–906. doi:10.1002/adfm.201102670

Krueger A, Stegk J, Liang Y, Lu L, Jarre G (2008) Biotinylated nanodiamond: simple and efficient functionalization of detonation diamond. Langmuir 24:4200–4204. doi:10.1021/la703482v

Krüger A, Kataoka F, Ozawa M, Fujino T, Suzuki Y, Aleksenskii AE, Vul' AY, Ōsawa E (2005) Unusually tight aggregation in detonation nanodiamond: identification and disintegration. Carbon 43:1722–1730. doi:10.1016/j.carbon.2005.02.020

Krüger A, Liang Y, Jarre G, Stegk J (2006) Surface functionalisation of detonation diamond suitable for biological applications. J Mater Chem 16:2322–2328. doi:10.1039/B601325B

Kuo Y, Hsu T-Y, Wu Y-C, Hsu J-H, Chang H-C (2013) Fluorescence lifetime imaging microscopy of nanodiamonds in vivo. In: Hasan ZU, Hemmer PR, Lee H, Santori CM (eds) pp 863503

Lee JW, Lee S, Jang S, Han KY, Kim Y, Hyun J, Kim SK, Lee Y (2013) Preparation of non-aggregated fluorescent nanodiamonds (FNDs) by non-covalent coating with a block copolymer and proteins for enhancement of intracellular uptake. Mol BioSyst 9:1004. doi:10.1039/c2mb25431j

Li Y, Zhou X (2010) Transferrin-coupled fluorescence nanodiamonds as targeting intracellular transporters: An investigation of the uptake mechanism. Diam Relat Mater 19:1163–1167

Li L, Davidson JL, Lukehart CM (2006) Surface functionalization of nanodiamond particles via atom transfer radical polymerization. Carbon 44:2308–2315. doi:10.1016/j.carbon.2006.02.023

Li J, Zhu Y, Li W, Zhang X, Peng Y, Huang Q (2010) Nanodiamonds as intracellular transporters of chemotherapeutic drug. Biomaterials 31:8410–8418. doi:10.1016/j.biomaterials.2010.07.058

Li X, Shao J, Qin Y, Shao C, Zheng T, Ye L (2011a) TAT-conjugated nanodiamond for the enhanced delivery of doxorubicin. J Mater Chem 21:7966. doi:10.1039/c1jm10653h

Li Y, Zhou X, Wang D, Yang B, Yang P (2011b) Nanodiamond mediated delivery of chemotherapeutic drugs. J Mater Chem 21:16406. doi:10.1039/c1jm10926j

Liang Y, Ozawa M, Krueger A (2009) A general procedure to functionalize agglomerating nanoparticles demonstrated on nanodiamond. ACS Nano 3:2288–2296. doi:10.1021/nn900339s

Liang Y, Meinhardt T, Jarre G, Ozawa M, Vrdoljak P, SchOll A, Reinert F, Krueger A (2011) Deagglomeration and surface modification of thermally annealed nanoscale diamond. J Colloid Interface Sci 354:23–30. doi:10.1016/j.jcis.2010.10.044

Lin Y-C, Perevedentseva E, Tsai L-W, Wu K-T, Cheng C-L (2012) Nanodiamond for intracellular imaging in the microorganisms *in vivo*. J Biophotonics 5:838–847. doi:10.1002/jbio.201200088

Liu KK, Cheng CL, Chang CC, Chao JI (2007) Biocompatible and detectable carboxylated nanodiamond on human cell. Nanotechnology 18:325102

Liu K-K, Chen M-F, Chen P-Y, Lee TJF, Cheng C-L, Chang C-C, Ho Y-P, Chao J-I (2008) Alpha-bungarotoxin binding to target cell in a developing visual system by carboxylated nanodiamond. Nanotechnology 19:205102. doi:10.1088/0957-4484/19/20/205102

Liu K-K, Wang C-C, Cheng C-L, Chao J-I (2009) Endocytic carboxylated nanodiamond for the labeling and tracking of cell division and differentiation in cancer and stem cells. Biomaterials 30:4249–4259. doi:10.1016/j.biomaterials.2009.04.056

Liu K-K, Zheng W-W, Wang C-C, Chiu Y-C, Cheng C-L, Lo Y-S, Chen C, Chao J-I (2010) Covalent linkage of nanodiamond-paclitaxel for drug delivery and cancer therapy. Nanotechnology 21:315106. doi:10.1088/0957-4484/21/31/315106

Liu J-H, Yang S-T, Chen X-X, Wang H (2012) Fluorescent carbon dots and nanodiamonds for biological imaging: preparation, application, pharmacokinetics and toxicity. Curr Drug Metab 13:1046–1056. doi:10.2174/1389200212802850083

Maitra U, Gomathi A, Rao CNR (2008) Covalent and noncovalent functionalisation and solubilisation of nanodiamond. J Exp Nanosci 3:271–278. doi:10.1080/17458080802574155

Man HB, Ho D (2012) Diamond as a nanomedical agent for versatile applications in drug delivery, imaging, and sensing. Phys Status Solidi sAppl Mater Sci 209:1609–1618. doi:10.1002/pssa.201200470

Marcon L, Riquet F, Vicogne D, Szunerits S, Bodart J-F, Boukherroub R (2010) Cellular and in vivo toxicity of functionalized nanodiamond in Xenopus embryos. J Mater Chem 20:8064. doi:10.1039/c0jm01570a

Martín R, Álvaro M, Herance JR, García H (2010a) Fenton-treated functionalized diamond nanoparticles as gene delivery system. ACS Nano 4:65–74. doi:10.1021/nn901616c

Martín R, Menchón C, Apostolova N, Victor VM, Álvaro M, Herance JR, García H (2010b) Nano-jewels in biology. Gold and platinum on diamond nanoparticles as antioxidant systems against cellular oxidative stress. ACS Nano 4:6957–6965. doi:10.1021/nn1019412

Maze JR, Stanwix PL, Hodges JS, Hong S, Taylor JM, Cappellaro P, Jiang L, Dutt MVG, Togan E, Zibrov AS, Yacoby A, Walsworth RL, Lukin MD (2008) Nanoscale magnetic sensing with an individual electronic spin in diamond. Nature 455:644–647. doi:10.1038/nature07279

McGuinness LP, Yan Y, Stacey A, Simpson DA, Hall LT, Maclaurin D, Prawer S, Mulvaney P, Wrachtrup J, Caruso F, Scholten RE, Hollenberg LCL (2011) Quantum measurement and orientation tracking of fluorescent nanodiamonds inside living cells. Nat Nanotechnol 6:358–363. doi:10.1038/nnano.2011.64

Mkandawire M, Pohl A, Gubarevich T, Lapina V, Appelhans D, ROdel G, Pompe W, Schreiber J, Opitz J (2009) Selective targeting of green fluorescent nanodiamond conjugates to mitochondria in HeLa cells. J Biophotonics 2:596–606. doi:10.1002/jbio.200910002

Mochalin V, Osswald S, Gogotsi Y (2009) Contribution of functional groups to the Raman Spectrum of nanodiamond powders. Chem Mater 21:273–279. doi:10.1021/cm802057q

Mochalin VN, Shenderova O, Ho D, Gogotsi Y (2011) The properties and applications of nanodiamonds. Nat Nanotechnol 7:11–23. doi:10.1038/nnano.2011.209

Mohan N, Chen C-S, Hsieh H-H, Wu Y-C, Chang H-C (2010) In vivo imaging and toxicity assessments of fluorescent nanodiamonds in caenorhabditis elegans. Nano Lett 10:3692–3699. doi:10.1021/nl1021909

Moore L, Chow EK-H, Osawa E, Bishop JM, Ho D (2013) Diamond-lipid hybrids enhance chemotherapeutic tolerance and mediate tumor regression. Adv Mater 25:3532–3541. doi:10.1002/adma.201300343

Neugart F, Zappe A, Jelezko F, Tietz C, Boudou JP, Krueger A, Wrachtrup J (2007) Dynamics of diamond nanoparticles in solution and cells. Nano Lett 7:3588–3591. doi:10.1021/nl0716303

Nguyen T, Chang HC, Wu VWK (2007) Adsorption and hydrolytic activity of lysozyme on diamond nanocrystallites. Diam Relat Mater 16:872–876

Opitz J, Mkandawire M, Sorge M, Rose N, Rudolph M, Krueger P, Hannstein I, Lapina VA, Appelhans D, Pompe W, Schreiber J, Roedel G (2010) Green fluorescent nanodiamond conjugates and their possible applications for biosensing In: Mohseni H, Razeghi M (eds) SPIE nano science + engineering pp 775914

Ozawa M, Inaguma M, Takahashi M, Kataoka F, Krüger A, Ōsawa E (2007) Preparation and behavior of Brownish, clear nanodiamond colloids. Adv Mater 19:1201–1206. doi:10.1002/adma.200601452

Perevedentseva E, Cheng C-Y, Chung P-H, Tu J-S, Hsieh Y-H, Cheng C-L (2007) The interaction of the protein lysozyme with bacteria E. coli observed using nanodiamond labelling. Nanotechnology 18:315102. doi:10.1088/0957-4484/18/31/315102

Perevedentseva E, Cai P-J, Chiu Y-C, Cheng C-L (2011) Characterizing protein activities on the lysozyme and nanodiamond complex prepared for bio applications. Langmuir 27:1085–1091. doi:10.1021/la103155c

Petrakova V, Taylor A, Kratochvilova I, Fendrych F, Vacik J, Kucka J, Stursa J, Cigler P, Ledvina M, Fiserova A, Kneppo P, Nesladek M (2012) Luminescence of nanodiamond driven by atomic functionalization: towards novel detection principles. Adv Funct Mater 22:812–819. doi:10.1002/adfm.201101936

Petráková V, Nesladek M, Taylor A, Fendrych F, Cigler P, Ledvina M, Vacik J, Stursa J, Kucka J (2011) Luminescence properties of engineered nitrogen vacancy centers in a close surface proximity. Phys Status Solidi 208:2051–2056. doi:10.1002/pssa.201100035

Philip J, Hess P, Feygelson T, Butler JE, Chattopadhyay S, Chen KH, Chen LC (2003) Elastic, mechanical, and thermal properties of nanocrystalline diamond films. J Appl Phys 93:2164–2171. doi:10.1063/1.1537465

Pinto H, Jones R, Palmer DW, Goss JP, Briddon PR, Oberg S (2011) Theory of the surface effects on the luminescence of the NV—defect in nanodiamond. Phys Status Solidi 208:2045–2050. doi:10.1002/pssa.201100013

Pinto H, Jones R, Palmer DW, Goss JP, Tiwari AK, Briddon PR, Wright NG, Horsfall AB, Rayson MJ, Oberg S (2012) First-principles studies of the effect of (001) surface terminations on the electronic properties of the negatively charged nitrogen-vacancy defect in diamond. Phys Rev B. doi:10.1103/PhysRevB.86.045313

Prabhakar N, Näreoja T, von Haartman E, Karaman DŞ, Jiang H, Koho S, Dolenko TA, Hänninen PE, Vlasov DI, Ralchenko VG, Hosomi S, Vlasov II, Sahlgren C, Rosenholm JM (2013) Core—shell designs of photoluminescent nanodiamonds with porous silica coatings for bioimaging and drug delivery II: application. Nanoscale 5:3713. doi:10.1039/c3nr33926b

Purtov KV, Burakova LP, Puzyr AP, Bondar VS (2008) The interaction of linear and ring forms of DNA molecules with nanodiamonds synthesized by detonation. Nanotechnology 19:325101. doi:10.1088/0957-4484/19/32/325101

Puzyr AP, Neshumayev DA, Tarskikh SV, Makarskaya GV, Dolmatov VY, Bondar VS (2004) Destruction of human blood cells in interaction with detonation nanodiamonds in experiments in vitro. Diam Relat Mater 13:2020–2023. doi:10.1016/j.diamond.2004.06.003

Puzyr AP, Baron AV, Purtov KV, Bortnikov EV, Skobelev NN, Mogilnaya OA, Bondar VS (2007) Nanodiamonds with novel properties: a biological study. Diam Relat Mater 16:2124–2128. doi:10.1016/j.diamond.2007.07.025

Puzyr' AP, Purtov KV, Shenderova OA, Luo M, Brenner DW, Bondar VS (2007) The adsorption of aflatoxin B1 by detonation-synthesis nanodiamonds. Dokl Biochem Biophys 417:299–301. doi:10.1134/S1607672907060026

Rehor I, Cigler P (2014) Precise Estimation of HPHT Nanodiamond Size Distribution Based on Transmission Electron Microscopy Image Analysis. Diam Relat Mater. doi:10.1016/j.diamond.2014.04.002

Rehor I, Mackova H, Filippov SK, Kucka J, Proks V, Slegerova J, Turner S, Vandeloo GV, Ledvina M, Hruby M, Cigler P (2014a) Fluorescent nanodiamonds with bioorthogonally reactive protein-resistant polymeric coatings. Chem Plus Chem 79:21–24. doi:10.1002/cplu.201300339

Rehor I, Slegerova J, Kucka J, Proks V, Petrakova V, Adam M-P, Treussart F, Turner S, Bals S, Sacha P, Ledvina M, Wen AM, Steinmetz NF, Cigler P (2014b) Fluorescent nanodiamonds embedded in biocompatible translucent shells. Small 10:1106–1115. doi:10.1002/smll.201302336

Resch-Genger U, Grabolle M, Cavaliere-Jaricot S, Nitschke R, Nann T (2008) Quantum dots versus organic dyes as fluorescent labels. Nat Methods 5:763–775. doi:10.1038/nmeth.1248

Rojas S, Gispert JD, Martín R, Abad S, Menchón C, Pareto D, Víctor VM, Álvaro M, García H, Herance JR (2011) Biodistribution of amino-functionalized diamond nanoparticles. *In Vivo* studies based on ^{18}F radionuclide emission. ACS Nano 5:5552–5559. doi:10.1021/nn200986z

Rondin L, Dantelle G, Slablab A, Grosshans F, Treussart F, Bergonzo P, Perruchas S, Gacoin T, Chaigneau M, Chang H-C, Jacques V, Roch J-F (2010) Surface-induced charge state conversion of nitrogen-vacancy defects in nanodiamonds. Phys Rev B. doi:10.1103/PhysRevB.82.115449

Sahoo H (2012) Fluorescent labeling techniques in biomolecules: a flashback. RSC Adv 2:7017–7029. doi:10.1039/C2RA20389H

Schietinger S, Barth M, Aichele T, Benson O (2009) Plasmon-enhanced single photon emission from a nanoassembled metal—Diamond hybrid structure at room temperature. Nano Lett 9:1694–1698. doi:10.1021/nl900384c

Schrand AM, Huang H, Carlson C, Schlager JJ, Osawa E, Hussain SM, Dai L (2007a) Are diamond nanoparticles cytotoxic? J Phys Chem B 111:2–7

Schrand AM, Dai L, Schlager JJ, Hussain SM, Osawa E (2007b) Differential biocompatibility of carbon nanotubes and nanodiamonds. Diam Relat Mater 16:2118–2123. doi:10.1016/j.diamond.2007.07.020

Schrand AM, Hens SAC, Shenderova OA (2009) Nanodiamond particles: properties and perspectives for bioapplications. Crit Rev Solid State Mater Sci 34:18–74. doi:10.1080/10408430902831987

Schrand AM, Lin JB, Hens SC, Hussain SM (2011) Temporal and mechanistic tracking of cellular uptake dynamics with novel surface fluorophore-bound nanodiamonds. Nanoscale 3:435. doi:10.1039/c0nr00408a

Shenderova OA, Zhirnov VV, Brenner DW (2002) Carbon nanostructures. Crit Rev Solid State Mater Sci 27:227–356. doi:10.1080/10408430208500497

Shimkunas RA, Robinson E, Lam R, Lu S, Xu X, Zhang X-Q, Huang H, Osawa E, Ho D (2009) Nanodiamond–insulin complexes as pH-dependent protein delivery vehicles. Biomaterials 30:5720–5728. doi:10.1016/j.biomaterials.2009.07.004

Slepetz B, Laszlo I, Gogotsi Y, Hyde-Volpe D, Kertesz M (2010) Characterization of large vacancy clusters in diamond from a generational algorithm using tight binding density functional theory. Phys Chem Chem Phys 12:14017–14022. doi:10.1039/c0cp00523a

Solarska K, Gajewska A, Skolimowski J, Woś R, Bartosz G, Mitura K (2010) Effect of non-modified and modified nanodiamond particles by Fenton reaction on human endothelial cells. Manuf Eng 43:603–607

Solarska K, Gajewska A, Bartosz G, Mitura K (2012a) Induction of apoptosis in human endothelial cells by nanodiamond particles. J Nanosci Nanotechnol 12:5117–5121. doi:10.1166/jnn.2012.4952

Solarska K, Gajewska A, Kaczorowski W, Bartosz G, Mitura K (2012b) Effect of nanodiamond powders on the viability and production of reactive oxygen and nitrogen species by human endothelial cells. Diam Relat Mater 21:107–113. doi:10.1016/j.diamond.2011.10.020

Solarska-Ściuk K, Gajewska A, Skolimowski J, Mitura K, Bartosz G (2013) Stimulation of production of reactive oxygen and nitrogen species in endothelial cells by unmodified and Fenton-modified ultradisperse detonation diamond. Biotechnol Appl Biochem 60:259–265. doi:10.1002/bab.1071

Sreenivasan VKA, Ivukina EA, Deng W, Kelf TA, Zdobnova TA, Lukash SV, Veryugin BV, Stremovskiy OA, Zvyagin AV, Deyev SM (2011) Barstar: barnase—a versatile platform for colloidal diamond bioconjugation. J Mater Chem 21:65. doi:10.1039/c0jm02819c

Thomas V, Halloran BA, Ambalavanan N, Catledge SA, Vohra YK (2012) In vitro studies on the effect of particle size on macrophage responses to nanodiamond wear debris. Acta Biomater 8:1939–1947. doi:10.1016/j.actbio.2012.01.033

Tisler J, Reuter R, Lämmle A, Jelezko F, Balasubramanian G, Hemmer PR, Reinhard F, Wrachtrup J (2011) Highly efficient FRET from a single nitrogen-vacancy center in nanodiamonds to a single organic molecule. ACS Nano 5:7893–7898. doi:10.1021/nn2021259

Tu J-S, Perevedentseva E, Chung P-H, Cheng C-L (2006) Size-dependent surface CO stretching frequency investigations on nanodiamond particles. J Chem Phys 125:174713. doi:10.1063/1.2370880

Tzeng Y-K, Faklaris O, Chang B-M, Kuo Y, Hsu J-H, Chang H-C (2011) Superresolution imaging of albumin-conjugated fluorescent nanodiamonds in cells by stimulated emission depletion. Angew Chem Int Ed 50:2262–2265. doi:10.1002/anie.201007215

Vaijayanthimala V, Chang H-C (2009) Functionalized fluorescent nanodiamonds for biomedical applications. Nanomed 4:47–55. doi:10.2217/17435889.4.1.47

Vaijayanthimala V, Tzeng Y-K, Chang H-C, Li C-L (2009) The biocompatibility of fluorescent nanodiamonds and their mechanism of cellular uptake. Nanotechnology 20:425103. doi:10.1088/0957-4484/20/42/425103

Vaijayanthimala V, Cheng P-Y, Yeh S-H, Liu K-K, Hsiao C-H, Chao J-I, Chang H-C (2012) The long-term stability and biocompatibility of fluorescent nanodiamond as an in vivo contrast agent. Biomaterials 33:7794–7802. doi:10.1016/j.biomaterials.2012.06.084

Vertegel AA, Siegel RW, Dordick JS (2004) Silica nanoparticle size influences the structure and enzymatic activity of adsorbed lysozyme. Langmuir 20:6800–6807. doi:10.1021/la0497200

Vial S, Mansuy C, Sagan S, Irinopoulou T, Burlina F, Boudou J-P, Chassaing G, Lavielle S (2008) Peptide-grafted nanodiamonds: preparation, cytotoxicity and uptake in cells. Chem Bio Chem 9:2113–2119. doi:10.1002/cbic.200800247

Villalba P, Ram MK, Gomez H, Bhethanabotla V, Helms MN, Kumar A, Kumar A (2012) Cellular and in vitro toxicity of nanodiamond-polyaniline composites in mammalian and bacterial cell. Mater Sci Eng C 32:594–598. doi:10.1016/j.msec.2011.12.017

von Haartman E, Jiang H, Khomich AA, Zhang J, Burikov SA, Dolenko TA, Ruokolainen J, Gu H, Shenderova OA, Vlasov II, Rosenholm JM (2013) Core—shell designs of photoluminescent nanodiamonds with porous silica coatings for bioimaging and drug delivery I: fabrication. J Mater Chem B 1:2358–2366. doi:10.1039/C3TB20308E

Wang H-D, Niu CH, Yang Q, Badea I (2011) Study on protein conformation and adsorption behaviors in nanodiamond particle–protein complexes. Nanotechnology 22:145703. doi:10.1088/0957-4484/22/14/145703

Wee T-L, Mau Y-W, Fang C-Y, Hsu H-L, Han C-C, Chang H-C (2009) Preparation and characterization of green fluorescent nanodiamonds for biological applications. Diam Relat Mater 18:567–573. doi:10.1016/j.diamond.2008.08.012

Wei L, Zhang W, Lu H, Yang P (2010) Immobilization of enzyme on detonation nanodiamond for highly efficient proteolysis. Talanta 80:1298–1304

Wei Q, Zhan L, Juanjuan B, Jing W, Jianjun W, Taoli S, Yi'an G, Wangsuo W (2012) Biodistribution of co-exposure to multi-walled carbon nanotubes and nanodiamonds in mice. Nanoscale Res Lett 7:1–9

Weissleder R, Ntziachristos V (2003) Shedding light onto live molecular targets. Nat Med 9:123–128. doi:10.1038/nm0103-123

Weng MF, Chiang SY, Wang NS, Niu H (2009) Fluorescent nanodiamonds for specifically targeted bioimaging: application to the interaction of transferrin with transferrin receptor. Diam Relat Mater 18:587–591

Weng M-F, Chang B-J, Chiang S-Y, Wang N-S, Niu H (2012) Cellular uptake and phototoxicity of surface-modified fluorescent nanodiamonds. Diam Relat Mater 22:96–104. doi:10.1016/j.diamond.2011.12.035

Williams DF (1987) Definitions in biomaterials: proceedings of a consensus conference of the European Society for Biomaterials, Chester. Elsevier, 3–5 Mar 1986

Williams OA (2011) Nanocrystalline diamond. Diam Relat Mater 20:621–640. doi:10.1016/j.diamond.2011.02.015

Wrachtrup J, Jelezko F, Grotz B, McGuinness L (2013) Nitrogen-vacancy centers close to surfaces. MRS Bull 38:149–154. doi:10.1557/mrs.2013.22

Wu VW-K (2010) Preparation for optimal conformation of lysozyme with nanodiamond and nanosilica as carriers. Chin J Chem 28:2520–2526

Xing Y, Xiong W, Zhu L, Osawa E, Hussin S, Dai L (2011) DNA damage in embryonic stem cells caused by nanodiamonds. ACS Nano 5:2376

Yan J, Guo Y, Altawashi A, Moosa B, Lecommandoux S, Khashab NM (2012) Experimental and theoretical evaluation of nanodiamonds as pH triggered drug carriers. New J Chem 36:1479. doi:10.1039/c2nj40226b

Yeap WS, Tan YY, Loh KP (2008) Using detonation nanodiamond for the specific capture of glycoproteins. Anal Chem 80:4659–4665. doi:10.1021/ac800009v

Yu S-J, Kang M-W, Chang H-C, Chen K-M, Yu Y-C (2005) Bright fluorescent nanodiamonds: no photobleaching and low cytotoxicity. J Am Chem Soc 127:17604–17605. doi:10.1021/ja0567081

Yuan Y, Chen Y, Liu J-H, Wang H, Liu Y (2009) Biodistribution and fate of nanodiamonds in vivo. Diam Relat Mater 18:95–100. doi:10.1016/j.diamond.2008.10.031

Yuan Y, Wang X, Jia G, Liu J-H, Wang T, Gu Y, Yang S-T, Zhen S, Wang H, Liu Y (2010) Pulmonary toxicity and translocation of nanodiamonds in mice. Diam Relat Mater 19:291–299. doi:10.1016/j.diamond.2009.11.022

Zhang B, Li Y, Fang C-Y, Chang C-C, Chen C-S, Chen Y-Y, Chang H-C (2009a) Receptor-mediated cellular uptake of folate-conjugated fluorescent nanodiamonds: a combined ensemble and single-particle study. Small 5:2716–2721. doi:10.1002/smll.200900725

Zhang X-Q, Chen M, Lam R, Xu X, Osawa E, Ho D (2009b) Polymer-functionalized nanodiamond platforms as vehicles for gene delivery. ACS Nano 3:2609–2616. doi:10.1021/nn900865g

Zhang X, Yin J, Kang C, Li J, Zhu Y, Li W, Huang Q, Zhu Z (2010) Biodistribution and toxicity of nanodiamonds in mice after intratracheal instillation. Toxicol Lett 198:237–243. doi:10.1016/j.toxlet.2010.07.001

Zhang X-Q, Lam R, Xu X, Chow EK, Kim H-J, Ho D (2011) Multimodal nanodiamond drug delivery carriers for selective targeting, imaging, and enhanced chemotherapeutic efficacy. Adv Mater 23:4770–4775. doi:10.1002/adma.201102263

Zhang X, Fu C, Feng L, Ji Y, Tao L, Huang Q, Li S, Wei Y (2012a) PEGylation and polyPEGylation of nanodiamond. Polymer 53:3178–3184. doi:10.1016/j.polymer.2012.05.029

Zhang X, Hu W, Li J, Tao L, Wei Y (2012b) A comparative study of cellular uptake and cytotoxicity of multi-walled carbon nanotubes, graphene oxide, and nanodiamond. Toxicol Res 1:62. doi:10.1039/c2tx20006f

Zhao L, Takimoto T, Ito M, Kitagawa N, Kimura T, Komatsu N (2011) Chromatographic separation of highly soluble diamond nanoparticles prepared by polyglycerol grafting. Angew Chem Int Ed 50:1388–1392. doi:10.1002/anie.201006310

Zhao N, Honert J, Schmid B, Klas M, Isoya J, Markham M, Twitchen D, Jelezko F, Liu R-B, Fedder H, Wrachtrup J (2012) Sensing single remote nuclear spins. Nat Nanotechnol 7:657–662. doi:10.1038/NNANO.2012.152

Zhu Y (2012) The Biocompatibility of nanodiamonds and their application in drug delivery systems. Theranostics 2:302–312. doi:10.7150/thno.3627

Intracellular Delivery of RNA via RNA-Binding Proteins or Peptides

Kazunori Watanabe and Takashi Ohtsuki

Abstract Methods that facilitate the cellular internalization of RNAs are widely utilized in medical and biological studies. Numerous small RNA carriers have been reported, many of which are comprised of synthetic polymers. Nanocarriers based on other materials, such as silica and metal nanoparticles, have also been developed. In this chapter, we describe methods for the intracellular delivery of small RNAs using carrier peptides, primarily cationic cell-penetrating peptides (CPPs) and carrier proteins, primarily CPP-RNA binding domain fusion proteins. In addition, we discuss the advantages and drawbacks of these carriers compared to other carriers.

Keywords Cell-penetrating peptide · RNA-binding protein · siRNA · shRNA · RNA delivery · Gene silencing · RNAi · Endosomal escape

Abbreviations

siRNA	Small interfering RNA
miRNA	Micro RNA
dsRNA	Double-stranded RNA
RBP	RNA-binding protein
RBD	RNA-binding domain
DRBD	DsRNA-binding domain
CPP	Cell-penetrating peptide
PTD	Peptide transduction domain
NLS	Nuclear localization signal
CLIP-RNAi	CPP-linked RBP-mediated RNA internalization and photoinduced RNAi

K. Watanabe · T. Ohtsuki (✉)
Department of Biotechnology, Okayama University, 3-1-1 Tsushima-naka, Kita-ku, Okayama 700-8530, Japan
e-mail: ohtsuk@okayama-u.ac.jp

A. Prokop et al. (eds.), *Intracellular Delivery II*,
Fundamental Biomedical Technologies 7, DOI: 10.1007/978-94-017-8896-0_19,

1 Introduction

Small RNAs, such as small interfering RNAs (siRNAs), short hairpin RNAs (shRNAs), and micro RNAs (miRNAs), have become an important tool for studying and controlling gene functions. The introduction of small double-stranded RNAs (dsRNAs), termed siRNA or shRNA, can lead to the down-regulation of specific genes through an evolutionarily conserved process known as RNA interference (RNAi) (Elbashir et al. 2001; Fire et al. 1998). The mechanism for miRNA-mediated gene silencing resembles that of RNAi-mediated silencing in that an Argonaute protein associated with an miRNA or an antisense strand of siRNA binds to a specific mRNA and interferes with its translation (Hutvagner and Simard 2008). In addition to studies of gene function, these small RNAs have been used in therapeutic studies and as potential therapeutic tools for the treatment of cancers and viral diseases (Gooding et al. 2012; Shim and Kwon 2010). For studies utilizing small RNAs, the development of an effective and non-toxic RNA delivery system is necessary.

Although viral vector technologies can stably introduce RNA genes into the host genome, there are many potential side effects, including immunotoxicity and mutagenesis (Raper et al. 2003; Check 2005; Guo et al. 2013). Therefore, in recent years, non-viral delivery systems, such as lipid-based agents (Foged 2012; Rehman et al. 2013), cationic polymers (Nimesh et al. 2011), and cell-penetrating peptides [CPPs, also called peptide transduction domains (PTDs)] (Fonseca et al. 2009; van den Berg and Dowdy 2011; Endoh and Ohtsuki 2009; Nakase et al. 2013) have been employed. Lipid-based agents and cationic polymers are used as standard tools *in vitro* due to their relatively low production cost and low immunogenecity (Akhtar and Benter 2007; Park et al. 2006), although serious cytotoxicity has been reported (Hunter 2006; Moghimi et al. 2005; Symonds et al. 2005).

The use of CPPs or carrier proteins is emerging as an effective approach for the intracellular delivery of RNA. CPPs are short peptides generally composed of 6–30 amino acids that can cross the cellular plasma membrane or be internalized by cells via endocytosis. CPPs are able to deliver biologically active macromolecules, such as proteins and nucleic acids, with high efficiency and low toxicity (van den Berg and Dowdy 2011; Schwarze et al. 1999; Jones and Sayers 2012). In addition to these peptides, carrier proteins for RNA delivery have recently been developed. These carrier proteins consist of an RNA-binding protein (or RNA-binding domain) linked to a cell-penetrating or cell-targeting peptide. In this chapter, we will discuss the use of peptides and proteins for the intracellular delivery of RNAs, mainly siRNAs. This review covers carrier peptides (mainly CPPs) and carrier proteins (mainly CPP-fusion RNA-binding proteins) that deliver RNA by themselves, it does not discuss CPP-modified nanocarriers including materials other than peptides, such as lipids, information on which may be found in other reports (Koren and Torchilin 2012).

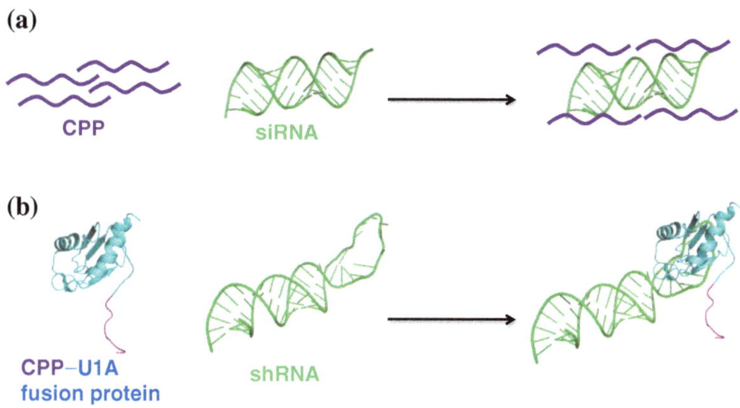

Fig. 1 CPP/siRNA (**a**) and CPP-U1A/shRNA (**b**) complexes. The accession numbers of the structures used here are 1R9F (siRNA), 1U6B (U1A RBD), and 1U6B+2F8S (shRNA bearing the U1A-binding sequence)

2 Use of CPPs for RNA Delivery

RNA delivery using CPP can be broken down into two main categories: (1) covalent conjugation of CPP to RNA, and (2) the formation of a non-covalent CPP-RNA complex. In 2004, siRNA delivery using covalently attached CPPs was demonstrated (Chiu et al. 2004; Davidson et al. 2004; Muratovska and Eccles 2004). However, it is not clear whether the CPP/siRNA conjugate was completely separated from free CPP in these studies. Turner et al. (2007) reported that "highly purified" CPP/siRNA conjugates induced RNAi, but the concentration of CPP/siRNA conjugates needed to achieve this was almost two orders of magnitude higher than in reports published in 2004, indicating that impurities (probably free CPP) enhanced cell-entry of the CPP/siRNA conjugate. Furthermore, Moschos et al. (2007) reported that covalently linked CPP/siRNA conjugates elicited immune responses. Positively charged CPPs and negatively charged RNAs can form complexes or aggregates through nonspecific electrostatic interactions (Law et al. 2008). Most studies utilizing CPPs for RNA delivery have employed these non-covalent CPP-RNA complexes (Table 1).

In 2003, Simeoni et al. (2003) demonstrated the successful delivery of siRNA into cells and subsequent gene silencing in HeLa and HS-68 cells using MPG and mutated MPG peptides. The amphipathic MPG peptide, an artificially constructed chimeric CPP, is a 27 amino acid peptide composed of a basic domain from the nuclear localization signal (NLS) of the SV40 large T antigen and a hydrophobic region derived from HIV gp41. The basic domain was thought to interact with siRNA while the hydrophobic region was believed to have membrane destabilizing properties. The mutated MPG peptide, in which a lysine residue in NLS domain was replaced with a serine (MPGΔNLS), was used to introduce siRNA into the HeLa and HS-68 cells. The MPGΔNLS/siRNA complexes induced gene silencing

Table 1 CPPs for RNA delivery

CPP	Sequence	Target	Cell line	RNAi efficiency	Notes	Ref.
MPG	GALFLGFLGAAGSTMGAWSQPKKKRKV	Luciferase	Cos7	~78%		Simeoni et al. 2003
		GAPDH	HS-68	~60%		Simeoni et al. 2003
MPGANLS	GALFLGFLGAAGSTMGAWSQPKSKRKV	Luciferase	Cos7	~90%		Simeoni et al. 2003
		GAPDH	HS-68	~80%		Simeoni et al. 2003
MPGa	GALFLAFLAAALSLMGLWSQPKKKRKV	Luciferase	ECV304	~83%		Veldhoen et al. 2006
Penetratin	RQIKIWFQNRRMKWKK	Luciferase	HeLa	0%		Lundberg et al. 2007
HA2-Penetratin	GLFGAIAGFIENGWEGMIDGRQIKIWFQNRRMKWKK	Luciferase	HeLa	~35%	Penetratin was co-added	Lundberg et al. 2007
MPGANLS	GALFLGFLGAAGSTMGAPKSKRKV	Luciferase	HeLa	~55%		Lundberg et al. 2007
TP-10	AGYLLGKINLKALAALAKKIL	Luciferase	HeLa	~18%		Lundberg et al. 2007
EB1	LIRLWSHLIHIWFQNRRLKWKKK	Luciferase	HeLa	~55%		Lundberg et al. 2007
Bovine PrP	GLWRALWRLLRSLWRLLWRA	Luciferase	HeLa	~48%		Lundberg et al. 2007
Polyarginine	RRRRRRRRR	EGFP	GC	~43%		Wang et al. 2007
CADY	GLWRALWRLLRSLWRLLWRA	GAPDH	HeLa	~95%		Crombez et al. 2008

(continued)

Table 1 (continued)

CPP	Sequence	Target	Cell line	RNAi efficiency	Notes	Ref.
		p53	HeLa	~97%		Crombez et al. 2008
LAH4	KKALLALALHHLAHLALHLALAKKA	Luciferase	Human embryonic retinoblasts	~80%		Langlet-Bertin et al. 2010
rPOA	CRRRRRRRRRC	VEGF	SCC	~50%		Won et al. 2010
PF14	stearyl-AGYLLGKLLOOLAAAALOOLL	Luciferase	BHK-21	~80%	N-terminal in PF14 was stearylated	Ezzat et al. 2012
TAT	RKKRRQRRR	Luciferase	A549-luc-C8	~92%	Calciumn was added	Baoum et al. 2011
dTAT	RKKRRQRRRHRRKKR	Luciferase	A549-luc-C8	~92%	Calciumn was added	Baoum et al. 2011

even more efficiently than the MPG/siRNA complexes. Veldhoen et al. (2006) demonstrated effective siRNA delivery using MPGα derived from MPG, and demonstrated that the uptake of MPGα/siRNA complexes into HeLa cells was mediated by endocytotic processes.

Lundberg et al. (2007) compared the siRNA delivery efficiencies of six CPPs, including penetratin, HA2-penetratin, MPGΔNLS, TP-10, EB1, and bovine PrP. Although the results showed that penetratin delivered larger amounts of siRNA than MPGΔNLS into cells, the penetratin/siRNA complexes did not induce gene silencing, while the MPGΔNLS/siRNA complexes did. These results seemed to be due to endosomal entrapment of the penetratin/siRNA complexes. Hence, EB1, a novel penetratin analog with endosomolytic properties, and HA2-penetratin, a chimeric peptide consisting of the HA2 influenza fusion peptide conjugated to penetratin, were designed for siRNA delivery. Indeed, the EB1/siRNA and HA2-penetratin/siRNA complexes efficiently induced gene silencing, in contrast to the results achieved with penetratin/siRNA complexes. The authors concluded that the increase in endosomal escape of the complexes was responsible for the significantly increased biological effect of the siRNA.

RNA delivery using non-covalent CPP/RNA complexes is a simple and effective strategy with low levels of cytotoxicity. Lundberg et al. (2007) reported that none of the cells transfected with peptides in complex with siRNA exhibited reduced proliferation, whereas the proliferation of cells treated with Lipofectamine 2000 was significantly reduced. However, an excess amount of CPP molecules were required for efficient RNA delivery. For example, MPGΔNLS and siRNA were mixed at a 25:1 molar ratio (2,500:100 nM) to obtain a 55 % suppression of luciferase protein expression in HeLa cells (Lundberg et al. 2007). In the case of MPG, the peptides and siRNA were used at an 84:1 molar ratio (4,200:50 nM) (Simeoni et al. 2003). Additionally, siRNA with Lipofectamine 2000, a lipid-based reagent, induced RNAi more efficiently compared to MPGΔNLS and EB1 (Lundberg et al. 2007).

To overcome these drawbacks, CPPs based on amphipathic peptides have been designed by several researchers (Crombez et al. 2008; Langlet-Bertin et al. 2010). Crombez et al. (2008) designed a secondary amphipathic peptide (CADY) of 20 amino acid residues containing aromatic tryptophan and cationic arginine residues. Although the CADY/siRNA complexes were used at a molar ratio of 40:1, the concentration of siRNA required was only 20 nM. Furthermore, GAPDH and p53 were efficiently silenced in human osteosarcoma U_2OS cells using the CADY/siRNA complexes. Langlet-Bertin et al. (2010) reported the development of a cationic amphipathic histidine-rich peptide (LAH4) and a mutated LAH4. LAH4 contains four histidines that are thought to enhance endosomal escape through acidification of the endosomal environment. This leads to the protonation of the histidines, which liberates a major fraction of peptides from the siRNAs that then interact with and destabilize the endosomal membrane. The LAH4/siRNA and mutated LAH4/siRNA complexes required 50 nM siRNA (vector/siRNA w/w ratio = 5:1–10.5:1) to induce a 70–80 % knockdown of luciferase in human embryonic retinoblasts.

CPP nanoparticles complexed with siRNA have also been reported recently. Won et al. (2010) demonstrated an effective RNA delivery method using a reducible poly(oligo-D-arginine) (rPOA), which forms nanoparticles with siRNAs that have a mean diameter of ~90 nm. rPOA is a polymer of Cys–(D-Arg)$_9$–Cys peptides linked together by disulfide bonds. The disulfide bonds are reduced when the rPOA/siRNA polyplex enters the cytoplasmic environment, and siRNA release is facilitated due to polyplex dissociation. Vascular endothelial growth factor (VEGF) was successfully knocked down using the rPOA/siRNA polyplex in squamous cell carcinoma (SCC) cells *in vitro* and *in vivo*. In addition, Ezzat et al. (2012) demonstrated the formation of CPP/siRNA nanocomplexes containing a novel CPP, PeptFect 14 (PF14), which were able to elicit efficient RNAi in various cell lines.

The use of carrier peptides is a technically simple method for RNA delivery, and generally, the cytotoxicity of carrier peptides is low compared with that of lipid-based agents. Many efforts to extend the applicability of CPPs are currently ongoing, including the introduction of chemical modifications to CPPs (Ezzat et al. 2012; Mae et al. 2009), the alteration of peptide design to facilitate endosomal escape (Varkouhi et al. 2010), and efforts to control the size of CPP/siRNA complexes (Baoum et al. 2011).

3 RNA Carrier Proteins: Fusion of an RNA-Binding Domain with a Cell-Penetrating or Cell-Targeting Peptide

Instead of using simple CPPs, some groups have recently developed CPP-fusion RNA-binding proteins (RBPs) as carrier proteins for RNA delivery (Eguchi et al. 2009; Michiue et al. 2009; Geoghegan et al. 2012; Endoh et al. 2008, 2009; Matsushita-Ishiodori et al. 2011). Many of these carrier proteins, mainly CPP-RBPs, are shown in Table 2.

In 2008, our group first demonstrated intracellular RNA delivery using TatU1A (Endoh and Ohtsuki 2010; Endoh et al. 2008). TatU1A consists of a Tat CPP and an RBP derived from the U1A RNA-binding domain (RBD), which recognizes a short RNA sequence (5′-CAUUGCACUCCG-3′) with high specificity and affinity (Oubridge et al. 1994). Although RNA bearing the U1A recognition sequence could be delivered by TatU1A, the RNA and TatU1A were trapped in the endosomes. To enhance escape from the endosomes, a photosensitizer (PS) was attached to the C-terminus of TatU1A. The PS enabled photoinduced release of RNA from endosomes and subsequent RNAi-mediated gene suppression. We designated this photoinduced RNAi strategy CLIP-RNAi (*C*PP-*l*inked RBP-mediated RNA *i*nternalization and *p*hotoinduced RNAi). Other CPPs and RBPs were also used to construct CPP-RBPs, and six additional CPP-RBP variants were constructed (Matsushita-Ishiodori et al. 2011). Among them, the most efficient

Table 2 RNA carrier proteins

CPP fusion RNAbinding domain (protein)	Target	Cell line	RNAi efficiency	Notes	Ref.
TAT–U1A	EGFP	CHO	~70%	Photo-irradiation was needed	Endoh et al.2008; Matsushita-Ishiodori et al. 2011
FHV–U1A	EGFP	CHO	~74%	Photo-irradiation was needed	Matsushita-Ishiodori et al. 2011
TAT–U1A–CL1	EGFP	CHO	~80%	Photo-irradiation was needed	Matsushita-Ishiodori et al. 2011
TAT–SxI	EGFP	CHO	~47%	Photo-irradiation was needed	Matsushita-Ishiodori et al. 2011
TAT–TAT–HA–TAT–DRBD1	GAPDH	H1299	~85%		Eguchi et al. 2009
DRBD1–DRBD2	GFP	Human breast carcinoma	~24%		Kim et al. 2009
			~57%	KALA was added	Kim et al. 2009
TAT–TAT–HA–TAT–DRBD1–DRBD2	HPRT	HeLa	~19%		Geoghegan et al. 2012
			~92%	Chloroquine was added	Geoghegan et al. 2012
TAT–TAT–HA–TAT–DRBD1	HPRT	HeLa	~82%		Geoghegan et al. 2012
			~92%	Chloroquine was added	Geoghegan et al. 2012
TAT–DRBD1–DRBD2	HPRT	HeLa	0%		Geoghegan et al. 2012
			~59%	Chloroquine was added	Geoghegan et al. 2012

(continued)

Table 2 (continued)

CPP fusion RNAbinding domain (protein)	Target	Cell line	RNAi efficiency	Notes	Ref.
B2–B2–B2–DFBD1–DRBD2	HPRT	HeLa	~30%		Geoghegan et al. 2012
			~65%	Chloroquine was added	Geoghegan et al. 2012
p19–YSA	RFP	SKOV3	~70%		Choi et al. 2013

carrier proteins were TatU1A-CL1 bearing the degradation signal peptide CL1, FHVU1A comprised of a flock house virus (FHV) coat peptide (as a CPP) and U1A RBD (RBP), and TatSxl comprised of Tat (CPP) and sex-lethal protein (Sxl) (RBP). Photoinduced RNAi was also achieved using TatU1A-CL1-PS, FHVU1A-PS, and TatSxl-PS. Attachment of CL1 to TatU1A accelerated RNAi-mediated gene silencing, suggesting that dissociation of the carrier protein and RNA due to the degradation of the carrier in the cytosol is important for RNAi activity. These CPP-RBP carriers cannot be used for cytoplasmic RNA delivery without photo-activation, but these light-dependent mechanisms for cytosolic RNA delivery and RNAi will open new possibilities for cellular and molecular biology and therapy (Matsushita-Ishiodori and Ohtsuki 2012).

Eguchi et al. (2009) utilized the double-stranded RNA-binding domain from the RNA-activated protein kinase PKR to construct the PTD-DRBD (CPP-DRBD) fusion protein. *In vitro* study showed that PTD-DRBD delivered siRNA into H1299 cells more rapidly than lipid-based agents. PTD-DRBD-delivered siRNA induced RNAi in various cell types, including T cells, human umbilical vein endothelial cells, and human embryonic stem cells. No evidence of cytotoxicity, off-target transcriptional changes, or induction of innate immune responses was observed in these cells. *In vivo* study confirmed that PTD-DRBD/siRNA complexes induced silencing of luciferase gene expression in mice, and treatment with PTD-DRBD/siRNA complexes targeting the EGF receptor and Act2 induced tumor-specific apoptosis and a significant increase in survival in an intracerebral glioblastoma mouse model (Michiue et al. 2009). Kim et al. (2009) also employed a DRBD derived from PKR as an siRNA delivery carrier without the addition of a CPP. The DRBD protected siRNA from degradation by ribonuclease, and silencing of GFP by 57 % was induced using DRBD/siRNA complexes with a cationic fusogenic peptide, KALA, to facilitate endosomal escape. Geoghegan et al. (2012) also used the DRBD derived from PKR to construct PTD-DRBD variants. PTD-2 × DRBD bound siRNA strongly compared to PTD-DRBD and delivered it into HeLa cells, but treatment with PTD-2 × DRBD/siRNA resulted in only 19 % knockdown of hypoxanthine-guanine phosphoribosyl transferase (HPRT). However, when the endosomolytic agent, chloroquine, was added, PTD-2 × DRBD/siRNA decreased HPRT expression by 92 %, indicating that PTD-2 × DRBD/siRNA was retained in the endosomes. They also designed a B2–2 × DRBD protein that included three repeats of the B2 peptide sequence (GHKVKRPKG), which they previously identified by phage display panning against the transferrin receptor. Treatment with B2–2 × DRBD/siRNA complexes resulted in a 30 % reduction in HPRT expression, an efficiency that was also enhanced by the addition of chloroquine to 65 %.

4 Targeted Delivery Using Carrier Peptides and Proteins

Many carrier peptides and carrier proteins have been reported as effective tools for RNA delivery. However, most of these carriers cannot target specific cell types, such as tumor cells. Therefore, the development of cell-specific carriers will be necessary for the effective therapeutic application of RNAi.

In 2007, Kumar et al. (2007) presented a new method using a peptide containing nonamer arginine residues at the carboxyl terminus of a short peptide derived from the rabies virus glycoprotein (RVG), which interacts specifically with the nicotinic acetylcholine receptor on neuronal cells. This RVG-9R peptide specifically introduced siRNA into neuronal cells *in vitro*, resulting in efficient gene silencing. Furthermore, *in vivo* analysis showed that RVG-9R delivered siRNA specifically to neuronal cells in mice and induced gene silencing within the brain. Targeted delivery of siRNA to macrophages using RVG-nona-D-arginine (RVG-9dR) was also reported by the same group (Kim et al. 2010). In addition, they reported an siRNA delivery system using an oligoarginine peptide conjugated to a CD7 specific single-chain antibody (scFvCD7-9R) (Kumar et al. 2008). Since the CD7 specific antibody bound to the T cell surface protein, CD7, scFvCD7-9R specifically delivered antiviral-siRNA into T cells and dramatically suppressed HIV infection.

Ren et al. (2012) developed a tumor-penetrating nanocomplex containing siRNA with a tandem tumor-penetrating and membrane-translocating peptide [LyP-1 and Transportan (TP)], which enabled the specific delivery of siRNA into the tumor parenchyma. The cyclic nonapeptide LyP-1 specifically bound the surface of stressed tumor and tumor-associated cells. TP-Lyp1/siRNA complexes suppressed the growth of tumors in mice and significantly improved their survival.

Choi et al. (2011, 2013) utilized p19 RNA-binding protein (p19), which specifically binds double-stranded siRNA in a sequence-independent manner, to promote cell-specific siRNA delivery by constructing a capsid (hepatitis B virus)-p19 fusion protein. Within the capsid moiety, the fusion protein included an integrin-binding peptide (RGD) sequence able to target tumor vasculature expressing $\alpha_v\beta_3$ integrins. The fusion proteins assembled into a capsid shell that encapsulated the siRNAs, and delivered the siRNA into B16F10 melanoma cells. They also designed a p19-ephrin mimetic peptide (YSA) fusion protein (p19-YSA) for siRNA delivery into tumor cells (Choi et al. 2013). YSA specifically targets the EphA2 receptor, which is highly expressed in many types of cancers and in tumor blood vessels. Efficient uptake of p19-YSA/siRNA was observed in human ovary cancer cells (SKOV3) expressing high levels of EphA2 receptors, while the nonmalignant normal cells showed relatively much lower uptake efficiency. Treatment of SKOV3 cells with the p19-YSA/siRNA complexes suppressed RFP expression by 68 %.

5 Conclusions

The development of efficient delivery systems is one of the most significant challenges for turning RNAs into clinically useful therapeutic drugs. One approach to overcoming this issue is the development of technologies designed to facilitate RNA delivery using carrier peptides or carrier proteins. To accomplish this, many research groups have devised carrier peptides, primarily CPPs bound non-covalently to siRNA, and carrier proteins that include an RNA-binding moiety, such as RBD from U1A and DRBDs from PKR and p19. In addition, various CPP-conjugates with targeting peptides, such as RVG and LyP-1, or a specific single antibody, have been devised for the delivery of siRNA into specific cell types, such as tumor and virus-infected cells. Several carriers, including RBPs and tumor-homing peptides, have also been developed. Generally, the cytotoxicities of carrier peptides and proteins are low compared to other RNA carrier molecules. Thus, these carrier peptides and proteins are promising options for gene therapy, and further advances in the design of carrier peptides and proteins are expected in future.

Acknowledgements We thank S. Matsumoto for help with preparation of Fig. 1.

References

Akhtar S, Benter IF (2007) Nonviral delivery of synthetic siRNAs in vivo. J Clin Invest 117:3623–3632

Baoum A, Ovcharenko D, Berkland C (2011) Calcium condensed cell penetrating peptide complexes offer highly efficient, low toxicity gene silencing. Int J Pharm 427:134–142

Check E (2005) Gene therapists urged to learn more immunology. Nature 434:812

Chiu YL, Ali A, Chu CY, Cao H, Rana TM (2004) Visualizing a correlation between siRNA localization, cellular uptake, and RNAi in living cells. Chem Biol 11:1165–1175

Choi KM, Choi SH, Jeon H, Kim IS, Ahn HJ (2011) Chimeric capsid protein as a nanocarrier for siRNA delivery: stability and cellular uptake of encapsulated siRNA. ACS Nano 5:8690–8699

Choi KM, Park GL, Hwang KY, Lee JW, Ahn HJ (2013) Efficient siRNA delivery into tumor cells by p19-YSA fusion protein. Mol Pharm 10:763–773

Crombez L, Aldrian-Herrada G, Konate K et al (2008) A new potent secondary amphipathic cell-penetrating peptide for siRNA delivery into mammalian cells. Mol Ther 17(1):95–103

Davidson TJ, Harel S, Arboleda VA et al (2004) Highly efficient small interfering RNA delivery to primary mammalian neurons induces microRNA-like effects before mRNA degradation. J Neurosci 24:10040–10046

Eguchi A, Meade BR, Chang YC et al (2009) Efficient siRNA delivery into primary cells by a peptide transduction domain-dsRNA binding domain fusion protein. Nat Biotechnol 27:567–571

Elbashir SM, Harborth J, Lendeckel W, Yalcin A, Weber K, Tuschl T (2001) Duplexes of 21-nucleotide RNAs mediate RNA interference in cultured mammalian cells. Nature 411:494–498

Endoh T, Ohtsuki T (2009) Cellular siRNA delivery using cell-penetrating peptides modified for endosomal escape. Adv Drug Deliv Rev 61:704–709

Endoh T, Ohtsuki T (2010) Cellular siRNA delivery using TatU1A and photo-induced RNA interference. Methods Mol Biol 623:271–281

Endoh T, Sisido M, Ohtsuki T (2008) Cellular siRNA delivery mediated by a cell-permeant RNA-binding protein and photoinduced RNA interference. Bioconjug Chem 19:1017–1024

Endoh T, Sisido M, Ohtsuki T (2009) Spatial regulation of specific gene expression through photoactivation of RNAi. J Control Release 137:241–245

Ezzat K, Zaghloul EM, El Andaloussi S et al (2012) Solid formulation of cell-penetrating peptide nanocomplexes with siRNA and their stability in simulated gastric conditions. J Control Release 162:1–8

Fire A, Xu S, Montgomery MK, Kostas SA, Driver SE, Mello CC (1998) Potent and specific genetic interference by double-stranded RNA in *Caenorhabditis elegans*. Nature 391:806–811

Foged C (2012) siRNA delivery with lipid-based systems: promises and pitfalls. Curr Top Med Chem 12:97–107

Fonseca SB, Pereira MP, Kelley SO (2009) Recent advances in the use of cell-penetrating peptides for medical and biological applications. Adv Drug Deliv Rev 61:953–964

Geoghegan JC, Gilmore BL, Davidson BL (2012) Gene silencing mediated by siRNA-binding fusion proteins is attenuated by double-stranded RNA-binding domain structure. Mol Ther Nucleic Acids 1:e53

Gooding M, Browne LP, Quinteiro FM, Selwood DL (2012) siRNA delivery: from lipids to cell-penetrating peptides and their mimics. Chem Biol Drug Des 80:787–809

Guo J, Evans JC, O'Driscoll CM (2013) Delivering RNAi therapeutics with non-viral technology: a promising strategy for prostate cancer? Trends Mol Med 19:250–261

Hunter AC (2006) Molecular hurdles in polyfectin design and mechanistic background to polycation induced cytotoxicity. Adv Drug Deliv Rev 58:1523–1531

Hutvagner G, Simard MJ (2008) Argonaute proteins: key players in RNA silencing. Nat Rev Mol Cell Biol 9:22–32

Jones AT, Sayers EJ (2012) Cell entry of cell penetrating peptides: tales of tails wagging dogs. J Control Release 161:582–591

Kim J, Lee SH, Choe J, Park TG (2009) Intracellular small interfering RNA delivery using genetically engineered double-stranded RNA binding protein domain. J Gene Med 11:804–812

Kim SS, Ye C, Kumar P et al (2010) Targeted delivery of siRNA to macrophages for anti-inflammatory treatment. Mol Ther 18:993–1001

Koren E, Torchilin VP (2012) Cell-penetrating peptides: breaking through to the other side. Trends Mol Med 18:385–393

Kumar P, Wu H, McBride JL et al (2007) Transvascular delivery of small interfering RNA to the central nervous system. Nature 448:39–43

Kumar P, Ban HS, Kim SS et al (2008) T cell-specific siRNA delivery suppresses HIV-1 infection in humanized mice. Cell 134:577–586

Langlet-Bertin B, Leborgne C, Scherman D, Bechinger B, Mason AJ, Kichler A (2010) Design and evaluation of histidine-rich amphipathic peptides for siRNA delivery. Pharm Res 27:1426–1436

Law M, Jafari M, Chen P (2008) Physicochemical characterization of siRNA-peptide complexes. Biotechnol Prog 24:957–963

Lundberg P, El-Andaloussi S, Sutlu T, Johansson H, Langel U (2007) Delivery of short interfering RNA using endosomolytic cell-penetrating peptides. FASEB J 21:2664–2671

Mae M, Andaloussi SE, Lehto T, Langel U (2009) Chemically modified cell-penetrating peptides for the delivery of nucleic acids. Expert Opin Drug Deliv 6:1195–1205

Matsushita-Ishiodori Y, Ohtsuki T (2012) Photoinduced RNA interference. Acc Chem Res 45:1039–1047

Matsushita-Ishiodori Y, Kuwabara R, Sakakoshi H, Endoh T, Ohtsuki T (2011) Photosensitizing carrier proteins for photoinducible RNA interference. Bioconjug Chem 22:2222–2226

Michiue H, Eguchi A, Scadeng M, Dowdy SF (2009) Induction of in vivo synthetic lethal RNAi responses to treat glioblastoma. Cancer Biol Ther 8:2306–2313

Moghimi SM, Symonds P, Murray JC, Hunter AC, Debska G, Szewczyk A (2005) A two-stage poly(ethylenimine)-mediated cytotoxicity: implications for gene transfer/therapy. Mol Ther 11:990–995
Moschos SA, Williams AE, Lindsay MA (2007) Cell-penetrating-peptide-mediated siRNA lung delivery. Biochem Soc Trans 35:807–810
Muratovska A, Eccles MR (2004) Conjugate for efficient delivery of short interfering RNA (siRNA) into mammalian cells. FEBS Lett 558:63–68
Nakase I, Tanaka G, Futaki S (2013) Cell-penetrating peptides (CPPs) as a vector for the delivery of siRNAs into cells. Mol Biosyst 9:855–861
Nimesh S, Gupta N, Chandra R (2011) Cationic polymer based nanocarriers for delivery of therapeutic nucleic acids. J Biomed Nanotechnol 7:504–520
Oubridge C, Ito N, Evans PR, Teo CH, Nagai K (1994) Crystal structure at 1.92 Å resolution of the RNA-binding domain of the U1A spliceosomal protein complexed with an RNA hairpin. Nature 372:432–438
Park TG, Jeong JH, Kim SW (2006) Current status of polymeric gene delivery systems. Adv Drug Deliv Rev 58:467–486
Raper SE, Chirmule N, Lee FS et al (2003) Fatal systemic inflammatory response syndrome in a ornithine transcarbamylase deficient patient following adenoviral gene transfer. Mol Genet Metab 80:148–158
Rehman Z, Zuhorn IS, Hoekstra D (2013) How cationic lipids transfer nucleic acids into cells and across cellular membranes: recent advances. J Control Release 166:46–56
Ren Y, Cheung HW, von Maltzhan G et al (2012) Targeted tumor-penetrating siRNA nanocomplexes for credentialing the ovarian cancer oncogene ID4. Sci Transl Med 4:147ra112
Schwarze SR, Ho A, Vocero-Akbani A, Dowdy SF (1999) In vivo protein transduction: delivery of a biologically active protein into the mouse. Science 285:1569–1572
Shim MS, Kwon YJ (2010) Efficient and targeted delivery of siRNA in vivo. FEBS J 277:4814–4827
Simeoni F, Morris MC, Heitz F, Divita G (2003) Insight into the mechanism of the peptide-based gene delivery system MPG: implications for delivery of siRNA into mammalian cells. Nucleic Acids Res 31:2717–2724
Symonds P, Murray JC, Hunter AC, Debska G, Szewczyk A, Moghimi SM (2005) Low and high molecular weight poly(L-lysine)s/poly(L-lysine)-DNA complexes initiate mitochondrial-mediated apoptosis differently. FEBS Lett 579:6191–6198
Turner JJ, Jones S, Fabani MM, Ivanova G, Arzumanov AA, Gait MJ (2007) RNA targeting with peptide conjugates of oligonucleotides, siRNA and PNA. Blood Cells Mol Dis 38:1–7
van den Berg A, Dowdy SF (2011) Protein transduction domain delivery of therapeutic macromolecules. Curr Opin Biotechnol 22:888–893
Varkouhi AK, Scholte M, Storm G, Haisma HJ (2010) Endosomal escape pathways for delivery of biologicals. J Control Release 151:220–228
Veldhoen S, Laufer SD, Trampe A, Restle T (2006) Cellular delivery of small interfering RNA by a non-covalently attached cell-penetrating peptide: quantitative analysis of uptake and biological effect. Nucleic Acids Res 34:6561–6573
Wang YH, Hou YW, Lee HJ (2007) An intracellular delivery method for siRNA by an arginine-rich peptide. J Biochem Biophys Methods 70:579–586
Won YW, Yoon SM, Lee KM, Kim YH (2010) Poly(oligo-D-arginine) with internal disulfide linkages as a cytoplasm-sensitive carrier for siRNA delivery. Mol Ther 19:372–380

Hyaluronic Acid Based Nanofibers for Wound Dressing and Drug Delivery Carriers

Jana Růžičková, Vladimír Velebný, Jindřich Novák, Katarzyna Szuszkiewicz, Kateřina Knotková, Marcela Foglarová and Marek Pokorný

Abstract Hyaluronic acid (HA) as one of the chief components of the human extracellular matrix is proven to be beneficial in the wound healing process, because hyaluronic acid contributes significantly to cell proliferation and migration. Electrospun hyaluronic acid nanofibers should mimic the natural architecture of tissue and so it should provide a wound dressing similar to the natural tissue. Addition of antimicrobial agents in HA nanofibers should eliminate infections and prevent occurrence of undesired tissue adhesions when applied to wounds in abdominal cavity. Drug release profile may be changed by the structure of the nanofibrous material.

Keywords Nanofibers · Electrospinning · Hyaluronic acid · 4Spin® · Wound dressing · Drug delivery · Release · Carrier

Abbreviations

DMF	*N, N*-Dimethylformamide
FDA	Food and drug administration
HA	Hyaluronic acid
PCL	Polycaprolactone
PEG	Polyethylene glycol
PEO	Polyethylene oxide
PLA	Polylactic acid
PLGA	Poly(lactic-co-glycolic) acid
PLLA	Poly-L-Lactide
IPA	Isopropyl alcohol
kDa	Kilodalton
$NaIO_3$	Sodium iodate

J. Růžičková (✉) · V. Velebný · J. Novák · K. Szuszkiewicz · K. Knotková · M. Foglarová · M. Pokorný
Contipro Biotech s.r.o, Dolní Dobrouč 401, 56102 Dolní Dobrouč, Czech Republic
e-mail: ruzickova@contipro.com

A. Prokop et al. (eds.), *Intracellular Delivery II*,
Fundamental Biomedical Technologies 7, DOI: 10.1007/978-94-017-8896-0_20,

$Na_2HPO_4 \cdot 12H_2O$	Disodium hydrogen phosphate dodecahydrate
KI	Potassium iodide
$NaH_2PO_4 \cdot 2H_2O$	Monosodium phosphate dihydrate
μl	Microliter
min	Minute
RH	Relative humidity
cm	Centimetre
nm	Nanometre
Nm	Newton metre
kV	Kilovolt
mg	Milligram
ml	Millilitre
SEM	Scanning electron microscopy
UV	Ultraviolet

1 Introduction

The use of nanofibers as a new form of the material is considered to be used in medicine for various purposes. Nanofibrous materials have advantages consisting in a small fiber diameter, a very large specific surface and a high amount of small inter-fiber pores. Nanofibrous layers produced of a suitable polymer may be beneficial for wound healing. Such wound dressing may release different types of incorporated substances improving the wound healing process. Nanofibrous drug carriers can deliver an incorporated drug to patients by physiologically more acceptable manner reducing the burden on the patient's organism during its administration. The drug is usually a component of the spinning solution. There are two possible forms of the drug incorporation. The drug is either dissolved or dispersed in the spinning solution. In both cases there are small areas of the drug in electrospun fibers. This system of drug delivery has a high rate of dissolution of incorporated drug areas caused by a large surface area of the drug given by the carrier form. The most desired drug release profile is constant and stable; such a profile can be achieved by incorporation of a lipophilic drug into a lipophilic polymer, while the hydrophilic polymer could serve as a carrier of hydrophilic drugs whereas the solvent used should be good for both drug and polymer (Zeng et al. 2005). Otherwise the drug cannot dissolve in the polymer solution which results in dispersion spinning. Electrospun drug dispersion causes the presence of the drug on the surface of the fibers or in the vicinity and consequently the "burst effect", the fast diffusion of the drug to the release media, occurs.

High reactivity of nanofibers is also affected by their small dimension causing a short diffusion pathway of incorporated additives releasing from nanofibers and their faster degradation, it makes nanofibers interesting only for certain types of drug releasing systems. But in the field of drug delivery systems and tissue engineering, electrospun polymeric nanofibers have already gained growing importance, mainly due to the following advantages: relatively easy incorporation of drugs during the electrospinning process, the possibility to incorporate high dosages of drugs, possible suppression of the "burst effect", stability and preservation of activity of drugs or growth factors, a large specific surface (improving the drug release) and specific morphology that may be controlled to a certain extent during the production process (Agarwal et al. 2008). The use of bicomponent fibers of the core-shell type or multilayer materials may contribute to the "burst effect" suppression and it is possible to control the kinetics by the thickness of the layers and fiber diameter (Okuda et al. 2010). On the contrary the "burst effect" is deepened by swelling of the carrier matrix used. One of the applications where the burst drug release profile seems to be ideal is the release of local antibiotics or antimicrobial agents from nanofibrous antiadhesive membranes, because most infections are related to contaminations introduced at the time of surgery when the skin barrier is violated (Zong et al. 2004).

The release of incorporated additives proceeds by two mechanisms, the diffusion and release given by the erosion of the matrix. Comparison of nanofibrous materials and films of the same polymer proved that the film releases the drug gradually mainly due to the dissolution of additive aggregates on the film surface, the release from nanofibers proceeds by the classical diffusion mechanism, because there were no aggregates formed on the surface of nanofibers (Taepaiboon et al. 2006). Nanofibers increase the efficiency of the additive utilization and in certain cases it may help to suppress side effects of the additive (Toshkova et al. 2010). Verreck et al. (2003) confirmed that electrospun materials in the studied cases reach a better applicability in pharmaceutical applications, as well as a controlled dissolution when compared with conventional forms like physical mixtures, solutions and extruded samples.

Another benefit of nanofibers doped by additives is a higher solubility bringing a possible interesting application in the field of delivery of drugs hardly soluble in aqueous media; these carriers could significantly improve their solubility in the organism due to a large surface area, even at oral administration (Leuner and Dressman 2000). In addition, this application form may be more advantageous than the conventional dispersions also contributing in the same way to increase the drug solubility (Chokshi et al. 2007).

2 Processing

There are several ways of nanofiber production; electrospinning is one of the few offering a high productivity especially in case of needleless systems (Lukas et al. 2008). Methods like drawing (Ondarcuhu and Joachim 1998), phase separation (Ma and Zhang 1999), template synthesis (Martin 1996) and self-assembly (Whitesides and Grzybowski 2002) are not able to satisfy the need of a massive production. On the other hand highly productive techniques like melt blown and centrifugal spinning (or force spinning) are hardly reaching submicron fibers, process primarily polymer melts or produce only randomly oriented samples of limited area and specific structure. The needleless and the nozzle-less electrospinning principle was denoted as probably the most productive one (Lukas et al. 2008) due to a high amount of Taylore cones that may be created on the surface of the large rotating electrodes, but it has not been proved. The needleless/nozzle-less electrospinning using different principles facilitating the Taylore cone formation like magnetic liquids (Yarin and Zussman 2004), bubbles (Forward and Rutledge 2012), wire spinning electrodes (Liu and He 2007) and also various types of rotating bodies capturing a polymer solution on its surface from which the spinning proceeds (Petrik and Maly 2009) are usually "open systems". In such systems there is generally a high amount of the polymer solution exposed to the air. This exposure causes an intensive evaporation of the solvent leading to the continual change of the solution properties and also the final product (Ruzickova et al. 2012a). On the contrary, the use of multiple jets necessary for the large scale production of nanofibers by the nozzles brings also some problems depending mostly on the distance between neighbouring nozzles. There are strong mutual interactions between neighbouring polymer jets caused by Coulombic forces leading to repulsion of the jets that results in the collection of a non-uniform nonwoven mat (Theron et al. 2004). It is possible to solve the problem by the use of jet matrices facilitating the uniformity of the resulting product but small inter-nozzle distances resulting in increased deposition rate over a smaller area prevent onset of the electrically driven bending instability which is the main factor responsible for the reduction of the fiber diameter. 4Spin® technology (Figs. 1 and 2) combines benefits of a classic electrospinning setup such as a closed feeding system and a needleless high production efficiency is approaching industrial production of nanofibrous materials for healthcare (Ruzickova et al. 2012a).

Presented numerical simulations (Fig. 3) show another disadvantage linked to a larger active surface of the nozzle-less spinning electrode on which the Taylor cones may originate, when compared with the needleless multijet described in this chapter leading to a weak electrostatic gradient and poor electrostatic forces (Pokorny et al. 2013b; Ruzickova et al. 2012a). Another interesting effect is related to the capillary forces acting in the solution and influencing the radius of curvature of the surface element area and thus the pressure in the solution droplet (Pokorny et al. 2011, 2013a, b). Reduction of the radius of curvature facilitates the Taylor cone formation so electrospinning devices using thin capillaries should be the most beneficial ones. In addition, 4Spin® technology enables to form unidirectionally oriented nanofibers (Pokorny et al. 2011, 2012a) and voluminous fluffy 3D nanofibrous structures (Ruzickova et al. 2013).

Petras et al. (2011) published a chapter focused on a comparison of the electrospinning process productivity of a multi-jet spinning electrode and nozzle-less wetted rotating electrodes of the different forms. He described that in the case of non-water soluble polymer solutions, such as polyurethane and polystyrene, the spinning process performed by the needleless technology fails. In the case of PVA

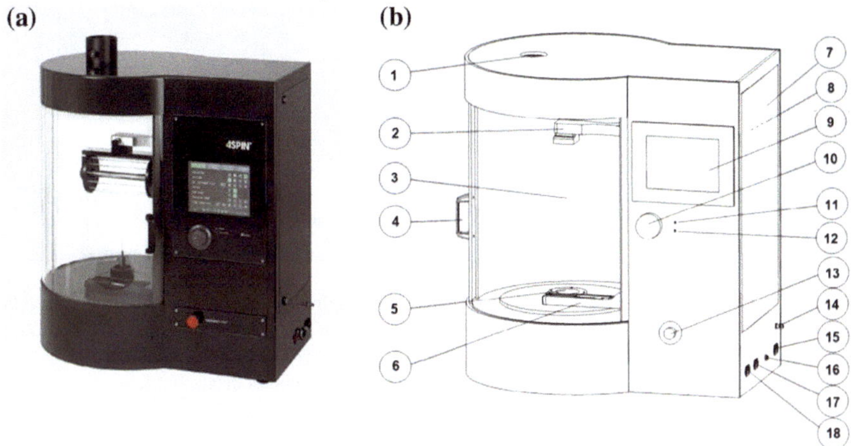

Fig. 1 4Spin® technology (Pokorny 2012b); **a** laboratory device, **b** laboratory device scheme *1* exhaust aperture, *2* collector bracket, *3* spinning chamber, *4* sliding door, *5* emitter and air connector, *6* dispenser, *7* fuse box door, *8* fuseboard and instrumentation inside the machine, *9* touch screen, *10* multifunction control, *11* standby signal, *12* on signal, *13* emergency stop button, *14* compressed air, *15* power, *16* safety clip, *17* peripheral ethernet connection, *18* power button

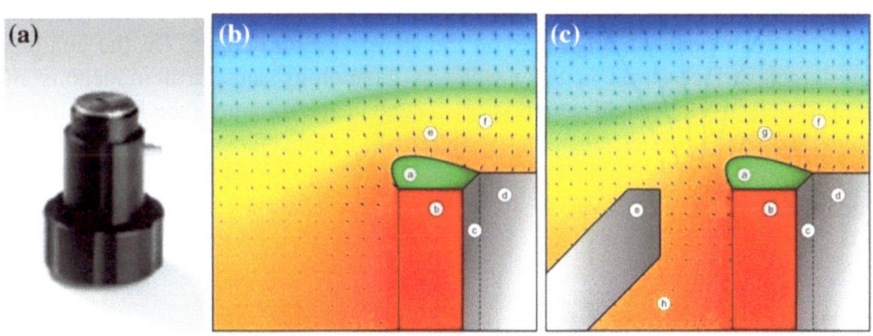

Fig. 2 Needleless multi-jet spinning electrode of the 4Spin® technology (Pokorny et al. 2013b); **a** electrode, **b** cross section of needleless multi-jet electrode showing numerically modelled electrostatic field, **c** cross section of needleless multi-jet electrode showing numerically modelled electrostatic field with additional hot air stream. *Notes a* a polymer solution droplet, *b* electrically conductive electrode connected to a high voltage source, *c* distribution channels of the polymer solution, *d* electrically nonconductive body, *e* electrically nonconductive contour shaping the air flow, *f* electric field strength vectors, *g* area of the high voltage gradient and tangential air forces acting to the polymer solution, and *h* idea of the air streamlines around the electrode (Ruzickova et al. 2012a)

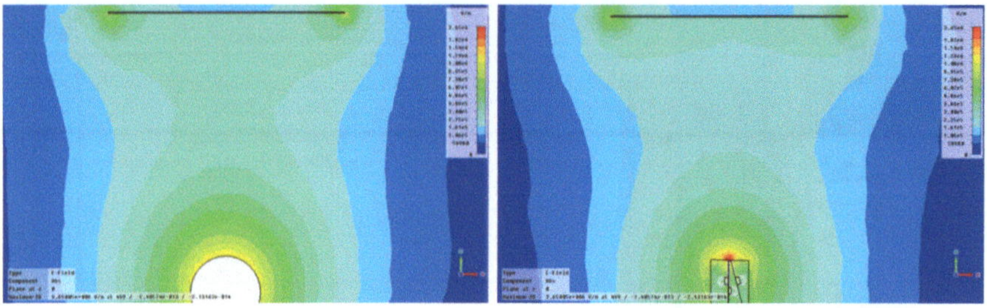

Fig. 3 The results of numerical simulation of electrostatic field intensity distribution between the emitter (*bottom side*) and the collector (*upper side*). There is visible the origin of the gradient field surrounding the thin emitter on the *left image*. The *right image* shows homogenous non-gradient field around the emitter with the large surface (Ruzickova et al. 2012a)

water solution, the jet system reached the same productivity when a smooth rotating cylinder was used and almost three times higher productivity when an electrode with tines was used. The use of a non-water soluble material brings limits of the free liquid surface electrospinning.

It was clearly demonstrated by the previously mentioned numerical simulations (Pokorny et al. 2012) that from the viewpoint of the functionality and the process efficiency it is beneficial to use the thin bodies as the spinning electrodes. The distribution of electrostatic field surrounding the conductive electrodes in the area of the Taylor cone formation is essential. In this area, sufficient electrostatic field gradient amplifying the action of its forces has to be created. Such gradient originates on the sharp and thin edges. On the contrary, the larger surfaces used in the needleless technology cause a homogenous and uniform electrostatic field which is inconvenient for spinning (see Fig. 3). It was proved (Ruzickova et al. 2012a) that production efficiency of the 4Spin® technology compared with the nozzle-less technology was higher in all three measured cases.

3 Electrospinning of Hyaluronic Acid

The use of hyaluronic acid (HA) as a carrier is also tempting because it is a natural biodegradable polymer improving migration and proliferation of cells and production of extracellular matrix (Schmidt and Friedl 2010). It was already proved that it is difficult to electrospin the hyaluronic acid as a polyanionic polymer because of its electrical characteristics and a high viscosity of the aqueous solution (Wang et al. 2005). The preparation of continuous fibers of polyelectrolytes by electrospinning of their aqueous solutions was considered to be impossible due to the repulsive forces between ionogenic groups (Park et al. 2004). The gelation of a HA solution usually occurs at very low polymer concentrations, so that the insufficient entanglement of polymer chains and high elasticity of the solution impede the generation of stable fibers (Kim et al. 2008). Nano-fibers from ionogenic polymers have been successfully electrospun from mixed solutions of an ionogenic polymer and a non-ionogenic polymer (Mincheva et al. 2005) and it was reported that HA nanofibers can be successfully electrospun by blending with other electro-spinnable polymers, such as poly(ethylene oxide) and gelatine, or by constructing a special device for example electro-blowing system or elevating the environmental temperature or by combination of the above mentioned (Ji et al. 2006; Li et al. 2006; Um et al. 2004). A few research groups improve the parameters of the HA spinning solution by the use of other solvents than water, solvents like *N,N*-dimethylformamide (Wang et al. 2005; Brenner et al. 2012; Fischer et al. 2012; Xu et al. 2009), formic acid (Liu et al. 2011, 2010; Ma et al. 2011), trifluoroethane (Soo et al. 2009; Mo et al. 2009) etc., that are absolutely inconvenient for the use in medical devices and may have a negative impact on the organism. Additionally, massive production of such material would lead to a necessity to dispose the large quantities of toxic waste. It is also possible to improve the solution parameters of HA solutions by surfactants but it is inconvenient from the biological viewpoint too (Uppal et al. 2011).

The utilization of HA nanofibers for medical purposes like wound dressing or scaffolds is highly anticipated. The use of classic nanofibrous materials for the mentioned applications is limited due to the small inter-fiber pore size making the migration of the cells to the scaffold impossible; the cells then grow on the scaffold surface and create 2D structure causing their dedifferentiation. The mentioned problem may be solved by the formation of tubular materials (Kuo et al. 2010), or by mutual interleaving of nanofibrous layers and cells (Patel et al. 2010; Wang 2008). Ideal solution may consist of the formation of 3D voluminous textiles. Another property limited by the classic nanofibrous structure is a sorption ability that is not bad in the case of hydrophilic polymers (Soo et al. 2013). But the amount of absorbed liquid decreases with increasing layer thickness (area weight), because nanofibrous layer is compact and more closed (Jun et al. 2011) and it does not allow massive swelling connected with the high sorption.

According to the mentioned problems with conventional nanofibrous textiles of hyaluronic acid the production of voluminous nanofibrous textiles of HA or its derivatives is very interesting. Production of such fluffy HA nanofibrous structure by electrospinning method is even more complicated and it has so far been described only by Kim et al. (2008). But the voluminous material described in the chapter was not stable and the structure collapsed immediately when the voltage was turned off. Therefore they poured the crystals of NaCl on the originating structure during electrospinning and subsequently washed them out, but the nanofibrous structure was swollen by the solvent and it had to be freeze-dried. Similar procedure used for the production of 3D nanofibrous structures of HA/collagen blend was described by Fischer et al. (2012) but he used DMF as a solvent. The mentioned washing and drying procedures are time consuming and expensive.

A stable fluffy voluminous nanofibrous material of hyaluronic acid has already been electrospun by Ruzickova et al. (2013), see in Fig. 9. Such a nanofibrous

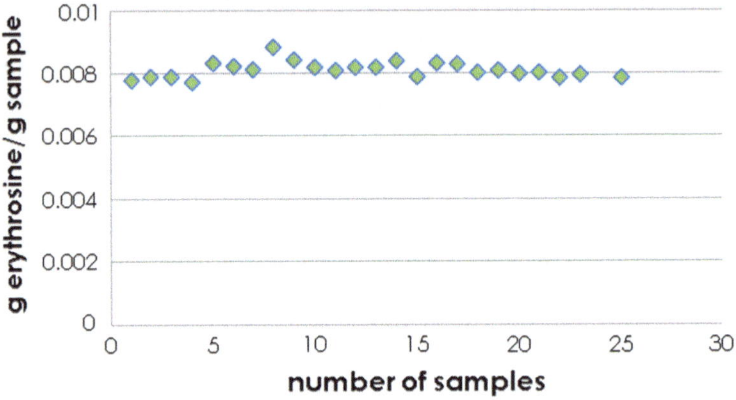

Fig. 4 Homogeneity of the additive content (Erythrosine) incorporated to the nanofibers produced by the 4Spin® technology (Knotkova et al. 2013). Erythrosine was homogeneously distributed in nanofibrous layers and the average concentration of this additive corresponded to concentration of erythrosine in original spinning solution

structure has great sorption properties so required for the moist wound healing and it may also contain additives improving the wound healing process. The release kinetics from the voluminous materials should proceed differently. From the viewpoint of a commercial production another important issue is the homogeneity of incorporated additives. It was already demonstrated (Knotkova et al. 2013) that 4Spin® technology processed the doped spinning solution into a homogenous product with uniform additive content (Fig. 4).

4 Drug Release from HA Nanofibers

Synthetic nanofibers are usually used for drug delivery carriers and they were studied in several cases, for example the nanofibrous carrier based on poly (ethylene-co-vinyl acetate), polylactic acid and their blends containing tetracycline hydrochloride (Kenawy et al. 2002) or nanofibrous membranes of polylactic acid used for delivery of antibiotic Mefoxin (Zong et al. 2002). Suppression of the aforementioned "burst effect" has been managed by electrospinning of poly (L-lactic acid) (PLLA) from a mixture of chloroform and acetone (Zeng et al. 2003) and encapsulation of additives has been supported by the use of surfactants, but the mentioned solvents or surfactants are not suitable for the use in healthcare. Further examples are heparin in PCL matrix (Zhang et al. 2006), paclitaxel in PLGA (Xie and Wang 2006) and doxorubicin hydrochloride in a copolymer of PEG-PLA (Xu et al. 2009).

A significant benefit of nanofibers can be the fact that nanofibrous implants made of polylactic acid blended with a quaternary chitosan and doxorubicin as a drug with anticancer activity, showed an increased efficiency of doxorubicin and concurrent suppression of side effects (Toshkova et al. 2010). This phenomenon is probably caused by synergistic effects of doxorubicin and quaternary chitosan. Combination of biopolymers and pharmaceuticals in the form of nanofibers may reach significantly different impacts in organism when compared to standard materials.

The benefits of HA nanofibers used for wound dressings and drug delivery have been already described in previous chapters. The truth is that hyaluronic acid in not a usual material for drug delivery carriers, because its use is limited by its solubility in aqueous media. Nanofibrous materials of hyaluronic acid release the incorporated substance immediately when they get in touch with humidity. The release is accelerated by the influence of dissolution and swelling and therefore it is very fast. Fast additive release from a carrier based on a native hyaluronic acid excludes it from use in drug carriers with a long duration of action. Hyaluronic acid based foils that serve as antiadhesive materials for prevention of post-surgical adhesions based on hyaluronic acid (HA) have already been used and some of them have already been approved by FDA for human use (Hooker et al. 1999), their disadvantage is a fast degradation and a fast absorption (Chang et al. 2012). In contrast, synthetic materials like PCL with a very slow degradation are also not suitable because their residues can cause an inflammatory reactions resulting in a higher amount of adhesions (Bölgen et al. 2007; Wallwiener et al. 2006). Therefore chemically crosslinked materials based on hyaluronic acid (HA) (Johns et al. 2001; DeIaco et al. 1998) with improved stability and slower degradation, eventually their blend with the native HA, may be very suitable for this type of application. Hyaluronic acid has also antiadhesive efficiency given by its ability to lubricate cells, maintain structural integrity of tissues, regulate liquid retention (Johns et al. 2001), stimulate mesothelium recovery (Wallwiener et al. 2006) etc. Therefore the nanofibrous materials based on HA should be useful also for the treatment of internal wounds.

Non water soluble derivatives of hyaluronan such as methacryloyl HA as a hydrophilic derivate and palmitoyl HA as a hydrophobic one, have been electrospun (Fig. 5) and the release of model incorporated substances of hydrophilic and hydrophobic nature have been measured (Ruzickova et al. 2012b) by a spectrophotometry and it is shown in Figs. 6, 7 and 8. Cumulative release profiles of studied samples exhibited similar release kinetics with a strong burst effect regardless of the hydrophilic or hydrophobic nature of the HA derivate or the drug incorporated to the structure (see Figs. 6 and 7). This fact may be explained by two possible reasons. Hydrophobized HA derivate with the degree of substitution 48 % contains certain amount of hydrophobic domains causing that the derivate is insoluble in the water but it is still partially hydrophilic. The difference in the polymer structure is not so dramatical to exhibit significantly different results of the release study. Hydrophobised derivate is water insoluble but it swells in the water such as the crosslinked hydrophilic derivate does. As the swelling facilitates the release and makes it faster, it suppresses the differencies in mutual compatibility of the drug and the polymer caused by its hydrophilic/hydrophobic nature

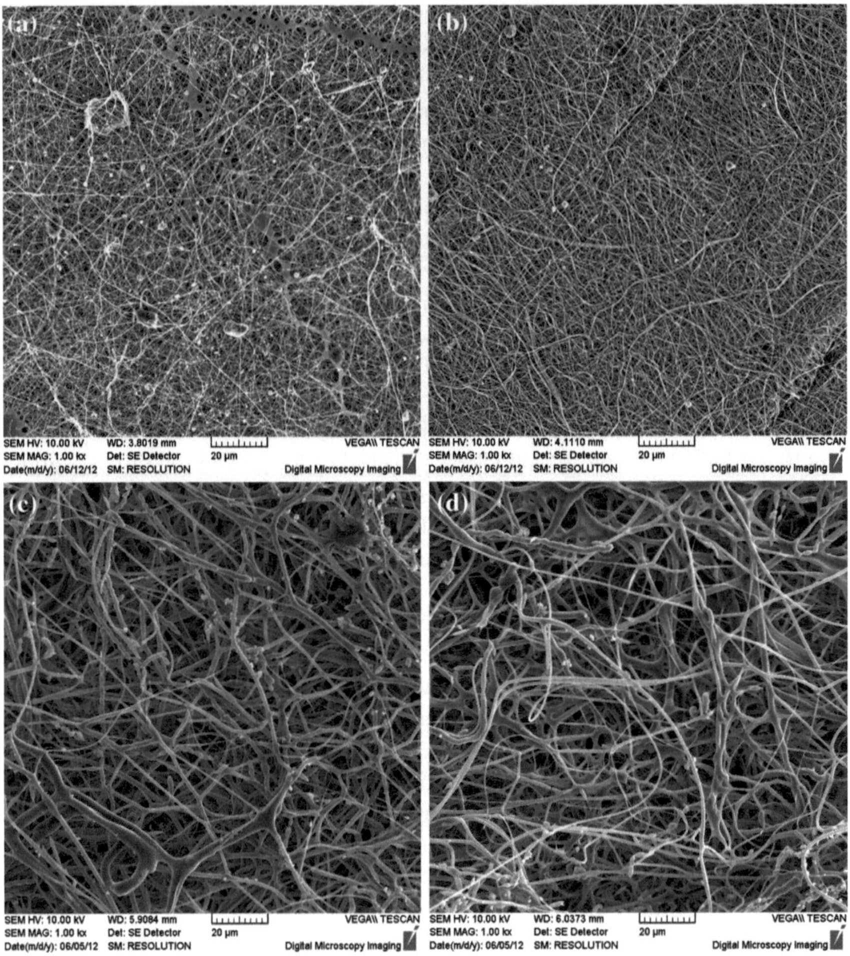

Fig. 5 SEM images of electrospun drug loaded nanofibers. **a** dexamethasone in hydrophilic HA derivate, **b** diclofenac in hydrophilic HA derivate, **c** dexamethasone in hydrophobised HA derivate, **d** diclofenac in hydrophobised HA derivate (Ruzickova et al. 2012b)

(Ruzickova et al. 2012b). The other reason may be insufficient solubility of the drug loaded causing dispersion processing and the drug crystallization on the surface of the nanofibrous layer, such crystals are visible on SEM images (see Fig. 5). A small amount of particles is visible in Fig. 5b in the case of a hydrophilic drug in a hydrophilic carrier electrospun from water solution that may indicate a good compatibility. Previous work (Ruzickova 2010) exhibited that even in the case of a good compatibility release of incorporated substance from hydrophilic carrier may be as fast as in the case of carrier dissolution when it dissolves.

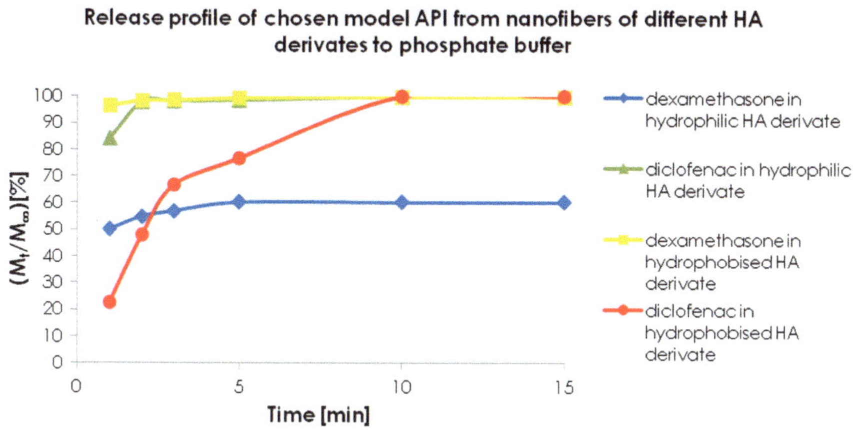

Fig. 6 Cumulative release profile of chosen model hydrophobic (dexamethasone) and hydrophilic (diclofenac) drug from nanofibers of hydrophobised HA derivate and hydrophilic derivate releasing to the distilled water (Ruzickova et al. 2012b)

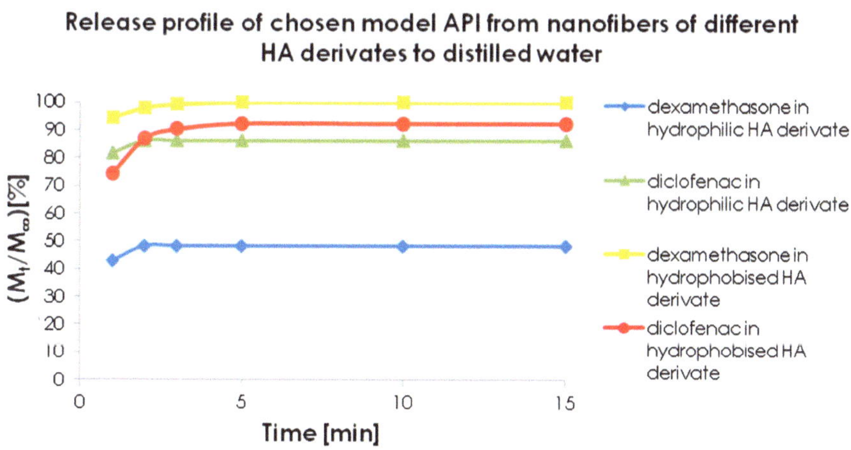

Fig. 7 Cumulative release profile of chosen model hydrophobic (dexamethasone) and hydrophilic (diclofenac) drug from nanofibers of hydrophobised HA derivate and hydrophilic derivate releasing to the PBF 7, 4 (Ruzickova et al. 2012b)

The wound environment does not necessarily contain a big amount of liquids. In the case of chronic wounds on the skin surface, the amount of liquid present in the wound is limited and the dissolution and release is much slower as demonstrates Figs. 10, 11 and 12 showing the release of iodine from native HA nanofibers measured *in vitro*. Nanofibers containing iodine generator were successfully electrospun and various types of nanofibrous structures including voluminous structure were created (Figs. 8 and 9).

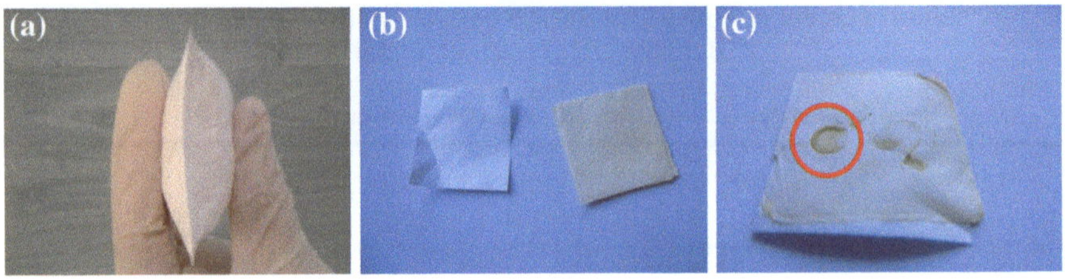

Fig. 8 Different nanofibrous structures: **a** fluffy; **b** flat and **c** compressed nanofibers- the designated place shows the area of compression (Ruzickova et al. 2012b)

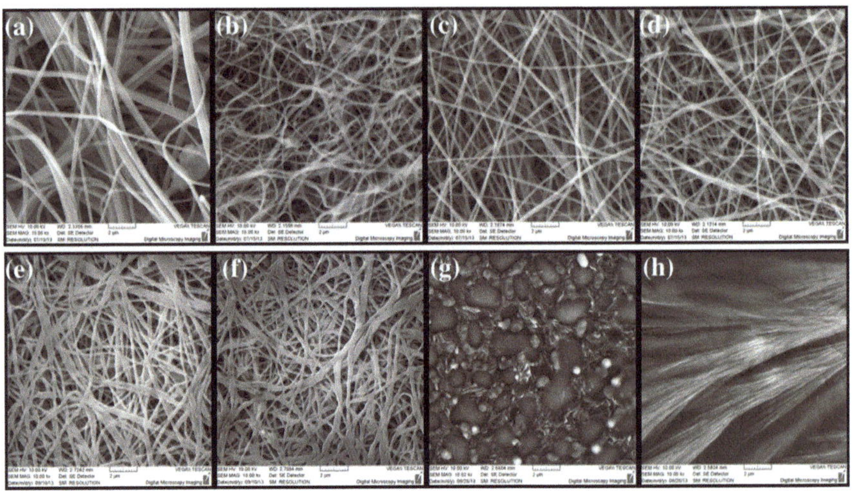

Fig. 9 SEM images of tested samples **a** fluffy nanofibers of solution A, **b** fluffy nanofibers of solution B, **c** flat nanofibers of solution A, **d** flat nanofibers of solution B, **e** compressed nanofibers of solution A, **f** compressed nanofibers of solution B, **g** foil of solution A, **h** foil of solution B

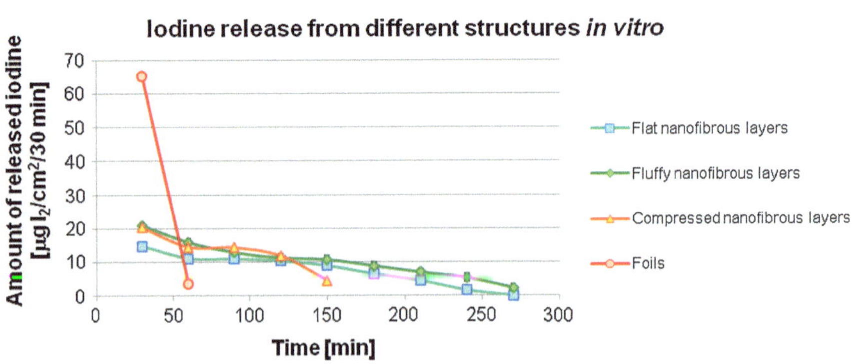

Fig. 10 Iodine release from carriers of different structure

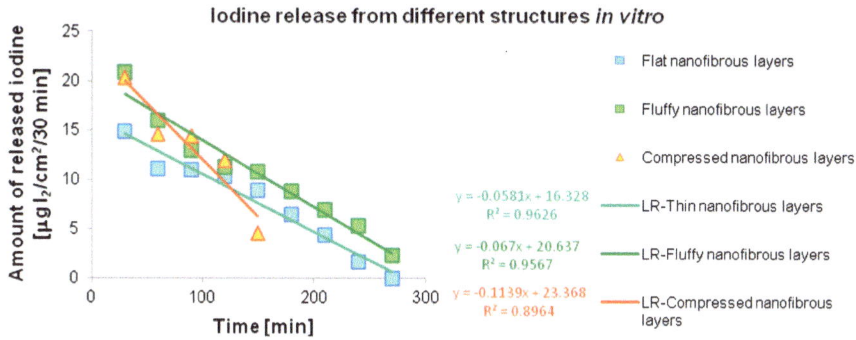

Fig. 11 Iodine release from nanofibrous carriers of different structure

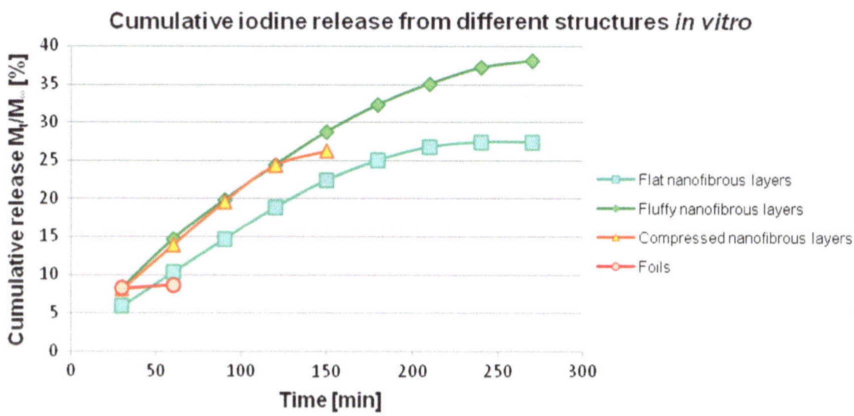

Fig. 12 Cumulative release of iodine from carriers of different structure *in vitro*

It was expected that more compact structures such as foils release slower whereas the voluminous structures like fluffy nanofibers release faster. It is obvious from Figs. 11 and 12 that iodine release from foils is much faster. It should be caused by a gradual release inflicted by the presence of additive aggregates on the surface of the films described also in (Taepaiboon et al. 2006). The mentioned aggregates or crystals are visible in Fig. 9g and h. Another problem related to the foils was their adhesion to the starch gel sometimes resulting in a destruction of the starch gel during sample removal. In this case nanofibers prolonged the additive release. Foils released iodine within 60 min whereas nanofibrous iodine generator carrier released for 270 min. The amount of released iodine was verified by spectrophotometry because the measurement was terminated when no more iodine was releasing but cumulative release curves did not reach 100 % (Fig. 12). It is apparently a consequence of iodine volatility. Iodine volatility of all iodine containing medical devices is a serious problem therefore the use of iodine generator that starts to release iodine at the moment of application and not before is very tempting. Such material doesn't need any special packages preventing iodine evaporation before application because the active layers containing iodine generator compounds are connected just before the application so the reaction cannot appear before. The use of the nanofibrous carriers is not limited only to the iodine generator but it is generally usable for other drug carriers.

It seems that the more compact form of nanofibrous structure releases iodine faster (Fig. 11) and in the case of more voluminous nanofibers a higher amount of iodine gets into the starch gel of *in vitro* model so a smaller amount evaporates (Figs. 11 and 12). This fact may be explained as follows. Nanofibrous layers made of hyaluronic acid in the used *in vitro* model absorb the liquid of the starch gel and they slowly dissolve and form a highly viscous gel-like layer. When the absorption takes place in a high amount of the surrounding liquid, absorption in more compact structures proceeds slower and absorbed amount of liquid is usually lower due to limited space usable for swelling. *In vitro* model modelling the chronic wound uses a limited amount of the liquid and in the case of voluminous and fluffy type of nanofibrous structure the absorption after the first contact with humidity proceeds selectively. Hyaluronic acid behaves as a supersorbing material. When it gets in touch with humidity it creates a gel. However, capillary actions do not occur in voluminous structures because humidity is captured in the swelled gel and due to higher amount of air pores the humidity is not transported so fast.

5 Concluding Remarks

Electrospinning is quite an old method of nanofiber production, it is highly productive and it is still being improved to reach even higher efficiency. There are plenty of patents being published every year but there are only few real applications and nanofibrous products already on the market although the potential of nanofibers for various applications is great.

There are a lot of benefits given by the structure that are contributing to increasing development of nanofibrous materials. In the field of wound healing and drug delivery releasing some active compound, the suitability of nanofibrous materials depends on the intended application. Nanofibers of hyaluronic acid and their massive production are even more difficult and their possible use in medicine is given by required kinetics of the substance release. Nanofibers of native hyaluronic acid or its swellable derivate are useful for the products releasing very fast. Non-swellable derivatives of hyaluronic acid should be useful for long-term release. Nanofibrous form may stabilize the release and prolong the release time when compared with foils.

Acknowledgement This research is conducted under financial support provided by Technology Agency of the Czech Republic (project TA02011238: Novel wound dressings based on nanofibers and staple microfibers of hyaluronan and chitin/chitosan-glucan complex).

References

Agarwal S, Wendorff JH, Greiner A (2008) Use of electrospinning technique for biomedical applications. Polymer 49:5603–5621

Bölgen N, Vargel I, Korkusuz P et al (2007) In vivo performance of antibiotic embedded electrospun PCL membranes for prevention of abdominal adhesions. J Biomed Mater Res B Appl Biomater 81(2):530–543

Brenner EK, Schiffman JD, Thompson EA et al (2012) Electrospinning of hyaluronic acid nanofibers from aqueous ammonium solutions. Carbohydr Polym 87(1):926–929

Chang JJ, Lee YH, Wu MH et al (2012) Electrospun anti-adhesion barrier made of chitosan alginate for reducing peritoneal adhesions. Carbohydr Polym 88:1304–1312

Chokshi RJ, Zia H, Sandhu HK et al (2007) Improving the dissolution rate of poorly water soluble drug by solid dispersion and solid solution-pros and cons. Drug Deliv 14:33–45

Delaco PA, Stefanetti M, Pressato D et al (1998) A novel hyaluronan-based gel in laparoscopic adhesion prevention: preclinical evaluation in an animal model. Fertil Steril 69(2):318–323

Fischer RL, McCoy MG, Grant SA (2012) Electrospinning collagen and hyaluronic acid nanofiber meshes. J Mater Sci Mater Med 23:1645–1654

Forward KM, Rutledge GC (2012) Free surface electrospinning from a wire electrode. Chem Eng J 183:492–503

Hooker GD, Taylor BM, Driman DK (1999) Prevention of adhesion formation with use of sodium hyaluronate–based bioresorbable membrane in a rat model of ventral hernia repair with polypropylene mesh—a randomized, controlled study. Surgery 125(2):211–216

Ji Y, Ghosh K, Shu XZ et al (2006) Electrospun three-dimensional hyaluronic acid nanofibrous scaffolds. Biomaterials 27:3782–3792

Johns DB, Keyport GM, Hoehler F et al (2001) Reduction of postsurgical adhesions with Intergel® adhesion prevention solution: a multicenter study of safety and efficacy after conservative gynecologic surgery. Fertil Steril 76(3):595–604

Jun HW, Tambralli A, Blakeney BA (2011) Electrospinning apparatus, methods of use, and uncompressed fibrous mesh. World Intellectual Property Organisation WO2011130110 (A2), 20 Oct 2011

Kenawy ER, Bowlin GL, Mansfield K et al (2002) Release of tetracycline hydrochloride from electrospun poly(ethylene-co-vinylacetate), poly(lactic acid), and a blend. J Controlled Release 81:57–64

Kim TG, Chung H, Park TG (2008) Macroporous and nanofibrous hyaluronic acid/collagen hybrid scaffold fabricated by concurrent electrospinning and deposition/leaching of salt particles. Acta Biomater 4(6):1611–1619

Knotkova K, Ruzickova J, Velebny V (2013) Homogeneity of additive content in the structure of electrospun nanofibers. In: 5th International conference Nanocon 2013, Brno, Oct 2013

Kuo CK, Zamarripa N, Thomas AH (2010) Scaffolds for tissue engineering and regenerative medicine. World Intellectual Property Organisation WO2010040129 (A2), 8 Apr 2010

Leuner C, Dressman J (2000) Improving drug solubility for oral delivery using solid dispersions. Eur J Pharm Biopharm 50:47–60

Li JX, He AH, Han CC et al (2006) Electrospinning of hyaluronic acid (HA) and HA/gelatin blends. Macromol Rapid Comm 27(2):114–120

Liu YL, He J (2007) Bubble electrospinning for mass production of nanofibers. Int. J. Nonlinear Sci Numer Simul 8:393–396

Liu Y, Ma G, Jun N et al (2010) Method for preparing pure hyaluronic acid nanofiber non-woven fabric. China Patent and Trademark Office CN101775704 (A), 14 Jul 2010

Liu Y, Ma G, Fang D et al (2011) Effects of solution properties and electric field on the electrospinning of hyaluronic acid. Carbohydr Polym 83:1011–1015

Lukas D, Sarkar A, Pokorny P et al (2008) Self-organisation of jets in electrospinning from free liquid surface: a generalized approach. J Appl Phys 103(8):0843091–0843097

Ma PX, Zhang R (1999) Synthetic nano-scale fibrous extracellular matrix. J Biomed Mat Res 46:60–72
Ma G, Jun N, Liu Y et al (2011) Preparation method of medicament-loaded biodegradable nanofiber medical dressing. China Patent and Trademark Office CN102068339 (A), 25 May 2011
Martin CR (1996) Membrane-based synthesis of nanomaterials. Chem Mater 8:1739–1746
Mincheva R, Paneva D, Manolova N et al (2005) Preparation of polyelectrolyte-containing nanofibers by electrospinning in the presence of a non-ionogenic water-soluble polymer. J Bioact Compat Polym 20:419–435
Mo X, Gao C, Liu W et al (2009) Preparation of bionic extracellular matrix hyaluronic acid and gelatin compound nanofiber membrane. China Patent and Trademark Office CN101581010 (A), 18 Nov 2009
Okuda T, Tominaga K, Kidoaki S (2010) Time-programmed dual release formulation by multilayered drug-loaded nanofiber meshes. J Control Release 143(2):258–264
Ondarcuhu T, Joachim C (1998) Drawing a single nanofibre over hundreds of microns. Europhys Lett 42(2):215–220
Park W, Jeong L, Yoo D et al (2004) Effect of chitosan on morphology and conformation of silk fibroin nanofibers. Polymer 45:7151–7157
Patel S, Wong C, Kurpinski K (2010) Multilayer fibrous polymer scaffolds, methods of production and methods of use. World Intellectual Property Organisation WO2010042651 (A1), 15 Apr 2010
Petrik S, Maly M (2009) Production nozzle-less electrospinning nanofiber technology. MRS Proc 1240:1
Petras D, Slobodian P, Klimmer D (2011) Comparison of productivity of different ways of electro spinning. In: 3rd International conference Nanocon 2011, Brno, Sept 2011
Pokorny M, Velebny V (2011) Apparatus for production of two-dimensional or three-dimensional fibrous materials of microfibres and nanofibres. World Intellectual Property Organisation WO2011095141 (A1), 5 Feb 2010
Pokorny M, Martincova L, Velebny V (2012a) Process for producing materials exhibiting anisotropic properties and composed of nanofibers or microfibers and apparatus for making the same. Czech Patent Office CZ303380 (B6), 27 Jun 2011
Pokorny M (2012b) Product design—4SPIN laboratory device for nanotechnologies. Europe Patent Community design CD002037176, 4 May 2012
Pokorny M, Rebicek J, Sukova L et al (2013a) Spinning nozzle for producing nano—and microfibrous materials composed of fibers with coaxial structure. Czech Patent Office CZ303780 (B6), 27 Jul 2012
Pokorny M, Sukova L, Rebicek J et al (2013b) Combined spinning nozzle for producing nanofibrous and microfibrous materials. Czech Patent Office CZ304097 (B6), 19 Jan 2012
Ruzickova J (2010) The release of incorporated substances from the nanofibrous structures and possibilities of application in medicine. Dissertation, Technical University of Liberec
Ruzickova J, Pokorny M, Sukova L et al (2012a) Alternative approach of large scale nanofibers production. Paper presented at International Nanofiber Symposium/N3 M—Nanofibers for the 3rd millenium—Nanofibers 2012, Tokyo Institute of Technology, Tokyo, 4–5 June 2012
Ruzickova J, Jouklova Z, Pravda M et al (2012b) Use of nanofibers for controlled release carriers. Paper presented at The 76th Prague Meeting on Macromolecules: polymers in Medicine, Institute of Macromolecular Chemistry, Prague, 1–5 July 2012
Ruzickova J, Novak J, Pravda M et al (2013) Objemný nanovlákenný materiál na bázi kyseliny hyaluronové, jejích solí nebo jejich derivátů, způsob jejich přípravy, způsob jejich modifikace, modifikovaný nanovlákenný materiál, nanovlákenný útvar a jejich použití (Voluminous nanofibrous material based on hyaluronic acid, its salts or their derivatives, the way of their preparation, the way of their modification, modified nanofibrous material, nanofibrous formation and use thereof). Czech Patent Office PV2013-913, 22 Nov 2013
Schmidt S, Friedl P (2010) Interstitial cell migration: integrin-dependent and alternative adhesion mechanisms. Cell Tissue Res 339:83–92

Soo SH, Moo LY, In JS et al (2009) Nano fiber for tissue regeneration and fabrication method thereof. Korean Intellectual Property Office KR20090071993 (A), 2 July 2009
Soo KI, Suhk KB, Oh KK et al (2013) Nanofiber having biocompatibility, method for producing the same and wound dressing. Japan Patent Office JP2013049927 (A), 26 Feb 2009
Taepaiboon P, Rungsardthong U, Supaphol P (2006) Drug-loaded electrospun mats of poly(vinyl alcohol) fibres and their release characteristics of four model drugs. Nanotechnology 17:2317–2329
Theron A, Zussman E, Yarin AL (2004) Experimental investigation of the governing parameters in the electrospinning of polymer solutions. Polymer 45:2017–2030
Toshkova R, Manolova N, Gardeva E et al (2010) Antitumor activity of quaternized chitosan-based electrospun implants against Graffi myeloid tumor. Int J Pharm 400(1–2):221–233
Um IC, Fang D, Hsiao BS (2004) Electro-spinning and electro-blowing of hyaluronic acid. Biomacromolecules 5:1428–1436
Uppal R, Ramaswamy GN, Arnold C et al (2011) Hyaluronic acid nanofiber wound dressing production, characterization, and in vivo behaviour. J Biomed Mater Res B Appl Biomater 97B(1):20–29
Verreck G, Chun I, Peeters J et al (2003) Preparation and characterization of nanofibers containing amorphous drug dispersions generated by electrostatic spinning. Pharm Res 20:810–817
Wang X, Um IC, Fang D et al (2005) Formation of water-resistant hyaluronic acid nanofibers by blowing-assisted electro-spinning and non-toxic post treatments. Polymer 46:4853–4867
Wang H (2008) Innovative bottom-up cell assembly approach to three-dimensional tissue formation using nano-or micro-fibers. The United States Patent and Trademark Office US20080112998 (A1), 14 Jan 2006
Wallwiener M, Brucker S, Hierlemann H et al (2006) Innovative barriers for peritoneal adhesion prevention: liquid or solid? a rat uterine horn model. Fertil Steril 86(4 Suppl):1266–1276
Whitesides GM, Grzybowski B (2002) Self-assembly at all scales. Science 295:2418–2421
Xie J, Wang CH (2006) Electrospun micro- and nanofibers for sustained delivery of paclitaxel to treat C6 glioma in vitro. Pharm Res 23:1817–1826
Xu S, Li J, He A (2009) Chemical crosslinking and biophysical properties of electrospun hyaluronic acid based ultra-thin fibrous membranes. Polymer 50:3762–3769
Yarin AL, Zussman E (2004) Upward needleless electrospinning of multiple nanofibers. Polymer 45(9):2977–2980
Zeng J, Xu X, Chen X et al (2003) Biodegradable electrospun fibers for drug delivery. J Controlled Release 92:227–231
Zeng J, Yang L, Liang Q et al (2005) Influence of the drug compatibility with polymer solution on the release kinetics of electrospun fiber formulation. J Controlled Release 105:43–51
Zhang YZ, Wang X, Feng Y et al (2006) Coaxial electrospinning of fluo-rescein isothiocyanate—conjugated bovine serum albu-min)–encapsulated poly(ε-caprolactone) nanofibers for sustained release. Biomacromolecules 7:1049–1057
Zong X, Kim K, Fang D et al (2002) Structure and process relationship of electrospun bioabsorbable nanofiber membranes. Polymer 43(16):4403–4412
Zong X, Li S, Chen E et al (2004) Prevention of post-surgery induced abdominal adhesions by electrospun bioabsorbable nanofibrous poly(lactide-co-glycolide) based membranes. Ann Surg 240:910–915

Potential of siRNA Therapy in Chronic Myeloid Leukemia

Juliana Valencia-Serna, Breanne Landry, Xiaoyan Jiang and Hasan Uludag

Abstract Leukemic cancers arise from genetic alterations in normal hematopoietic stem or progenitor cells, leading to impaired regulation of proliferation, differentiation, apoptosis and survival of the malignant cells. A range of molecular alterations is beginning to be elucidated in specific types of leukemias, providing potential targets for molecular modulation as the basis of a therapy. With the advent of RNA interference (RNAi) and, in particular, the short interfering RNA (siRNA) as its pharmacological mediator, it is becoming possible to specifically modulate desired leukemic targets at will. This chapter will summarize the current attempts to utilize siRNAs in leukemic therapy using chronic myeloid leukemia (CML) as a prototypical disease model. We first provide a brief background on the CML disease with particular emphasis on molecular mediators critical in this disease and the current drug therapy. The limitations of current drugs and potential of RNAi are presented. We then provide a summary of delivery efforts employed to deliver siRNA to CML cells, with emphasis on non-viral delivery approach due to its better safety profile for utility in a clinical setting. Important factors involved in intracellular delivery of siRNA are highlighted, emphasizing features critical for non-viral delivery. We conclude with a perspective on the future of siRNA therapy for the CML disease.

J. Valencia-Serna · H. Uludag (✉)
Department of Biomedical Engineering, Faculty of Medicine and Dentistry,
University of Alberta, Edmonton, AB T6G 2V2, Canada
e-mail: huludag@ualberta.ca

B. Landry · H. Uludag
Department of Chemical and Materials Engineering, Faculty of Engineering,
University of Alberta, Edmonton, AB T6G 2G6, Canada

X. Jiang
Terry Fox Laboratories, British Columbia Cancer Agency, Department of Medical Genetics,
University of British Columbia, Vancouver, BC V5Z 1L3, Canada

H. Uludag
Faculty of Pharmacy and Pharmaceutical Sciences, University of Alberta, Edmonton,
AB T6G 2G6, Canada

A. Prokop et al. (eds.), *Intracellular Delivery II*,
Fundamental Biomedical Technologies 7, DOI: 10.1007/978-94-017-8896-0_21,

Keywords siRNA · Drug delivery · Chronic myeloid leukemia · Non-viral delivery · RNAi · Molecular therapy · Controlled delivery systems

Abbreviations

RNAi	RNA interference
dsRNA	Double stranded RNA
CML	Chronic myeloid leukemia
AML	Acute myeloid leukemia
LSC	Leukemic stem cells
RISC	RNA-induced silencing complex
siRNA	Short interfering RNA
mRNA	Messenger RNA
shRNS	Short hairpin RNA
TKI	Tyrosine-kinase inhibitors
SDF-1	Stromal cell-derived factor-1
CXCR4	C-X-C chemokine receptor type 4
IC50	Concentration required for 50 % loss of cell viability
PRAME	Preferentially expressed antigen of melanoma
STAT	Signal transducer and activator of transcription
PPP2R5C	Protein phosphatase 2, Regulatory subunit B′, gamma
PEI	Polyethylenimine
PLL	Poly-l-lysine
MW	Molecular weight
kDa	Kilo Dalton
PEI2	2 kDa PEI
PEI2LA	Linoleic acid substituted 2 kDa PEI
PEI1.2PA	Palmitic acid substituted 1.2 kDa PEI
GFP	Green fluorescent protein
MSC	Mesenchymal stem cells
ECM	Extracellular matrix
CPP	Cell penetrating peptides
PTD-DRBD	Peptide transduction domain and double-stranded RNA-binding domain

1 Introduction

Leukemic cancers arise from genetic alterations in normal hematopoietic stem or progenitor cells, leading to impaired regulation of proliferation, differentiation, apoptosis and survival of malignant cells. The US National Cancer Institute calculated an overall 5-year relative survival (between 2003 and 2009) rate of 56.0 %

for various leukemias combined (Fast Stats 2013). The front line therapy in leukemia is chemo (drug) therapy (Estey 2013; Stefanachi et al. 2012); current therapeutic approaches include broad-spectrum drugs against fast-proliferating cells and small-molecule inhibitors targeting specific signal transduction pathways, so called molecular therapies (Ferrara 2012). Leukemic cells generally respond well to drug therapy at the onset of the treatment, but the drugs lose their effectiveness over a period of 6–12 months. It is well recognized now that the resistance to conventional (broad-spectrum) therapeutic agents is inevitable, but recent evidence also indicated that even the most advanced molecularly-targeted drugs lose their efficacy as a result of resistance development in a relatively short time. The inherent plasticity of the cells combined with diverse resistance mechanisms make malignant cells naturally adapt by mounting an effective resistance against the drugs. The high relapse rate in leukemia patients has been additionally attributed to existence of a rare population of leukemia stem cells (LSC) resistant to current drug therapies (Ishikawa et al. 2007; Mikkola et al. 2010).

With better understanding of molecular changes in malignant transformations, treatments that target tumor-specific changes are expected to lead to more effective therapies as the normal cells transform into malignant cells. Towards this end, a highly specific leukemia therapy can be developed by exploiting the RNA interference (RNAi) mechanism to silence the aberrant protein(s) responsible for the disease (Iorns et al. 2007; Rossbach 2010). There are two main approaches for RNAi, using either a plasmid encoding for short hairpin RNA (shRNA) or delivering small interfering RNA (siRNA) where the shRNA transcription and processing steps can be omitted (Guo 2010). The use of siRNA is a more practical approach bypassing the need to express the shRNA at sufficient quantities in hard-to-transfect primary cells. The siRNA essentially acts as a pharmaceutical 'drug' in this respect. In cytosol, the siRNA duplexes assemble into a pre-RISC (RNA-induced silencing complex) containing specific proteins, including argonaute proteins (AGO1, 3 or 4) (Yoda et al. 2010; Wang et al. 2009), which is subsequently guided to target desired mRNA based on complementary base pairing (Yoda et al. 2010). Endonucleoyltic cleavage and/or translational repression of the mRNA (Yoda et al. 2010; Wang et al. 2009) then silences the protein target. Delivery systems, however, are an absolute necessity for effective use of siRNA since the molecules are highly sensitive to serum nucleases and their large ($\sim$13 kDa) and anionic nature (due to its phosphodiesterase backbone) prevents the siRNA to traverse cellular membranes. Viral means to deliver siRNA has been emphasized initially, but the undesirable side-effects of viral delivery in a clinical setting makes this approach highly risky for therapeutic use. Alternatively, cationic biomolecules capable of binding and neutralizing the anionic charges of siRNA and packaging the siRNA into nano-sized complexes can serve as effective siRNA carriers (Abbasi et al. 2013). Formulations of cationic biomolecules, such as lipids, small amines or polymers, with siRNA, typically results in nano-sized particles that are suitable for systemic administration and cellular uptake.

In this chapter, we will summarize the attempts to deliver siRNA molecules using non-viral carriers in leukemia. We will focus on a particular type of leukemic cancers, namely chronic myeloid leukemia (CML), since it is one of the major classes of leukemia and it is well understood at molecular level. The siRNA therapy has not reached clinical setting in CML treatment, so that we will review the literature on preclinical studies exploring the potential of siRNA therapy in this disease. In conjunction, we will explore the technology of siRNA delivery, investigating the critical issues pertinent to effective siRNA delivery.

2 Chronic Myeloid Leukemia and Current Drug Therapies

Myeloid leukemias (46 % of all leukemias) affect the myeloid cells of the bone marrow, which normally go on to form the blood cells. Thirteen percent of those cases account for the CML. Approximately 350,000 people worldwide are diagnosed with leukemia annually, with ~250,000 death resulting from leukemia each year. Most leukemia occur in the elderly and peaks between the ages of 75 and 79 (Elert 2013). CML is a myeloproliferative disease initialized at the hematopoietic stem cells that is thought to arise due to translocation of chromosomes 9 and 22, which results in a fusion between *ABL* and *BCR* genes, or in the so-called Philadelphia (Ph) Chromosome (Sloma et al. 2010; Kumar et al. 2009; Bocchia et al. 2005). Once the normally regulated tyrosine kinase of the ABL protein is permanently activated by the juxtaposition of the BCR sequence, it leads initially to a chronic phase characterized by myeloid cell expansion, while allowing differentiation of expanded cells in the peripheral blood. As the disease progresses, either by increased *BCR-ABL* expression or activation of other pathways, patients enter a more aggressive disease phase (blast crisis), which is characterized by a progressive loss of the capacity of hematopoietic cells to differentiate and increased expansion and accumulation of immature blast cells in the bone marrow and spread to the bloodstream (Ito 2013; Melo and Barnes 2007). The Philadelphia chromosome is not specific for CML, since it can also be found in ~5 % of children with acute lymphoblastic leukemia (ALL), the most common childhood cancer (DeWeerdt 2013).

Current therapies for CML are based on the use of tyrosine-kinase inhibitors (TKIs) and allogeneic hematopoietic stem-cell transplantation. Stem-cell transplantation therapy is an option when the treatment with TKIs fails; however, this therapy has substantial risk of mortality due to chronic graft-versus-host disease (Goldman and Melo 2003; Baccarani et al. 2009). TKIs, such as Imatinib, have revolutionized CML therapy. Imatinib binds to the ABL kinase domain with the formation of a bond that impedes ATP binding, subsequently blocking or preventing the interaction of the ABL kinase with substrates and therefore from activating its oncogenic pathways (Zhang and Li 2013; Deininger et al. 2000). Targeted therapy with Imatinib has transformed the survival potential for patients with chronic phase of CML; it has significant impact on patients with accelerated phase but a minimal impact for those patients at the blastic phase stage (Giles 2006). However, CML patients, especially those in advance-stage disease, can develop TKI resistance leading to relapse (Baccarani et al. 2009). This acquired drug resistance could be due to the amplification of *BCR-ABL* gene, and overexpression of *BCR-ABL* mRNA and protein (Weisberg et al. 2007). However, resistance most often results from point mutations in the kinase domain of BCR-ABL protein that affect drug binding to the protein, thereby reducing the ability of Imatinib to block the tyrosine kinase activity. More than 50 distinct *BCR-ABL* mutations have been reported to-date and the current repertoire of TKIs can cover all known mutations leading to resistance; however, no single drug can prevent all forms of resistance (Zhang and Li 2013), necessitating the use of TKI cocktails to overcome any possible resistance.

Next-generation drugs such as Dasatinib and Nilotinib, are more potent TKIs and produce more rapid declines in CML disease burden than the Imatinib, which translates into more durable cytogenetic (absence of metaphase Ph^+ cells) and hematological remissions (achievement of normal white blood and platelet cell counts and, no signal of CML symptoms) (Weisberg et al. 2007; Sawyers 2013). Nilotinib is $\sim$30 fold more potent than Imatinib as an ABL inhibitor. Dasatinib is a potent inhibitor of ABL kinase Src-family kinases, which are know to be involved in multiple intracellular signal transduction pathways including oncogenesis and disease progression (Weisberg et al. 2007). Ponatinib is a newer drug that has the unique property of inhibiting both the native (un-mutated) and mutated BCR-ABL proteins, specially those including the T315I mutation, which confers resistance to all other CML drugs (including Nilotinib and Dasatinib) and seems to translate in worst overall survival compared with other mutations found in Imatinib-treated patients (Weisberg et al. 2007; Sawyers 2013; Cortes et al. 2012). A strategy of combing two or three ABL inhibitors with non-overlapping *BCR-ABL* mutations resistance profiles, such the example of Ponatinib exemplified above, could prevent the emergence of drug resistance (Sawyers 2013). However, it is expected that treatment with these new ABL inhibitors could also lead to new point mutations that overcome the resistance of these new drugs (Weisberg et al. 2007), given the plasticity of leukemic stem cells.

3 Insensitivity of CML Stem Cells to TKIs

Although Imatinib inhibits the production of $\sim$99 % of differentiated leukemic cells, it fails to deplete the LSCs (Weisberg et al. 2007). Studies have shown that despite the complete depletion of *BCR-ABL* transcript levels in these LSCs with TKIs, the cells remain viable. These data indicate that even in the presence of Imatinib, especially in accelerated and advanced states of the disease, CML has progressed and evolved so these LSCs no longer require BCR-ABL activity to maintain their viability (Kumar et al. 2009; Zhang and Li 2013; Corbin et al. 2011; Muvarak et al. 2012), and anti-apoptotic and pro-survival signals are provided by alternate pathways. Thus, the drivers of cell proliferation and survival, probably influenced by BCR-ABL in the early stages of CML, now operate autonomously and could lead to CML relapse (Savona and Talpaz 2008). Thus, it is clear that not only BCR-ABL inhibition is needed for the eradication of progenitor CML cells, but BCR-ABL-independent survival mechanisms of LSCs also needs to be targeted for a complete CML eradication.

One characteristic common to all LSCs is that they require the particular microenvironment of bone marrow, the stroma, to thrive. The stroma bathes the leukemic cells in growth factors, chemical signals and cell-surface ligands that keep the cells in a dormant phase resistant to drug therapy. Part of what keeps these cells entrenched in the bone marrow is a chemical signal sent by the stroma, called stromal cell-derived factor (SDF-1). This signal binds to a protein located at the surface of the stem cells, called C-X-C chemokine receptor type 4 (CXCR4) (Willyard 2013). In the case of CML, BCR-ABL protein seems to be involved in the inhibition of SDF-1-induced migration and signaling which allows an abnormal release of immature myeloid cells from the bone marrow into the circulation (Jin et al. 2008). On the other hand, it has been shown that under Imatinib, CXCR4 expression in $CD34^+$ can be reversibly up-regulated, hence allowing these cells to home to bone marrow, helping them to become quiescent and to become insensitive to TKIs (Jin et al. 2008; Copland 2009). Down-regulation of CXCR4 expression along with TKIs therapy could enhance the eradication of LSCs in CML.

Researchers are also developing drugs that target a key property of stem cells, namely their self-renewal potential. One signaling pathway that seems to play an important role in self-renewal of CML LSCs hinges on two proteins: Wnt and beta-catenin (Copland 2009; Zhao et al. 2007). In 2012, Armstrong et al. reported that a small molecule that inhibits beta-catenin, given in combination with Imatinib, reduces CML survival and eliminates leukemia stem cells in a CML mouse model (Willyard 2013; Heidel et al. 2012). Finally, *AHI-1* is a newly discovered oncogene that is highly expressed in primitive hematopoietic CML stem and progenitor cells, and whose overexpression has been shown to promote abnormal differentiation and proliferative activity of myeloid cells in CML. Zhou et al. showed that *AHI-1* overexpressing $BCR\text{-}ABL^+$ cells (CML cells transduced with an *AHI-1* construct) showed greater resistance to growth inhibition effects of Imatinib in comparison to control cells. Suppression of *AHI-1* by transduction of an AHI-1 silencing construct (AHI-1/sh4) resulted in increased sensitivity to Imatinib. AHI-1 was also found to significantly increase or reduce protein expression and phosphorylation of BCR-ABL, JAK2 and STAT5 once *AHI-1* is overexpressed or suppressed, respectively. Suppression of *AHI-1* in primary $CD34^+$ CML cells was also shown to increase Imatinib sensitivity especially in Imatinib resistant and blast crisis patients who express relatively higher levels of AHI-1 (Zhou et al. 2008).

4 Down-Regulation of Protein Targets by RNAi in CML

RNA interference (RNAi) is a process by which double-stranded small interfering RNA (siRNA) induces sequence-specific, post-transcriptional gene silencing (De Paula et al. 2007). Endogenous RNAi is triggered by the transcription of long pieces of double-stranded RNA (dsRNA), which are cleaved into the smaller (21–23 nucleotides long) fragments by the enzyme Dicer. In practice, siRNA is synthetically produced and then directly introduced into the cell, thus circumventing Dicer mechanics (Whitehead et al. 2009). Once siRNA is present in the cytoplasm of the cell, it is incorporated into the protein complex RISC (RNA-induced silencing complex). Thereafter, Argonaute, a protein contained within RISC, cleaves the sense strand of the siRNA, thereby releasing it from RISC. The now activated RISC, which contains the antisense strand of the siRNA, selectively seeks out and cleaves mRNA that is complementary to the antisense strand (De Paula et al. 2007; Whitehead et al. 2009). The activated RISC complex is not affected by this reaction and can move on to destroy additional mRNA targets, which further propagates the silencing of gene expression. In mammalian cells, RNAi persists effectively only for an average of 66 h due to its dilution during cell divisions (De Paula et al. 2007), and so repeated administration is necessary to achieve a persistent effect if needed (Whitehead et al. 2009).

The shortcomings of current leukemia treatments call for development of new strategies to deliver more efficacious drugs into CML cells. Owing to increasing knowledge of CML at a molecular level, RNAi is a promising approach for leukemia treatment. To control the expression of *BCR-ABL* and other genes involved in these cellular malfunctioning processes, synthetic small interfering RNA (siRNA) can be delivered into diseased cells to interact with the target mRNA of aberrant genes and silence their protein expression. However, a delivery carrier that helps these siRNA moieties to reach the mRNA in the cytoplasm is needed. In order to achieve this purpose, carriers need to interact with the siRNA moieties to form siRNA nanoparticles that protect the siRNA from serum nucleases and facilitate their cell membrane interaction, internalization via endocytosis and escape from endosomes, in order to ultimately be released in the cytoplasm (Prokop 2011; Dominska and Dykxhoorn 2010; Mintzer and Simanek 2009).

Several potential targets have been pursued for siRNA therapy of CML cells (Table 1). Silencing specific targets has been used as a tool to elucidate their functional role in CML and the biological outcome upon depleting the selected target. The main aim of these studies was identification of novel targets to decrease proliferation rates and induce programmed cell death that can be used in combination with conventional drugs to improve drug sensitivity. In one of the first studies to explore siRNA therapy, Wohlbold and co-workers targeted *BCR-ABL* expression in *BCR-ABL*-transduced cells. This siRNA treatment resulted in a significant reduction of BCR-ABL protein, which led to a reduced regulatory effect of its substrates, reducing the expression of antiapoptotic Bcl-X_L protein and increasing the expression of cell cycle inhibitor p27. BCR-ABL silencing led to a

Table 1 siRNA targets shown to be beneficial in CML. siRNA studies which reported significant anti-survival effects in CML cells were selected from a PubMed 'CML siRNA' keyword search

References	Target	Rationale	siRNA carrier (concentration)	Outcome
Wilda et al. (2002)	BCR-ABL	Compare efficiency of cell killing by Imatinib to that of silencing of *BCR-ABL* with siRNA	Oligofectamine (unknown)	Reduction of mRNA and protein were found with apoptosis levels 2.5x higher than controls. Apoptosis rate of BCR-ABL siRNA treated cells was at the same level as cells treated with Imatinib or ~5 times more than control cells
Wohlbold et al. (2003)	BCR-ABL	Inhibit *BCR-ABL* expression and evaluate sensitization to imatinib	Electroporation (200–800 nM)	Decreased cell viability and sensitization of imatinib-resistant K562 cells to imatinib
Rangatia and Bonnet (2006)	BCR-ABL	Study anti-leukemic properties of BCR-ABL by RNAi	Electroporation (1 μg per 5×10^5 cells)	60 % reduction of *BCR-ABL* mRNA expression. Slight increase of apoptosis. 2-fold increase of DNA fragmentation. Caspase-7 and -9 activated. Cells unable to actively divide for at least 2 weeks after silencing
Arthanari et al. (2010)	BCR-ABL	To assess efficacy of Tat-LK15 peptide in delivering siRNA to target *BCR-ABL*	Tat-LK15 peptide: fusion of HIV-Tat-derived peptide to cationic peptide LK15 (1 to 30 μg siRNA/mL—24–729 nM calculated)	Expression of p210 BCR-ABL was reduced for all concentrations. Cytotoxicity due to siRNA nanoparticles ranging from 0 % (10 μg) to 30 % (30 μg). No silencing detected after 48 h
Wang et al. (2008)	Cyclin A2	Deliver cyclin A2 siRNA with SWNTs and evaluation of cyclin A_2 role upon doxorubicin treatment.	Functionalized single wall carbon nanotubes (f-SWNTs) (25 nM in culture)	A positive correlation between ability of doxorubicin to induce apoptosis and up-regulation of cyclin A_2
Gioia et al. (2011)	Syk and Axl	Identify downstream effectors of Lyn involved in resistance to nilotinib	Nucleofection (unknown)	Silencing Lyn's downstream effectors Syk and Axl restored capacity of nilotinib to inhibit cell proliferation

(continued)

Table 1 (continued)

References	Target	Rationale	siRNA carrier (concentration)	Outcome
Tanaka et al. (2011)	PRAME	Investigate function of PRAME in CML progression by RNAi in K562 cells	Nucleofection (1.5 μM siRNA/1.5 × 10^6 cells)	70 % knockdown of *PRAME* mRNA. Significant inhibition of cell proliferation and decrease of clonogenic growth. 60 % of apoptotic cells in comparison with 15 % of controls
Kosova et al. (2010)	STAT5A	Effects of STAT5A siRNA knockdown on cell growth and apoptosis induction	HiPerFect (unknown)	~75 % suppression of *STAT5A* mRNA. Resistant K562 cells became ~4 times more sensitive to Imatinib. An increase in caspase-3 activation was seen
Zhang et al. (2011)	GCS or MDR1	Relation of GCS and MDR1 to regulation P-gp gene expression and function activity in drug retention	Lipofectamine 2000™ (unknown)	Silencing of GCS can affect MDR1 expression and inhibit P-gp efflux. Silencing of GCS or MDR1 sensitized drug-resistant cells to chemotherapy and increased drug retention
Koldehoff et al. (2013)	BCR-ABL and GFI1B	Anti-leukemic additive effect of co-silencing of BCR-ABL and GFI1B	DOTAP, liposomal transfection (175 pM for GFI1B and 54 pM for BCR-ABL)	Additive effect in the inhibition of cell growth and in the increase of apoptosis in comparison with transfection of either siRNA alone
Shen et al. (2013)	PPP2R5C	Effect of PPP2R5C down-regulation in imatinib-sensitive and –resistance CML cells	Nucleofection (3 μg)	Inhibition of the proliferation of CML cells. Rendered imatinib-resistant cells more sensitive to TKIs

significant reduction of cell viability in a dose-dependent manner. A significant drop in the IC_{50} values of Imatinib (3.4-fold drop) was observed in K562 cells transfected BCR-ABL siRNA but not in untreated K562 cells (Wohlbold et al. 2003). Effective *BCR-ABL* silencing was also obtained by Rangatia et al, where ~50 % decrease at the mRNA level was found after 72 h of siRNA delivery. This silencing resulted in a two-fold increase of sub-G1 cell population as well as an increase of DNA fragmentation and mitochondrial-induced apoptosis. Although only a transient mRNA reduction was seen with siRNA treatment, a long-term effect was observed in proliferation of targeted cells: cells were unable to actively divide for at least 2 weeks in comparison with untreated cells. The reason for this was the cell cycle arrest in G1 phase, which is observed by decrease in cyclin D1 and increase in p21 and p27 cell cycle inhibitors (Rangatia and Bonnet 2006).

The related cell cycle mediator cyclin A_2 was targeted by RNAi in CML K562 cells. Silencing cyclin A_2 in doxorubicin-treated K562 cells led to a significant decrease in growth inhibition, apoptosis induction and increased erythroid differentiation. This suppression also caused a small fraction of K562 cells to differentiate along megakaryocytic and monocyte-macrophage pathways upon doxorubicin treatment. A positive correlation between the ability of doxorubicin to induce apoptosis in K562 cells and upregulation of cyclin A_2 was seen; the higher the cyclin A_2 expression, the higher the sensitivity to doxorubicin was. These results indicated a pro-apoptotic role of cyclin A_2 and its ability to regulate cell differentiation in CML (Wang et al. 2008).

Upregulated expression of Lyn has been suggested as an additional mechanism of cell resistance to nilotinib (Mahon et al. 2008). Gioia et al. investigated the role of Lyn kinase signalling as a mediator of resistance to nilotinib. Tyrosine kinase Lyn was overexpressed ~8 times more in nilotinib-resistant K562 cells in comparison with TKI-sensitive K562 cells. The proteins spleen tyrosine kinase Syk, UFO receptor Axl, and an adaptor protein CDCP-1 were found to have increased tyrosine phosphorylation in Lyn-overexpressing cells. Co-immunoprecipitation studies showed that Lyn interacted with Syk and Axl proteins in both cell lines. An increase in Syk phosphorylation was detected in nilotinib-resistant cells (with no significant difference of expression). Inhibition of Syk either by addition of Syk inhibitor R406, Syk shRNA or Syk siRNA increased (or restored) the sensitivity to nilotinib. Silencing of Axl and CDCP-1 by siRNA increased the sensitivity to nilonitib partially, suggesting that Axl and CDCP-1 may be mediators of Syk/Lyn signaling pathways. An overexpression of Lyn, CDCP-1 and Axl was also detected in nilotinib-resistant $CD34^+$ patient cells. The role played by Syk and Axl in the nilotinib resistance identifies these genes as potential targets as a combinatorial therapy for CML (Gioia et al. 2011).

Tanaka et al. investigated the function of preferentially expressed antigen of melanoma (PRAME) in leukemia. PRAME acts as a repressor of retinoic acid receptor (RAR) signalling and thus the functional repression of PRAME was investigated in K562 cell line in the absence of retinoic acid. Three days after siRNA delivery, ~70 % knockdown of *PRAME* mRNA and a complete inhibition of protein expression was achieved, which resulted in a significant inhibition of proliferation and clonogenic growth. This PRAME knockdown also lead to a significant increase of cells in G_0/G_1 phase and a related decrease in cells in S phase in comparison with the control group, which suggests a relationship between PRAME and cell cycle arrest in the G_0/G_1 phase. This cell cycle arrest was followed by a gradual increase in apoptotic cells and caspase-3 activation. Overexpression of PRAME was also found to prevent the cells from erythroid differentiation (Tanaka et al. 2011).

Kosova et al. studied the effect of STAT (signal transducer and activator of transcription) knockdown in apoptosis and proliferation in sensitive and Imatinib-resistant K562 cells. STAT5 is involved in the development of myeloproliferative diseases while STAT3 is implicated in malignant transformation; both STAT5 and STAT3 are constitutively expressed in haematological malignancies (Bromberg 2002; Turkson 2004). Quantification of mRNA levels revealed a significant increase in *STAT5B*, and *STAT5A* (>50 %), but not *STAT3* level (4 %) in Imatinib-resistant cells as compared to Imatinib-sensitive cells. Transient knockdown of STAT5A by siRNA led to sensitization of Imatinib-resistant and Imatinib-sensitive cells by 4.5 and 1.2 times to Imatinib treatment (Kosova et al. 2010). When Imatinib-resistant cells were treated with 5 μM Imatinib, cell viability was decreased by ~20 %, while the same concentration of Imatinib with STAT5A-siRNA-treated cells decreased cell viability by ~60 % (Kosova et al. 2010).

The sphingolipid ceramide plays an important role in apoptotic signalling in response to anticancer drugs. Intracellular levels of pro-apoptotic ceramide were shown to increase when cells respond to drugs, contributing to their anti-cancer efficacy. However, multidrug-resistant cells accumulate ceramide due to an enhanced activity of glugosyceramide synthase (GCS), which converts the available ceramide in glucosylceramide (GlcCer). This conversion impedes ceramide from being involved in activation of apoptosis (Baran et al. 2011; Zhang et al. 2011). Although controversial, down-regulation of GCS has been shown to down-regulate expression of P-glycoprotein (P-gp) (Zhang et al. 2011), an efflux pump that decreases intracellular levels of drugs. With the aim of decreasing the multidrug resistance in doxorubicin-resistance K562 cells, Zhang and co-workers targeted GCS or P-gp by siRNA. Upon silencing GCS or P-gp with specific siRNAs, the transporter activity was significantly decreased, suggesting a linkage between GCS and P-gp expression, and providing potential therapeutic targets in CML therapy (Zhang et al. 2011).

Growth factor independence-1B (GFI-1B) is a transcription factor that controls the development and differentiation of erythroid cells and megakaryocytes at the erythro-megakaryocytic progenitor stage (Randrianarison-Huetz et al. 2010). *GPI1B* mRNA expression was found to be overexpressed in leukemic cells (Elmaagacli et al. 2007). Koldehoff et al. investigated whether antileukemic effect of *BCR-ABL* silencing can be further increased by co-silencing of *GFI1B* (Koldehoff et al. 2013). A significant drop in cell viability was evident with the combination of GFI1B and BCR-ABL siRNAs, as well as *BCR-ABL* mRNA levels after co-silencing. An additive induction of apoptosis after co-silencing was observed. Similar results of the inhibition of mRNA levels of *BCR-ABL* and

GFI1B were found in advanced CML patient cells. The co-silencing led to a significant reduction of *MDR1* (P-gp) and *c-Myc* mRNA levels, suggesting *BCR-ABL* and *GFI1B* to be connected to other critical mediators involved in cancer transformation (Koldehoff et al. 2013).

Protein Phosphatase 2, Regulatory Subunit B', Gamma (PPP2R5C) levels were over-expressed in peripheral blood mononuclear cells from chronic phase CML patients, and *PPP2R5C* expression was significantly decreased in patients undergoing remission (Zheng et al. 2011). *PPP2R5C* plays an important role in cell proliferation, differentiation, and transformation based on its induction of the dephosphorylation of p53 at various residues, which negatively modulates its apoptotic activities, and thus promoting cell survival (Zheng et al. 2011). It was possible to reduce *PPP2R5C* mRNA and protein levels in K562 and resistant-K562 cells with specific siRNA treatment. *PPP2R5C* mRNA levels in CML primary cells was also reduced with specific siRNA treatment, leading to reduced proliferation rate in both K562 and CML primary cells. An increase in apoptosis rate in K562 cells was also evident. These results indicate that down-regulation of *PPP2R5C* could significantly inhibit the proliferation of CML cells and more importantly, could render imatinib-resistant cells more sensitive to TKIs (Shen et al. 2013).

Taken together, it is evident that several promising protein targets are available for siRNA-mediated silencing. Effective functional responses have been obtained, in the form of reduced proliferation, apoptosis induction as well as sensitization to CML drugs, after targeting individual or combination of the appropriate targets. Whether this approach could be applied clinically remains to be seen and it is generally assumed that effective delivery of siRNA is the limiting step.

5 Biomaterials in siRNA Delivery

Biomaterials such as cationic lipids and polyamines have been used to condense nucleic acids for delivery into cells. However, existing transfection and delivery methods are more suitable for attachment-dependent cells (e.g., breast cancer cells) rather than attachment-independent CML cells. Physical treatments such as electroporation on the other hand, although helpful to investigate the effect of gene depletion by RNAi, especially on difficult-to-transfect cells such as primary or suspension-growing cells (Gioia et al. 2011; Tanaka et al. 2011; Kosova et al. 2010), cannot be translated in vivo because of the significant toxicity they induce to cells after transfection, and because they have only been designed for an in vitro setting (Rangatia and Bonnet 2006; Merkerova et al. 2007). Electroporation (and related 'nucleofection') is the most common method to deliver siRNA for experimental purposes. Viral vector have been effectively used but these present significant safety risk because they can integrate to the host's genome or cause lethal immune responses and inflammation (Mintzer and Simanek 2009; Arthanari et al. 2010).

Cationic polyamines are desirable for siRNA delivery because they are capable of condensing anionic siRNAs into spherical, stable nano-particles (De Paula et al.

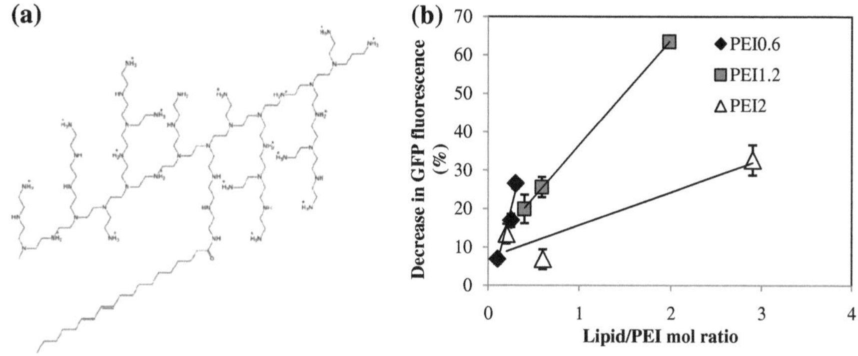

Fig. 1 **a** LA-substituted 2 kDa PEI. **b** GFP silencing in GFP-K562 cells with PA-substituted polymers (72 nM siRNA concentration). Decrease in mean GFP fluorescence was evaluated by flow cytometry 3 days after transfection. PEI1.2PA (lipid substitution of 1.98 PEI/PA) stands out showing a 63 % decrease in the mean GFP fluorescence and a milder effect on the cell counts decreasing it by 40 %. A strong correlation between the extent of lipid substitution and the GFP silencing was found with all the PEIs used. GFP silencing was also found to be dependent on the molecular weight (MW) of PEI used: the lower the MW of the PEI, the greater the increase in GFP silencing at increasing lipid substitutions

2007). Polyethylenimines (PEIs) with various molecular weights (MWs) and modifications have been used for transfection of nucleic acids in different cell lines and live animals (De Paula et al. 2007; Aliabadi et al. 2011, 2012). The high transfection efficiency of PEI is attributed to its "proton-sponge" effect, by which PEI once in the endosome attracts ions that lead to swelling and bursting of the endosome (Whitehead et al. 2009), which results in the release of the nucleic acids into the cytoplasm. This high transfection efficiency is mostly seen with high MW ($\sim$25 kDa) PEI where cellular delivery of nucleic acid cargo is efficient (unlike low MW PEIs). However, excessive endosome rupture leads to cell toxicity, thus limiting the dose of siRNA that can be delivered (Aliabadi et al. 2011; Wang et al. 2010). Moreover, an inverse relationship between cytotoxicity and transfection is observed in PEI, such that low MW (2–5 kDa) PEIs are considered to have better safety profiles due to non-toxicity, but are ineffective for nucleic acid delivery (Neamnark et al. 2009). By using the amine groups of the PEI that allow conjugation with other ligands, the Uludag group investigated the effect of lipid substitutions on 2 kDa PEI (PEI2) in an attempt to increase the polymer interaction with the cell membrane and nucleic acids delivery (Fig. 1a). It was found that the relatively nontoxic but ineffective PEI2 polymer carrier can be transformed into an effective delivery agent by grafting a lipid molecule onto the polymer (Aliabadi et al. 2011; Neamnark et al. 2009). Although generally effective, the gene delivery efficiency of these modified polymers can vary among cell lines (Aliabadi et al. 2011; Abbasi et al. 2008; Farrell et al. 2007; Alshamsan et al. 2009). Our recent studies on AML cells indicate that LA substitution (and to a lesser extent caprylic acid substitution in recent studies) sustained most silencing among the lipid-substituted 2 kDa PEIs (PEI2LA) for down-regulation of Green Fluorescent Protein (GFP) reporter gene and endogenous CXCR4 gene (Landry et al. 2012). Similarly, PEI2LA polymer also showed the highest silencing performance with GAPDH and P-glycoprotein expression when targeted in the MDA-MB-435 breast cancer cells among the different lipid-substituted PEI2s utilized (Aliabadi et al. 2011). However, the polymers that were effective in CML cells were different; we found a particular polymer (1.2 kDa PEI) substituted with a relatively high amount of palmitic acid (PEI1.2PA; 2.0 PA per PEI1.2) to be effective. The ability of PEI1.2PA to deliver siRNA intracellularly was high (Fig. 1b), explaining its relative efficiency. The oncogene *BCR-ABL* was also effectively silenced with this polymer, resulting in the expected apoptosis induction in the targeted cells (Valencia Serna et al. 2013). It is presently not known if this is a unique combination, or other molecular weight PEIs and/or lipids can substitute for its efficiency. The liposomal agent Lipofectamine^{TM} 2000, however, seems to be equally effective in the K562 cell model of CML.

To better explore the range of delivery systems suitable for CML, we can also inspect the siRNA delivery attempts in a related leukemia, namely acute myeloid leukemia (AML). These studies are summarized in Table 2, where the promising siRNA targets involved in cell survival (66) were categorized based on the physiological role of the target chosen (e.g., proteins involved in cell cycle regulation, bone marrow microenvironment interactions, drug sensitization, regulating transcription, phosphorylation as well as common AML mutations and related proteins). One can again see a range of non-viral carriers that have been used for delivering different siRNAs, but electroporation, as an experimental approach, has again dominated the delivery attempts (Fig. 2). Both polymeric and liposomal reagents were successfully employed for siRNA delivery in this leukemia and one can see the relatively limited studies involving non-viral carriers. The major focus in AML studies remained on elucidating suitable targets, which is expected to be analogues in the case of CML as well. As expected, there was a general increase in these attempts in recent years (Fig. 3a) but the effective dose of the siRNA delivery formulations (whether formulated with a carrier or delivered naked typically with electroporation) did not significantly change over time (Fig. 3b). One would have liked to see an improved effective dose (i.e., lower effective dose) with recently developed delivery systems, but this did not appear to be the case. This issue is discussed in more detail below.

6 Mechanism of Uptake and Intracellular Processing of siRNA Nanoparticles in Leukemic Cells

Cell membrane is the first interface that siRNA nanoparticles need to interact with for internalization. The lipid bilayer acts as an impermeable membrane to entry of unwanted materials from the external environment (including siRNA nanoparticles) and as selectively permeable, by the control of protein channels and pores, to the entry of nutrients and exit of metabolites (De Paula et al. 2007; Whitehead

Table 2 siRNA targets shown to be beneficial in AML models

References	Target (role/pathway)	Rationale	siRNA carrier (concentration)	Outcome
Cell cycle				
Yang et al. (2013)	SGOL1 (cell cycle—mitosis)	SGOL1 is a centromeric protein overexpressed in leukemia's including AML	Electroporation (unknown)	Decreased proliferation; mitotic arrest, intrinsic apoptosis
Tibes et al. (2012)	WEE1, CHEK1, PKMYT, ATR (cell-cycle check-points and DNA-damage repair proteins)	siRNA kinase/cytarabine screen to determine chemosensitizing targets to use with cytarabine	A cationic lipid-based transfection reagent (unknown)	Increased cytarabine efficacy
Bone marrow (BM) microenvironment interaction				
Fernandez-Vidal et al. (2006)	CDC25A (cell cycle)	Effects on cell adhesion and proliferation	Electroporation (8 pmol per 6 × 10^6 cells)	Decreased adhesion-dependent proliferation
Landry et al. (2013)	CXCR4 (BM micro-environment interaction)	CXCR4 mediated adhesion of AML cells	CA-PEI 2 kDa (25–50 nM)	Decreased proliferation
De Toni et al. (2006)	GSK3beta, p65 subunit NF-κB	Resistance due to adhesion molecules/integrin and morphogen Wnt soluble factors in AML	Electroporation (200 nM)	Restored chemosensitivity (daunorubicin)
Hu et al. (2011)	IGFBP7 (tumor suppressor)	To investigate the role of the known solid tumor suppressor (IGFBP7), in childhood AML	Lipofectamine 2000 (unknown)	Decreased adhesion, migration, invasion, proliferation. Role in BM microenvironment interaction was apparent
Despeaux et al. (2012)	FAK (stem cell pathway)	Over expression; FAK pathway deregulated in cancers (replaces Wnt3a-controlled canonical pathway)	Electroporation (200 nM)	Decreased survival

(continued)

Table 2 (continued)

References	Target (role/pathway)	Rationale	siRNA carrier (concentration)	Outcome
Recher et al. (2004)	FAK (cell motility and survival)	FAK involvement in AML	Electroporation (200 nM)	Decreased migration, increased chemosensitivity to daunorubicin, decreased FAK did not improve resistance due to fibronectin adhesion
Sansonetti et al. (2012)	MCL-1 (anti-apoptosis)	Survival effects of adhesion interactions with BMSCs. (Induced CD44 expression upregulated MCL-1)	Lipofectamine RNAiMax (50 nM)	Increased apoptosis
Kim et al. (2013)	SDF1 (BM micro-environment interaction)	Role of SDF-1 in survival and proliferation in AML	HiPerFect (5–25 nM)	Decreased proliferation
Drug sensitizing targets				
Konopleva et al. (2004)	BCL2 (anti-apoptosis)	Determining CDDO (novel synthetic triterpenoid 2-cyano-3,12-dioxooleana-1,9-dien-28-oic acid) mechanisms in AML	Electroporation (100–500 nM)	Decreased cell proliferation and increased apoptosis with co-treatment of CDDO (but not without) in CDDO resistant cells
Rao et al. (2011)	BCL2 (anti-apoptosis)	Involvement in curcumin action in daunorubicin insensitive CD34+ AML	Lipofectamine 2000 (50 nM)	Increased chemosensitivity of daunorubicin in CD34+ AML
Cluzeau et al. (2012)	BCL2L10 (anti-apoptosis)	BCL210 over-expression in azacitidine resistant cells	Electroporation (50 nM)	Sensitized cells to azacitidine
McLornan et al. (2013)	C-FLIP$_L$ (anti-apoptosis)	Higher expression of C-FLIP (drug resistance role) correlated with decreased patient survival	Electroporation (1.5 μg/1–2.5 $\times$ 10^6 cells)	Increased apoptosis, sensitization to rTRAIL induced apoptosis

(continued)

Table 2 (continued)

References	Target (role/pathway)	Rationale	siRNA carrier (concentration)	Outcome
Wang et al. (2010)	COT1	COT1 increases effect of silibinin/1,25-dihydroxyvitamin D3 combinations	Electroporation (5,000 nM)	Increased G1 arrest and differentiation caused by Silibinin/1, 25-dihydroxyvitamin D3 combinations
Kasper et al. (2012)	MCL-1 (anti-apoptosis)	MCL-1 is over expressed in FLT3-ITD cell lines	Electroporation (unknown)	Increased chemosensitivity in FLT3-ITD+ AML
Wang et al. (2013)	MCL-1 (anti-apoptosis)	Involvement in arsenic trioxide effect in AML	Unknown	Increased arsenic trioxide-induced mitochondrial apoptosis (chemosensitivity)
Nishioka et al. (2009)	MEK-1	Study of 5-AzadC (DNA methyltransferase inhibitor) and AZD6244 (MEK inhibitor) in AML	Electroporation (unknown)	Decreased viability with 5-AzadC co-treatment but not without
Altman et al. (2010)	Mnk1, Mnk2	Involvement in cytarabine mechanism of action	Unknown	Decreased leukemic colony formation with cytarabine treatment but not without
Nishioka et al. (2010)	4E-BP1 (MEK/ERK pathway) and MCL-1 (anti-apoptotic)	AZD6244 causes apoptosis and suppresses 4E-BP1 and MCL-1 in HL-60 cells but not in EOL-1 and MOLM13 cells	Electroporation (300 nM)	Decreased MCL-1 expression and increased apoptosis with AZD6244 (4E-BP1). Increased apoptosis with/out AZD6244 (MCL-1)
Transcription factor				

(continued)

Table 2 (continued)

References	Target (role/pathway)	Rationale	siRNA carrier (concentration)	Outcome
Pan et al. (2012)	Gli1 (transcription factor—activator/hedgehog pathway)	Effects of aberrant expression and inhibition of Gli	Jet-PEI (100 nM)	Decreased proliferation and decreased survival
Elmaagacli et al. (2007)	GFI1B	Role of GF1IB in erythropoietic and megakaryocytic malignancies	TransMessenger (0.1–2.4 μg/24-well plate well)	Decreased proliferation in AML, CML and normal CD34+ cells and increased apoptosis in AML and CML cells
Rushworth and MacEwan (2008)	HO-1, Nrf2, c-FLIP	Involvement in NF-κB and TNF-induced apoptosis in AML	Electroporation (30 nM)	Susceptible to TNF-induced cell death (HO-1, Nrf2), Susceptible to TNF but not with NF-κB inhibitor BAY 11–7082 (c-FLIP)
Rushworth et al. (2010)	NF-κB and HO-1	Inhibition of highly expressed NF-κB did not cause apoptosis due to HO-1	Electroporation (30 nM)	Increased apoptosis after targeting both HO-1 and NF-κB in AML cells but not in CD34+ non-malignant cells
Carvalho et al. (2007)	NF-κB subunit p65, IKK1, IKK2, NEMO	Understanding the role of NF-κB activation in AML	Electroporation (unknown)	Increased apoptosis
Braun et al. (2006)	NF-κB subunit p65	NF-κB is continuously activated in P39 MDS/AML cells	Electroporation (unknown)	Increased apoptosis
Zhang et al. (2013)	STAT3	Development of targeted STAT3 (role in cancers) siRNA delivery in TLR9+ hematopoietic cells	TLR9 antagonist CpG(A)-siRNA (100–500 nM in vitro, 100 μg (5 mg/kg) every 24 h IT for in vivo)	In vivo, decreased tumor growth. The delivery system is immunostimulatory and can contribute to overall anti-cancer effects

(continued)

Table 2 (continued)

References	Target (role/pathway)	Rationale	siRNA carrier (concentration)	Outcome
Gao et al. (2011)	WT1 (transcription factor and tumor suppressor)	Involvement in miR-15a and miR-16-1 tumor suppressors	HiPerFect (50 nM)	Decreased proliferation
Elmaagacli et al. (2005)	WT1 (transcription factor and tumor suppressor)	WT1 is overexpressed in leukemia	TransMessenger (0.8 μg, 1 × 10^5 cells/well)	Decreased proliferation, increased apoptosis in AML and CML (but not in normal CD34+ cells). Increased anti-survival effects when WT1 and BCR-ABL were targeted in K562 cells
Tyrosine kinase related				
Park et al. (2013)	Axl (receptor tyrosine kinase, various pathways)	Determine role of Axl in FLT3 signaling in AML	Electroporation (unknown)	Inhibited cell growth, arrested cell-cycle, induced apoptosis and differentiation in FLT3-ITD+ AML
Gu et al. (2007)	CSFIR	Identification of tyrosine-phosphorylated proteins in AML M7 (AMKL)	Electroporation (unknown)	Decreased proliferation and increased apoptosis in AML M7 MKPL1 cells but not in CML K562 cells. (C-KIT siRNA did not decrease proliferation)
Tyner et al. (2008)	EPHA4, JAK1, JAK3, LTK, LYN, PTK2, PTK2B PTK6, PTK9, and SRC (all cytosolic kinases except EPHA4 and LTK)	siRNA screen of tyrosine kinase proteins in AML cells	Electroporation (1,000 nM)	Decreased viability

(continued)

Table 2 (continued)

References	Target (role/pathway)	Rationale	siRNA carrier (concentration)	Outcome
Voisset et al. (2010)	FES, FER	Investigation of FES and FER in AML	Electroporation (0.4–0.8 nmol in 0.2–0.5 ml; 800–4,000 nM estimate)	Decreased proliferation (FER) and decreased survival (FES) in FLT3-ITD+ AML but not in non-mutated tested cells
Walters et al. (2006)	JAK2	To determine kinases that cause STAT5 phosphorylation in AML	Electroporation (unknown)	Decreased proliferation and viability in AML HEL cells but not in CML K562 cells. Decreased phosphorylation of STAT1/3/5 and Erk1/2. JAK1, JAK3 and TYK2 had no effect
Walters et al. (2006)	JAK3	To identify activated tyrosine kinases in AMKL cells without FLT3 and KIT mutations	Electroporation (unknown)	Decreased proliferation, inhibition of STAT5 tyrosine phosphorylation, increased apoptosis in AMKL. JAK2 and TYK2 had no effect
Dos Santos et al. (2008)	Lyn (a Src family kinase)	Lyn is highly activated. PP2 (SRK inhibitor) caused decreased proliferation and increased apoptosis	Electroporation (3 μg/100 μl for 2×10^6 cells, 2,143 nM estimated)	Decreased leukemic colony formation, linked to mTOR pathway
Okamoto et al. (2007)	Lyn (a Src family kinase)	Lyn and FLT-ITD interactions in AML	Electroporation (3 μg)	Decreased proliferation in FLT3-IDT+ 32D cells. Decreased STAT5 phosphorylation

(continued)

Table 2 (continued)

References	Target (role/pathway)	Rationale	siRNA carrier (concentration)	Outcome
Common mutations and related targets				
Gessner et al. (2010)	AML/MTG8 (transcription factor)	AML/MTG8 fusion gene found in AML. (Also studied MLL/AF4 found in ALL)	Electroporation (unknown)	Reduced clonogenicity, induction of replicative senescence (also decreased TERT expression and increased telomere shortening)
Caligiuri et al. (2007)	c-CBL (mutation)	Identification and study of c-CBL and CBL-b mutations in AML	Electroporation (unknown)	Decreased cell proliferation
Wang et al. (2011)	CIP2A (oncoprotein)	Determine role of CIP2A (involved in cancers) in AML	Electroporation (unknown)	Decreased proliferation, decreased clonogenic activity and increased differentiation
Wang et al. (2011)	FLT3 (AML mutation)	FLT3 over-expressed/mutated in AML	sc-29528 reagent, Santa Cruz (in vitro unknown, 5 × 100 μg/kg every 72 h IP for in vivo without reagent)	Arrested in G0/G1 phase, decreased proliferation in vivo and in vitro, increased apoptosis
Walters et al. (2005)	FLT3 (AML mutation)	FLT3 over-expressed/mutated in AML. Developing multiple methods for inhibiting FLT3 due to need for better specificity and resistance development	Electroporation (1 μg/1 × 10^7 cells)	Decreased proliferation, increased apoptosis, and increased sensitivity to MLN518 (a FLT3 inhibitor)
Spirin et al. (2011)	KIT (c-kit oncogene)	siRNA and shRNA studies target c-kit (over-expressed / mutations in AML)	Lipofectamine 2000 (50–200 nM)	Effects were not studied for siRNA transfections. (shRNA studies)

(continued)

Table 2 (continued)

References	Target (role/pathway)	Rationale	siRNA carrier (concentration)	Outcome
Balusu et al. (2011)	NPM1 (molecular chaperone—proteins and nucleic acids)	Common mutation in AML	Electroporation (100 nM)	Chemosensitizes (ATRA and cytarabine), decreased cells in S-phase, induced differentiation, increased apoptosis (NPM1 mutant+ AML)
Fernandes et al. (2010)	Rho, Rac, Cdc4 (Rho family GTPases)	Understanding AML Casitas B lineage lymphoma (CBL) mutations	Electroporation (unknown)	Decreased proliferation (CBL mutation+ AML)
Geletu et al. (2007)	Ubc9 (small ubiquitin-related modifier conjugation)	To identify target proteins of C/EBPalphap30 (mutation that occurs in 10 % of AML)	Electroporation (500 ng)	Prevents differentiation block caused by C/EBPalphap30 (co-transfected) when CD34+/U937 cells go through granulocytic differentiation
Various others				
Fiskus et al. (2006)	EZH2	Effect of EZH2 on AML cells	Electroporation (100 nM)	Co-treatment with LBH589 (inhibitor) decreased colony formation (HL-60 and U937) and increased differentiation (U937)
Gao et al. (2009)	hnRNP K	Role of hnRNP K in drug induced suppression and apoptosis induction	DharmaFECT-4 (100 nM)	Induced apoptosis
Miyazaki et al. (2010)	HO-1	Involvement in AML	Electroporation (unknown)	Reduced cell survival with and without cytarabine
Schepers et al. (2005)	HSP27 (Heat shock protein family, stress response)	Involvement in AML	Oligofectamine 25 nM	Increased VP-16 mediated apoptosis but not CD95/Fas mediated apoptosis

(continued)

Table 2 (continued)

References	Target (role/pathway)	Rationale	siRNA carrier (concentration)	Outcome
Muranyi et al. (2010)	ILK (PI3K-dependent signalling pathway)	Investigation of ILK and FLT3 as targets in AML (inhibitors used for FLT3 suppression)	Accell modified siRNA (unknown)	Decreased leukemic colony formation
Muranyi et al. (2009)	ILK (PI3K-dependent signalling pathway)	ILK role in AML. Possible benefit in targeting both ILK and FLT-3	Electroporation (50 μg per 5×10^6 cells)	Decreased colony formation, increased cell death
Wang et al. (2010)	MMP-2, MT1-MMP, and TIMP-2	Role of MMP-2, MT1-MMP, or TIMP-2 in AML extramedullary infiltration	Lipofectamine 2000 (400 nM estimated)	Decreased invasion
Lu et al. (2008)	NRP-1 (VEGF receptor)	Involvement in AML	Lipofectamine 2000 (unknown)	Decreased proliferation and chemotaxis of leukemic cells
Powell et al. (2009)	OPN (Ser585-survival pathway, cytokine and chemoattractant))	Investigated Ser585-survival pathway. OPN is a secreted protein and therapeutically accessible	Unknown (50–150 nM)	Increased cell death and decreased survival in AML blasts and leukemic stem and progenitor cells
Yang et al. (2012)	S100A8 (autophagy)	To determine S100A8 role in autophagy in AML	Lipofectamine RNAiMAX (unknown)	Increased chemosensitivity, increased arsenic trioxide induced apoptosis, decreased autophagy

siRNA studies which reported significant anti-survival effects in AML cells were selected from a PubMed 'AML siRNA' keyword search. The targets were segregated based on their physiological mechanism (or action)

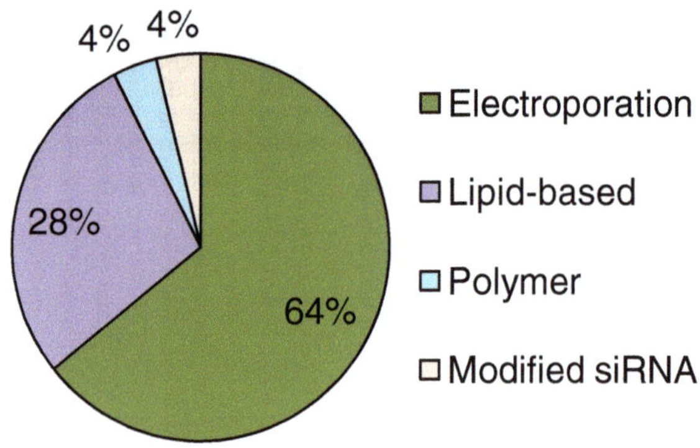

Fig. 2 Different delivery approaches used for siRNA delivery in AML models. The delivery approaches were obtained from the studies summarized in Table 2

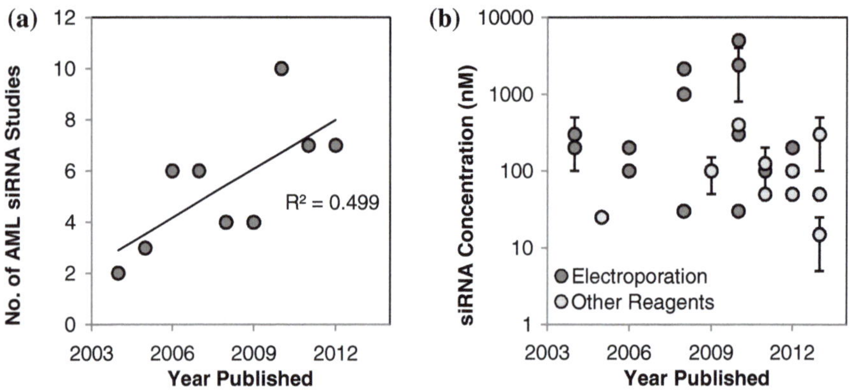

Fig. 3 **a** Number of studies published between 2004–2012 involving siRNA delivery in AML (original data from Table 2). **b** Effective in vitro siRNA concentrations utilized (if reported) in the studies outlined in Table 2. For clinical purposes, one would like to have an effective concentration less than 50 nM. Error bars are displayed to indicate the siRNA concentration range used in a given study

et al. 2009; Lodish et al. 2013). The heterogeneous lipid composition and distribution of hundreds of lipid species present in cell membrane influences the degree of lipid diffusion in the membrane as well, as the thickness and shape (architecture) of the cell membrane. These characteristics are not conserved among the cells and are dependent on the cell type, cellular activity and the constant changes in signaling with the external environment (Lodish et al. 2013; Janmey and Kinnunen 2006). The heterogeneity of these components and the affinity among some of them, such as the affinity between cholesterol and sphingolipids, lead to the formation of clusters along the membrane that are known as lipid rafts (Brown and London 2000; van Blitterswijk 1988). These clusters have their own charge, which can make a siRNA nanoparticle more or less interactive with specific regions on cell surface, leading to different type of interactions along the membrane. The successful integration of the siRNA nanoparticles with the membrane will depend on the ability to interact to this area by charge affinity or to move to another area of more affinity (Nel et al. 2009). These factors make the cell membrane a dynamically uneven surface with unstable characteristic features for interaction with siRNA nanoparticles (Nel et al. 2009).

6.1 Differences in Delivery Between Suspension-Growing Versus Attachment-Dependent Cells

The delivery of siRNA nanoparticles is challenging when the target cells are suspension-growing (i.e., attachment-independent) cells. The interaction of suspended particles with cells growing in suspension is expected to be different from cells that are attached to other cells or to tissue culture plastic. The siRNA nanoparticles need to be designed in such a way that they are able to bind to suspended cells and promote their entrance into the cells for delivery of nucleic acids. Uptake of siRNA nanoparticles in suspension-growing cells such as leukemic cells has been studied and it has been found that the delivery of nucleic acid moieties is more difficult to achieve in these cells (Valencia Serna et al. 2013).

Initial studies performed in Uludag lab compared siRNA uptake between the adherent breast cancer MDA-MB-231 cells and suspension-growing K562 CML cells with a generally-effective PEI2LA. A 15-fold reduction in siRNA uptake was found in K562 cells in comparison to MDA-MB-231 cells, showing a considerable reduction in the siRNA delivery efficiency. In addition, despite a 29-fold increase in the siRNA uptake with these carriers (in comparison to non-carrier groups), silencing of the green fluorescent protein (GFP) in GFP-positive K562 cells was proven to be ineffective (Valencia Serna et al. 2013). The relatively small amount of siRNA may have not reached its target (RISC for degradation of the mRNA), possibly due to incomplete internalization or endosomal entrapment of the particles (Valencia Serna et al. 2013). Similarly, Lorenz et al. evaluated the interaction of polymeric particles with different cell types and found that when the amount of amino groups of the particles was increased, a greater amount of particles interacted with cell membranes (Lorenz et al. 2006). Moreover, although it was found that the interaction between the cells and nanoparticles was the same with all the cell lines tested, whether attachment-dependent (HeLa and mesenchymal stem cells (MSC), or suspension-growing (KG1a as a model for CD34+ hematopoietic stem cells and Jurkat as model for T cells) cells, the internal location of these particles differed among the cells: particles that interacted with MSC and HeLa (attachment-dependent) cells were located in intracellular compartments, most likely located inside endosomes; while particles that interacted with KG1a and Jurkat (suspension-growing) cells were found at the cell membrane or periphery of the cells (Lorenz et al. 2006), suggesting that these particles were not internalized in the cells and were not able to overcome the cell membrane barrier. An active endocytosis was perhaps limited in the latter cells. Zhao et al. (1996) also compared the oligonucleotide uptake between leukemic and the different types of normal (from bone marrow or peripheral blood) human hematological cells and found that the uptake differed among the hematopoietic cell types: the uptake was highest in myeloid/macrophages, followed by B-cells, T-cells and finally with neutrophils having the lowest uptake level. On the other hand, human leukemic cells were also found to take up more oligonucleotides than normal or residual cells from the same patient and, this uptake was increased or decreased in leukemic and normal cells upon cell growth factor stimulation and cell growth inhibition, respectively. This led the authors to conclude that the uptake in leukemic cells was greater due to their higher cell growth and activation (Zhao et al. 1996). These observations indicated that the uptake of nucleic acid nanoparticles is dependent; not only on the cell type (i.e., attachment-dependent vs. suspension-growing or attachment-independent), but also on the internalization pathway that could be variable among the attachment independent cells. Thus, not only cationic charges seem to be important for high affinity interactions with cell membranes of suspension-growing cells, there might be also a need of specific ligands (as will be discussed later) for particles to display direct interaction with cell membrane components that lead to a complete cell internalization and continuation with the silencing of the targeted mRNA.

Among the explanations for less effective transfection in leukemic or suspension-growing cells, Labat-Moleur et al. (1996) suggested that the poor transfection ability of cationic vectors in lymphocytes, and other non-adherent cells might be attributed to weak interaction of these vectors due to the lack of Ca^{2+}-dependent cell surface extracellular matrix (ECM) ligands, such as, proteoglycans and cadherins, that are only present in adherent cells. Others have explored further the presence of binding proteins in leukemic cells. Rainaldi et al. for example, had cell culture flasks coated with polylysine and added foetal bovine serum to surfaces so that polylysine could immobilize proteins necessary for binding of K562 cells. The expressions of the most common ECM proteins, including fibronectin (VLA-5), vitronectin ($\alpha_v\beta_3$), collagen (VLA-2) and, hyaluronan (CD44), were then evaluated and, fibronectin was found to be the only cell membrane protein that was expressed in K562 cells attached to a surface under these conditions. The authors concluded that facilitated-adhesion of K562 cells onto polylysine did not occur directly between these two, but rather between the fibronectin receptors on the surface of K562 cells and the fibronectin absorbed onto the polylysine-coated surface by the addition of FBS (Rainaldi et al. 2001).

Sun et al. (2013) investigated the expression of CD44 in different cell lines (SHI-1, THP-1, NB4 and K562) and K562 cells were found to differ from the rest of cells lines: the expression of CD44 in SHI-1, THP-1 and NB4 cells (all cell models of acute myeloid leukemia, AML) was shown to be significantly higher than that of K562 cells. Moll et al. (1998) have also shown that K562 cells do not normally express CD44 protein but that these cells can express this molecule when they are stimulated to differentiate towards the myeloid lineage. The expression of CD44 receptors on the cell surface seems to be a characteristic related to the adhesion and migration of AML cells. Suspension-growing and leukemic cells seems to not express most of the cell membrane receptors that are involved in the cell adhesion. How the lack of these receptors reduces the interaction with the binding of nucleic acids and their carriers still remains to be clarified.

6.2 Structure-Function Relationships in Cellular Delivery

In order for the uptake to take place, particle binding and engulfment at the adhesion site require specific and nonspecific interactions to overcome the resistive forces that hinder particle uptake (Nel et al. 2009). There are two types of strategies that aim to deliver nucleic acid to the suspension-growing cells, mediated by either unspecific binding or specific binding. Here, we review the latest studies and delivery systems that are being used for improving the delivery efficiency in the suspension-growing leukemic cells. The challenges for the optimization of the design of the nucleic acid delivery carriers are additionally discussed.

6.2.1 Delivery Based on Unspecific Binding to Cells

Calcium phosphate transfection method consists of addition of a co-precipitate containing calcium phosphate and nucleic acids to cell culture. The sedimented co-precipitates are then taken up by cells after non-specific binding of the co-precipitates to the cell membrane (Ravid and Freshney 1998). Jordan and Wurm (2004) used calcium phosphate to transfect attachment-dependent and suspension-growing cells, emphasizing some of their transfection differences. They mention that since adherent cells are at the bottom of the plate, they do not interact with the particles present in the medium. To increase this interaction, the particles should be large enough to settle down by gravity, but the bigger the size the lower the transfection efficiency (expected). When the particle is small, it is unlikely to settle, but its inherent transfection ability would be higher due to its smaller size. The suspension-growing cells are more likely to interact with particles because they are suspended in the medium, however the affinity plays an important role for the particles to tightly interact with the cells (Jordan and Wurm 2004).

Cell penetrating peptides (CPPs) can be up to 30 amino acids long and are inherently able to translocate cell membranes. Overall cationic charge of CPPs confers them the ability to interact electrostatically with the phosphate backbone of nucleic acids to form stable nanoparticles, while allowing them to interact with cell membranes as well. Arthanari et al. (2010) used the cationic Tat-derived CPPs (aminoacids 49–57 of HIV-1 TAT protein) covalently attached to cationic membrane active peptide LK15 (Tat-LK15 peptide) for the delivery of siRNA and plasmid DNA encoding for a short hairpin RNA (shRNA) in K562 cells. The authors found that the combination of these two peptides increased the transfection efficiency by two folds in comparison with Tat peptide alone in several cell lines. With a dose of 1–30 μg of siRNA in 1 ml (24–729 nM based on our calculation), expression of p210 BCR-ABL was reduced to at least ~70 % 48 h post-transfection for all concentrations. High density of positive charges of siRNA nanoparticles led to cytotoxicity ranging from 0 % (10 μg siRNA) to 30 % cell death (30 μg siRNA) (Arthanari et al. 2010).

Wang et al. (2008) prepared siRNA nanoparticles with functionalized single wall carbon nanotubes (f-SWNTs) and siRNA against cyclin A2, which is involved in cell cycle regulation and associated with proliferation in leukemic K562 cells. These cells were treated with particles formed of f-SWNTs and Cyclin A2 siRNA (25 nM), which resulted in a reduction of cell numbers of up to ~70 % 60 h after treatment in comparison with cells treated with f-SWNTs and control siRNA. No significant toxicity was found with cells treated with f-SWNTs alone or in combination with control siRNA. Both a reduction of the *Cyclin A2* mRNA was found 32 h after treatment as well as a 70 % reduction in the colony formation assay 3 weeks post-transfection (Wang et al. 2008).

6.2.2 Delivery Based on Specific Binding to Cells

Among the most effective specific-binding interaction are those ligands or antibodies coupled onto nanoparticles that allow them to interact with complementary molecules (or receptors) on cell membranes (Nel et al. 2009; Wang et al. 2011). These interactions result in either receptor-mediated endocytosis or receptor-mediated direct penetration in the absence of endocytosis, for example when gold nanoparticles and cell-penetrating peptides are used as delivery carriers (Nel et al. 2009). For nanoparticle adherence and engulfment to take place at an adhesion site, ligand-receptor interactions need to overcome the resistive forces that prevent the nanoparticle uptake. Examples of these resistive forces are the memory of the cell membrane to return to its original form and the hydrophobic exclusion of polar surfaces by the cell membrane (Nel et al. 2009).

Antibody-mediated attachment does not necessarily induce internalization of nanoparticles. For example, doxorubicin-loaded liposomes were attached to anti-CD34 monoclonal antibody that targets CD34+ KG-1a cells. The cell-associated level of Dox in KG-1a cells was found to be lower in comparison with free Dox exposure. Since 40 % of the drug seemed to be released after 2 h of incubation for both systems, this targeted-drug delivered system would need to be used in combination with inhibitors of the P-gp efflux pump—whose overexpression in cancerous cells is known to act as a drug resistant method—so that the Dox extrusion is inhibited and the drug can exert its cytotoxic effect inside the cell. The IC_{50} of Dox-loaded CD34+ liposomes was similar to that of free Dox, but 8 times higher than the non-targeted liposome. Based on confocal and flow cytometry studies that show that neither the liposome nor the antibody were internalized by the cells, the authors suggested that the delivery mechanism of this immunoliposome is binding to targeted cells and Dox at the vicinity of the cells, where the drug is consequently internalized as free Dox. The CD34 cell binding did not seem to be capable of triggering the liposome cell entry (Carrion et al. 2004) and/or the cells were not able to internalize the Dox-loaded liposomes, probably because of the nature of the liposome used.

Thus, immuno-targeting with antibodies by themselves does not necessarily correspond to high internalization since the carrier needs to also play an important role with cell membrane interactions and internalization. Guillem et al. constructed immunopolyplexes, using 25 kDa PEI-based polyplexes attached via a streptavidin bridge to biotin-labeled antibodies that target cell membrane proteins. A significant selectivity in delivery was observed: an anti-CD3 immunopolyplex was functional only in Jurkat T-cells (CD3+/CD19-), while an anti-CD19 immunopolyplex was functional only in Granta B-cell line (CD3-/CD19+). However, only ~11 % of Jurkat cells and ~2 % for Granta cells were transfected, showing a dependency of the transfection efficiency on the cell type used. Transfection of a mixture of Jurkat and J.RT3/T3.5 cells (a CD3-/CD19- T-cell line) with anti-CD3 immunopolyplexes showed that >80 % of transfected cells were CD3+, indicating the selectivity of the delivery system in a heterogeneous cell population. Viability studies showed a decrease in cell viability to 50 % for Jurkat cells and to 90 % for J.RT3/T3.5 (Guillem et al. 2002), which shows an association of transfection with significant cytotoxicity in this system. Poor transfection was shown with naked 25-PEI (5 % cells were positive for transfection) in comparison with anti-CD3 immunopolyplex (10 % of cells were positive).

In an attempt to develop a delivery system for T-cell leukemias, more specifically against JL1-positive cells, an antibody-coupled CPP (oligo-arginine$_9$) complex was developed (Lee et al. 2010). Uptake studies showed a higher binding affinity of JL1-CPP nanoparticles to JL1-overexpressing cells than Jurkat cell with low JL1 expression, when a 200-pmol siRNA was used and after 2 h of transfection 96 % of the cell population was already FITC-positive. The uptake was shown to be 5.7 % for JL1- negative H9 cells (Lee et al. 2010). No toxicity studies or silencing experiments were performed. Major limitation of peptide delivery systems is their excessive positive charge and lack of target cell specificity, which may result in non-specific tissue distribution and aberrant immunogenic toxicity (reviewed in Lee et al. 2010).

Therefore, using an antibody seems to increase selectivity and enhance the efficacy of the carrier in the last two cases. However, this targeted systems seems to be limited by the efficacy of the carrier used, therefore, a more efficacious carrier could enhance transfection even further.

6.3 Secondary Effects of siRNA Silencing: Off-Target Effects and Cytotoxicity

Transferrin receptor (TrfR) is a cell membrane-associated glycoprotein that is overexpressed in cancer cells, can be easily accessible and can promote endocytosis once its ligand transferrin (Trf) is bound at the cell surface (Mendonça et al. 2010). Mendonça et al. developed a transferrin receptor (TrfT)-targeted sterically stabilized liposomes encapsulating BCR-ABL siRNA. K562 and LAMA-84 cells were transfected twice every 2 days with a siRNA concentration ranging from 100 to 2,000 nM. Efficacy experiments showed a dose-dependent reduced viability when BCR-ABL siRNA was delivered to these cells: cell viability was reduced up to ~47 % for LAMA-84 cells and up to ~24 % for K562 cells. However, a dose-dependent toxicity (58 % with 2 μM of siRNA) was also seen with scrambled siRNA with LAMA-84 cells, but to a much lower extent with K562 cells. Levels of *BCR-ABL* mRNA were reduced in a dose-dependent manner up to 1 μM (~60 %) with specific siRNA in LAMA-84 cells while no significant reduction was found with scrambled siRNA. Similar results were found for the oncoprotein as well, except that the scrambled sequence reduced non-specifically the protein levels at the highest siRNA concentration (2 μM) (Mendonça et al. 2010). High toxicity and off-targets effect in this system was probably due to high siRNA dose ($2 \times 1 - 2$ μM of siRNA) and the highly cationic carrier.

Eguchi et al. generated a carrier composed of a TAT-peptide transduction domain (PTD) and double-stranded RNA-binding domain (PTD-DRBD); RNAi induction was evaluated in difficult-to-transfect cells, including T cells. GFP siRNA delivered with PTD-DRBD (siRNA concentration of 100–400 nM) in Jurkat T-cells containing an integrated GFP reporter gene resulted in a reduction of the mean GFP fluorescence to ~10 % and also showed an mRNA reduction to 10 %, while Lipofection (100 nM for Lipofectamine 2000™ and 10–50 nM with Lipofectamine RNAiMAX™) reduced the mean fluorescence to 50–60 % and mRNA levels to ~50 %. Similar results were found when targeting CD4 and CD8 in primary murine T cells with PTD-DRBD with specific siRNAs. No toxicity was found in primary human umbilical cord vein endothelial cells (HUVEC) cells treated with siRNA and PTD-DRBD. About 20 % mRNA reduction or off-target effect was seen when scrambled (negative control) siRNA were used either in Jurkat or HUVEC cells (Eguchi et al. 2009). This system is highly effective in silencing however; off-target effects are probably common when high siRNA concentrations are used.

7 Perspective on the Future of siRNA Delivery in CML

New functional carriers that promote efficient delivery of gene-based agents (i.e., siRNA) in a controlled and non-toxic way, are motivating researchers to find physiological solutions for treatment of CML. A better understanding of the clues that lay behind the uptake and intracellular trafficking of siRNA nanoparticles in the challenging suspension-growing leukemic cells will further help in this endeavor. The effect of carrier characteristics such as molecular size, degree of substitution (or modification) and optimal balance of the lipophilic-cationic moieties should be better understood not only on siRNA delivery efficiency, but also on toxicity, intracellular trafficking and cell specificity. This together with the identification of novel siRNA targets that can be used in combination with classical siRNA targets in CML, such as *BCR-ABL*, to silence gene combinations involved in the activation of different survival pathways in CML should prove beneficial. The combinational delivery, where multiple targets are silenced simultaneously, is likely going to yield more efficacious therapy, and possibly more specific outcomes. Irrespective of the target, however, non-viral siRNA delivery is more likely to be the clinically acceptable approach, given the relatively safe nature of such a delivery mode. The siRNA therapy could act in conjunction with the current clinically-employed drugs to improve their effectiveness or re-sensitize the cells to current drugs. However, the siRNA therapy could also serve as a stand-alone therapy if LSC could be specifically targeted. There is no reason why the delivery methods used for CML cells could not be applied to other types of leukemias, but this will most likely require a different set of biomaterials effective in a particular types of leukemia. Very little information exists on the molecular details for effective carriers in different leukemias so that this should be a fruitful avenue of exploration in the future.

Since the suspension-growing cells tend to be more difficult to transfect than the attachment-dependent cells, added pressure exists for non-viral delivery to be functional for leukemic diseases. The siRNA nanoparticles need to be effective at 20–50 nM range in culture for a practical translation to preclinical animal models. It is typical for reported delivery system to employ concentrations beyond this range, including our own work (Valencia Serna et al. 2013). Concerted effort to lower efficacious doses will be beneficial in this regard. In addition to efficacy, specificity is important not to down-regulate targets critical in normal physiology. Given the cationic nature of these nanoparticles, they could theoretically bind to a multitude of cells in vivo. However, 'biochemical' targeting could alleviate this limitation to some extent: only those genes that are aberrantly expressed in CML cells, such as the *BCR-ABL* or other supporting mediators, could be the target of RNAi, so that nanoparticles penetrating 'normal' cells might not lead to silencing important targets. On the other hand, in order to increase the specificity of siRNA delivery, polymers could be coupled with CML-specific ligands, such as antibodies, to deliver the siRNA to only certain cell populations. For example, carriers could be coupled with an anti-CD34 antibody to target at least most of the CML stem cell portion. However, these antibodies need to be chosen with care so the delivery system is not too limited to certain cell populations. These antibody ligands need to be also exclusively or substantially over-expressed in the target cells to minimize nanoparticle binding to normal cells. A modular design could be envisioned where a delivery system optimized for general cellular uptake is further functionalized with leukemic cell specific cell surface binding molecules.

Finally, little information is available on siRNA delivery to primary cells, either healthy or malignant cells from CML patients. It is critical not only to evaluate the efficacy in human cells, but also evaluating off-target effects of the siRNA delivered and cytotoxic effect of the polymers. While cell lines are preferred (due to practical reasons) to optimize cellular delivery, endocytosis rate and intracellular trafficking pathways are expected to be significantly different from cell lines. Misleading directions could be avoided by employing primary cells early on in development process.

References

Abbasi M, Uludağ H, Incani V, Hsu CYM, Jeffery A (2008) Further investigation of lipid-substituted poly (L-Lysine) polymers for transfection of human skin fibroblasts. Biomacromolecules 9(6):1618–1630

Abbasi M, Lavasanifar A, Uludağ H (2013) Recent attempts at RNAi-mediated P-glycoprotein downregulation for reversal of multidrug resistance in cancer. Med Res Rev 33(1):33–53

Aliabadi HM, Landry B, Bahadur RK, Neamnark A, Suwantong O, Uludağ H (2011) Impact of lipid substitution on assembly and delivery of siRNA by cationic polymers. Macromol Biosci 11(5):662–672

Aliabadi HM, Landry B, Sun C, Tang T, Uludağ H (2012) Supramolecular assemblies in functional siRNA delivery: where do we stand? Biomaterials 33(8):2546–2569

Alshamsan A, Haddadi A, Incani V, Samuel J, Lavasanifar A, Uludağ H (2009) Formulation and delivery of siRNA by oleic acid and stearic acid modified polyethylenimine. Mol Pharm 6(1):121–133

Altman JK, Glaser H, Sassano A, Joshi S, Ueda T, Watanabe-Fukunaga R, Fukunaga R, Tallman MS, Platanias LC (2010) Negative regulatory effects of Mnk kinases in the generation of chemotherapy-induced antileukemic responses. Mol Pharmacol 78(4):778–784

Arthanari Y, Pluen A, Rajendran R, Aojula H, Demonacos C (2010) Delivery of therapeutic shRNA and siRNA by Tat fusion peptide targeting BCR-ABL fusion gene in Chronic Myeloid Leukemia cells. J Control Release 145(3):272–280

Baccarani M, Cortes J, Pane F, Niederwieser D, Saglio G, Apperley J, Cervantes F, Deininger M, Gratwohl A, Guilhot F, Hochhaus A, Horowitz M, Hughes T, Kantarjian H, Larson R, Radich J, Simonsson B, Silver RT, Goldman J, Hehlmann R, European LeukemiaNet (2009) Chronic myeloid leukemia: an update of concepts and management recommendations of European LeukemiaNet. J Clin Oncol 27(35):6041–6051

Balusu R, Fiskus W, Rao R, Chong DG, Nalluri S, Mudunuru U, Ma H, Chen L, Venkannagari S, Ha K, Abhyankar S, Williams C, McGuirk J, Khoury HJ, Ustun C, Bhalla KN (2011) Targeting levels or oligomerization of nucleophosmin 1 induces differentiation and loss of survival of human AML cells with mutant NPM1. Blood 118(11):3096–3106

Baran Y, Bielawski J, Gunduz U, Ogretmen B (2011) Targeting glucosylceramide synthase sensitizes imatinib-resistant chronic myeloid leukemia cells via endogenous ceramide accumulation. J Cancer Res Clin Oncol 137(10):1535–1544

Bocchia M, Gentili S, Abruzzese E, Fanelli A, Iuliano F, Tabilio A, Amabile M, Forconi F, Gozzetti A, Raspadori D, Amadori S, Lauria F (2005) Effect of a p210 multipeptide vaccine associated with imatinib or interferon in patients with chronic myeloid leukaemia and persistent residual disease: a multicentre observational trial. Lancet 365(9460):657–662

Braun T, Carvalho G, Coquelle A, Vozenin M-C, Lepelley P, Hirsch F, Kiladjian J-J, Ribrag V, Fenaux P, Kroemer G (2006) NF-kappaB constitutes a potential therapeutic target in high-risk myelodysplastic syndrome. Blood 107(3):1156–1165

Bromberg J (2002) Stat proteins and oncogenesis. J Clin Invest 109(9):1139–1142

Brown DA, London E (2000) Structure and function of sphingolipid- and cholesterol-rich membrane rafts. J Biol Chem 275(23):17221–17224

Caligiuri MA, Briesewitz R, Yu J, Wang L, Wei M, Arnoczky KJ, Marburger TB, Wen J, Perrotti D, Bloomfield CD, Whitman SP (2007) Novel c-CBL and CBL-b ubiquitin ligase mutations in human acute myeloid leukemia. Blood 110(3):1022–1024

Carrion C, de Madariaga MA, Domingo JC (2004) In vitro cytotoxic study of immunoliposomal doxorubicin targeted to human CD34(+) leukemic cells. Life Sci 75(3):313–328

Carvalho G, Fabre C, Braun T, Grosjean J, Ades L, Agou F, Tasdemir E, Boehrer S, Israel A, Véron M, Fenaux P, Kroemer G (2007) Inhibition of NEMO, the regulatory subunit of the IKK complex, induces apoptosis in high-risk myelodysplastic syndrome and acute myeloid leukemia. Oncogene 26(16):2299–2307

Cluzeau T, Robert G, Mounier N, Karsenti JM, Dufies M, Puissant A, Jacquel A, Renneville A, Preudhomme C, Cassuto JP, Raynaud S, Luciano F, Auberger P (2012) BCL2L10 is a predictive factor for resistance to azacitidine in MDS and AML patients. Oncotarget 3(4):490–501

Copland M (2009) Chronic myelogenous leukemia stem cells: What's new? Curr Hematol Malig Rep 4(2):66–73

Corbin AS, Agarwal A, Loriaux M, Cortes J, Deininger MW, Druker BJ (2011) Human chronic myeloid leukemia stem cells are insensitive to imatinib despite inhibition of BCR-ABL activity. J Clin Invest 121(1):396–409

Cortes JE, Kantarjian H, Shah NP, Bixby D, Mauro MJ, Flinn I, O'Hare T, Hu S, Narasimhan NI, Rivera VM, Clackson T, Turner CD, Haluska FG, Druker BJ, Deininger MWN, Talpaz M (2012) Ponatinib in refractory Philadelphia chromosome-positive leukemias. N Engl J Med 367(22):2075–2088

De Paula D, Bentley MVLB, Mahato RI (2007) Hydrophobization and bioconjugation for enhanced siRNA delivery and targeting. RNA 13(4):431–456

De Toni F, Racaud-Sultan C, Chicanne G, Mas VM-D, Cariven C, Mesange F, Salles J-P, Demur C, Allouche M, Payrastre B, Manenti S, Ysebaert L (2006) A crosstalk between the Wnt and the adhesion-dependent signaling pathways governs the chemosensitivity of acute myeloid leukemia. Oncogene 25(22):3113–3122

Deininger MW, Goldman JM, Melo JV (2000) The molecular biology of chronic myeloid leukemia. Blood 96(10):3343–3356

Despeaux M, Chicanne G, Rouer E, De Toni-Costes F, Bertrand J, Mansat-De Mas V, Vergnolle N, Eaves C, Payrastre B, Girault J-A, Racaud-Sultan C (2012) Focal adhesion kinase splice variants maintain primitive acute myeloid leukemia cells through altered Wnt signaling. Stem Cells 30(8):1597–1610

DeWeerdt S (2013) Genetics: written in blood. Nature 498(7455):S4–S6

Dominska M, Dykxhoorn DM (2010) Breaking down the barriers: siRNA delivery and endosome escape. J Cell Sci 123(8):1183–1189

Dos Santos C, Demur C, Bardet V, Prade-Houdellier N, Payrastre B, Recher C (2008) A critical role for Lyn in acute myeloid leukemia. Blood 111(4):2269–2279

Eguchi A, Meade BR, Chang Y-C, Fredrickson CT, Willert K, Puri N, Dowdy SF (2009) Efficient siRNA delivery into primary cells by a peptide transduction domain-dsRNA binding domain fusion protein. Nat Biotechnol 27(6):567–571

Elert E (2013) Living with leukaemia. Nature 498(7455):S2–S3

Elmaagacli AH, Koldehoff M, Peceny R, Klein-Hitpass L, Ottinger H, Beelen DW, Opalka B (2005) WT1 and BCR-ABL specific small interfering RNA have additive effects in the induction of apoptosis in leukemic cells. Haematologica 90(3):326–334

Elmaagacli AH, Koldehoff M, Zakrzewski JL, Steckel NK, Ottinger H, Beelen DW (2007) Growth factor-independent 1B gene (GFI1B) is overexpressed in erythropoietic and megakaryocytic malignancies and increases their proliferation rate. Br J Haematol 136(2):212–219

Estey EH (2013) Acute myeloid leukemia: 2013 update on risk-stratification and management. Am J Hematol 88(4):318–327

Farrell L-L, Pepin J, Kucharski C, Lin X, Xu Z, Uludağ H (2007) A comparison of the effectiveness of cationic polymers poly-L-lysine (PLL) and polyethylenimine (PEI) for non-viral delivery of plasmid DNA to bone marrow stromal cells (BMSC). Eur J Pharm Biopharm 65(3):388–397

Fast Stats: An interactive tool for access to SEER cancer statistics. Surveillance Research Program. National Cancer Institute. [Online]. Available http://seer.cancer.gov/faststats. Accessed 20 Oct 2013

Fernandes MS, Reddy MM, Croteau NJ, Walz C, Weisbach H, Podar K, Band H, Carroll M, Reiter A, Larson RA, Salgia R, Griffin JD, Sattler M (2010) Novel oncogenic mutations of CBL in human acute myeloid leukemia that activate growth and survival pathways depend on increased metabolism. J Biol Chem 285(42):32596–32605

Fernandez-Vidal A, Ysebaert L, Didier C, Betous R, De Toni F, Prade-Houdellier N, Demur C, Contour-Galcéra M-O, Prévost GP, Ducommun B, Payrastre B, Racaud-Sultan C, Manenti S (2006) Cell adhesion regulates CDC25A expression and proliferation in acute myeloid leukemia. Cancer Res 66(14):7128–7135

Ferrara F (2012) New agents for acute myeloid leukemia: is it time for targeted therapies? Expert Opin Investig Drugs 21(2):179–189

Fiskus W, Pranpat M, Balasis M, Herger B, Rao R, Chinnaiyan A, Atadja P, Bhalla K (2006) Histone deacetylase inhibitors deplete enhancer of zeste 2 and associated polycomb repressive complex 2 proteins in human acute leukemia cells. Mol Cancer Ther 5(12):3096–3104

Gao F-H, Wu Y-L, Zhao M, Liu C-X, Wang L-S, Chen G-Q (2009) Protein kinase C-delta mediates down-regulation of heterogeneous nuclear ribonucleoprotein K protein: involvement in apoptosis induction. Exp Cell Res 315(19):3250–3258

Gao S-M, Xing C-Y, Chen C-Q, Lin S-S, Dong P-H, Yu F-J (2011) miR-15a and miR-16-1 inhibit the proliferation of leukemic cells by down-regulating WT1 protein level. J Exp Clin Cancer Res 30:110

Geletu M, Balkhi MY, Peer Zada AA, Christopeit M, Pulikkan JA, Trivedi AK, Tenen DG, Behre G (2007) Target proteins of C/EBPalphap30 in AML: C/EBPalphap30 enhances sumoylation of C/EBPalphap42 via up-regulation of Ubc9. Blood 110(9):3301–3309

Gessner A, Thomas M, Castro PG, Büchler L, Scholz A, Brümmendorf TH, Soria NM, Vormoor J, Greil J, Heidenreich O (2010) Leukemic fusion genes MLL/AF4 and AML1/MTG8 support leukemic self-renewal by controlling expression of the telomerase subunit TERT. Leukemia 24(10):1751–1759

Giles FJ (2006) Clone wars in CML. Leukemia 20(6):939–940

Gioia R, Leroy C, Drullion C, Lagarde V, Etienne G, Dulucq S, Lippert E, Roche S, Mahon F-X, Pasquet J-M (2011) Quantitative phosphoproteomics revealed interplay between Syk and Lyn in the resistance to nilotinib in chronic myeloid leukemia cells. Blood 118(8):2211–2221

Goldman JM, Melo JV (2003) Chronic myeloid leukemia—advances in biology and new approaches to treatment. N Engl J Med 349(15):1451–1464

Gu T-L, Mercher T, Tyner JW, Goss VL, Walters DK, Cornejo MG, Reeves C, Popova L, Lee K, Heinrich MC, Rush J, Daibata M, Miyoshi I, Gilliland DG, Druker BJ, Polakiewicz RD (2007) A novel fusion of RBM6 to CSF1R in acute megakaryoblastic leukemia. Blood 110(1):323–333

Guillem VM, Tormo M, Revert F, Benet I, García-Conde J, Crespo A, Aliño SF (2002) Polyethyleneimine-based immunopolyplex for targeted gene transfer in human lymphoma cell lines. J Gene Med 4(2):170–182

Guo P (2010) The emerging field of RNA nanotechnology. Nat Nanotechnol 5(12):833–842

Heidel FH, Bullinger L, Feng Z, Wang Z, Neff TA, Stein L, Kalaitzidis D, Lane SW, Armstrong SA (2012) Genetic and pharmacologic inhibition of β-catenin targets imatinib-resistant leukemia stem cells in CML. Cell Stem Cell 10(4):412–424

Hu S, Chen R, Man X, Feng X, Cen J, Gu W, He H, Li J, Chai Y, Chen Z (2011) Function and expression of insulin-like growth factor-binding protein 7 (IGFBP7) gene in childhood acute myeloid leukemia. Pediatr Hematol Oncol 28(4):279–287

Iorns E, Lord CJ, Turner N, Ashworth A (2007) Utilizing RNA interference to enhance cancer drug discovery. Nat Rev Drug Discov 6(7):556–568

Ishikawa F, Yoshida S, Saito Y, Hijikata A, Kitamura H, Tanaka S, Nakamura R, Tanaka T, Tomiyama H, Saito N, Fukata M, Miyamoto T, Lyons B, Ohshima K, Uchida N, Taniguchi S, Ohara O, Akashi K, Harada M, Shultz LD (2007) Chemotherapy-resistant human AML stem cells home to and engraft within the bone-marrow endosteal region. Nat Biotechnol 25(11):1315–1321

Ito T (2013) Stem cell maintenance and disease progression in chronic myeloid leukemia. Int J Hematol 98(6):641–647

Janmey PA, Kinnunen PKJ (2006) Biophysical properties of lipids and dynamic membranes. Trends Cell Biol 16(10):538–546

Jin L, Tabe Y, Konoplev S, Xu Y, Leysath CE, Lu H, Kimura S, Ohsaka A, Rios M-B, Calvert L, Kantarjian H, Andreeff M, Konopleva M (2008) CXCR4 up-regulation by imatinib induces

chronic myelogenous leukemia (CML) cell migration to bone marrow stroma and promotes survival of quiescent CML cells. Mol Cancer Ther 7(1):48–58
Jordan M, Wurm F (2004) Transfection of adherent and suspended cells by calcium phosphate. Methods 33(2):136–143
Kasper S, Breitenbuecher F, Heidel F, Hoffarth S, Markova B, Schuler M, Fischer T (2012) Targeting MCL-1 sensitizes FLT3-ITD-positive leukemias to cytotoxic therapies. Blood Cancer J 2(3):e60
Kim H-Y, Oh Y-S, Song I-C, Kim S-W, Lee H-J, Yun H-J, Kim S, Jo D-Y (2013) Endogenous stromal cell-derived factor-1 (CXCL12) supports autonomous growth of acute myeloid leukemia cells. Leuk Res 37(5):566–572
Koldehoff M, Zakrzewski JL, Beelen DW, Elmaagacli AH (2013) Additive antileukemia effects by GFI1B- and BCR-ABL-specific siRNA in advanced phase chronic myeloid leukemic cells. Cancer Gene Ther 20(7):421–427
Konopleva M, Tsao T, Estrov Z, Lee R-M, Wang R-Y, Jackson CE, McQueen T, Monaco G, Munsell M, Belmont J, Kantarjian H, Sporn MB, Andreeff M (2004) The synthetic triterpenoid 2-cyano-3,12-dioxooleana-1,9-dien-28-oic acid induces caspase-dependent and -independent apoptosis in acute myelogenous leukemia. Cancer Res 64(21):7927–7935
Kosova B, Tezcanli B, Ekiz HA, Cakir Z, Selvi N, Dalmizrak A, Kartal M, Gunduz U, Baran Y (2010) Suppression of STAT5A increases chemotherapeutic sensitivity in imatinib-resistant and imatinib-sensitive K562 cells. Leuk Lymphoma 51(10):1895–1901
Kumar C, Purandare AV, Lee FY, Lorenzi MV (2009) Kinase drug discovery approaches in chronic myeloproliferative disorders. Oncogene 28(24):2305–2313
Labat-Moleur F, Steffan AM, Brisson C, Perron H, Feugeas O, Furstenberger P, Oberling F, Brambilla E, Behr JP (1996) An electron microscopy study into the mechanism of gene transfer with lipopolyamines. Gene Ther 3(11):1010–1017
Landry B, Aliabadi HM, Samuel A, Gül-Uludag H, Jiang X, Kutsch O, Uludağ H (2012) Effective non-viral delivery of siRNA to acute myeloid leukemia cells with lipid-substituted polyethylenimines. PLoS ONE 7(8):e44197
Landry B, Gül-Uludag H, Uludag H (2013) Lipid-polymer mediated siRNA therapy for silencing CXCR4 in acute myeloid leukemia. In: Presented at the ASGCT 16th Annual Meeting 2013, Salt Lake City
Lee YK, Kim KS, Kim JS, Baek JE, Park SI, Jeong HY, Yoon SS, Jung KC, Song HG, Park YS (2010) Leukemia-specific siRNA delivery by immunonanoplexes consisting of anti-JL1 minibody conjugated to oligo-9 Arg-peptides. Mol Cells 29(5):457–462
Lodish H, Berk A, Kaiser CA, Krieger M, Bretscher A (2013) Molecular cell biology, 5th edn. W. H. Freeman & Company, New York
Lorenz MR, Holzapfel V, Musyanovych A, Nothelfer K, Walther P, Frank H, Landfester K, Schrezenmeier H, Mailänder V (2006) Uptake of functionalized, fluorescent-labeled polymeric particles in different cell lines and stem cells. Biomaterials 27(14):2820–2828
Lu L, Zhang L, Xiao Z, Lu S, Yang R, Han ZC (2008) Neuropilin-1 in acute myeloid leukemia: expression and role in proliferation and migration of leukemia cells. Leuk Lymphoma 49(2):331–338
Mahon F-X, Hayette S, Lagarde V, Belloc F, Turcq B, Nicolini F, Belanger C, Manley PW, Leroy C, Etienne G, Roche S, Pasquet J-M (2008) Evidence that resistance to nilotinib may be due to BCR-ABL, Pgp, or Src kinase overexpression. Cancer Res 68(23):9809–9816
McLornan D, Hay J, McLaughlin K, Holohan C, Burnett AK, Hills RK, Johnston PG, Mills KI, McMullin MF, Longley DB, Gilkes A (2013) Prognostic and therapeutic relevance of c-FLIP in acute myeloid leukaemia. Br J Haematol 160(2):188–198
Melo JV, Barnes DJ (2007) Chronic myeloid leukaemia as a model of disease evolution in human cancer. Nat Publishing Group 7(6):441–453
Mendonça LS, Firmino F, Moreira JN, Pedroso de Lima MC, Simões S (2010) Transferrin receptor-targeted liposomes encapsulating anti-BCR-ABL siRNA or asODN for chronic myeloid leukemia treatment. Bioconjug Chem 21(1):157–168

Merkerova M, Bruchova H, Kracmarova A, Klamova H, Brdicka R (2007) Bmi-1 overexpression plays a secondary role in chronic myeloid leukemia transformation. Leuk Lymphoma 48(4):793–801

Mikkola HKA, Radu CG, Witte ON (2010) Targeting leukemia stem cells. Nat Publishing Group 28(3):237–238

Mintzer MA, Simanek EE (2009) Nonviral vectors for gene delivery. Chem Rev 109(2):259–302

Miyazaki T, Kirino Y, Takeno M, Samukawa S, Hama M, Tanaka M, Yamaji S, Ueda A, Tomita N, Fujita H, Ishigatsubo Y (2010) Expression of heme oxygenase-1 in human leukemic cells and its regulation by transcriptional repressor Bach1. Cancer Sci 101(6):1409–1416

Moll J, Khaldoyanidi S, Sleeman JP, Achtnich M, Preuss I, Ponta H, Herrlich P (1998) Two different functions for CD44 proteins in human myelopoiesis. J Clin Invest 102(5):1024–1034

Muranyi AL, Dedhar S, Hogge DE (2009) Combined inhibition of integrin linked kinase and FMS-like tyrosine kinase 3 is cytotoxic to acute myeloid leukemia progenitor cells. Exp Hematol 37(4):450–460

Muranyi AL, Dedhar S, Hogge DE (2010) Targeting integrin linked kinase and FMS-like tyrosine kinase-3 is cytotoxic to acute myeloid leukemia stem cells but spares normal progenitors. Leuk Res 34(10):1358–1365

Muvarak N, Nagaria P, Rassool FV (2012) Genomic instability in chronic myeloid leukemia: targets for therapy? Curr Hematol Malig Rep 7(2):94–102

Neamnark A, Suwantong O, Bahadur RKC, Hsu CYM, Supaphol P, Uludağ H (2009) Aliphatic lipid substitution on 2 kDa polyethylenimine improves plasmid delivery and transgene expression. Mol Pharm 6(6):1798–1815

Nel AE, Mädler L, Velegol D, Xia T, Hoek EMV, Somasundaran P, Klaessig F, Castranova V, Thompson M (2009) Understanding biophysicochemical interactions at the nano-bio interface. Nat Mater 8(7):543–557

Nishioka C, Ikezoe T, Yang J, Komatsu N, Koeffler HP, Yokoyama A (2009) Blockade of MEK signaling potentiates 5-Aza-2′-deoxycytidine-induced apoptosis and upregulation of p21(waf1) in acute myelogenous leukemia cells. Int J Cancer 125(5):1168–1176

Nishioka C, Ikezoe T, Yang J, Yokoyama A (2010) Inhibition of MEK/ERK signaling induces apoptosis of acute myelogenous leukemia cells via inhibition of eukaryotic initiation factor 4E-binding protein 1 and down-regulation of Mcl-1. Apoptosis 15(7):795–804

Okamoto M, Hayakawa F, Miyata Y, Watamoto K, Emi N, Abe A, Kiyoi H, Towatari M, Naoe T (2007) Lyn is an important component of the signal transduction pathway specific to FLT3/ITD and can be a therapeutic target in the treatment of AML with FLT3/ITD. Leukemia 21(3):403–410

Pan D, Li Y, Li Z, Wang Y, Wang P, Liang Y (2012) Gli inhibitor GANT61 causes apoptosis in myeloid leukemia cells and acts in synergy with rapamycin. Leuk Res 36(6):742–748

Park I-K, Mishra A, Chandler J, Whitman SP, Marcucci G, Caligiuri MA (2013) Inhibition of the receptor tyrosine kinase Axl impedes activation of the FLT3 internal tandem duplication in human acute myeloid leukemia: implications for Axl as a potential therapeutic target. Blood 121(11):2064–2073

Powell JA, Thomas D, Barry EF, Kok CH, McClure BJ, Tsykin A, To LB, Brown A, Lewis ID, Herbert K, Goodall GJ, Speed TP, Asou N, Jacob B, Osato M, Haylock DN, Nilsson SK, D'Andrea RJ, Lopez AF, Guthridge MA (2009) Expression profiling of a hemopoietic cell survival transcriptome implicates osteopontin as a functional prognostic factor in AML. Blood 114(23):4859–4870

Prokop A (2011) Intracellular delivery: fundamentals and applications, vol 5. Springer, Berlin

Rainaldi G, Filippini P, Ferrante A, Indovina PL, Santini MT (2001) Fibronectin facilitates adhesion of K562 leukemic cells normally growing in suspension to cationic surfaces. J Biomed Mater Res 55(1):104–113

Randrianarison-Huetz V, Laurent B, Bardet V, Blobe GC, Huetz F, Duménil D (2010) Gfi-1B controls human erythroid and megakaryocytic differentiation by regulating TGF-beta signaling at the bipotent erythro-megakaryocytic progenitor stage. Blood 115(14):2784–2795

Rangatia J, Bonnet D (2006) Transient or long-term silencing of BCR-ABL alone induces cell cycle and proliferation arrest, apoptosis and differentiation. Leukemia 20(1):68–76

Rao J, Xu D-R, Zheng F-M, Long Z-J, Huang S-S, Wu X, Zhou W-H, Huang R-W, Liu Q (2011) Curcumin reduces expression of Bcl-2, leading to apoptosis in daunorubicin-insensitive CD34+ acute myeloid leukemia cell lines and primary sorted CD34+ acute myeloid leukemia cells. J Transl Med 9:71

Ravid K, Freshney RI (1998) DNA transfer to cultured cells. Wiley, Hoboken

Recher C, Ysebaert L, Beyne-Rauzy O, Mansat-De Mas V, Ruidavets J-B, Cariven P, Demur C, Payrastre B, Laurent G, Racaud-Sultan C (2004) Expression of focal adhesion kinase in acute myeloid leukemia is associated with enhanced blast migration, increased cellularity, and poor prognosis. Cancer Res 64(9):3191–3197

Rossbach M (2010) Small non-coding RNAs as novel therapeutics. Curr Mol Med 10(4):361–368

Rushworth SA, MacEwan DJ (2008) HO-1 underlies resistance of AML cells to TNF-induced apoptosis. Blood 111(7):3793–3801

Rushworth SA, Bowles KM, Raninga P, MacEwan DJ (2010) NF-kappaB-inhibited acute myeloid leukemia cells are rescued from apoptosis by heme oxygenase-1 induction. Cancer Res 70(7):2973–2983

Sansonetti A, Bourcier S, Durand L, Chomienne C, Smadja-Joffe F, Robert-Lézénès J (2012) CD44 activation enhances acute monoblastic leukemia cell survival via Mcl-1 upregulation. Leuk Res 36(3):358–362

Savona M, Talpaz M (2008) Getting to the stem of chronic myeloid leukaemia. Nat Rev Cancer 8(5):341–350

Sawyers CL (2013) Perspective: combined forces. Nature 498(7455):S7

Schepers H, Geugien M, van der Toorn M, Bryantsev AL, Kampinga HH, Eggen BJL, Vellenga E (2005) HSP27 protects AML cells against VP-16-induced apoptosis through modulation of p38 and c-Jun. Exp Hematol 33(6):660–670

Shen Q, Liu S, Chen Y, Yang L, Chen S, Wu X, Li B, Lu Y, Zhu K, Li Y (2013) Proliferation inhibition and apoptosis induction of imatinib-resistant chronic myeloid leukemia cells via PPP2R5C down-regulation. J Hematol Oncol 6(1):64

Sloma I, Jiang X, Eaves AC, Eaves CJ (2010) Insights into the stem cells of chronic myeloid leukemia. Leukemia 24(11):1823–1833

Spirin PV, Nikitenko NA, Lebedev TD, Rubtsov PM, Stocking C, Prasolov VS (2011) Modulation of activated oncogene c-kit expression with RNA-interference. Mol Biol (Mosk) 45(6):1036–1045

Stefanachi A, Leonetti F, Nicolotti O, Catto M, Pisani L, Cellamare S, Altomare C, Carotti A (2012) New strategies in the chemotherapy of leukemia: eradicating cancer stem cells in chronic myeloid leukemia. Curr Cancer Drug Targets 12(5):571–596

Sun Y-H, Sun Y-L, Ran X-H, Cui J-Y, Zhang H-L, Chen Z-X (2013) The influence of CD44 on the adhesive, migratory and infiltrative abilities of leukemia cells. Zhonghua Xue Ye Xue Za Zhi 34(1):60–63

Tanaka N, Wang Y-H, Shiseki M, Takanashi M, Motoji T (2011) Inhibition of PRAME expression causes cell cycle arrest and apoptosis in leukemic cells. Leuk Res 35(9):1219–1225

Tibes R, Bogenberger JM, Chaudhuri L, Hagelstrom RT, Chow D, Buechel ME, Gonzales IM, Demuth T, Slack J, Mesa RA, Braggio E, Yin HH, Arora S, Azorsa DO (2012) RNAi screening of the kinome with cytarabine in leukemias. Blood 119(12):2863–2872

Turkson J (2004) STAT proteins as novel targets for cancer drug discovery. Expert Opin Ther Targets 8(5):409–422

Tyner JW, Walters DK, Willis SG, Luttropp M, Oost J, Loriaux M, Erickson H, Corbin AS, O'Hare T, Heinrich MC, Deininger MW, Druker BJ (2008) RNAi screening of the tyrosine kinome identifies therapeutic targets in acute myeloid leukemia. Blood 111(4):2238–2245

Valencia Serna J, Gül-Uludag H, Mahdipoor P, Jiang X, Uludağ H (2013) Investigating siRNA delivery to chronic myeloid leukemia K562 cells with lipophilic polymers for therapeutic BCR-ABL down-regulation. J Control Release 172(2):495–503 van Blitterswijk WJW (1988) Structural basis and physiological control of membrane fluidity in normal and tumor cells. Sub-Cell Biochem 13:393–413

Voisset E, Lopez S, Chaix A, Georges C, Hanssens K, Prébet T, Dubreuil P, De Sepulveda P (2010) FES kinases are required for oncogenic FLT3 signaling. Leukemia 24(4):721–728

Walters DK, Stoffregen EP, Heinrich MC, Deininger MW, Druker BJ (2005) RNAi-induced down-regulation of FLT3 expression in AML cell lines increases sensitivity to MLN518. Blood 105(7):2952–2954

Walters DK, Goss VL, Stoffregen EP, Gu T-L, Lee K, Nardone J, McGreevey L, Heinrich MC, Deininger MW, Polakiewicz R, Druker BJ (2006a) Phosphoproteomic analysis of AML cell lines identifies leukemic oncogenes. Leuk Res 30(9):1097–1104

Walters DK, Mercher T, Gu T-L, O'Hare T, Tyner JW, Loriaux M, Goss VL, Lee KA, Eide CA, Wong MJ, Stoffregen EP, McGreevey L, Nardone J, Moore SA, Crispino J, Boggon TJ, Heinrich MC, Deininger MW, Polakiewicz RD, Gilliland DG, Druker BJ (2006b) Activating alleles of JAK3 in acute megakaryoblastic leukemia. Cancer Cell 10(1):65–75

Wang X, Ren J, Qu X (2008) Targeted RNA interference of cyclin A2 mediated by functionalized single-walled carbon nanotubes induces proliferation arrest and apoptosis in chronic myelogenous leukemia K562 cells. ChemMedChem 3(6):940–945

Wang B, Li S, Qi HH, Chowdhury D, Shi Y, Novina CD (2009) Distinct passenger strand and mRNA cleavage activities of human Argonaute proteins. Nat Struct Mol Biol 16(12):1259–1266

Wang J, Lu Z, Wientjes MG, Au JLS (2010a) Delivery of siRNA therapeutics: barriers and carriers. AAPS J 12(4):492–503

Wang X, Gocek E, Novik V, Harrison JS, Danilenko M, Studzinski GP (2010b) Inhibition of Cot1/Tlp2 oncogene in AML cells reduces ERK5 activation and up-regulates p27Kip1 concomitant with enhancement of differentiation and cell cycle arrest induced by silibinin and 1,25-dihydroxyvitamin D(3). Cell Cycle 9(22):4542–4551

Wang C, Chen Z, Li Z, Cen J (2010c) The essential roles of matrix metalloproteinase-2, membrane type 1 metalloproteinase and tissue inhibitor of metalloproteinase-2 in the invasive capacity of acute monocytic leukemia SHI-1 cells. Leuk Res 34(8):1083–1090

Wang W, Li W, Ou L, Flick E, Mark P, Nesselmann C, Lux CA, Gatzen H-H, Kaminski A, Liebold A, Lützow K, Lendlein A, Li R-K, Steinhoff G, Ma N (2011a) Polyethylenimine-mediated gene delivery into human bone marrow mesenchymal stem cells from patients. J Cell Mol Med 15(9):1989–1998

Wang J, Li W, Li L, Yu X, Jia J, Chen C (2011b) CIP2A is over-expressed in acute myeloid leukaemia and associated with HL60 cells proliferation and differentiation. Int J Lab Hematol 33(3):290–298

Wang CM, Sheng GY, Lu J, Xie L, Bai ST, Xu XJ, Liu YF (2011c) Effect of small interfering RNA targeting wild-type FLT3 in acute myeloid leukaemia cells in vitro and in vivo. J Int Med Res 39(5):1661–1674

Wang R, Xia L, Gabrilove J, Waxman S, Jing Y (2013) Downregulation of Mcl-1 through GSK-3β activation contributes to arsenic trioxide-induced apoptosis in acute myeloid leukemia cells. Leukemia 27(2):315–324

Weisberg E, Manley PW, Cowan-Jacob SW, Hochhaus A, Griffin JD (2007) Second generation inhibitors of BCR-ABL for the treatment of imatinib-resistant chronic myeloid leukaemia. Nat Publishing Group 7(5):345–356

Whitehead KA, Langer R, Anderson DG (2009) Knocking down barriers: advances in siRNA delivery. Nat Rev Drug Discov 8(2):129–138

Wilda M, Fuchs U, Wössmann W, Borkhardt A (2002) Killing of leukemic cells with a BCR/ABL fusion gene by RNA interference (RNAi). Oncogene 21(37):5716–5724

Willyard C (2013) Stem cells: bad seeds. Nature 498(7455):S12–S13

Wohlbold L, van der Kuip H, Miething C, Vornlocher H-P, Knabbe C, Duyster J, Aulitzky WE (2003) Inhibition of bcr-abl gene expression by small interfering RNA sensitizes for imatinib mesylate (STI571). Blood 102(6):2236–2239

Yang L, Yang M, Zhang H, Wang Z, Yu Y, Xie M, Zhao M, Liu L, Cao L (2012) S100A8-targeting siRNA enhances arsenic trioxide-induced myeloid leukemia cell death by down-regulating autophagy. Int J Mol Med 29(1):65–72

Yang J, Ikezoe T, Nishioka C, Yokoyama A (2013) A novel treatment strategy targeting shugoshin 1 in hematological malignancies. Leuk Res 37(1):76–82

Yoda M, Kawamata T, Paroo Z, Ye X, Iwasaki S, Liu Q, Tomari Y (2010) ATP-dependent human RISC assembly pathways. Nat Struct Mol Biol 17(1):17–23

Zhang H, Li S (2013) Molecular mechanisms for survival regulation of chronic myeloid leukemia stem cells. Protein Cell 4(3):186–196

Zhang Y-Y, Xie K-M, Yang G-Q, Mu H-J, Yin Y, Zhang B, Xie P (2011) The effect of glucosylceramide synthase on P-glycoprotein function in K562/AO2 leukemia drug-resistance cell line. Int J Hematol 93(3):361–367

Zhang Q, Hossain DMS, Nechaev S, Kozlowska A, Zhang W, Liu Y, Kowolik CM, Swiderski P, Rossi JJ, Forman S, Pal S, Bhatia R, Raubitschek A, Yu H, Kortylewski M (2013) TLR9-mediated siRNA delivery for targeting of normal and malignant human hematopoietic cells in vivo. Blood 121(8):1304–1315

Zhao Q, Song X, Waldschmidt T, Fisher E, Krieg AM (1996) Oligonucleotide uptake in human hematopoietic cells is increased in leukemia and is related to cellular activation. Blood 88(5):1788–1795

Zhao C, Blum J, Chen A, Kwon HY, Jung SH, Cook JM, Lagoo A, Reya T (2007) Loss of beta-catenin impairs the renewal of normal and CML stem cells in vivo. Cancer Cell 12(6):528–541

Zheng H, Chen Y, Chen S, Niu Y, Yang L, Li B, Lu Y, Geng S, Du X, Li Y (2011) Expression and distribution of PPP2R5C gene in leukemia. J Hematol Oncol 4(21):1756–8722

Zhou LL, Zhao Y, Ringrose A, DeGeer D, Kennah E, Lin AEJ, Sheng G, Li X-J, Turhan A, Jiang X (2008) AHI-1 interacts with BCR-ABL and modulates BCR-ABL transforming activity and imatinib response of CML stem/progenitor cells. J Exp Med 205(11):2657–2671

Index

A. Prokop et al. (eds.), *Intracellular Delivery II*,
Fundamental Biomedical Technologies 7, DOI: 10.1007/978-94-017-8896-0,

MIX
Papier aus verantwortungsvollen Quellen
Paper from responsible sources
FSC® C105338

If you have any concerns about our products,
you can contact us on
ProductSafety@springernature.com

In case Publisher is established outside the EU,
the EU authorized representative is:
Springer Nature Customer Service Center GmbH
Europaplatz 3, 69115 Heidelberg, Germany

Printed by Libri Plureos GmbH
in Hamburg, Germany